Smart Innovation, Systems and Technologies

Volume 184

The Smart Innovation, Systems and Technologies book series encompasses the topics of knowledge, intelligence, innovation and sustainability. The aim of the series is to make available a platform for the publication of books on all aspects of single and multi-disciplinary research on these themes in order to make the latest results available in a readily-accessible form. Volumes on interdisciplinary research combining two or more of these areas is particularly sought.

The series covers systems and paradigms that employ knowledge and intelligence in a broad sense. Its scope is systems having embedded knowledge and intelligence, which may be applied to the solution of world problems in industry, the environment and the community. It also focusses on the knowledge-transfer methodologies and innovation strategies employed to make this happen effectively. The combination of intelligent systems tools and a broad range of applications introduces a need for a synergy of disciplines from science, technology, business and the humanities. The series will include conference proceedings, edited collections, monographs, handbooks, reference books, and other relevant types of book in areas of science and technology where smart systems and technologies can offer innovative solutions.

High quality content is an essential feature for all book proposals accepted for the series. It is expected that editors of all accepted volumes will ensure that contributions are subjected to an appropriate level of reviewing process and adhere to KES quality principles.

**** Indexing: The books of this series are submitted to ISI Proceedings, EI-Compendex, SCOPUS, Google Scholar and Springerlink ****

More information about this series at http://www.springer.com/series/8767

Anna Esposito · Marcos Faundez-Zanuy ·
Francesco Carlo Morabito · Eros Pasero
Editors

Progresses in Artificial Intelligence and Neural Systems

Springer

Editors
Anna Esposito
Dipartimento di Psicologia and IIASS
Università della Campania "Luigi Vanvitelli"
Caserta, Italy

Marcos Faundez-Zanuy
Fundació Tecnocampus
Pompeu Fabra University
Mataró, Barcelona, Spain

Francesco Carlo Morabito
Department of Civil, Environmental, Energy, and Material Engineering
University Mediterranea of Reggio Calabria
Reggio Calabria, Italy

Eros Pasero
Laboratorio di Neuronica, Dipartimento Elettronica e Telecomunicazioni
Politecnico di Torino
Torino, Italy

ISSN 2190-3018 ISSN 2190-3026 (electronic)
Smart Innovation, Systems and Technologies
ISBN 978-981-15-5092-8 ISBN 978-981-15-5093-5 (eBook)
https://doi.org/10.1007/978-981-15-5093-5

This Springer imprint is published by the registered company Springer Nature Singapore Pte Ltd.
The registered company address is: 152 Beach Road, #21-01/04 Gateway East, Singapore 189721, Singapore

Program Committee

Amorese Terry, Università della Campania "Luigi Vanvitelli" and IIASS
Apolloni Bruno, Università di Milano
Buonanno Michele, Università della Campania "Luigi Vanvitelli" and IIASS
Buono Carmela, Università della Campania "Luigi Vanvitelli" and IIASS
Cordasco Gennaro, Università della Campania "Luigi Vanvitelli" and IIASS
Cuciniello Marialucia, Università della Campania "Luigi Vanvitelli" and IIASS
Damasevicius Robertas, Kaunas University of Technology
Esposito Anna, Università della Campania "Luigi Vanvitelli" and IIASS
Esposito Antonietta Maria, Osservatorio Vesuviano sezione di Napoli
Esposito Marilena, International Institute for Advanced Scientific Studies (IIASS)
Faundez-Zanuy Marcos, Tecnocampus Universitat Pompeu Fabra
Gabrielli Leonardo, Università Politecnica delle Marche
Koutsombogera Maria, Trinity College Dublin
Cosimo Ieracitano, Università degli Studi Mediterranea Reggio Calabria
Maiorana Emanuele, Roma TRE University
Mekyska Jiri, Brno University
Scarpiniti Michele, Università di Roma "La Sapienza"
Schaust Jonathan, University of Applied Science in Koblenz
Senese Vincenzo Paolo, Università degli Studi della Campania "Luigi Vanvitelli"
Severini Marco, Università Politecnica delle Marche
Squartini Stefano, Università Politecnica delle Marche
Troncone Alda, Università degli Studi della Campania "Luigi Vanvitelli" and IIASS
Tschacher Wolfgang, Universität Bern
Uncini Aurelio, Università di Roma "La Sapienza"
Vitabile Salvatore, Università degli Studi di Palermo
Vogel Carl, Trinity College Dublin

Sponsoring Institutions

International Institute for Advanced Scientific Studies (IIASS) of Vietri S/M, Italy
Department of Psychology, Università della Campania "Luigi Vanvitelli", Italy
Provincia di Salerno, Italy
Comune di Vietri sul Mare, Salerno, Italy
International Neural Network Society (INNS)
Università Mediterranea di Reggio Calabria, Italy

Preface

This book provides an overview of the current progress in Artificial Intelligence and Neural Nets. Artificial Intelligence and Neural Nets have shown great capabilities in modeling, prediction, and recognition tasks for social signal processing and big data mining. The adoption of such intelligent and advanced computational tools has achieved a mature degree of understanding in many application areas, in particular, in complex multimodal systems supporting human–machine or human–human interaction, a field that is broadly addressed by the scientific communities and has a strong commercial impact.

At the same time, the emotional issue, i.e., the need to make ICT interfaces emotionally and socially believable, has gained increasing attention in the implementation of such complex systems due to the relevance of emotional aspects in everyday human functional abilities (like cognitive processes, perception, learning, communication, and even "rational" decision-making). The real challenge is taking advantage of the emotional characterization of humans' emotions to make the computers interfacing with them more natural and therefore useful.

The volume proposed assesses to what extent and how sophisticated computational intelligence tools developed so far might support the multidisciplinary research on the characterization of appropriate systems' reactions to human emotions and expressions in interactive scenarios.

Interdisciplinary aspects are taken into account, and research is proposed from different fields including mathematics, computer vision, speech analysis and synthesis, machine learning, signal processing, telecommunication, human–computer interaction, psychology, anthropology, sociology, neural networks, machine learning, and advanced sensing in order to provide contributions on the most recent trends, innovative approaches, and future challenges.

The contributions reported in the book cover different scientific areas and are grouped according to the thematic classification reported below; even though these areas are closely connected in the themes, they afford and provide fundamental insights for the cross-fertilization of different disciplines:

- Neural Networks and Related Applications,
- Neural Networks and Pattern Recognition in Medicine,
- Computational and Methodological Intelligence in Economics and Finance,
- Advanced Smart Multimodal Data Processing, and
- Dynamics of Signal Exchanges and Empathic Systems.

The chapters composing this book were first discussed at the international workshop on neural networks (WIRN 2019) held in Vietri Sul Mare from 12 to 14 June 2019, in regular and special sessions. The workshop hosted four special sessions. The first special session on *Computational and Methodological Intelligence in Economics and Finance* organized by Marco Corazza includes contributions discussing of bio-inspired optimizers used to solve complex financial/economic decision-making problems. The second special session on *Neural Networks and Pattern Recognition in Medicine* organized by Giansalvo Cirrincione, Vitoantonio Bevilacqua, reports on the most recent AI techniques on the processing of biomedical images, medical classification, and gene expression analysis. The third special session on *Advanced Smart Multimodal Data Processing* organized by Francesco Camastra, Angelo Ciaramella, Michele Scarpiniti, and Antonino Staiano collects contributions on recent research advances and the state-of-the-art methods in the fields of Soft Computing, Machine Learning, and Data Mining methodologies. The fourth special session on *Dynamics of Signal Exchanges and Empathic System* gives emphasis to contributions devoted to the implementation of Empathic Systems, considering that empathy is central for successful social interactional exchanges. The session organized by Anna Esposito, Anna Sorrentino, Antonietta M. Esposito, Gennaro Cordasco, Nelson Mauro Maldonato, Francesco Carlo Morabito, Maria Ines Torres, Stephan Schlögl, and Zoraida Callejas Carrión was sponsored by two H2020 funded projects: *Empathic* (empathic-project.eu/) and "Menhir" (menhir-project.eu/), aiming to implement socially and emotionally believable automatic systems, the Italian Government funded project SIROBOTICS (https://www.istitutomarino.it/project/si-robotics-social-robotics-for-active-and-healthy-ageing/) aiming to implement social robot assistants for supporting elderly everyday independent living, and the ANDROIDS project funded by the program V:ALERE 2019 Università della Campania "L. Vanvitelli", D. R. 906 del 4/10/2019, prot. n. 157264,17/10/2019. This particular special session was also intended to celebrate professor Anna Costanza Baldry, Ufficiale al Merito della Repubblica Italiana and Full Professor of Social Psychology at Department of Psychology, Università della Campania "Luigi Vanvitelli". Her unexpected loss left a huge emptiness and shattered all of her colleagues and friends. Her research topics were about fighting violence against women. For her, empathy was a blessing since it suggested the right way to help abused women, and a curse since empathy forced her to share their suffering.

The scientists contributing to this book are specialists in their respective disciplines, and through their contributions have made this volume a significant scientific effort. The coordination and production of this book have been brilliantly conducted by the Springer Project Coordinator Mr. **Ramamoorthy Rajangam**, the

Springer Executive Editor Dr. **Thomas Ditzinger**, and the Editor Assistant Mr. **Holger Schaepe**. They are the recipients of our deepest appreciation. This initiative has been skillfully supported by the Editors in chief of the Springer series Smart Innovation, Systems and Technologies, Professors **Jain Lakhmi C.** and **Howlett Robert James**, to whom goes our deepest gratitude.

The Editors

Caserta, Italy — Anna Esposito
Mataró, Spain — Marcos Faundez-Zanuy
Reggio Calabria, Italy — Francesco Carlo Morabito
Torino, Italy — Eros Pasero

Contents

About the Editors

Anna Esposito received her Laurea degree summa cum laude in Information Technology and Computer Science from Salerno University (1989), and her Ph.D. degree in Applied Mathematics and Computer Science from Napoli University Federico II (1995) with a thesis developed at MIT, Boston, USA. She was a postdoc at IIASS, Lecturer at Salerno University Department of Physics (1996–2000), and Research Professor (2000–2002) at WSU Department of Computer Science and Engineering, Ohio, USA. She is currently an Associate Professor at Campania University L. Vanvitelli. She has published over 240 peer-reviewed papers in journals, books, and conference proceedings.

Marcos Faundez-Zanuy received his B.Sc. degree (1993) and Ph.D. (1998) from the Polytechnic University of Catalunya. He is a Full Professor at ESUP Tecnocampus Mataro, where he also heads the Signal Processing Group. His research focuses on biometrics applied to security and health. He was the initiator and Chairman of the EU COST action 277 "Nonlinear speech processing" and secretary of COST action 2102 "Cross-Modal Analysis of Verbal and Non-verbal Communication". He is the author of over 50 papers indexed in ISI Journal citation report, over 100 conference papers and several books, and is the PI of 10 national and EU funded projects.

Francesco Carlo Morabito joined the University of Reggio Calabria, Italy, in 1989, and has been a Full Professor of Electrical Engineering there since 2001. He served as President of the Electronic Engineering Course, a member of the University's Inner Evaluation Committee, Dean of the Faculty of Engineering and Deputy Rector, and is currently Vice-Rector for Internationalization. He is a member of the Italian Society of Electrical Engineering's Steering Committee.

Eros Pasero has been a Professor of Electronics at Politecnico of Turin since 1991. He was a visiting Professor at ICSI Berkeley (1991), Tongji University Shanghai (2011, 2015), and Tashkent Politechic University, Uzbekistan. His interests include artificial neural networks and electronic sensors. He heads the Neuronica Lab, which develops wired and wireless sensors for biomedical, environmental and automotive applications, and neural networks for sensor signals processing. Professor Pasero is President of the Italian Society for Neural Networks (SIREN) and was General Chair of IJCNN2000, SIRWEC2006, and WIRN 2015. He has received several awards and holds 5 international patents, and is the author of over 100 international publications.

Introduction

Towards Socially and Emotionally Believable ICT Interfaces

Anna Esposito, Marcos Faundez-Zanuy, Francesco Carlo Morabito, and Eros Pasero

Abstract In order to realize an artificial intelligence focused on human needs, it is necessary to identify the interactional characteristics that describe human mood, social behavior, beliefs, and experiences. The cross-modal analysis of communicative macro-signals represents the first step in this direction. The second step requires the definition of adequate mathematical representations of these signals to validate them perceptively (on the human side) and computationally.

1 Introduction

The human beings have always aimed to conceive tools that would allow exceeding their limits and improving their skills. In the past, the exploitation of these tools required long training acquired skills favoring the establishment of dedicated expertise. This trend changed radically few decades ago in computer science when automatic systems have been proposed which are more and more capable of operating autonomously and intelligently. Behind this radical change there is the need to exploit automatic systems in everyday living environments to simplify communications,

A. Esposito (✉)
Università della Campania "Luigi Vanvitelli", Dipartimento di Psicologia and IIASS, Caserta, Italy
e-mail: iiass.annaesp@tin.it

M. Faundez-Zanuy
Tecnocampus Universitat Pompeu Fabra, Barcelona, Spain
e-mail: faundez@tecnocampus.cat

F. C. Morabito
Università degli Studi "Mediterranea" di Reggio Calabria, Reggio Calabria, Italy
e-mail: morabito@unirc.it

E. Pasero
Politecnico di Torino, Dipartimento di Elettronica e Telecomunicazioni, Turin, Italy
e-mail: eros.pasero@polito.it

A. Esposito et al. (eds.), *Progresses in Artificial Intelligence and Neural Systems*, Smart Innovation, Systems and Technologies 184,
https://doi.org/10.1007/978-981-15-5093-5_1

exchanges, and services, facilitate social inclusion and improve the quality of life and wellbeing of all citizens, and particularly vulnerable people.

There is a huge demand to develop complex autonomous systems capable of assisting people with different needs, ranging from different physical illnesses (such as limited motor skills) to psychological and communicative disorders. Key services required to these systems include: (a) the automatic monitoring of daily functional activities; (b) the abilities to detect subtle changes in individuals' functional, physical, social and psychological activities; (c) the identification of appropriate actions enabling individuals, particularly vulnerable ones such as seniors, to appropriately operate in society, and (d) the offering of therapeutic interventions and rehabilitation tools.

Currently one category of the world population particularly in needs is seniors. The world population is aging rapidly (https://www.un.org/en/sections/issues-depth/ageing/) and the World Health Organization consider the consequences of this process very difficult to handle, since there is a great inhomogeneity among individuals with regard to their physical and cognitive functionality while aging. The consequences of this process are far-reaching, since aging leads to physical and cognitive declines, mental disorders with varying degrees of severity, such as such as Parkinson's disease and dementia, and anxiety and depression. The increasing number of seniors who need support to carry out their daily activities will lead to a future lack of professional and/or informal caregivers to meet their needs and a limited access to therapeutic interventions.

As a result of this situation, national health institutions will be exposed to considerable burdens in terms of health care costs, and care associated with the medical cures. One solution is to develop complex autonomous systems, in the form of socially and emotionally believable ICT interfaces capable of detecting the onset of such disorders, providing, where possible, initial on-demand support to patients, offering doctors' sustenance for diagnoses and treatments, suggesting strategies for favoring social inclusion and wellbeing.

In addition, such socially and emotionally believable ICT interfaces can be embedded in several promising applications such as:

- Platforms for measuring creative learning, mobile learning, experiential learning, social learning and collaborative learning (multi-dimensional learning)
- Integrated technological platforms for early interventions and managements of cognitive and physical diseases
- Responsive and symbiotic personal assistants
- Real-time cognitive brain-like computing systems simulating perception, attention, inner speech, imagination and emotions.

The efforts made so far have not been satisfactory due to a lack of attention to users' needs and expectations and the difficulty of understanding how to model interactions taking into account contextual and environmental (including people and objects) effects on individuals' actions/reactions, social perception, and practices of attributing meanings/sensations in long-term relationships. To understood human needs in context more investigations are needed, aimed at analyzing human-machine interaction in the domestic and social spheres in order to develop complex autonomous systems capable of establishing with their users trustworthy relationships, and taking appropriate actions to arouse empathic feelings and esteem.

In addition, there are no standard procedures for the structured development of complex autonomous systems "meeting" users' expectations and demands. Although cognitive psychology and neuroscience have offered suggestions to model these behaviors [1–6] and potential solutions have been proposed [7] to avoid triggering repulsive reactions by users towards such systems (the "uncanny valley" effect, Moore [8] the research is still at a seminal state.

To implement friendly and socially believable human-machine interactions would require accounting of research aspects such as:

- How communication practices are transformed in different contexts;
- Which are the user's cognitive and emotional consequences when interacting with machines;
- How endorse machines of an effective ability to process behavioral and contextual information.

As a consequence it would be necessary to:

- Explore new data to gather models of behaviours in a multimodal communication environment;
- Elaborate new mathematical models accounting of contextual, social, cognitive and emotional effects.

2 Content

The themes of this book tackled aspects of dynamics of signal exchanges either processed by natural or artificial systems attempting to contribute to the above research in four fields organized in sections. Each section contains contributes aiming to progresses AI and neural nets toward socially and emotionally believable human-machine interactions. These contributes were initially discussed at the 29th edition of the International Workshop on Neural Networks (WIRN 2019) held in Vietri sul Mare, Italy, from the 12th to the 14th of June 2019. Particularly:

Section I contain this introductory paper [9], this volume.
Section II contains contributes devoted to applications of neural networks in seismic signal classification, sentiment analysis, image classification, behavioral traits, stress, and emotions from music, images and handwriting. It includes 20 original contributes.
Section III includes 5 papers dedicated to exploit neural networks and pattern recognition techniques in medicine.

Section IV contains 6 papers reporting on computational and methodological intelligent approaches in economics and finance, such as bio-inspired optimizers to solve complex financial optimization problems and machine learning techniques to model the behavior of economic agents.
Section V contains 6 original papers on recent research advances in multimodal signal processing that process and combine information from a variety of modalities (e.g., speech, language, text) in order to enhance the performance of human-computer interaction devices.
Section VI discusses themes on dynamics of signal exchanges and empathic systems and related applications of ICT interfaces able to detect health and affective states of their users, and interpret their psychological and behavioral patterns. The section includes 13 chapters.

3 Conclusions

A change is needed to approach the implementation of socially and emotionally believable cognitive systems (no matter if they are or not physical embodied) that must reflect on:

A deep investigations on the relevant consequences that occur at the cognitive and emotional level of the final user;
An efficient *modeling of users' communicative signals, competencies, beliefs and environmental information* in order to provide relevant feedbacks and services.

In these contexts artificial intelligence is the theoretical structure from which to derive algorithms for information processing able to produce new representations of such problems and generalized solutions for nonstationary and non-linear input-output relations. Limitations due to the fact that algorithms of this type may not converge towards adequate solutions and produce meaningless results are surpassed by biologically inspired machine learning models. These new AI ICT Interfaces will deliver innovative sets of cognitive/physical care services, personalized and cooperative interactions to boost social inclusion and engage user intimate companionship, preserving privacy and offering safety and security by-design.

Acknowledgments European Commission The research leading to these results has received funding from the EU H2020 research and innovation program under grant agreement No. 769872 (EMPATHIC) and N. 823907 (MENHIR), the project SIROBOTICS that received funding from Italian MIUR, PNR 2015–2020, D.D. 1735, 13/07/2017, and the project ANDROIDS funded by the program V: ALERE 2019 Università della Campania "Luigi Vanvitelli", D.R. 906 del 4/10/2019, prot. no. 157264,17/10/2019.7.

References

1. Troncone, A., Amorese, T., Cuciniello, M., Saturno, R., Pugliese, L., Cordasco, G., Vogel, C., Esposito, A.: Advanced assistive technologies for elderly people: a psychological perspective on seniors' needs and preferences (Part A). Acta Polytech. Hung. **17** (2), 163–189 (2020)
2. Esposito, A., Cuciniello, M., Amorese, T., Esposito, A.M., Troncone, A., Maldonato, M.N., Vogel, C., Bourbakis, N., Cordasco, G.: Seniors' appreciation of humanoid robots. In: Esposito A., Faundez-Zanuy M., Morabito F., Pasero, E. (eds.) Neural Approaches to Dynamics of Signal Exchanges. Smart Innovation, Systems and Technologies, vol. 151, pp. 331–345. Springer, Singapore (2020)
3. Maskeliunas, R., Damaševicius, R., Lethin, C., Paulauskas, A., Esposito, A., Catena, M., Aschettino, V.: Serious game iDO: towards better education in dementia care. Information **10**(355), 1–15
4. Esposito, A., Amorese, T., Cuciniello, M., Riviello, M.T., Esposito, A.M., Troncone, A., Torres, M.I., Schlögl, S., Cordasco, G.: Elder user's attitude toward assistive virtual agents: the role of voice and gender. J. Ambient Intell. Human Comput. (2019). https://doi.org/10.1007/s12652-019-01423-x
5. Esposito, A., Esposito, A.M., Vogel, C.: Needs and challenges in human computer interaction for processing social emotional information. Pattern Recogn. Lett. **66**, 41–51 (2015)
6. Esposito, A., Fortunati, L., Lugano, G.: Modeling emotion, behaviour and context in socially believable robots and ICT interfaces. Cogn. Comput. **6**(4), 623–627 (2014)
7. Buendia, A., Devillers, L.: From informative cooperative dialogues to long-term social relation with a robot. In Mariani et al. (eds.) Natural Interaction with Robots, Knowbots and Smartphones. Springer, New York, NY, pp. 135–151 (2014)
8. Moore, R.: A Bayesian explanation of the "Uncanny Valley" effect and related psychological phenomena. Nat. Sci. Rep. **2**(864) (2012)
9. Esposito, A., Faundez-Zanuy, M., Morabito, F.C., Pasero, E.: Towards socially and emotionally believable ICT interfaces. This volume (2020)

Neural Networks and Related Applications

The Simplification Conspiracy

Bruno Apolloni, Aamna Al Shehhi, and Ernesto Damiani

Abstract We study in a quantitative way the efficacy of a social intelligence scheme that is an extension of Extreme Learning Machine paradigm. The key question we investigate is whether and how a collection of elementary learning parcels can replace a single algorithm that is well suited to learn a relatively complex function. Per se, the question is definitely not new, as it can be met in various fields ranging from social networks to bio-informatics. We use a well known benchmark as a touchstone to contribute its answer with both theoretical and numerical considerations.

1 Introduction

Simplification is a keyword of modern social life covering the general aim of removing superfluous rules and actions from customary interactions among people and between people and institutions. Drawbacks may arise when we transfer this philosophy to the interaction between people and natural phenomena, e.g. on scientific matters. Undoubtedly, facing complex phenomena such as those involving biological organisms, we are compelled to simplify observations in order to obtain suitable, though approximate, explanations. Using the paradigmatic framework of artificial neural networks, we adopt a first simplification consisting of abandoning the search for a formal (symbolic) explanation of the physics of the phenomenon in favor of

B. Apolloni (✉)
Department of Computer Science, via Comelico 39/41, 20135 Milano, Italy
e-mail: apolloni@di.unimi.it

A. A. Shehhi
Khalifa University & Massachusetts Institute of Technology, Cambridge, MA, USA
e-mail: alshehhi@mit.edu

E. Damiani
Center on Cyber-Physical Systems, Khalifa University, Abu Dhabi, United Arab Emirates
e-mail: ernesto.damiani@ku.ac.ae

A. Esposito et al. (eds.), *Progresses in Artificial Intelligence and Neural Systems*, Smart Innovation, Systems and Technologies 184,
https://doi.org/10.1007/978-981-15-5093-5_2

a suitable simulation of it. To this end, one may consider using a neural network to combine symbolic functions that locally *understand* the phenomenon under study. This approach is known as the *hybrid ANN* paradigm [1]. In this paradigm, the neural network acts as *glue* connecting *parcels of formal knowledge*. Still in the simplification direction, we may replace the local symbolic functions with simple neural networks called *Gossiping Parcels* (GPs) to be combined with either majority voting or a decision tree on the parcel outputs when the overall explanation is a classification rule, or with a linear combination of the outputs in the case of a continuous function. All these schemes ask for a learning phase where either the glue and/or the parcels must be trained to simulate the phenomenon under study. This training can be regarded as *ensemble learning* [2], i.e. using multiple learning algorithms to obtain better accuracy than could be obtained from any of the constituent learning algorithms alone.

Looking at classification algorithms, in the past we developed a "learning by gossip" model where the opinions of different agents are variously combined to empower the conviction of any of them [3]. In this paper we focus on an ensemble of knowledge parcels that we combine via a linear model to produce a continuous function simulating the phenomenon under study. The basic scheme is the following: knowledge parcels are gossip producers that we can listen to, yet we can not modify them; we can only train a linear regression to find the best way to combine their outputs. This paradigm has been developed in terms of standard *Extreme Learning Machines* [4] when the parcels are simply nonlinear neurons (the hidden layer of a three-layers Forward Neural Network), and more in general of *Reservoir Computing* (Echo State Networks, Liquid State Machines, etc.) [5–7] when the parcels are more complex networks. Many results are available about the learning capabilities of these machines. Learning by gossip, as an implementation of ensemble learning, extends the learning facility from a "divide et impera" perspective, that in a modern acceptation could be reformulated as "Object Oriented Learning": in place of a monolithic code focused to solve the learning problem in toto, we fragment it in subproblems whose solutions are properly combined to get the final answers. The ways of fragmenting are various, the common goal is to deal with weak learners to combine in an optimal ensemble. The general idea is that weak learners are more simple to train, though less accurate, meanwhile the inaccuracy can be statistically recovered by the combiner (a perceptron, an SVM, a majority voter, etc.). In this paper we focus on an ensemble of trainable elementary neural networks on the same input, in the role of gossip producers, and a linear function of their outputs as their combiner. The latter is identified in one shot as an MSE regressor on the entire gossip ouput landscape.

We discuss the sample complexity of this contrivance contrasted with the one of the target function and elaborate on some numerical results of its implementation. As a result, we provide some hints to appreciate the benefits of our ensemble approach, as a function of the degree of *competence* of the GPs and of empowering expedients that are peculiar of this framework.

2 The Statistical Framework

We frame our approach in the Algorithmic Inference framework where random variable are described through a sampling mechanism as follows.

Definition 1 Denoting with X the random variable (r.v.), a sampling mechanism $\mathcal{M}(X)$ is a pair $(g_{\boldsymbol{\theta}}, z)$, where z is a specification of a completely known r.v. Z and $g_{\boldsymbol{\theta}}$ is a function mapping from $\mathfrak{Z}$ to $\mathfrak{X}$ such that, for each sample realization $\{z_1, \ldots, z_m\}$ randomly drawn from Z, the set $\{x_1 = g_\theta(z_1), \ldots, x_m = g_\theta(z_m)\}$ is a sample realization of X.

An example of universal sampling mechanism, suggested by the probability integral transform theorem [8], is represented by $\mathcal{M}_X = (U, \widetilde{F}_{X_\theta}^{-1})$, where U is a [0, 1] uniform variable and $\widetilde{F}_{X_\theta}^{-1}$ is a generalized inverse function of the CDF F_{X_θ} of the r.v. X we have selected. Namely, $\widetilde{F}_{X_\theta}^{-1}(u) = \min\{x | F_{X_\theta}(x) \geq u\}$.

In this scenario, an entire observation history, consisting of both the actual prefix—the *sample*—and the data we will observe in the future—the *population*—looks like a sequence of seeds z (in turn partitioned into those referring to the sample and those referring to the population) mapped into the history through an *explaining function* $g_{\boldsymbol{\theta}}$. As for the former, we can say nothing new, since they are randomly extracted from a perfectly known distribution; hence they are completely unquestionable as to the single value, and completely known as for their ensemble properties. *Vice versa*, the explaining function has $\boldsymbol{\theta}$ as a free parameter that we want to infer from the sample. We cannot say which is the value of $\boldsymbol{\theta}$ in the sampling mechanism, as we do not know the related seeds. Rather, we may transfer the probability mass of the seed from the sample to the parameter value realizing the sample. This is a well-known strategy, adopted in many approaches, such as [9–11], which misses, however, the *univocity issue*.

In terms of the gossip framework outlined in the previous Section, seeds are multivariate, as they are represented by the outputs of the GPs ensemble (GPE). This is not a drawback for our sampling mechanism definition, apart for the knowledge of their distribution that we may normally achieve only by simulation. A further drift from the Algorithmic Inference approach may concern the independence of the seeds. We start our analysis with a basic version of the model where GPs are *three-layers perceptrons* whose hidden layer is activated by a sigmoidal-logistic function. The output layer is linearly activated and uniformly random bound weights are used in all layers (the feed-forward neural networks (FNN) in Fig. 1). In this model, seeds comply with their definition in Definition 1, where a multivariate X is the output of the neural networks that in turn is the seed Z of the varible Y produced by the combiner. Different sampling mechanisms, hence different X distributions, correspond to different weights extractions, and, thank to the weights independence, their ensemble is a suitable Z for Y . Things become more complicated if we decide to train the GP. Complexity does not derive from a reduction of the parameters' randomness *per se*; actually function $g_{\boldsymbol{\theta}}$ in Definition 1 may be completely deterministic

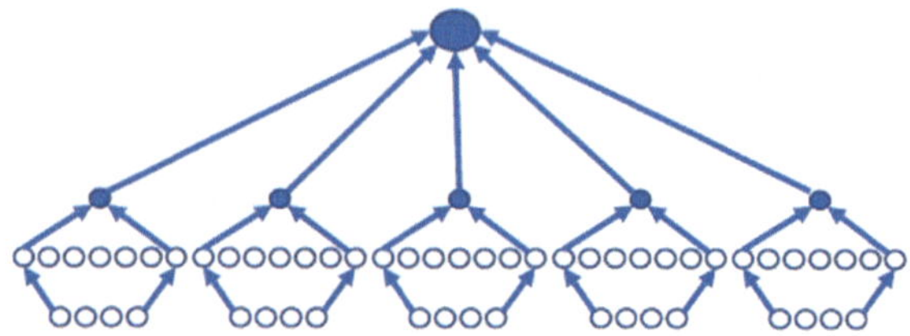

Fig. 1 The basic model

too. Rather, it derives from the dependency of the learned parameters on the training set that determines a dependency between seeds as well.

3 Learning by Gossip

Let us consider a typical inference instance of the AI framework

Example 1 Let X be a negative exponential r.v. whose CDF and sampling mechanism are defined as:

$$F_X(x) = 1 - \mathrm{e}^{-\lambda x} I_{[i,\infty)}(x); \qquad x = \frac{-\log u}{\lambda} \tag{1}$$

where u is a seed uniformly drawn in $[0, 1]$ and $\lambda \in [0, \infty)$ is the unknown parameter to be inferred on the basis of an m-sized sample $\boldsymbol{x}$. If we identify a suitable (well-behaving in [12]) statistic the sufficient statistic $s_\Lambda = \sum_{i=1}^{m} x_i$, a monotonic non increasing relationship reads:

$$\left(\lambda \leq \widetilde{\lambda}\right) \Leftrightarrow \left(s_\lambda \geq s_{\widetilde{\lambda}}\right) \tag{2}$$

where $s_{\widetilde{\lambda}} = \sum_{i=1}^{m} \widetilde{x}_i$ and $\widetilde{x}_i$ is the value into which u_i would map if we substitute λ with $\widetilde{\lambda}$ in the explaining function. As S_λ follows a Gamma distribution law with shape and scale parameters respectively m and $1/\lambda$ [8], we have that Λ as well follows a Gamma distribution of parameters m and $1/s_\Lambda$, thanks to the commuting roles of variable specification and parameter between s_λ and λ.

Using the gossip responses Z as seeds has the drawback of working with a unknown seed distribution law that my be recovered only by simulation. As a counterpart, computing the output of the questioned function f as a linear function of Z makes the statistic mean square error $S = \sum(t - y)^2$—with $t =$ the target value and $y =$ the value actually computed by the learned function—to be individually sufficient [8] w.r.t. the coefficient of the linear function. Rather than on the values of the regression coefficients, we focus directly on S (or analogous ones) to appreciate the quality of our inference. Denoting with Σ the extension of the above sum over the entire Y population, i.e. to the entire input to the learned function, S is an estimate of Σ that we may parametrize as follows

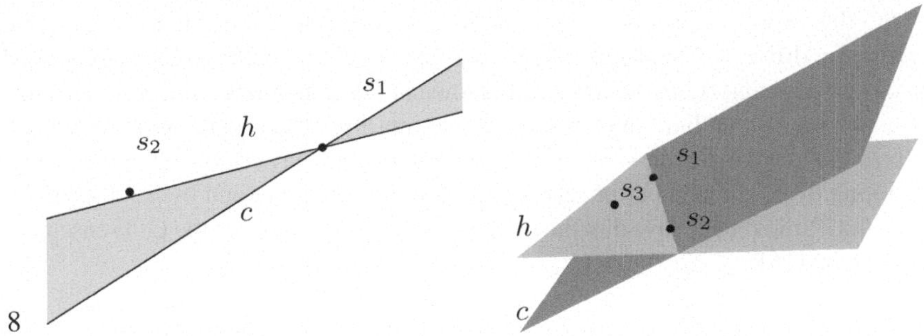

(a) (b)

Fig. 2 Sentry points needed in the worst case to bind the symmetric difference between the hypothesis h and the target concept c in **a** two-dimensional and **b** three-dimensional spaces

$$S = h(\mathbf{X}, \mathbf{W}) \tag{3}$$

where $\mathbf{X}$ is a sample in input to f and $\mathbf{W}$ is the set of the three-layer perceptron weights. For fixed $\mathbf{w}$ we face the usual bias-variance trade off. For any training set—test set split $\mathbf{x}$, different $\mathbf{w}$s denote different representations of the learning problem, among which we may screen the more efficient ones (see Sect. 3.2). In accordance to our model, we read the entire GPE contrivance as a subsymbolic kernelization of a linear regression problem. Kernels are originated by a mapping function ϕ so that the Kernel matrix at the bassis of the support vector discrimination is

$$K_{ij} = \phi(\mathbf{x}_i) \cdot \phi(\mathbf{x}_i)$$

where ϕ is decided a priori. In our case, we aim to learn ϕ exactly.

To clarify our approach, let us formalize a binary version of the problem as follows. Let us consider a straight line in the two-dimensional case and a plane in the three-dimensional one, as linear separator templates dividing positive from negative points as in picture (Fig. 2). One of the linear delimiters is the true divider—the concept c, the second is a hypothesis h about it. To avoid the growth of the symmetric difference between concept and hypothesis, we need at most 2 or 3 points (we call them *sentry points*) that bar a rotation of the hypothesis so as to have another symmetric difference *completely* including the current one. Sentry poits provide a way of characterizing the sample complexity of a class of concepts [13] (hyperplanes, in ouer case), that is dual to the Vapnick Cervonenkis complexity [14]. There are theorems in the literature that establish the equivalence between these two notions of complexity, but the notion adopted here allows to visualize the role of the points determining the complexity [15]. In our case, the points to be divided are the results of (weakly trained) GPs . This is another way of kernelizing the X space where the original points lie.

In other words, sentry points are minimal sets of points capable of binding the symmetric difference between concepts in a given class and hypotheses generated by a consistent algorithm. Sentry points shown in Fig. 2 are representatives of groups of sentry points that bind in own turn the concept-hypothesis symmetric difference related to the weak learners. In terms of our model, we may establish the following. Denoting by n_c the number of sentry points of our combiner and by n_{w_i} the one of the i-th GP, the theory says that the number n_t of sentry points of the GPE ensemble is given by [15]

$$n_t \leq n_c \times n_w \tag{4}$$

Equation 4 represents a more binding inequality of the analogous one which holds on the growth functions [16]. In our case the common value n_w of n_{w_i}s is $O(NLog(N))$, where N is the total number of training parameters of the multi-layer perceptron realizing the weakly trained GPs [17]. The detail n_c of a hyperplane (the target of our learning) is equal to the hyperplane dimension. Within this framework, we will now investigate the efficiency of our learning paradigm and its exploitability beyond the usual statistic properties.

3.1 Efficiency Versus Size

From (4) we see that the sample complexity of GPEs is independent from the target function f and training time as well. Vice versa, the approximation of the trained function g to f depends on the gap between the sample complexity of f and the one of g. In synthesis, we can modulate our learning effort by enriching the complexity of our contrivance as much as f is complex. This can be done, for instance, by enlarging the number of GPs or by increasing their hidden layer size. As much as n_t increases, so much the training time increases. Binding the training time of the GPs provides poor hypotheses that revert in a smaller n_w.

Training GPs has the effect of inducing dependency between the seeds Z corresponding to the items of **X**, in spite of the Independence of the **W**. This behavior, that we may quantize in terms of correlations between the differences of the target vector minus its version leaned by the various GPs, is another ambivalent aspect of our training procedure. On the one hand a high correlation denotes that we are uniformly training the various GPs, in spite of their random initialization. On the other hand, too similar GPs are useless, since they provide a similar information to the combiner. The last aspect induces a virtual reduction of n_{w_i}s.

3.2 Beyond Probability

Probability looks the most comfortable tool for describing uncertain situations. Provided you know your sample space and you have enough observations of it, then you can organize the observations into statistics so as to reliably infer general properties of the sample space that may prove useful in future occasions. The key is to relate the statistics to the properties in a way that the frequency with which these properties are falsified is asymptotically as small as you want. This is the basis strategy of Algorithmic Inference [18].

The current framework is notably more complex: we have two families of seeds, one underlying the random input **X**, another the random weights **W** and a subsymbolic function (the neural network) relating them to the questioned property, i.e. the MSE Σ of the inferred g. This foils any effort to infer the Σ distribution law. Rather, we may try conditioning our operational framework so as to favor low values of Σ. We acquaint this problem from the *ergodic processes* perspective.

Let us consider a sequence of random variables **X** on the same discrete and finite probability space. It means that the set $\mathfrak{X}$ of values the random variable may assume along the sequence is the same. What changes is the probability distribution on them. Thus, the probability mass of a value is a function of two parameters: the specific value concerned and the step along the sequence at which we are questioning the probability. In the Markov process this step is denoted as a time clock, where the time progress is beaten by the transform applied to the probability distribution over X. It is a transition matrix M so that $P_{t+1} = M P_t$. In our case **W** randomly generated at each clock affects the Σ distribution in a way that is independent of the randomness of **X** (since we train on a given **x**) . In general terms, a process is denoted as ergodic if for any *regular* function ψ the sample mean of ψ along an infinite sampled trajectory of **X** equals the expected value of $\psi(\mathbf{X})$ at any clock time. Our goal is to enforce a similar property on the random process of our computation, so that the observation of a sequence of MSE for a fixed **X** sample along a long sequence of **W** will give us insight about properties of Σ for a fixed **W** along a long sequence of **X** samples. In this way we aim at identifying an optimal **W** which minimizes the generalization MSE, expecting that this optimality remains also when further **X** samples will be considered. Hence our first question is about the ergodicity of our pseudo-process. From the algorithmic perspective, **X** and **W** play a symmetric role in the input a to the FNNs hidden layer. For instance, let consider a three (input, hidden, output) layer GP and split **W** into $\mathbf{W}_{01}$ weights between input and hidden layer and $\mathbf{W}_{12}$ between hidden and output layer. We simply obtain:

$$\mathbf{a} = \mathbf{W}_{01} \cdot \mathbf{x}$$

Any subsequent computation depends on **a**, apart from the final linear regression which depends separately on **x**. Moreover, a proper rescaling of both variables induces similar ranges in the two directions **X** and $\mathbf{W}_{01}$, while the prevalence of linear operators leads to Normal distributions in both cases. Numerical experiments discussed in the next section will confirm this analysis. Hence, given this directions' *twisting* for grant, on a given sample we simulate many GPEs and focus on the one computing the minimum S. On the one hand we remark that the cumulative distribution of minimal Σ goes to be biased toward to 0 with the number of ensembles—a condition that makes the observed minimum close to the optimum. On the other hand our question is: who tell us that this optimality condition is preserved on a new instance of **X**? Actually, our inference problem is to learn a hyperplane, that is the most elementary function to be inferred. Indeed, VC dimension and detail of a hyperplane are equal to the hyperplane dimensionality d. This makes S extremely close to the Σ with high probability $(1-\eta)$—that meets our wish, namely:

$$\Sigma \leq S + \sqrt{\frac{d(\log(2N/d) + 1 - log(\eta/4)}{N}} \tag{5}$$

where N is the sample size.

4 Preliminary Results

We started with an elementary case study where we aim to learn the well-knowm Pumadyn benchmark `pumadyn8-nm`. This benchmark is drawn from a family of datasets which are synthetically generated from a Matlab simulation of a robot arm [19]. It contains 4, 500 samples, each constituted by 8 inputs and one output. The former record the angular positions and velocities of three arm joints plus the value of two applied torques. The latter is the resulting angular acceleration of one of the joints. This acceleration is a nonlinear function of the inputs which is affected by moderate noise as well. Our reference result is the one obtained some years ago through a special 5 layers FFN where the neurons of a layer are allowed to move inside it to get the most rewarding position in respect to he neurons of the upper layer [20]. We appreciate the generalization capability of the network in Fig. 3a, where we represent in gray the sorted test set targets and in black the corresponding values computed by our network for one 512 both training and testing size replica of a Delve testing scheme [21].

We replace this complex neural network with our contrivance getting results like in Fig. 3b. Namely, we span a set configurations where we stressed:

- the GP architecture: either 3Layer $(8, n_h, 1)$ or 5L$(8, n_h, \lfloor n_h/2 \rfloor, \lfloor n_h/3 \rfloor, 1)$

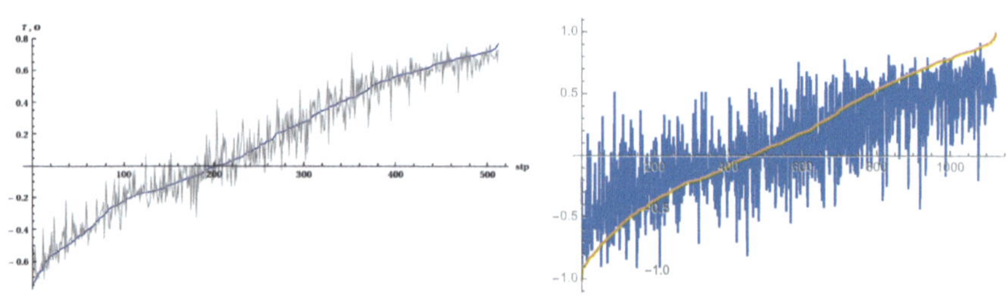

Fig. 3 Errors on Pumadyn regression. Course of the network output with sorted target patterns achieved by **a** a complex neural network, **b** one of the most performing GPE

Table 1 Features and conditions

Minima and extremes

Experiment	Training		Testing	
	Min	Exteme	Min	Exteme
$3L\alpha = 0.00005$	0.0738381	0.0755341	0.0818115	0.0885704
$5L\alpha = 0.00005$	0.063183	0.063183	0.0740216	0.0750288
$5L\alpha = 0.0005$	0.063165	0.0634284	0.0656679	0.0694787

- number n_h of neurons in the main hidden layer, ranging from 60 to 160
- number of GPs, ranging from 30 to 90.

We recall that on the one hand we train the GPs in parallel using a rather regular backpropagation and sigmoidal-logistic activation function—except the output layer where the activation function is linear. Then we solve in one shot the identification of the regression coefficients of the combiner neuron. Hence, another pair of operational parameters we investigate are

- number of training epochs, ranging from 200 to 500, and
- learning rate, ranging from 10^{-2} to 10^{-7}.

Figure 4 reports the course of the GPE MSE in training and testing in the various operational conditions.

From these pictures we see that enriching the GP architecture generally pays in terms of training and testing error. The smaller learning rate (5×10^{-6}) insures a regular descent of these errors that prevents overfitting. Table 1 denotes that a more aggressive learning (learning rate = 5×10^{-5}) may provide somehow smaller errors with a non clear parametrization that produces overfittinfg, since the min is not in correspondence of the maximal degrees of freedom and training epochs (where we registered the extreme value). The MSE trends raise some questions about the gain of ensemble learning versus single GP learning. The graph in Fig. 5 shows a gain around 5, with obviously better benefits with the growth of the training degrees of freedom. This trend reassures us on the advantage of having a GP team in spite of the high correlation of GPs' results we mention in Sect. 3.1. Actually the trend in 5 L architecture is shown in Fig. 6a, whereas we expect a similar asymptote with the 3 L architecture as in Fig. 6b. However, the spread around the true values is well distributed among the GPs, as it emerges from the pictures in Fig. 7.

Thus, a first lesson we may learn from this experiment is that adding knowledge to GPE by training its GPs pays in terms of MSE performance. A second lesson concerns *meritocracy*. Does it make sense to focus on the best performing GPEs? Fig. 8a fosters a positive answer to this question. Red graph and blue graph represent the empirical generalization MSE CDFs from a population generated from a single sample and different replicas of the GP set and from a single GP set and different samples, respectively. Since hyperplanes are driven by the sampled values, the second population has a greater number of degrees of freedom that reflects into a better

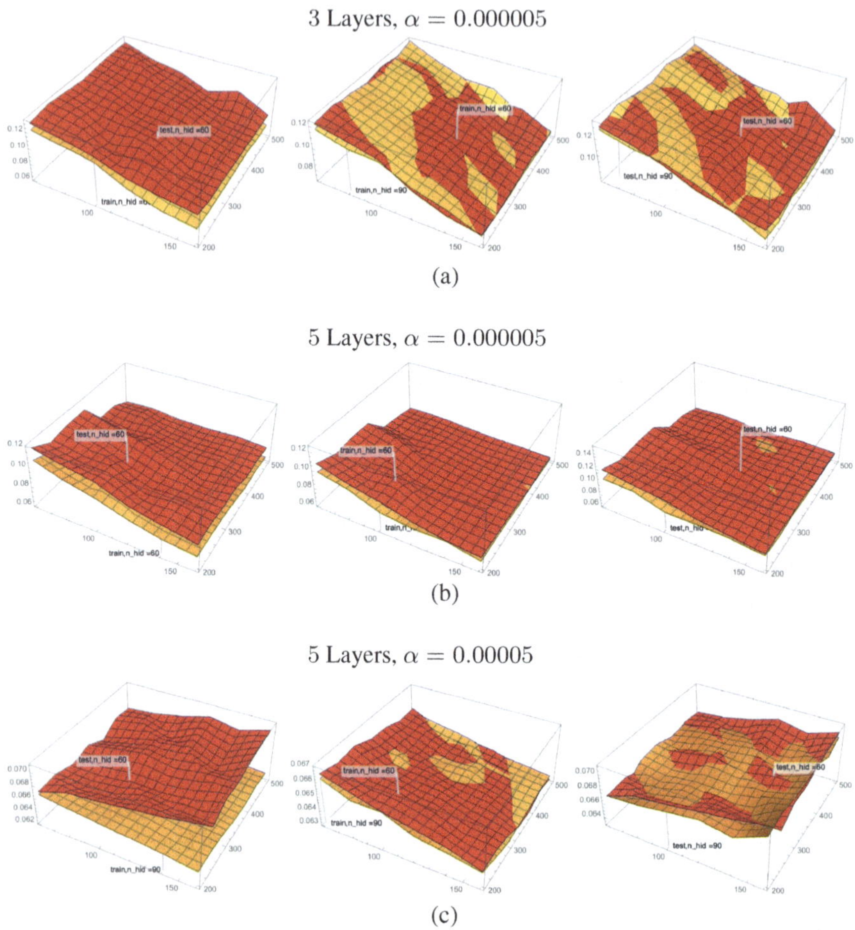

Fig. 4 MSE trend with number of hidden nodes varying from 60 to 160 and number of epochs varying from 200 to 500. Labels on the graph specify the number of GPs, the number of layers and the learning rate

descent toward minimal MSE—that implies some drift from the ergodicity assumption in Sect. 3.2. However, our operational scenario is: given a sample, look for the most convenient GPE. Hence we selected a GPE close to the optimal, namely the one inducing a generalization error S= 0.0327, and used it on 150 more samples. The red graph in Fig. 8 shows that this GPC maintains its value on average, with obvious spreads due to the sample variations. As a matter of fact, there is a high correlation between training set and test set MSE. For instance in Fig. 8a we followed the evolution of these values on the experiment 5 L, $\alpha = 10^{-6}$ which denote a correlation equal 0.992.

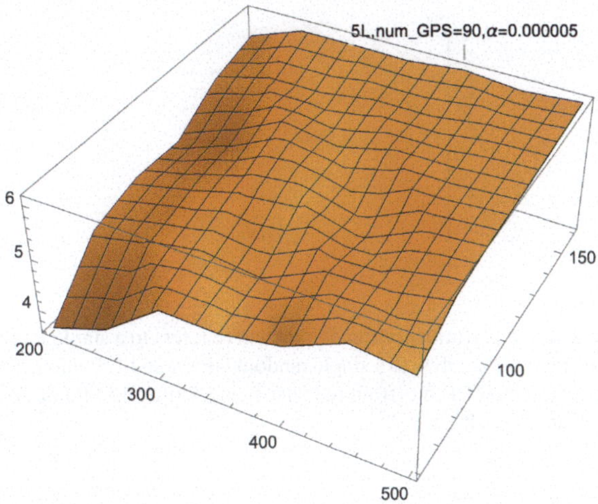

Fig. 5 Ratio of Mean GP training MSE over ensemble training MSE with the number of hidden nodes and of training epochs

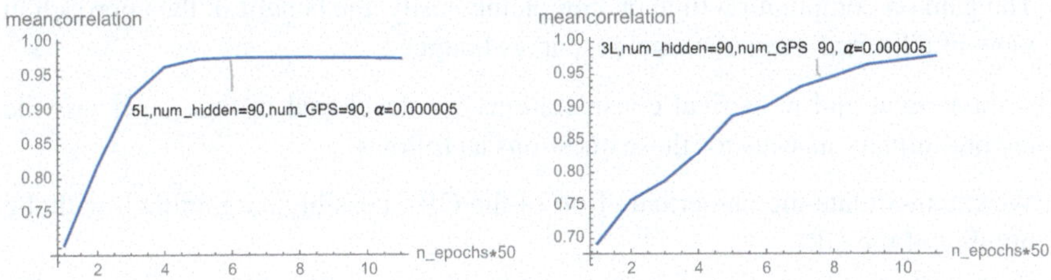

Fig. 6 Growth of the correlation between the errors of the various GPs with their training

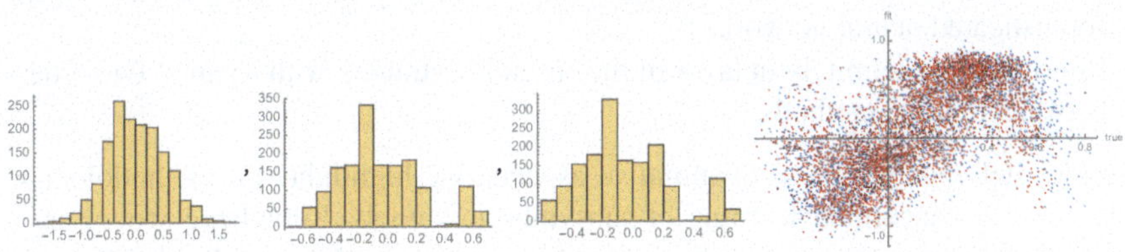

Fig. 7 Hitrogram of the GPS spreads around three true points and true-fitted values scatterplot of two (blue, red) GPs

5 Conclusions

In this paper we formalized an ensemble learning setting, based on our Learning by Gossip approach, and analyzed its properties. Our results help the user to understand the potential benefits of this approach. The parameters that are taken into consideration generally are:

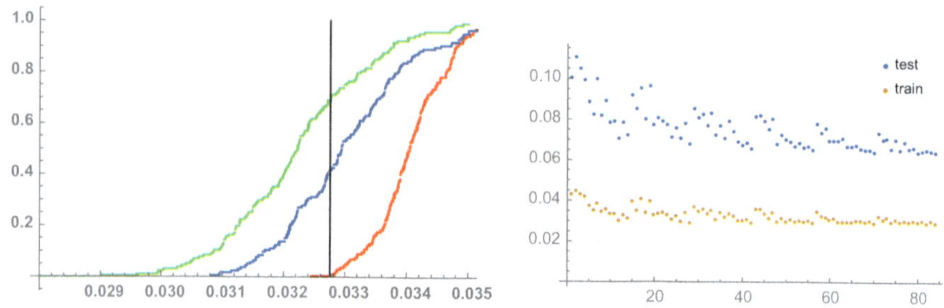

Fig. 8 Left: Generalization performance trends. Red curve refers to a single Pumadyn sample and 150 random GP sets; blue curve refers to a single random GP set and 150 Pumadyn samples. Green curve refers to a quasi-optimal GP set (from red curve) and further 150 Pumadyn samples. Right: Joint course of Train and test MSEs

- The sample complexity of the learning problem demanded to the weak learners (the one of the combiner is relatively low by definition)
- The accuracy of the final solution
- The gain on computation time or, complementarily, the benefit of the approach in view of a limited computational resources budget

The theoretical and numerical considerations we developed in this paper provide some preliminary answes tor these questions as follows.

- We can modulate the numerical effort of the GPs, possibly bargaining it with the number of the GPs.
- Both pushing the single GP to improve its learning capabilities, and increasing their number pays off in terms of accuracy. However a gap of around half order exists between our best MSE and the one we achieved with a single but highly sophisticated neural network.
- Runtime goes definitely in favor of the ensemble strategy with a gap of two orders between the two methods.

Besides delving into these quantitative aspects, we highlighted a methodological one as well, concerning learning of subsymbolic kernels. Kernelized spaces are a formidable resource to get ride of some learning problems. However the identification of a proper kernel remains in the domain of intuition and trial-and-error methods, with the further problem of rendering kernel implementation as less costly as possible [22]. Our approach provides a way of *learning* the mapping functions originating the kernels with a computational effort that remains up to us. What we propose in this paper is just the start of a research line that we plan to develop in the future with more extensive both theoretical and numerical investigations.

References

1. Apolloni, B., Kurfess, F. (eds.): From Synapses to Rules—Discovering Symbolic Rules from Neural Processed Data. Kluwer Academic/Plenum Publishers, New York (2002)
2. Dietterich T.G.: Ensemble methods in machine learning. In: Multiple Classifier Systems, pp. 1–15. Springer Berlin Heidelberg, Berlin, Heidelberg (2000)
3. Apolloni, B., Malchiodi, D., Taylor, J.: Learning by gossip: a principled information exchange model in social networks. Cogn. Comput. **5**, 327–339 (2013). https://doi.org/10.1007/s12559-013-9211-6
4. Huang, G.-B., Chen, L.: Convex incremental extreme learning machine. Neurocomputing **70**(16), 3056–3062 (2007)
5. Lukoševičius, M., Jaeger, H.: Survey: reservoir computing approaches to recurrent neural network training. Comput. Sci. Rev. **3**(3), 127–149 (2009)
6. Lukoševičius, M.: A Practical Guide to Applying Echo State Networks, pp. 659–686. Springer, Berlin Heidelberg, Berlin, Heidelberg (2012)
7. Zhang, Y., Li, P., Jin, Y., Choe, Y.: A digital liquid state machine with biologically inspired learning and its application to speech recognition. IEEE Trans. Neural Networks Learn. Syst. **26**(11), 2635–2649 (2015)
8. Wilks, S.S.: Mathematical Statistics, Wiley Publications in Statistics. Wiley, New York (1962)
9. Hannig, J.: On generalized fiducial inference. Statistica Sinica **19**(2), 491–544 (2009)
10. Iyer, H.K., Patterson, P.: A recipe for constructing generalized pivotal quantities and generalized confidence intervals, Tech. Rep. 2e002/10, Department of Statistics, Colorado State University (2002)
11. Martin, R., Liu, C.: Inferential models: a framework for prior-free posterior probabilistic inference. J. Am. Stat. Assoc. **108**(501), 301–313 (2013)
12. Apolloni, B., Pedrycz, W., Bassis, S., Malchiodi, D.: The Puzzle of Granular Computing. Springer, Berlin (2008). https://doi.org/10.1007/978-3-540-79864-4
13. Apolloni, B., Bassis, S., Malchiodi, D.: Compatible worlds. Nonlinear Anal.: Theory, Methods Appl. **71**(12), e2883–e2901 (2009)
14. Vapnik, V.: The Nature of Statistical Learning Theory. Springer, New York (1995)
15. Apolloni, B., Malchiodi, D.: Gaining degrees of freedom in subsymbolic learning. Theoret. Comput. Sci. **255**, 295–321 (2001)
16. Abu-Mostafa, Y.S.: Hints and the vc dimension. Neural Comput. **5**(2), 278–288 (1993)
17. Baum, E.B., Haussler, D.: What size net gives valid generalization? Neural Comput. **1**(1), 151–160 (1989)
18. Apolloni, B., Malchiodi, D., Gaito, S.: Algorithmic Inference in Machine Learning, vol. 5. Advanced Knowledge International Pty, ADELAIDE—AUS (2006)
19. Corke, P.I.: A robotics toolbox for matlab. IEEE Robot. Autom. Mag. **3**(1), 24–32 (1996)
20. Apolloni, B., Bassis, S., Valerio, L.: Training a network of mobile neurons. In: The 2011 International Joint Conference on Neural Networks, pp. 1683–1691 (2011)
21. Rasmussen, C.E., Neal, R.M., Hinton, G.E., van Camp, D., Revow, M., Ghahramani, Z., Kustra, R., Tibshirani, R.J.: The delve manual (1996). http://www.cs.toronto.edu/~delve/
22. Cesa-Bianchi, N., Mansour, Y., Shamir, O.: On the complexity of learning with kernels, CoRR abs/1411.1158. http://arxiv.org/abs/1411.1158

Passengers' Emotions Recognition to Improve Social Acceptance of Autonomous Driving Vehicles

Jacopo Sini, Antonio Costantino Marceddu, Massimo Violante, and Riccardo Dessì

Abstract Autonomous driving cars hopefully could improve road safety. However, they pose new challenges, not only on a technological level but also from ethical and social points of view. In particular, social acceptance of those vehicles is a crucial point to obtain a widespread adoption of them. People nowadays are used to owning manually driven vehicles, but in the future, it will be more probable that the autonomous driving cars will not be owned by the end users, but rented like a sort of driverless taxis. Customers can feel uncomfortable while riding an autonomous driving car, while rental agencies will need to differentiate the services offered by their fleets of vehicles. If people are afraid to travel by these vehicles, even if from the technological point of view they are safer with respect to the manually driven ones, customers will not use them, making the safety improvements useless. To prevent the occupants of the vehicle from having bad feelings, the proposed strategy is to adapt the vehicle driving style based on their moods. This requires the usage of a neural network trained by means of facial expressions databases, of which there are many freely available online for research purposes. These resources are very useful, but it is difficult to combine them due to their different structures. To overcome this issue, a tool designed to uniform them, in order to use the same training scripts, and to simplify the application of commonly used postprocessing operations, has been implemented.

J. Sini (✉) · A. C. Marceddu · M. Violante
Politecnico di Torino, Turin, Italy
e-mail: jacopo.sini@polito.it

M. Violante
e-mail: massimo.violante@polito.it

R. Dessì
General Motors – Global Propulsion Systems, Turin, Italy
e-mail: riccardo.dessi@gm.com

A. Esposito et al. (eds.), *Progresses in Artificial Intelligence and Neural Systems*, Smart Innovation, Systems and Technologies 184,
https://doi.org/10.1007/978-981-15-5093-5_3

1 Introduction

The development of autonomous driving vehicles is a challenging activity, with technological, ethical, and social implications. The main driver of this development is an expected increase in road safety [1]: the idea is that such vehicles, equipped with complex sensors like LIDAR, RADAR, cameras and so on, will be more able, with respect to human drivers, to avoid (or at least to mitigate the consequences if inevitable) crashes.

People used cars for over a hundred years, and during all this time period the vehicles have been operated by human drivers. Due to this habit, it is simpler for people to trust another person with respect to a computer to drive their vehicles. Moreover, we expect that drivers adopt different styles depending on their emotions [2]. So, it is important to take into account their social implications in order to avoid making the expected safety improvement useless due to a lack of people's trust on those vehicles. These regards ethic issues, social trust on autonomous driving capabilities, and novel commercialization model for the car manufacturers. In this paper, we would like to analyze the problem and propose an idea to improve the trust in these new driverless cars.

We wish to draft a preliminary workflow to perform autonomous vehicles passengers' emotions detection, in order to improve their comfort and so their trust. From a technical perspective, we would like to present a software tool to simplify the training of a neural network that, starting from the passengers' faces pictures, recognize their emotions. It could be used to adapt consequentially the autonomous cars driving style. As evidenced by various studies [3], positive emotions would consider faster or riskier driving as more favorable than people with negative emotions so, regarding the driving style adaptation, we can make more precise clarifications about what we wish to achieve:

- if the people on the car are scared or sad, the car adopts a caring driving style, that slows the speed and takes the curve more accurately (lowering the lateral accelerations);
- if the people on the car are neutral, the car adopts a normal driving style;
- if the people on the car are happy or over enjoyed, the car adopts a sportive driving style (steeper acceleration/braking ramps and curve trajectory with a higher level of lateral accelerations).

To obtain these results, as the first step in the development of the approach, a novel software has been developed. It is designed to allow usage of various third-party facial expressions databases for the training, the validation, and the testing of the emotions recognition neural network. It also simplifies the applications of commonly used postprocessing operations on the images, like grayscale conversion and histogram equalization.

2 State of Art

In recent years the miniaturization, which allows the creation of even smaller devices, the spread of smartphones with better and better cameras, the continuous improvement of wireless connections, and the advent of neural networks and other technologies have opened the door to the use of these techniques for different purposes. These devices can be easily embedded inside vehicles in a cost-effective way.

Neural networks have been used in recent years to solve various problems. For example, the LeNet-5 network [4] proposed by Yann LeCun, Joshua Bengio and Patrick Haffner has been used to automatically read the digits on bank cheques in the United States and is still used nowadays in the industry. Other examples of these technologies are the artificial intelligence developed by DeepMind [5], a subsidiary of Alphabet Inc.'s Google, like AlphaGo, AlphaZero, and the last AlphaStar, that was capable of beating different professional StarCraft players.

Around the 1970s, psychologist Paul Ekman [6] identified six basic emotional states common to all cultures: anger, disgust, fear, happiness, sadness, and surprise. Further emotional states were subsequently theorized and then extended by Paul Ekman himself and other researchers [7–9]. For the sake of this paper, we considered these six states with the addition of the neutrality and contempt states.

Starting from the 2000s, several databases depicting human faces have been published with the aim of improving the algorithms for identifying facial expressions [18]. These databases are composed of posed and non-posed photos, and sometimes show additional data such as, for example, the Facial Action Coding System (FACS) coding [10, 11] and the Active Appearance Model (AAM) facial landmarks [12, 13].

Various authors [14, 15] described in detail that the interactions between cars, drivers, passengers and the ownership of cars are really complex [16]. Autonomous driving will put in place a new era in car ownership and usage, so car manufacturers need to understand how passengers will want to use that kind of vehicles. Other than that, to make things even more complex, it is reasonable to expect a time (from 10 to 30 years) in where driver-owned manually driven cars coexist and compete on the market with rented autonomous driving vehicles.

A lot of effort has been spent in the development of autonomous driving cars, considered in both owners, regulatory bodies and passengers' perspective [1]. But some questions remain open. In the future, the customers will want to own a car or they will become an on-demand transportation service only like taxis, trains or airplanes? If the answer is the second one, how can different carmakers and their new business-to-business customers, in this case transport companies and not the passengers, who will hold autonomous cars fleet, differentiate from each other in order to reach more final users? And finally, how much the customers be willing to pay this kind of service? At the moment, it is difficult to forecast answers to these questions, so they need to invest into the user experience of passengers, as much as airlines and railways companies did.

We claim that an enabling technology to improve the customers experience could be a properly trained neural network able to recognize the passengers' emotions. In this way, it will be possible to modify the vehicle behavior based on the passengers' feelings detected during the ride.

3 Proposed Approach

Our proposal is to use cameras mounted in front of each seat of the car, and a properly trained neural network in order to adapt the vehicle driving style basing on the passengers' emotions. We claim that in this way it will be possible to improve the passengers' trustiness on these vehicles, hence their social acceptance. To avoid privacy issues, all the related emotions detection computations will be executed onboard thanks to a dedicated ECU, without any transmission to the external world. Moreover, the algorithm does not need to store into a permanent memory the frames representing the passengers' faces. This device is not directly involved in safety-critical tasks, so it can be implemented in a cost-effective way. The onboard system (Fig. 1) will be composed of:

- micro-cameras, placed in front of each seat to take pictures of the passengers' faces;
- a neural network, to recognize the emotions of each passenger;
- a decision algorithm, that determines the vehicle configuration on the base of the recognized passengers' emotions;
- a centralized emotions detector and control management computer, to tune vehicle parameters dynamically to improve the passengers' experiences.

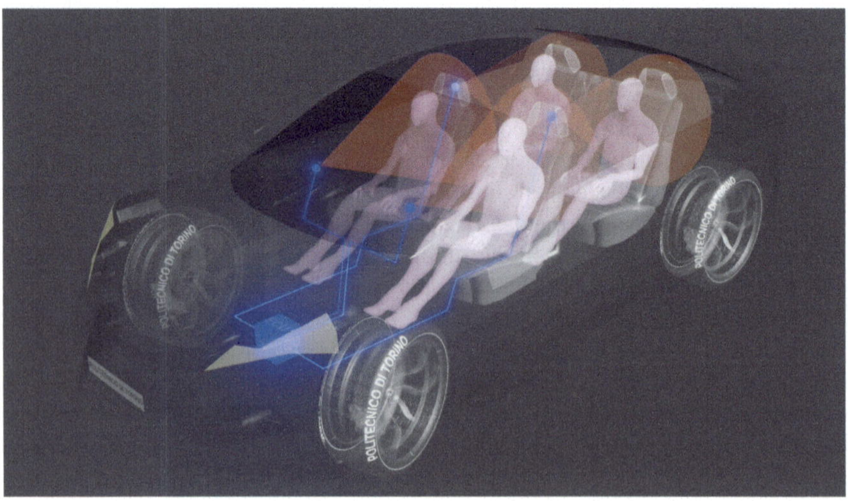

Fig. 1 Graphical representation of the proposed

The workflow to obtain the dataset needed to train this neural network is the following. Firstly, it is necessary to obtain the databases needed to train the neural network. There are many databases containing facial expressions that can be used for research purposes, which will be discussed later.

After this step, it is possible to start with the training, validation, and testing of the neural network. After that a sufficient emotions recognition accuracy has been achieved, a test campaign on a real vehicle can be started, in order to verify the effectiveness of the network in different lighting conditions (sun position, shadows due to the frame of the car, night driving) and in more realistic scenarios.

The last part of the work will regard how to properly manage the driving style depending on the detected feelings.

For the sake of this paper, we want to focus on the first phase, regarding the datasets preparation activities. To improve the effectiveness of this phase, we chose to develop a novel software that we called Facial Expressions Databases Classifier (FEDC).

FEDC is a program able to automatically classify images of some of the most used databases, depicting posed human faces:

- Extended Cohn-Kanade Database (CK+) [17, 18];
- Facial Expression Recognition 2013 Database (FER2013) [19];
- Japanese Female Facial Expression (JAFFE) [20];
- Multimedia Understanding Group Database (MUG) [21];
- Radboud Faces Database (RaFD) [22].

In all databases, emotions were classified according to Ekman's framework, but from a technical point of view, each author has adopted a different structure to organize the images. Since we would like to use pictures from multiple databases, without having to adapt the neural network training script for each of these, we have chosen to develop a novel program to uniform the classifications.

Since FEDC exploits the labels provided by the databases' creators, the accuracy relies only upon the classifications performed by them. In addition to this, it is also able to do several useful operations on the images, in order to simplify the neural network training operations and enhance its performances:

- scaling of the horizontal and vertical resolutions;
- conversion in grayscale color space;
- histogram equalization;
- face detection to crop the images to face only.

This allows, for the people who make use of these databases, to minimize the time necessary for their classification, so that they can dedicate directly to other tasks, such as training of a neural network.

3.1 FEDC Description

FEDC has a clean and essential user interface (Fig. 2), consisting of four macro areas:

- in the left column, it is possible to choose the database to be classified;
- in the right column, it is possible to select the operations to be performed on the photos: those available have already been mentioned previously;
- in the lower part of the window, there are the buttons to select the input file, the output folder, and to start and cancel the classification;
- finally, above the buttons, there is the progress bar, that indicates the progress of the current operation.

It should be noted that:

- the user must choose a size for the photos to be classified: it must be between 48×48 and 1024×1024 pixels. For the FER2013 database, since the starting images have a 48×48 pixels resolutions, this possibility, alongside to the face cropping feature, is not available;
- the JAFFE and the FER2013 databases contain grayscale-only images;

Fig. 2 Graphical User Interface of FEDC

- the RaFD database also contains photos taken in profile: the program excels in the recognition of frontal photos and allows recognition to be made even for profile photos, although it is likely that it will not be able to classify all these photos.

The images automatically classified with this program can be used, for example, for training a neural network with Keras [23] or similar frameworks: using Python with the Scikit-learn library [24], these images can be subdivided in the training, validation and test datasets or cross-validated.

Antonio Costantino Marceddu, that is one of the authors of this paper, developed FEDC resorting to Eclipse, with Java and the addition of the OpenCV framework [25], and released his software under the MIT license on GitHub [26].

4 Conclusions

Social acceptance implications on autonomous driving cars have to be considered in autonomous driving vehicles development. To achieve this goal, customers' trustiness plays a key role. In this paper, a possible improving methodology focused on the adaptation of the vehicle driving style, based on the passengers' emotions have been proposed.

At the moment, the effort of the authors is directed on the first stage of the methodology development, which is the data preparation for the training, validation, and test of the neural network, using pictures from the various databases of facial expressions available online for research purposes. Once that the neural network is trained, it will be possible to perform tests on cars in order to verify the emotions detection algorithm capabilities from the pictures captured with onboard cameras, in grayscale, in daylight or IR at night. When an acceptable matching rate will be reached, the next steps will be the development of the algorithms for car parameters tuning and for choosing the prevailing emotion, since the passengers can have conflicting moods. The last point will be the finding of an acceptable frequency to update the driving style.

References

1. Burns L.D.: Sustainable mobility: a vision of our transport future. In: Nature 497(7448), 181–182 (2013)
2. Trògolo, A.M., Melchior, F., Medrano, L.A.: The role of difficulties in emotion regulation on driving behavior. J. Behav. Health Soc. Issues (2014). https://doi.org/10.5460/jbhsi.v6.1.47607
3. Hu T, Xie X., Lee, J.: Negative or positive? The effect of emotion and mood on risky driving (2013). https://doi.org/10.1016/j.trf.2012.08.009
4. LeCun, Y., Bottou, L., Bengio, Y., Haffner, P.: Gradient-based learning applied to document recognition. Proc. IEEE **86**(11), 2278–2324 (1998)
5. DeepMind company website, https://deepmind.com/. Last accessed 10 May 2019

6. Ekman, P., Friesen, W.V.: Constants across cultures in the face and emotion. J. Pers. Soc. Psychol. **17**(2), 124–129 (1971)
7. Ekman, P.: Basic Emotions. In: Dalgleish, T., Power, M.J. (eds.) Handbook of Cognition and Emotion, pp. 45–60. Wiley, New York, NY (1999)
8. Ekman, P., Cordaro, D.: What is meant by calling emotions basic. Emot. Rev. **3**(4), 364–370 (2011)
9. Cordaro, D.T., Sun, R., Keltner, D., Kamble, S., Huddar, N., McNeil, G.: Universals and cultural variations in 22 emotional expressions across five cultures. Emotion **18**(1), 75–93 (2018)
10. Ekman, P., Friesen, W.: Facial Action Coding System: A Technique for the Measurement of Facial Movement. Consulting Psychologists Press, Palo Alto (1978)
11. Ekman, Paul, Friesen, Wallace V., Hager, Joseph C.: Facial Action Coding System: The Manual on CD ROM. A Human Face, Salt Lake City (2002)
12. Edwards, G.J., Taylor, C.J., Cootes, T.F.: Interpreting face images using active appearance models. Proceedings Third IEEE International Conference on Automatic Face and Gesture Recognition, pp. 300–305 (1998)
13. Cootes, T.F., Edwards, G.J., Taylor, C.J.: Active Appearance Models. IEEE Trans. Pattern Anal. Mach. Intell. **23**(6), 681–685 (2001)
14. Fraedrich, E., Lenz, B.: Societal and individual acceptance of autonomous driving. In: Autonomous Driving (2016). https://doi.org/10.1007/978-3-662-48847-8_29
15. Fraedrich, E., Lenz, B.: Taking a drive, hitching a ride: autonomous driving and car usage. In: Autonomous Driving (2016). https://doi.org/10.1007/978-3-662-48847-8_31
16. Woisetschläger, D.M.: Consumer perceptions of automated driving technologies: an examination of use cases and branding strategies. In: Autonomous Driving (2016). https://doi.org/10.1007/978-3-662-48847-8_32
17. Kanade, T., Cohn, J. F., Tian, Y.: Comprehensive database for facial expression analysis. In: Proceedings of the Fourth IEEE International Conference on Automatic Face and Gesture Recognition (FG'00), Grenoble, France, pp. 46–53 (2000)
18. Lucey, P., Cohn, J.F., Kanade, T., Saragih, J., Ambadar, Z., Matthews, I.: The extended Cohn-Kanade Dataset (CK+): a complete expression dataset for action unit and emotion-specified expression. In: Proceedings of the Third International Workshop on CVPR for Human Communicative Behavior Analysis (CVPR4HB 2010), San Francisco, USA, pp. 94–101 (2010)
19. Goodfellow, I., Erhan, D., Carrier, P.L., Courville, A., Mirza, M., Hamner, B., Cukierski, W., Tang, Y., Lee, D.H., Zhou, Y., Ramaiah, C., Feng, F., Li, R., Wang, X., Athanasakis, D., Shawe-Taylor, J., Milakov, M., Park, J., Ionescu, R., Popescu, M., Grozea, C., Bergstra, J., Xie, J., Romaszko, L., Xu, B., Chuang, Z., Bengio, Y.: Challenges in representation learning: a report on three machine learning contests. arXiv (2013)
20. Lyons, M., Akamatsu, S., Kamachi, M., Gyoba, J.: Coding Facial Expressions with Gabor Wavelets. In: 3rd IEEE International Conference on Automatic Face and Gesture Recognition, pp. 200–205 (1998)
21. Aifanti, N., Papachristou, C., Delopoulos, A.: The MUG facial expression database. In: Proceedings of the 11th International Workshop on Image Analysis for Multimedia Interactive Services (WIAMIS), Desenzano, Italy, April 12–14 (2010)
22. Langner, O., Dotsch, R., Bijlstra, G., Wigboldus, D.H.J., Hawk, S.T., van Knippenberg, A.: Presentation and validation of the Radboud faces database. Cogn. Emot. **24**(8), 1377–1388 (2010). https://doi.org/10.1080/02699930903485076
23. Chollet, F., Keras, et al.: https://keras.io (2015). Last accessed 10 May 2019
24. Scikit-learn: Machine Learning in Python, Pedregosa et al., JMLR 12, pp. 2825–2830 (2011)
25. Bradski, G.: The OpenCV Library. Dr. Dobb's Journal of Software Tools **120**, 122–125 (2000)
26. Facial Expression Database Classifier (FEDC), https://github.com/AntonioMarceddu/Facial_Expression_Database_Classifier. Last accessed 10 May 2019

Road Type Classification Using Acoustic Signals: Deep Learning Models and Real-Time Implementation

Giovanni Pepe, Leonardo Gabrielli, Emanuele Principi, Stefano Squartini, and Luca Cattani

Abstract Nowadays, cars host an increasing number of sensors to improve safety, efficiency and comfort. Acoustic sensors have been proposed, in recent works, to acquire information related to the road conditions. Thanks to effectiveness of Deep Learning techniques in analyzing audio data, new scenarios can be envisioned. Based on previous works employing Convolutional Neural Networks (CNN) trained specifically for either of the two tasks, we compare the performance of a CNN trained to jointly classify both wetness and roughness and a transfer learning approach where two CNN, specialized for one task singularly, are joined in a single network. Then we investigate several issues related to the deployment of a classification system able to detect road wetness and roughness on an embedded processor. The first approach seems to score better in our tests and is, thus, selected for the deployment to an ARM-based embedded processor. The computational cost, Real-Time Factor and memory requirements are discussed, as well as the degradation related to the extraction of the features and the weights quantization. Results are promising and show that such an application can be readily deployed on off-the-shelf hardware.

1 Introduction

In today's society, road conditions are the main factors that cause accidents and driving annoyances. The weather is the major factor affecting car accident numbers [1], with a doubling of the accidents on wet pavements [2]. For what concerns the cabin comfort, tyre-road noise is one of the main factor on vehicle noise emissions, causing annoyance on passengers and affecting driver's concentration [3].

G. Pepe (✉) · L. Gabrielli · E. Principi · S. Squartini
Department of Information Engineering, Università Politecnica delle Marche, Via Brecce Bianche, 60131 Ancona, AN, Italy
e-mail: g.pepe@pm.univpm.it

G. Pepe · L. Cattani
ASK Industries S.p.A., Viale Ramazzini, 42124 Reggio Emilia, RE, Italy

A. Esposito et al. (eds.), *Progresses in Artificial Intelligence and Neural Systems*, Smart Innovation, Systems and Technologies 184,
https://doi.org/10.1007/978-981-15-5093-5_4

The automated detection of the road conditions is playing an important role on intelligent cars safety systems and on autonomous vehicles [4]. This kind of systems, based on sensors like cameras, optical sensors and radar/lidar, is increasing in the last years. Acoustic sensors have been employed to detect various features of the road surface [5, 6], suggesting that combining them with machine learning techniques is possible to replace expensive optical sensors with inexpensive ones, integrating them with the infotainment system for automatic equalization and speech enhancement [7]. Automotive companies are starting employing these safety systems in their cars.[1]

In previous works microphone recordings were processed by Support Vector Machine (SVM) for road roughness [8] and wetness [9, 10]. One of the first deep learning approach [5] employed Bidirectional Long-Short Term Memory neural networks (BLSTM) [11, 12] for the road wetness. Other works used CNN for road roughness [6] and road wetness detection [13]. In the latter, CNN and BLSTM are compared both in terms of performance and computational cost, showing that CNN achieves similar results at a lower computational cost.

1.1 Scope of the Work

Considering the good performance and scalability of CNN architectures, it is worth verifying the real-world applicability of the Deep Learning approach to road classification. The goal of this work is twofold: we want to compare approaches able to conduct joint classification of both roughness and wetness, and verify their scalability to embedded processors, for possible integration in an automotive data processing system.

In our previous works, CNN were tested for an individual binary classification task: rough versus smooth (R/S) and dry versus wet (D/W). Unfortunately, employing two separate networks with the same architecture trained on one task each, implies a doubling of the computational resources. Having one network performing both the tasks could reduce the computational burden and the memory required to store network weights and data.

The paper is structured as follows: we first describe the proposed method in Sect. 2, then we discuss how the experiments are carried out in Sect. 3. Finally in Sect. 4 results are presented. Section 5 concludes the paper.

2 Proposed Method

The work is based on supervised training of deep learning architectures using Auditory Spectral Features (ASF) [14] as input features, extracted from microphones. The dataset contains both roughness and wetness label, manually annotated during the

[1] See e.g. https://www.autonews.com/blogs/porsche-says-new-wet-mode-will-solve-911s-aquaplaning.

recording sessions. Two microphones are employed, one positioned on the bottom of the trunk, below the spare wheel (T) and one close to the head of the passenger in the back right seat (IB). The microphone positions are selected based on the results of a thorough analysis of the best positions [13]. Moreover these two specific locations protect the sensors from water and wind.

2.1 *Auditory Spectral Features*

ASF are acoustic features widely used in audio applications and sound event recognition [14], since they can simulate human auditory perception [15]. They are extracted from audio samples applying the Short Time Fourier Transform (STFT) and using a frame size of 30 ms and a frame step of 10 ms [15]. Each STFT provides the power spectrogram which is converted to the Mel-Frequency scale using 26 triangular filters, obtaining the Mel spectrograms $M^{30}(n, m)$. To match the human perception of loudness, Mel spectrograms are transformed to a logaritmic scale:

$$M_{log}^{30}(n, m) = log_{10}(M^{30}(n, m) + 1.0) \tag{1}$$

For each Mel spectrogram, it has been calculated the positive first order differences $D^{30}(n, m)$:

$$D^{30} = M_{log}^{30}(n, m) - M_{log}^{30}(n - 1, m) \tag{2}$$

The frame energy in logaritmic scale and its derivative have been included, for a total of 54 features. In order to feed the CNN, audio chunks of 1 s are obtained from 98 feature vectors.

2.2 *Learning Approaches*

Two approaches are compared in this work:

- a CNN (from now on called joint-CNN) trained for joint classification of wetness and roughness (Fig. 1),
- an architecture (from now on called TL-CNN) created following a Transfer Learning approach from two specialized CNNs trained separately (Fig. 2).

The joint-CNN network has two binary outputs for joint classification. The TL-CNN employs a transfer learning strategy. Transfer learning is an approach where an architecture is trained on different tasks in order to improve generalization, sharing informations from different data [16]. In this work we exploit Transfer Learning to improve generalization of road classification from networks that have been separately trained on the two tasks [6, 7, 13]. The best networks emerging from individual evaluations are merged adding one dense layer that is trained to optimize performance

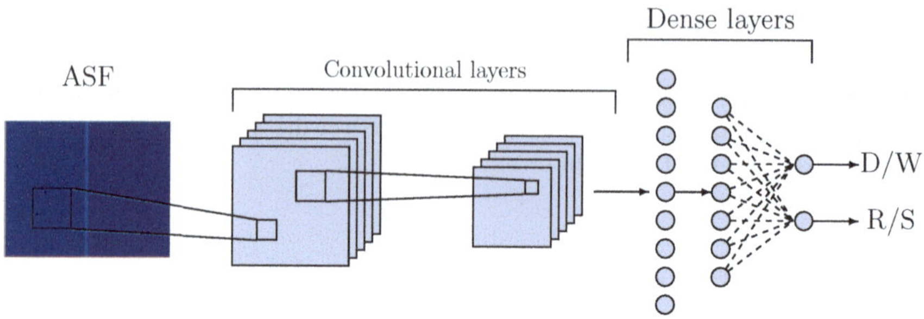

Fig. 1 Joint-CNN for roughness and wetness classification

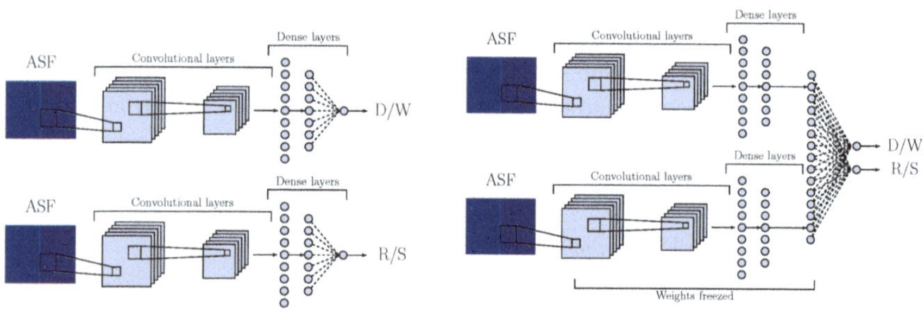

Fig. 2 Left figure represents the networks that separately perform wetness and roughness detection, right figure represents the Transfer Learning approach adding one dense layer

with the joint classification problem, while the CNN are not re-trained. The merged networks have the same architectures but one is used for wetness classification, while the other for roughness classification.

3 Experiments

Experiments are performed according to Fig. 3. We first train and compare the joint-CNNs and the TL-CNNs on a GPU, with features extracted with OpenSmile, an open-source audio analysis toolkit [17] (see Fig. 3a). Feature extraction is also implemented in C++ on the embedded processor, to assess the performance variation implied by a different implementation of the extraction process. Then the features are transferred from the processor to a computer and used to run a second batch of experiments (see Fig. 3b) and assess the performance variation. Finally, the best configurations from (a) are used to implement the network in an embedded system, that is trained on GPU but run on the board using the testing set (see Fig. 3c).

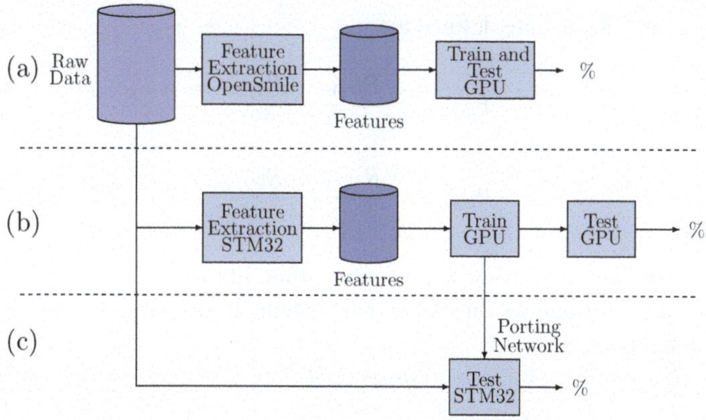

Fig. 3 Experiments overview: **a** feature extraction using OpenSmile [17], train and test of networks using GPU; **b** feature extraction using STM32 board, train and test of networks using GPU; **c** importing of network trained by GPU on board and test of networks using STMicroelectronics (STM) board

Computer experiments were performed using Keras[2] with Tensorflow[3] as backend. Parallel computing is performed using cuDNN[4] libraries and Nvidia GeForce GTX 970. Embedded experiments were performed using the ST Microelectronics tools and an STM32 board.

Classification results are reported using the F_1-score, defined as:

$$F_1 = 2 \times \frac{P \times R}{P + R} \tag{3}$$

where P and R stand for precision and recall, respectively:

$$P = \frac{TP}{TP + FP} \tag{4}$$

$$R = \frac{TP}{TP + FN} \tag{5}$$

where TP, FP, FN are the true positive, false positive and false negative rate respectively. For the double classification, we have calculated the macro average F_1-score.

$$F_{1_{macro}} = 2 \times \frac{P_{macro} \times R_{macro}}{P_{macro} + R_{macro}} \tag{6}$$

[2]https://keras.io.

[3]https://www.tensorflow.org/.

[4]https://developer.nvidia.com/cudnn.

where P_{macro} and R_{macro} are defined as:

$$P_{macro} = \frac{P_{D/W} + P_{R/S}}{2} \quad (7)$$

$$R_{macro} = \frac{R_{D/W} + R_{R/S}}{2} \quad (8)$$

The metrics are evaluated using scikit-learn python library.[5]

To analyze performances on embedded system, Ratio TIme Factor (RTF) was analyzed, defined as:

$$RTF = \frac{t_p}{t_a} \quad (9)$$

t_a is the time duration of audio frame where t_p is the time required to process the audio chunk.

3.1 Dataset

The Dataset is the same used in [13]: audio recording was performed using PCB Piezotronics model 378C20 microphones and Squadriga II multichannel audio recording device running at 44100 Hz in an A-Class Mercedes. Differently from [13], we do not perform cross-validation, in order to analyze the feasibility of the approach in a real scenario. The dataset is split according to recorded audio: 64% of recordings were used for the train set, 16 % for validation and 20% for test set. In total 5675 s are used for training, 1418 s seconds for validating and 1773 s seconds for testing. For the purpose of training, the number of dry and wet samples were balanced, thus, leaving the rough/smooth samples unbalanced.

3.2 STM32CubeMX and X-CUBE-AI Expansion Package

The STM32 embedded platform from ST Microelectronics is chosen for the evaluation as it provides tools to run neural networks. The networks need to be trained on a GPU and can be later exported to the platform. A graphical tool, STM32CubeMX allows to configure the microprocessor and generate C code for ARM® Cortex®-M cores.

The tool provides an expansion package called X-CUBE-AI that automatically converts pre-trained neural networks optimized in C code in order to import them into STM32 microcontrollers, compressing the networks in order to reduce the size of weights and biases. As reported by the manufacturer, compression is applied to

[5]scikit-learn.org.

Table 1 Best performing joint-CNN for T and IB microphone for wetness and roughness classification. The F1-score is the average-macro

Mic	CNN layer size	Kernel shape	Strides shape	Dense layer size	$F1_{macro}$ (%)
T	20, 25	[[1, 2], [1, 7]]	[[2, 2], [3, 6]]	900, 300	**94.10**
IB	25, 20	[[6, 7], [2, 8]]	[[2, 2], [6, 3]]	900, 400	**91.56**

dense layers, applying K-means clustering on weights and biases, and concatenating all layers [18]. Three compression factors are available: none, ×4 and ×8. Depending on used compression factor, the tool can compress weights and biases quantizing them on 4 (×4) or 8 bit (×8). Tool with expansion package was used to import best models trained on GPU on STM32 microcontroller.

4 Results

4.1 Experiments Using GPU

The experiments were performed using a random search. All configurations have a max pooling layer of dimension 2 × 2 and strides 1 × 1.

In Table 1 the best results using CNN for wetness and roughness classification are presented, while in Tables 2 and 3 the F_1-score are presented separately for the two tasks together with the final macro average score. Finally in Table 4 the results

Table 2 Results obtained with the separated networks using T microphone

CNN layer size	Kernel shape	Strides shape	Dense layer size	D/W	R/S	$F1_{macro}$ (%)
				F1-score (%)	F1-score (%)	
25, 30	[[5, 3], [5, 2]]	[[4, 5],[5, 2]]	200, 800	98.31	82.70	**88.10**
20, 20	[[8, 3], [6, 6]]	[[3, 1], [7, 10]]	900, 900	97.73	81.28	87.67
30, 20	[[9, 6], [2, 2]]	[[6, 2], [2, 9]]	200, 1000	90.59	85.59	87.62
20, 25	[[1, 2], [2, 3]]	[[4, 3], [2, 5]]	200, 500	97.29	77.05	85.42
30, 25	[[2, 8], [5, 3]]	[[4, 2], [2, 8]]	700, 400	97.42	80.24	84.87
30, 30	[[7, 5], [1, 2]]	[[6, 1], [4, 8]]	600, 700	97.45	75.42	81.74
20, 30	[[4, 4], [4, 5]]	[[4, 2], [6,]]	500, 200	95.62	77.77	81.68
25, 30	[[8, 3], [5, 3]]	[[3, 2], [9, 5]]	200, 200	90.66	64.82	80.49
30, 20	[[3, 2], [8, 2]]	[[2, 1], [3, 10]]	300, 1000	82.08	80.55	80.32
25, 20	[[4, 7], [6, 1]]	[[4, 3], [5, 4]]	500, 800	89.77	67.12	79.95

Table 3 Results obtained with the separated networks using IB microphone

CNN layer size	Kernel shape	Strides shape	Dense layer size	D/W	R/S	$F1_{macro}$ (%)
				F_1 (%)	F_1 (%)	
20, 20	[[2, 6], [2, 6]]	[[3, 1], [10, 6]]	500, 200	94.56	74.22	**81.16**
25, 20	[[4, 7], [6, 1]]	[[4, 3], [5, 4]]	500, 800	81.40	67.23	81.11
30, 20	[[7, 3], [6, 4]]	[[9, 2], [1, 7]]	500, 300	89.64	77.98	80.81
20, 25	[[2, 8], [1, 8]]	[[5, 2], [8, 3]]	800, 600	92.21	75.92	79.49
30, 20	[[9, 6], [2, 2]]	[[6, 2], [2, 9]]	200, 1000	89.03	72.11	78.96
30, 20	[[5, 8], [4, 10]]	[[4, 2], [7, 3]]	600, 700	95.37	61.08	78.42
20, 25	[[8, 2], [1, 6]]	[[3, 3], [3, 2]]	100, 100	95.17	61.08	78.39
20, 30	[[5, 10], [7, 8]]	[[3, 2], [8, 3]]	500, 1000	92.90	48.84	77.63
30, 20	[[7, 7], [2, 2]]	[[5, 4], [5, 1]]	500, 800	93.41	60.07	77.47
20, 30	[[10, 10], [9, 6]]	[[3, 2], [7, 4]]	600, 300	96.02	59.65	77.30

Table 4 Results obtained with the merged networks training the new layers. The best performance is obtained using CNN composed by 2 layers of 20 and 30 kernels respectively, dimensions of kernel are [[4, 4], [4, 5]], strides equal to [[4, 2], [6, 1]] and two dense layers of 500 and 200 units respectively for microphone T and 2 layers of 20 kernels each, with dimensions [[2, 6], [2, 6]] and strides [[3, 1], 10, 6] and two dense layers of 500 and 200 units respcetively for microphone IB

T microphone		IB microphone	
Dense layer size	$F1_{macro}$ (%)	Dense layer size	$F1_{macro}$ (%)
20	93.40	20	90.15
40	**94.01**	40	90.03
60	92.71	60	90.18
80	93.69	80	90.30
100	93.67	100	**90.38**
120	93.65	120	90.29
140	93.70	140	90.34
160	93.75	160	90.36
180	93.73	180	90.23
200	93.72	200	90.20

with the merged networks are presented. Best results are obtained by the joint-CNN, however the TL-CNN shortly follows. Both approaches bring a notable improvement of the results. This is due to the fact that separate networks excel on the wetness task, but fail to provide remarkable performances on the second task.

4.2 Experiments Using STM32 Board

The best configurations of Sect. 4.1 are used to carry out the experiments on STM32H743ZI board. The embedded system has a 32-bit ARM® processor with frequency up to 480 MHz, 2 MB of Flash Memory and 1 MB of RAM.

To compare the experiments using the network on STM32 board and GPU, audio data has been transferred using UART communication to the board, where feature extraction was performed. Finally the extracted features are transferred back to the PC where training is accomplished running on GPUs. First experiments were performed testing networks on GPU using the features extracted on the board. In Table 5 the results are presented: comparing the performance with Table 1, $F1_{macro}$ is 3.89% and 6.35% lower for T and IB respectively. This is caused by differences in the feature extraction algorithms as implemented in OpenSmile and on the board. This problem can be alleviated by performing a random search with the features extracted by the board. The trained network was also deployed on STM32 board using the STM32CubeMX tool. None of the networks could be imported on the board without applying compression, due to Flash Memory limitations. A factor ×4 compression was employed. The results obtained on GPUs and on the board are presented in Table 5. Considering the performance degradation of the feature extraction on the board, the embedded processor and the GPU are able to achieve similar results, with a $F1_{macro}$ bearing as little as 0.27% degradation on the IB microphone only.

Regarding the computational complexity, the performance has been evaluated using Multiply-and-Accumulate Complexity (MACC), RTF and RAM size. The MACC index indicates the complexity of a model including multiply-and-accumulate instructions and an estimate of the activation functions computational cost [18]. Features extraction comes in 1 ms for Mel spectrograms in logarithmic scale and Energy processing for each frame and 1 ms for the first order derivate. The network processes input data in 178 ms (best network for T microphone) and 235 ms (best network for IB microphone). In both cases the RTF is lower than 1, 27.7% and 33.4%, respectively.

Table 5 Results obtained with the same architecture used in Table 1 but trained with features extracted from ST board

Mic	GPU		STM32					
	$F1_{macro}$ (%)	Memory size (MB)	$F1_{macro}$ (%)	RTF (%)	Compression factor	Memory size (MB)	RAM (kB)	Complexity (MACC)
T	90.21	19.09	90.22	27.7	×4	1.57	200.86	2099885
IB	85.21	16.21	84.94	33.4	×4	1.35	251.08	3510810

5 Conclusions

In the present work, a deep learning approach for wetness and roughness road classification is discussed. Two learning approaches are compared for this task. The best model found during the evaluation has been used to evaluate the performance degradation implied by the deployment to an embedded system and its computational cost.

Regarding the comparison of learning approaches, it was seen that training CNN with joint classification gives better results than merging two specialized CNN using a standard transfer learning approach. Moreover, from an engineering perspective, the deployment of the joint-CNN is more efficient than the transfer learning model, given the reduced complexity of the model.

The deployment has been carried on an STM32 board. The performance was evaluated, showing results comparable with GPUs. The extraction of the features and the processing are computationally feasible, not exceeding 33.4% of the overall available time. This shows the feasibility of audio processing by deep learning on embedded processors for audio classification tasks with lightweight architectures such as CNN.

Acknowledgements The authors would like to thank ASK industries S.P.A. for financial support and technical assistance. This work is supported by the Italian Ministry of Economic Development (MISE)'s fund for the sustainable growth (F.C.S.) under grant agreement (CUP) B48I15000130008, project VASM ("Vehicle Active Sound Management"). We gratefully acknowledge the support of NVIDIA Corporation forp the donation of the GPU used for this research.

References

1. Mondal, P., Sharma, N., Kumar, A., Bhangale, U., Tyagi, D., Singh, R.: Effect of rainfall and wet road condition on road crashes: a critical analysis. Tech. Rep., SAE Technical Paper (2011). https://doi.org/10.4271/2011-26-0104
2. Gothié, M.: The contribution to road safety of pavement surface characteristics. Bulletin des Laboratoires des Ponts et Chaussees, pp. 5–12 (January 2000)
3. Junoh, A.K., Muhamad, W., Fouladi, M.: A study on the effects of tyre vibration to the noise in passenger car cabin. Adv. Model. Optim. **13**(3), 567–581 (2011)
4. Zhang, X., Gao, H., Guo, M., Li, G., Liu, Y., Li, D.: A study on key technologies of unmanned driving. CAAI Trans. Intell. Technol. **1**(1), 4–13. https://doi.org/10.1016/j.trit.2016.03.003
5. Abdić, I., Fridman, L., Brown, D.E., Angell, W., Reimer, B., Marchi, E., Schuller, B.: Detecting road surface wetness from audio: a deep learning approach. In: 23rd International Conference on Pattern Recognition (ICPR), pp. 3458–3463. IEEE (2016). https://doi.org/10.1109/ICPR.2016.7900169
6. Ambrosini, L., Gabrielli, L., Vesperini, F., Squartini, S., Cattani, L.: Deep neural networks for road surface roughness classification from acoustic signals. In: Audio Engineering Society Convention 144 (2018). http://www.aes.org/e-lib/browse.cfm?elib=19451
7. Gabrielli, L., Ambrosini, L., Vesperini, F., Bruschi, V., Squartini, S., Cattani, L.: Processing acoustic data with siamese neural networks for enhanced road roughness classification. In: Proceedings of IJCNN. IEEE, Budapest, Hungary (14–19 July (Accepted) 2019). https://doi.org/10.1109/SIU.2017.7960154

8. Doğan, D.: Road-types classification using audio signal processing and svm method. In: 25th Signal Processing and Communications Applications Conference (SIU), pp. 1–4 (2017). https://doi.org/10.1109/SIU.2017.7960154
9. Alonso, J., Lopez, J., Pavón, I., Recuero, M., Asensio, C., Arcas, G., Bravo, A.: On-board wet road surface identification using tyre/road noise and support vector machines. Appl. Acoust. **76**, 407–415 (2014). https://doi.org/10.1016/j.apacoust.2013.09.011
10. Kanarachos, S., Blundell, M., Kalliris, M., Kotsakis, R.: Speed-dependent wet road surface detection using acoustic measurements, octave-band frequency analysis and machine learning algorithms. In: ISMA Conference on Noise and Vibration Engineering (9 2018). https://doi.org/10.13140/RG.2.2.30521.01121
11. Schuster, M., Paliwal, K.K.: Bidirectional recurrent neural networks. IEEE Trans. Signal Process. **45**(11), 2673–2681 (1997). https://doi.org/10.1109/78.650093. Nov
12. Hochreiter, S., Schmidhuber, J.: Long short-term memory. Neural Comput. **9**(8), 1735–1780 (1997). https://doi.org/10.1162/neco.1997.9.8.1735
13. Pepe, G., Gabrielli, L., Ambrosini, L., Squartini, S., Cattani, L.: Detecting road surface wetness using microphones and convolutional neural networks. In: Audio Engineering Society Convention 146 (Mar 2019). http://www.aes.org/e-lib/browse.cfm?elib=20326
14. Marchi, E., Ferroni, G., Eyben, F., Gabrielli, L., Squartini, S., Schuller, B.: Multi-resolution linear prediction based features for audio onset detection with bidirectional lstm neural networks. In: 2014 IEEE International Conference on Acoustics, Speech and Signal Processing (ICASSP), pp. 2164–2168 (May 2014). https://doi.org/10.1109/ICASSP.2014.6853982
15. Eyben, F., Böck, S., Schuller, B., Graves, A.: Universal onset detection with bidirectional long-short term memory neural networks. In: Proceedings of the 11 th. International Society for Music Information Retrieval Conference, ISMIR, Utrecht, The Netherlands. pp. 589–594 (2010). http://ismir2010.ismir.net/proceedings/ismir2010-101.pdf
16. Goodfellow, I., Bengio, Y., Courville, A.: Deep Learning. MIT Press (2016). http://www.deeplearningbook.org
17. Eyben, F., Weninger, F., Gross, F., Schuller, B.: Recent developments in openSMILE, the munich open-source multimedia feature extractor. In: Proceedings of the ACM Multimedia 2013, pp. 835–838. ACM, Barcelona, Spain (2013). https://doi.org/10.1145/2502081.2502224
18. © STMicroelectronics: UM2526 : Getting started with X-CUBE-AI Expansion Package for Artificial Intelligence (AI) (January 2019)

Emotional Content Comparison in Speech Signal Using Feature Embedding

Stefano Rovetta, Zied Mnasri, and Francesco Masulli

Abstract Expressive speech processing has been improved in the recent years. However, it is still hard to detect emotion change in the same speech signal or to compare emotional content of a pair of speech signals, especially using unlabeled data. Therefore, feature embedding has been used in this work to enhance emotional content comparison for pairs of speech signals, cast as a classification task. Actually, feature embedding was proved to reduce the dimensionality and the intra-feature variance in the input space. Besides, deep autoencoders have recently been used as a feature embedding tool in several applications, such as image, gene and chemical data classification. In this work, a deep autoencoder is used for feature embedding before performing classification by vector quantization of the emotional content of pairs of speech signals. Autoencoding was performed following two schemes, for all features and for each group of features. The results show that the autoencoder succeeds (a) to reveal a more compact and a clearly separated structure of the mapped features, and (b) to improve the classification rates for the similarity/dissimilarity of all emotional content aspects that were compared, i.e neutrality, arousal and valence; in order to calculate the emotion identity metric.

S. Rovetta · Z. Mnasri (✉) · F. Masulli
DIBRIS, Università Degli Studi di Genova, Genoa, Italy
e-mail: zied.mnasri@enit.utm.tn

S. Rovetta
e-mail: stefano.rovetta@unige.it

Z. Mnasri
Electrical Engineering Department, ENIT, University Tunis El Manar, Tunis, Tunisia

F. Masulli
Sbarro Inst. for Cancer Research and Molecular Medecine, Temple University, Philadelphia, PA, USA
e-mail: francesco.masulli@unige.it

A. Esposito et al. (eds.), *Progresses in Artificial Intelligence and Neural Systems*, Smart Innovation, Systems and Technologies 184,
https://doi.org/10.1007/978-981-15-5093-5_5

1 Introduction

Speech technology is widely used in interactive applications. However, expressive speech still poses significant challenges. Emotional content analysis would be very useful for sophisticated man-machine interaction, with possible applications even beyond efficient vocal interfaces, for instance in collaborative robotics. However, detection of emotional content and its characteristics from speech signals is still inaccurate.

Since emotion recognition is a pattern recognition problem, data-driven models have been usually used for that. Indeed, supervised learning techniques like neural networks [5, 10], SVM [15], or generative models like HMM-GMM [11, 14] have been classically utilized. More recently, deep learning models have also been developed for that purpose using feedforward, recurrent or convolutional neural networks [7].

However, to analyse the emotional content in a huge database of speech signals, supervised learning would require tedious labeling, with the associated cost and the underlying risk of mistakes. Therefore, an unsupervised approach would be a suitable alternative. In this scope, there have been some successful works like in [19] where SOM were used to detect emotions from audiobooks, and in [3] where hierarchical k-means were applied to detect emotions from a corpus to build a model for expressive speech synthesis. However, to the best of our knowledge, unsupervised learning hasn't been used for emotional content comparison in speech signal so far.

Being able to compare two data items is a fundamental ability for machine learning methods ranging from clustering to kernel-based classification. Change point detection, one-class classification, novelty detection, outlier analysis, concept drift tracking are also made possible by the availability of similarity indexes. However, in the case of emotional content in speech, this task is challenging, since emotions are at an intermediate level between structural properties and semantic content.

Therefore, this work aims to find a better feature embedding which allows enhancing either clustering or classification results for emotional content comparison of speech signals. To achieve this goal, a deep autoencoder has been applied as a tool for feature embedding. More particularly, this work addresses the problem of emotional content analysis from speech independently from speaker or text. The similarity of the following expressive speech characteristics is modeled for each pair of speech signals: (a) neutrality of speech, (b) arousal and (c) valence. The input features undertake two types of preprocessing: normalization and/or embedding using the autoencoder. Finally, the results of vector quantization are aggregated to calculate a metric for emotion identity similarity.

The paper is organized as follows: Sect. 2 reviews the related work, including the standard feature sets for expressive speech analysis and the main feature extraction techniques used in expressive speech processing, Sect. 3 presents the feature embedding technique used in this work, i.e. deep autoencoder, Sect. 4 describes the speech material used in this work, whereas Sect. 5 details the experiments and discusses the obtained results.

2 Related Work

Since emotion recognition is a pattern recognition task, data-driven models have been looking for the feature set presenting the closest correlation to emotion classes. However, the usefulness of features used for emotion recognition has not been proved for emotional content comparison. Nevertheless, some combinations of speech parameters have been used for this purpose with a relative success.

Classical signal-extracted features have been proved to be extremely efficient for supervised emotion recognition, such as acoustic parameters like Mel-frequency cepstral coefficients (MFCC), prosodic parameters like fundamental frequency (F_0) and energy, or signal-related parameters like harmonic-to-noise ratio (HNR) and zero-crossing rate (ZCR). Such features, and others, have been grouped into standard feature sets for expressive speech analysis and/or recognition, such as Interspeech emotion and paralinguistic challenges [16, 17], and the Geneva minimalistic acoustic parameter set (GeMAPS) [4].

Though the aforementioned features have reached outstanding performance in emotion recognition using supervised learning, they haven't been quite efficient while using clustering techniques [13]. Besides, such an important quantity of features induces a high dimensionality of the input space. Therefore, feature extraction should be studied for such data sets in order to improve the performance.

The aim of feature analysis is to optimize the feature space so that only the most relevant features are selected or extracted. Several techniques based on ANOVA (Analysis of variance) and mutual information or cross-validation have been used for input selection, to keep only the most contributory features. An alternative way to reduce the feature space dimensionality consists in applying feature transformation methods such as PCA (Principal component analysis) and LDA (Linear discriminant analysis).

Also, to deal with feature sparseness, autoencoding neural networks were used [8]. An autoencoder is a neural network, which outputs are the same than its inputs. It is generally used to discover latent data structures in the inputs. In [2], an autoencoder was used to resolve the problem of feature transfer learning in emotion recognition. Actually, an emotion classifier trained on some kind of data, e.g. adult speech, wouldn't be efficient when tested on another kind of data, e.g. children voices. The technique consisted in applying a single autoencoder for each class of targets. The reconstructed data was then used to build the emotion recognition system.

3 Autoencoders for Feature Embedding

The autoencoder is an unsupervised learning algorithm, which is basically used for automatic extraction of features from unlabeled data. Therefore, the autoencoder can be used for feature extraction either for classification or for clustering.

3.1 Deep Autoencoder

An autoencoder is a neural network which approximates the identity function, i.e. the output is the same as the input. The autoencoder optimizes the weights (W) and the biases (b) of the neural network, such that $y_i = h_{W,b}(x_i) = x_i \; \forall x_i \in X = (x_1, x_2, ..., x_n) \subset R^n$, where x_i, y_i and $h_{W,B}$ are respectively the inputs, the outputs and the hidden layer code [9].

For real-valued data, the objective function is the mean square error $E = \sum_{i=1}^{N} ||x_i - h_{W,b}(x_i)||^2$ where $||.||$ denotes the Euclidean norm; whereas for binary data, the objective function is the cross-entropy $E = -\sum_{i=1}^{N}(x_i \log h_{W,b}(x_i) + (1 - x_i)\log(1 - h_{W,b}(x_i))$. The weights W_i and biases b_i are updated using a gradient descent algorithm, such as SGD (Stochastic gradient descent). To calculate the gradient of the objective function $J_{W,b} = (\frac{\partial E(W,b)}{\partial W}, \frac{\partial E(W,b)}{\partial b})$, the backpropagation algorithm is used [6].

The deep autoencoder is composed of two parts, namely the encoder and the decoder. Both parts consist of hidden layers, usually stacked in a mirror symmetry, with a bottleneck layer in the middle, i.e. the code layer. Then the encoded data are the output of the code layer. The usefulness of such an architecture consists in the structure of the encoded data. Actually it has been shown that the code layer can (a) reveal a hidden structure of the input features, discovered through the encoding process, (b) reduce the dimensionality of the input space. Then the encoded data will be used as an input for classifiers or clustering algorithms, in order to improve their accuracy.

3.2 Feature Embedding Using Autoencoders

Very often, original input features have a large variance between each other, which yields a complex distribution. To cope with this issue, non-linear mapping can be used to reduce the intra-feature variance in the input space. Such a mapping can be performed either by kernel methods, such as kernel k-means which applies a non-linear transformation using fixed kernel function [18], or by autoencoders, as it has been proved recently in several works [9, 18, 20, 21].

Since the autoencoder aims to learn a new representation of the input features, supplied by the code layer, then the use of a smaller number of nodes in this layer helps obtaining a new feature space with a smaller dimension. Furthermore, the new mapping often reveals a new structure where the input features (or their coded images) are grouped into compact regions, which would be more helpful for clustering tasks. However, autoencoding hasn't been widely used for clustering so far.

In [18], an autoencoder used for clustering was trained with a new objective function where the centroids are updated at the same time as the networks parameters, i.e. weights and biases. In [21], autoencoders were used for deep embedded clustering. The approach consists in a two-step algorithm by (a) initializing the parameters with an autoencoder, (b) optimizing both the centroids and the autoencoder parameters. Optimization is performed with KL divergence as the objective function (cf. Sect. 3.1), to maximize the similarity between the distribution of the embedded features and the centroids.

4 Speech Material

This work was performed using a standard emotional speech database, i.e. EMO-DB [1], which has been widely used for emotion recognition and analysis. The feature set was selected among the standard ones (cf. Sect. 2). In particular, the Interspeech 2009 emotion challenge feature set has been proved to be highly efficient in emotion recognition.

4.1 Speech Database

EMO-DB is an acted speech database specifically designed for emotional speech processing. It has been known for providing the best emotion recognition rates using supervised classifiers such as SVM, and generative models like HMM-GMM [16]. It includes 5 short and 5 long sentences in German, uttered by 5 male and 5 female speakers. Each sentence is uttered in 7 emotional states (neutrality, anger, boredom, fear, disgust, happiness and sadness). The signals were registered at 16-KHz sampling rate in an anechoic chamber.

4.2 Feature Set

The Interpeech'2009 emotion challenge feature set, proposed by Schuller et al. [16] contains prosodic, acoustic and signal-related LLD's (low-level descriptor) required for emotion recognition. Each LLD is presented as a vector of 12 coefficients or functionals, including its most relevant statistics, calculated on the whole signal. Besides, each LLD vector is duplicated using its Δ value, i.e. temporal difference. Finally, each signal is represented by (16 LLD+ 16 Δ-LLD) $\times$ 12 functionals, thus by 384 coefficients (cf. Table 1). Then each pair of signals is represented by 768 coefficients.

Table 1 Interspeech 2009 emotion challenge LLD's and functionals [16]

Groups	LLD's	Functionals (for all LLD's)
Prosodic	(Δ) RMS energy	Min, max, range
	(Δ) ln F_0	Min rel. position, max rel. position
Signal-related	(Δ) ZCR	Kurtosis, skewness
	(Δ) HNR	Standard deviation, arithmetic mean
Spectral	(Δ) MFCC 1–12	Linear regression (offset, slope, MSE)

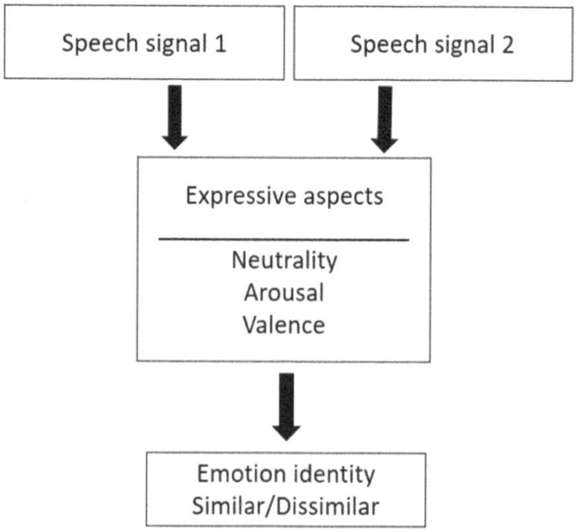

Fig. 1 Emotional content aspects and labels

4.3 Classes Related to Emotional Content

Since this work is interested in emotional content comparison, the signals of the database were grouped into pairs, of which 89386 were selected. Each pair has been assigned four labels regarding the similarity/dissimilarity of the following aspects: identity of emotions, neutrality, arousal and valence (cf. Fig. 1). It should be noted that except for valence, which is similar only for 40% of the pairs, 50% of them have similar emotions identity, neutrality and arousal.

5 Experiments

The experiments led in this work aim to evaluate (a) the effect of feature embedding on kmeans-based vector quantization of the emotional content of speech, and (b) the different ways of feature embedding using autoencoders.

5.1 Feature Embedding and Representation

Before applying vector quantization, the set of 768 features of each pair of signal (cf. Table 1) is preprocessed. Three types of preprocessing are achieved: (i) Normalization, to get zero-mean and unit-variance features, (ii) Normalization and application of the autoencoder on the whole feature set, so that the dimension of the feature vector, i.e. 768, is reduced to a lower value, (iii) Normalization and application of the autoencoder to the joint subsets of 12 coefficients, i.e. LLD, for each pair of signals (cf. Table 1). In this way, the 24-dimension of each pair of subsets is reduced to 1-dimension.

To apply the autoencoder, a deep neural network was implemented, where the output is the same as the input. The autoencoder architectures used in (ii) and (iii) are described in Table 2. In all the experiments, the training options were set as follows: ADAM optimizer, 50 epochs at maximum, a minibatch size of 32, a gradient threshold of 1, and a sigmoid transfer function.

The weights and biases of the code layer are utilized to calculate its output, using the sigmoid function. In the case of (LLD+ Δ-LLD) input features, the final embedded vector consists of the concatenated outputs of each autoencoder, i.e. a 32-coefficient vector.

The embedded data are finally represented by applying a vector quantization step [12], practically done with the kmeans algorithm using kmeans++ initialization. Classes are attributed to codevectors by majority voting. The codebook size was set to 100. The result is a smoothed-out representation of class distributions.

5.2 Feature Visualization

The autoencoder reveals the intrinsic structure of the input data, which helps in codevectors optimization. Figures 2, 3, 4, 5 show a comparison between the original input features and the autoencoded features, either all together or by LLD-group. It looks obvious that autoencoding allows visualizing (a) a compact structure, where features are more tightly distributed in the input space, (b) a clearer separation between the original classes. Therefore, the autoencoder, especially when applied to each LLD-group, seems to provide a good representation of the embedded features.

Table 2 Autoencoder architectures (layers and number of nodes)

Input features	Input layer	Hidden layer 1	Code layer	Hidden layer 3	Output layer
All features	768	500	32	500	768
LLD features	24	100	1	100	24

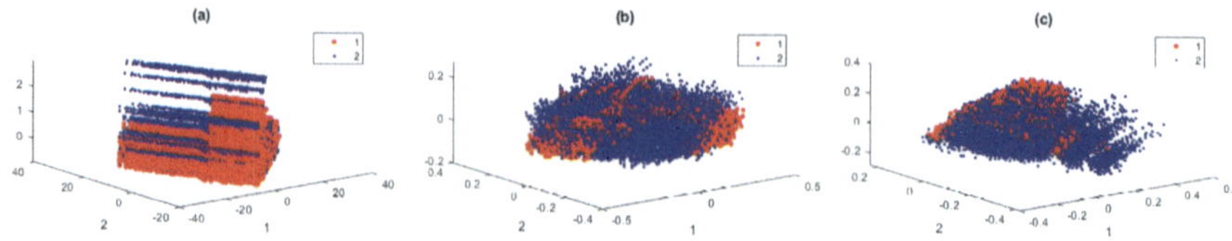

Fig. 2 Neutrality classes distribution: **a** original features, **b** autoencoded features, **c** autoencoded features by LLD-group

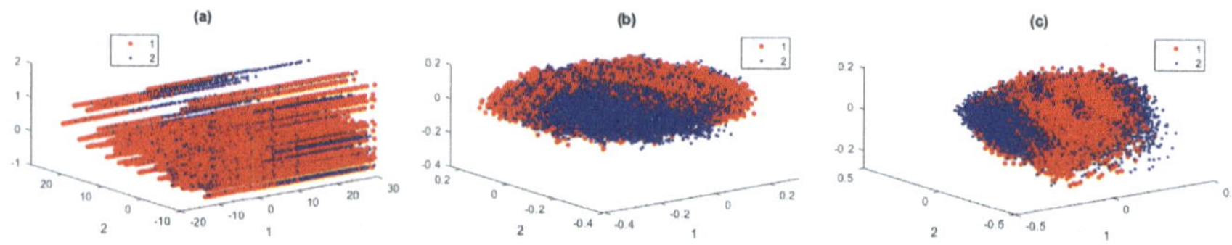

Fig. 3 Arousal classes distribution

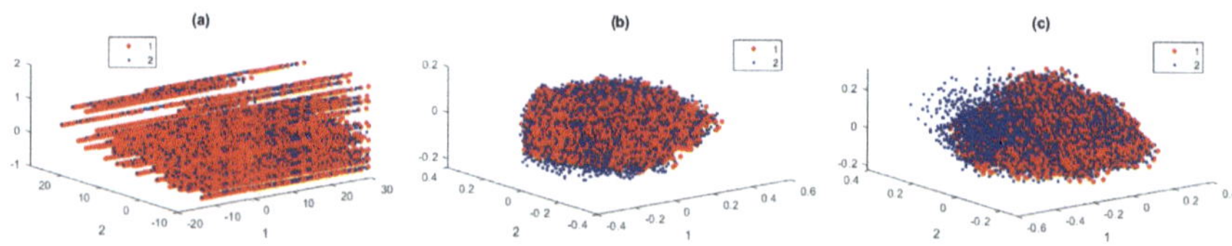

Fig. 4 Valence classes distribution

5.3 *Metric for Emotion Identity*

The three emotional aspects binary labels collected from vector quantization, i.e. neutrality, valence and arousal are aggregated to yield a metric for similarity measure of the emotion identity (μ_{id}). To calculate this metric, the mean value was used, i.e. $\mu_{id} = \frac{N+V+A}{3}$ where N, V and A are respectively neutrality, valence and arousal similarity/dissimilarity labels for each pair of signals. Then the identity metric is located into [0,1] interval.

As an application example, we cluster pairs of signals with similar emotions by using agglomerative hierarchical clustering to represent a dendrogram, where the leaves (x-axis) represents signals grouped using the distance between clusters calculated using the metric μ_{id} (cf. Fig. 5).

5.4 *Results and Discussion*

Table 3 shows the results of vector quantization using the aforementioned feature transformations (cf. Sect. 5.1). The following notes and interpretations can be drawn: (i) The effect of feature embedding on the classification results is clear, which may

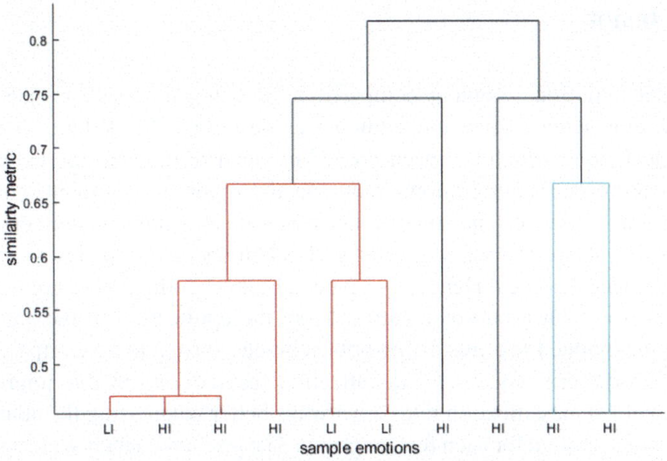

Fig. 5 Dendrogram of the hierarchical clustering of emotion identity similarity label based on the aggregated metric (HI: High-intensity emotions, LI: Low intensity emotions)

Table 3 Vector quantization accuracy using a codebook of 100 codevectors (NF: Normalized features, AF: Autoencoded features, AFL: Autoencoded features by LLD)

Expressive aspect	Total accuracy (%)			Similarity accuracy (%)			Dissimilarity accuracy (%)		
	NF	AF	AFL	NF	AF	AFL	NF	AF	AFL
Neutrality	67.9	75.2	77.5	75.3	81.0	83.9	60.5	69.3	71.1
Arousal	58.9	67.4	60.3	57.1	65.4	64.7	60.6	69.4	56.1
Valence	60.6	60.9	62.4	9.3	12.8	23.5	94.9	93.3	88.4

be explained by the improvement of features distribution, thanks to the autoencoder (cf. Figures 2, 3, 4, 5). (ii) The autoencoder applied on each LLD-group seems to improve, though slightly, the classification results. This may be due to the fact that this strategy allows selecting only one feature per LLD-group, thus avoiding redundant LLD information. (iii) The classification results using the autoencoder are more balanced between both classes, than those using only normalized data. This could be explained by the effect of compacting data, which allows detecting the codevectors more easily. (iv) Increasing the size of the codebook improves the classification results. However, it should be reasonably adapted to the number of samples, therefore we opted for a maximum of 100 codevectors for ca. 90,000 samples. (v) Using an aggregation metric for agglomerative hierarchical clustering allows grouping samples with similar emotions. However, in this case the result depends on the accuracy of vector quantization applied on the emotional aspects used to calculate the metric.

6 Conclusion

In this paper, emotional content comparison for pairs of speech signals by vector quantization using feature embedding was described. Embedding is a feature extraction technique which has been proved to enhance the learning performance. Actually, feature embedding allows reducing the input space dimensionality and the intra-feature variance. The autoencoder was used to achieve feature embedding, through the use of deep neural networks, with a bottleneck middle layer, which provides the encoded features. Hence, two types of autoencoding were applied, for all features, and for each group of features. First, the application of the autoencoder shows that the mapped features, using both schemes, have a more compact and distinguishable structure. The vector quantization results confirm this improvement, since the obtained classification rates are always better when using the autoencoder, especially when applied for each feature group. The predicted labels were aggregated to calculate a metric to compare emotion identity in speech. As an outlook, such a metric would form the basis for higher-level tasks, such as clustering utterances by emotional content, or applying kernel methods for expressive speech analysis.

Acknowledgments This work was supported by the research grant funded by "Fondi di Ricerca di Ateneo 2016" of the University of Genova.

References

1. Burkhardt, F., Paeschke, A., Rolfes, M., Sendlmeier, W. F., & Weiss, B.: A database of German emotional speech. In: Ninth European Conference on Speech Communication and Technology (2005)
2. Deng, J., Zhang, Z., Marchi, E., Schuller, B.: Sparse autoencoder-based feature transfer learning for speech emotion recognition. In: 2013 IEEE Humaine Association Conference on Affective Computing and Intelligent Interaction, pp. 511–516 (2013)
3. Eyben, F., Buchholz, S., Braunschweiler, N., Latorre, J., Wan, V., Gales, M. J., & Knill, K.: Unsupervised clustering of emotion and voice styles for expressive TTS. In: 2012 IEEE International Conference on Acoustics, Speech and Signal Processing (ICASSP), pp. 4009–4012 (2012)
4. Eyben, F., Scherer, K.R., Schuller, B.W., Sundberg, J., Andre, E., Busso, C., Truong, K.P.: The Geneva minimalistic acoustic parameter set (GeMAPS) for voice research and affective computing. IEEE Trans. Affect. Comput. **7**(2), 190–202 (2016)
5. Hozjan, V., Kacic, Z.: Context-independent multilingual emotion recognition from speech signals. Int. J. Speech Technol. **6**(3), 311–320 (2003)
6. Huang, P., Huang, Y., Wang, W., & Wang, L.: Deep embedding network for clustering. In: IEEE 2014 22nd International Conference on Pattern Recognition, pp. 1532–1537 (2014)
7. Kim, J., Saurous, R.A.: Emotion recognition from human speech using temporal information and deep learning. Proc. Interspeech **2018**, 937–940 (2018)
8. Moneta, C., Parodi, G., Rovetta, S., Zunino, R.: Automated diagnosis and disease characterization using neural network analysis. In: Proceedings of 1992 IEEE International Conference on Systems, Man, and Cybernetics, pp. 123–128 (1992)
9. Ng, A.: Sparse autoencoder. CS294A Lecture notes. http://web.stanford.edu/class/cs294a/sparseAutoencoder2011.pdf

10. Nicholson, J., Takahashi, K., Nakatsu, R.: Emotion recognition in speech using neural networks. Neural Comput. Appl. **9**(4), 290–296 (2000)
11. Nwe, T.L., Foo, S.W., De Silva, L.C.: Speech emotion recognition using hidden Markov models. Speech Commun. **41**(4), 603–623 (2003)
12. Ridella, S., Rovetta, S., Zunino, R.: K-winner machines for pattern classification. IEEE Trans. Neural Networks **12**(2), 371–385 (2001)
13. Rovetta, S., Mnasri, Z., Masulli, F., & Cabri, A.: Emotion recognition from speech signal using fuzzy clustering. In:: EUSFLAT Conference (2019) (to appear)
14. Schuller, B., Rigoll, G., Lang, M.: Hidden Markov model-based speech emotion recognition. In: 2003 IEEE International Conference on Acoustics, Speech, and Signal Processing—ICASSP 2003, vol. 2, pp. II-1 (2003)
15. Schuller, B., Rigoll, G., Lang, M.: Speech emotion recognition combining acoustic features and linguistic information in a hybrid support vector machine-belief network architecture. In: 2004 IEEE International Conference on Acoustics, Speech, and Signal Processing, vol. 1, pp. I-577 (2004)
16. Schuller, B., Steidl, S., Batliner, A.: The interspeech 2009 emotion challenge. In: Tenth Annual Conference of the International Speech Communication Association (2009)
17. Schuller, B., Steidl, S., Batliner, A., Vinciarelli, A., Scherer, K., Ringeval, F., Mortillaro, M.: The INTERSPEECH 2013 computational paralinguistics challenge: Social signals, conflict, emotion, autism. In: Proceedings INTERSPEECH 2013, 14th Annual Conference of the International Speech Communication Association, Lyon, France (2013)
18. Song, C., Liu, F., Huang, Y., Wang, L., Tan, T.:Auto-encoder based data clustering. In: Iberoamerican Congress on Pattern Recognition, pp. 117–124. Springer, Berlin, Heidelberg (2013)
19. Szekely, E., Cabral, J. P., Cahill, P., Carson-Berndsen, J.: Clustering expressive speech styles in audiobooks using glottal source parameters. In: Twelfth Annual Conference of the International Speech Communication Association (2011)
20. Tian, F., Gao, B., Cui, Q., Chen, E., Liu, T.Y.: Learning deep representations for graph clustering. In: Twenty-Eighth AAAI Conference on Artificial Intelligence (2014)
21. Xie, J., Girshick, R., Farhadi, A.: Unsupervised deep embedding for clustering analysis. In: International Conference on Machine Learning, pp. 478–487 (2016)

Efficient Data Augmentation Using Graph Imputation Neural Networks

Indro Spinelli, Simone Scardapane, Michele Scarpiniti, and Aurelio Uncini

Abstract Recently, data augmentation in the semi-supervised regime, where unlabeled data vastly outnumbers labeled data, has received a considerable attention. In this paper, we describe an efficient technique for this task, exploiting a recent framework we proposed for missing data imputation called graph imputation neural network (GINN). The key idea is to leverage both supervised and unsupervised data to build a graph of similarities between points in the dataset. Then, we augment the dataset by severely damaging a few of the nodes (up to 80% of their features), and reconstructing them using a variation of GINN. On several benchmark datasets, we show that our method can obtain significant improvements compared to a fully-supervised model, and we are able to augment the datasets up to a factor of $10\times$. This points to the power of graph-based neural networks to represent structural affinities in the samples for tasks of data reconstruction and augmentation.

1 Introduction

Semi-supervised learning (SSL) studies how to exploit vast amounts of unlabeled data to improve the performances of a model trained on a smaller number of labeled data points [5]. Over the years, a large number of solutions were devised, ranging from manifold regularization [3] to label propagation [8]. With the emergence of deep learning techiques and the increase in datasets' size, several new methods were

I. Spinelli · S. Scardapane (✉) · M. Scarpiniti · A. Uncini
Department of Information Engineering, Electronics and Telecommunications (DIET), "Sapienza" University of Rome, Via Eudossiana 18, 00184 Rome, Italy
e-mail: simone.scardapane@uniroma1.it

M. Scarpiniti
e-mail: michele.scarpiniti@uniroma1.it

A. Uncini
e-mail: aurelio.uncini@uniroma1.it

A. Esposito et al. (eds.), *Progresses in Artificial Intelligence and Neural Systems*, Smart Innovation, Systems and Technologies 184,
https://doi.org/10.1007/978-981-15-5093-5_6

also proposed to leverage end-to-end training of deep networks in the SSL regime, e.g., pseudo-labels [12] and ladder networks [13].

An interesting line of research recently concerns the exploitation of unsupervised information to perform data augmentation [7]. Augmented datasets can then be used directly, or indirectly through the imposition of one or more regularizers on the learning process, by enforcing the model to have stable outputs with respect to the new points [8]. Several papers have shown that semi-supervised data augmentation has the potential to provide significant boosts in accuracy. for deep learning models compared to standard SSL algorithms or equivalent fully-supervised solutions [4, 15].

In this light, an essential research question concerns how to devise efficient and general-purpose strategies for data augmentation. A popular data augmentation strategy, mixup [17], builds new datapoints by taking linear combinations of points in the dataset, but it requires knowledge of the underlying labels and has shown mixed results when dealing with non-image data. Other popular techniques, such as AutoAugment [6], are instead especially designed to work in structured domains such as images or audio samples.

In this paper, we propose a new method to perform data augmentation from general vectorial inputs. Our starting point is the graph imputer neural network (GINN) [14], an algorithm we proposed recently for multivariate data imputation in the presence of missing features. GINN works by building a graph of similarities from the points in the dataset (similarly to manifold regularization [3]). Missing features are then estimated by applying a customized graph autoencoder [11] to reconstruct the full dataset starting from the incomplete dataset and the graph information [14]. GINN has shown to outperform most state-of-the-art methods for data imputation, leading us to investigate whether its performance can be extended to the case of data augmentation.

The algorithm proposed in this paper is an application of GINN showing its effectiveness for data augmentation. After constructing the similarity graph starting from both labeled and unlabeled data, we corrupt some of the nodes by removing up to 80% of their features. After recomputing on-the-fly their connections with the neighboring nodes, we apply the previously trained GINN architecture for performing imputation, effectively generating a new data point that can be added to the dataset. Despite its conceptual simplicity, we show through an extensive experimental evaluation that augmenting the dataset in this fashion (up to 5/10 times) leads to significant improvements in accuracy when training standard supervised learning algorithms.

The rest of the paper is structured as follows. In Sect. 2 we describe our proposed technique for data augmentation. Section 3 shows an experimental evaluation on a range of different datasets. We conclude with a few open research directions in Sect. 4.

2 Semi-supervised Data Augmentation with GINN

Figure 1 shows a high-level overview of our framework. As we stated before, we first build a similarity graph from the data, used to train a GINN model [14]. Then, we employ GINN on highly-corrupted points from the dataset to generate new data-points. We first describe GINN in Sect. 2.1, before describing how we use it for data augmentation in Sect. 2.2.

2.1 *Graph Imputer Neural Network (GINN)*

GINN [14] is a graph-based neural network that takes a damaged dataset as input (i.e., a dataset with a few missing entries), and is able to reconstruct the missing features by constructing a similarity graph and exploiting the graph convolution operation [2, 11]. Here we summarize the GINN architecture in broad terms, while for a more in-depth description we refer the interested reader to [14].

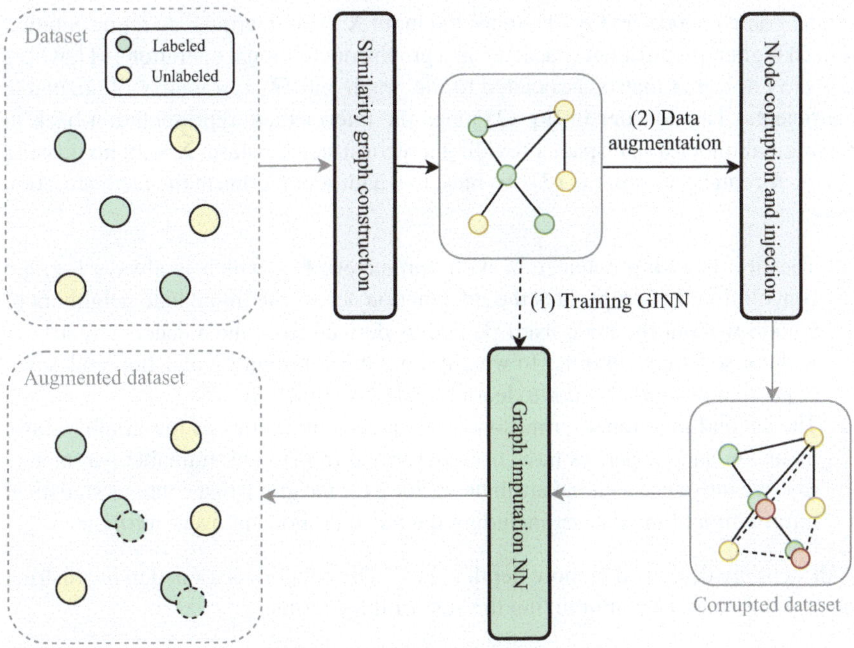

Fig. 1 Overall schema of the proposed framework for data augmentation. In green we show the GINN method which is optimized from the dataset

Consider a generic dataset $\mathbf{X} \in \mathbb{R}^{n \times d}$, where each row encodes one example, defined by a vector of d features. We would like a model that can reconstruct the original $\mathbf{X}$ even when a few of its elements are missing, i.e., an algorithm performing missing data imputation.

In order to train GINN to this end, we first augment the matrix information in $\mathbf{X}$ with a graph describing the structural proximity between points. In particular, we encode each feature vector as a node in a graph $\mathbf{G}$. The adjacency matrix $\mathbf{A}$ of the graph is derived from a similarity matrix $\mathbf{S}$ of the feature vectors, containing the pairwise Euclidean distances of the feature vectors. In order to keep only the most relevant edges, we apply a two-step pruning on $\mathbf{S}$. We compute the $97.72nd$ percentile for each row, corresponding to $+2\sigma$ under a Gaussian assumption, and we use this as a threshold, discarding all the connection below this value. The second step replicates the first one but acts on all the surviving elements of the matrix $\mathbf{S}$ (see the original paper [14] for a rationale of this technique).

The core of GINN is a graph-convolutional autoencoder described as follows:

$$\mathbf{H} = \text{ReLU}\left(\mathbf{L}\mathbf{X}\boldsymbol{\Theta}_1\right) , \tag{1}$$

$$\widehat{\mathbf{X}} = \text{Sigmoid}\left(\mathbf{L}\mathbf{H}\boldsymbol{\Theta}_2 + \widetilde{\mathbf{L}}\mathbf{X}\boldsymbol{\Theta}_3 + \boldsymbol{\Theta}_4\mathbf{g}\right) . \tag{2}$$

The graph encoder in Eq. (1), maps the input $\mathbf{X}$ to an intermediate representation $\mathbf{H}$ in an higher dimensional space, using a graph convolutional operation [11], where $\mathbf{L}$ is the Laplacian matrix associated to the graph and $\boldsymbol{\Theta}_1$ is a matrix of adaptable coefficients. The decoder in Eq. (2) maps the intermediate representation back to the original dimensional space providing a reconstructed dataset $\widehat{\mathbf{X}}$ with no missing values. As can be seen in Eq. (2), we have two additional terms in the reconstruction process:

1. The first is a skip connection with parameters $\boldsymbol{\Theta}_2$, which is always a graph convolution that propagates the information across the immediate neighbors of a node without the node itself. $\widetilde{\mathbf{L}}$ is thus derived from the adjacency matrix $\mathbf{A}$ without self-loops in order to weight more the contribution of 1-hop nodes and to prevent the autoencoder to learn the identity function.
2. The second additional term models the global properties of the graph, whose inclusion and properties have been described in [2], with trainable parameters $\boldsymbol{\Theta}_3$. We introduce a global attribute vector $\mathbf{g}$ for the graph that contains statistical information of the dataset including the mean or mode of every attribute.

Both steps are described in more depth in [14]. The network is trained to reconstruct the original dataset by minimizing the sum of three terms:

$$L_A = \alpha \text{MSE}(\mathbf{X}, \widehat{\mathbf{X}}) + (1-\alpha)\text{CE}(\mathbf{X}, \widehat{\mathbf{X}}) + \gamma\, \text{MSE}(\text{global}(\widehat{\mathbf{X}}), \text{global}(\mathbf{X})) . \tag{3}$$

The loss function in Eq. (3) minimizes the reconstruction error over the non-missing elements combining the mean squared error (MSE) for the numerical variables and the cross-entropy (CE) for the categorical variables. In addition, since the computation of the global information is differentiable, we compute a loss term with respect to the global attributes of the original dataset. Here, α and γ are additional hyperparameters, the first is initialized as the ratio between numerical and categorical columns, the second is a weighting factor. Practically, at every iteration of optimization we randomly drop a given percentage of elements in $\mathbf{X}$ in order to learn to reconstruct the original matrix irrespective of which elements are missing.

This model can be trained as standalone or paired with a critic network C in an adversarial fashion [16]. In the latter case we have a 3-layer feed-forward network that learns to distinguish between imputed ($\hat{\mathbf{x}} \sim \mathbb{P}_{imp}$) and real ($\mathbf{x} \sim \mathbb{P}_{real}$) data. To train the models together we use the Wasserstein distance [1]. The objective function of the critic is:

$$\min_{A} \max_{C \in \mathscr{D}} \mathop{\mathbb{E}}_{\mathbf{x} \sim \mathbb{P}_{real}} [C(\mathbf{x})] - \mathop{\mathbb{E}}_{\hat{\mathbf{x}} \sim \mathbb{P}_{imp}} [C(\hat{\mathbf{x}}))] , \tag{4}$$

where $\mathscr{D}$ is the set of 1-Lipschitz functions, $\mathbb{P}_{imp}$ is the model distribution implicitly defined by our GCN autoencoder A, and $\mathbb{P}_{real}$ is the unknown data distribution.

In addition, we use a gradient penalty introduced in [9] to enhance training stability and enforcing the Lipschitz constraint, obtaining the final critic loss:

$$L_C = \mathop{\mathbb{E}}_{\hat{\mathbf{x}} \sim \mathbb{P}_{imp}} [C(\hat{\mathbf{x}})] - \mathop{\mathbb{E}}_{\mathbf{x} \sim \mathbb{P}_{real}} [C(\mathbf{x})] + \lambda \mathop{\mathbb{E}}_{\tilde{\mathbf{x}} \sim \mathbb{P}_{\tilde{\mathbf{x}}}} \left[(\|\nabla_{\tilde{\mathbf{x}}} C(\tilde{\mathbf{x}})\|_2 - 1)^2 \right] , \tag{5}$$

where λ is an additional hyper-parameter. We define $\mathbb{P}_{\tilde{\mathbf{x}}}$ as sampling uniformly from the combination of the real distribution $\mathbb{P}_{imp}$ and from the distribution resulting from the imputation $\mathbb{P}_{imp}$. This means that the feature vector $\tilde{\mathbf{x}}$ will be composed by both real and imputed elements in almost equal size.

The total loss of the autoencoder becomes:

$$L_T = L_A - \mathop{\mathbb{E}}_{\hat{\mathbf{x}} \sim \mathbb{P}_{imp}} [C(\hat{\mathbf{x}})] , \tag{6}$$

since it must fool the critic and minimize the reconstruction error at the same time. In our implementation the GCN autoencoder is trained once for every five optimization steps of the critic and they will influence each other during the whole process.

2.2 *Semi-supervised Data Augmentation*

In order to augment the dataset in the semi-supervised setting, we build the graph with labeled and unlabeled features and let GINN learn its representation. Rather than learning the labels, like in a transductive setting [11], we want to generate completely new nodes in the graph and thus new feature vectors with labels that can be used later for other objectives. To do so we take all the labeled data and use it to generate a new data matrix of the size of the desired augmentation level. This means that features vectors can be repeated in this matrix. We then damage the matrix in a MCAR (Missing Completely At Random) fashion removing 80% of its elements. These severely damaged vectors will be the starting point of our data augmentation strategy. We note that in [14] we have shown already the ability of GINN in imputing vectors having few non-zero elements.

We inject these new damaged feature vectors in the graph. To recompute their connections on-the-fly, we follow the same procedure described above, without considering the second pruning step in order to guarantee that every node will have at least one neighbor. The connections will be computed only with unlabeled nodes in order to prevent the new nodes to be too similar to the ones they originated from. This time the similarity has to take into account the fact that the new vectors will have only few non-zero elements. For this reason we formulate the distance as follows:

$$S_{ij} = d(\mathbf{x}_i \odot (\mathbf{m}_i \odot \mathbf{m}_j), \mathbf{x}_j \odot (\mathbf{m}_i \odot \mathbf{m}_j)), \tag{7}$$

where $\odot$ stands for the Hadamard product between vectors, $\mathbf{m}_i$ and $\mathbf{m}_j$ are binary vectors describing the missing elements in $\mathbf{x}_i$ and $\mathbf{x}_j$ respectively, and d is the Euclidean distance.

Once the new data is in the graph, we use GINN to impute all missing elements, and add the resulting vectors to our original dataset. The resulting imputed vectors will have the label of the node they have been generated from and will be influenced by the unlabeled nodes in the graph sharing similar features.

3 Experimental Evaluation

In this section, we analyze the data augmentation performance of our framework. For the evaluation we used 6 classification datasets from the UCI Machine Learning Repository,[1] and their characteristics are described in Table 1. These datasets contain numerical, categorical and mixed features in order to show that our framework is capable of generating realistic feature vectors composed of different types of attributes.

We tracked the performances of 5 different classifiers when using the default and the augmented datasets. The algorithms used are a k-NN classifier with $k = 5$, regularized logistic regression (LOG), C-Support Vector Classification with an RBF kernel (SVC), a random forest classifier with 10 estimators (RF) and a maximum depth of 10 and a feedforward neural network composed of six layers with an embedding dimension of 128 (MLP). All hyper-parameters are initialized with their default values in the scikit-learn implementation.

Concerning our framework, we use an embedding dimension of the hidden layer of 128, sufficient for an overcomplete representation for all the datasets involved. We trained GINN for a maximum of 3000 epochs with an early stopping strategy for the reconstruction loss over the known elements. The critic used is a simple 3-layer

[1] http://archive.ics.uci.edu/ml/index.php.

Table 1 Datasets used in our experiments. All datasets were downloaded from the UCI repository

Name	Observations	Numerical attr.	Categorical attr.
Abalone	4177	8	0
Heart	303	8	5
Ionosphere	351	34	0
Phishing	1353	0	9
Tic-tac-toe	958	9	9
Wine-quality-red	1599	11	0

feed-forward network trained 5 times for each optimization step of the autoencoder. We used the Adam optimizer [10] for both networks with a learning rate of 1×10^{-3} and 1×10^{-5} respectively for autoencoder and critic. All other hyper-parameters are taken from the original GINN paper [14], whose implementation is available as an open-source package.[2]

3.1 Semi-supervised Classification

In the semi-supervised setting, we divided our data between the training set, 70%, and test set, 30%. Only 10% of the training set has labels associated with feature vectors. We augment the dataset with GINN creating 3 different versions of the training set, respectively having $2\times$, $5\times$ and $10\times$ more labeled data. We then train the classifiers on this 4 different training sets and compute the accuracy, averaged over 5 different trials. Each trial has different splits of training and test sets.

In Table 2 we show the classification results. As can be seen, our method consistently improves classification accuracy. These improvements range from less than a percentage point up to an increment of 24% points. When we do not improve the results, our decrease in performance regards at most 6% points for a single classifier (e.g., LOG in the hearth dataset). In Fig. 2 we summarize the results of Table 2 and propose a comparison of the times our augmented datasets allows to achieve a better classification accuracy to the classifiers with respect to the original dataset. As can be clearly seen this difference increases with the size of the augmentation, showing that our framework is capable of generating good feature vectors even when creating a data matrix 10 times the size of the original dataset.

[2]https://github.com/spindro/GINN.

Table 2 Mean classification accuracy over 5 different trials on different splits of data obtained using the standard dataset **X** and the augmented versions with GINN. In Fig. 2 we summarize the results of this table

Dataset	Classifier	Baseline	Augmented (2×)	Augmented (5×)	Augmented (10×)
Abalone	LOG	52.87	52.54	52.47	54.50
	k-NN	52.07	52.07	52.07	52.07
	SVC	52.87	52.87	52.87	52.87
	RF	51.53	51.18	52.38	53.67
	MLP	50.53	52.66	52.03	54.78
Heart	LOG	76.92	70.77	70.77	66.59
	k-NN	58.24	64.40	64.40	62.56
	SVC	55.88	56.04	56.04	56.04
	RF	79.56	81.10	80.44	78.02
	MLP	65.71	66.81	63.96	61.32
Ionosphere	LOG	78.30	80.94	79.06	78.49
	k-NN	66.04	90.57	90.57	90.57
	SVC	64.15	85.09	84.34	85.28
	RF	83.96	87.92	85.47	86.23
	MLP	90.57	88.87	86.04	86.04
Phishing	LOG	83.50	82.07	82.02	81.67
	k-NN	82.76	84.04	83.55	84.19
	SVC	82.51	82.17	81.92	81.92
	RF	81.33	81.58	81.08	81.82
	MLP	81.38	81.97	83.05	82.91
Tic-tac-toe	LOG	77.43	73.89	69.86	70.07
	k-NN	73.61	74.10	75.00	78.13
	SVC	65.28	64.65	67.85	68.47
	RF	71.94	75.76	72.36	74.38
	MLP	78.40	73.06	78.75	82.01
Wine-quality	LOG	53.75	53.25	53.96	54.21
	k-NN	45.83	46.21	49.67	49.79
	SVC	45.63	47.67	52.42	52.38
	RF	50.58	50.92	54.96	54.21
	MLP	48.83	50.83	51.04	50.38

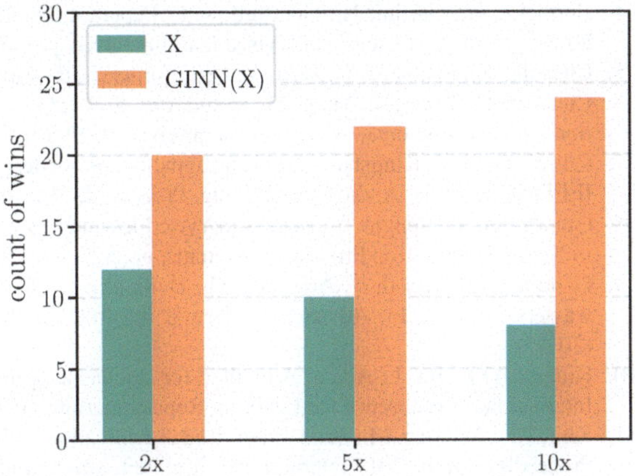

Fig. 2 Number of times the default and the augmented datasets with GINN had a better classification performances over 5 different trials considering all datasets and classifiers in the benchmark

4 Conclusions

In this paper, we proposed an approach to data augmentation in the semi-supervised regime by reformulating it as a problem of extreme missing data imputation. To this end, we employ a novel algorithm for missing data imputation built on top of a graph autoencoder. Our results on a set of standard vectorial benchmarks show that the method can significantly improve over using only the labeled information, even when the dataset is augmented up to ten times its original size.

The method lends itself to a variety of improvements. First of all, we are interested in evaluating the augmentation strategies not only by directly retraining a supervised algorithms, but also in the context of several regularization strategies commonly used today [15]. We would also like to test the algorithm on non-vectorial domains, including images and audio, where the challenge is to define a proper metric to build the similarity graph. As a final remark, we note that the experiments presented here open the way to a set of interesting additional questions. In particular, viewing data augmentation as an extreme case of data imputation bridges two different fields with high potential for cross-fertilization.

References

1. Arjovsky, M., Chintala, S., Bottou, L.: Wasserstein generative adversarial networks. In: Proceeding of the 34th International Conference on Machine Learning (ICML), vol. 70, pp. 214–223 (2017)
2. Battaglia, P.W., Hamrick, J.B., Bapst, V., Sanchez-Gonzalez, A., Zambaldi, V., Malinowski, M., Tacchetti, A., Raposo, D., Santoro, A., Faulkner, R., et al.: Relational inductive biases, deep learning, and graph networks. arXiv preprint arXiv:1806.01261 (2018)
3. Belkin, M., Niyogi, P., Sindhwani, V.: Manifold regularization: a geometric framework for learning from labeled and unlabeled examples. J. Mach. Learn. Res. 7(Nov), 2399–2434 (2006)

4. Berthelot, D., Carlini, N., Goodfellow, I., Papernot, N., Oliver, A., Raffel, C.: Mixmatch: a holistic approach to semi-supervised learning. arXiv preprint arXiv:1905.02249 (2019)
5. Chapelle, O., Schlkopf, B., Zien, A.: Semi-Supervised Learning, 1st edn. The MIT Press (2010)
6. Cubuk, E.D., Zoph, B., Mane, D., Vasudevan, V., Le, Q.V.: Autoaugment: learning augmentation policies from data. arXiv preprint arXiv:1805.09501 (2018)
7. Cui, X., Goel, V., Kingsbury, B.: Data augmentation for deep neural network acoustic modeling. IEEE/ACM Trans. Audio Speech Lang. Process. **23**(9), 1469–1477 (2015)
8. Grandvalet, Y., Bengio, Y.: Semi-supervised learning by entropy minimization. In: Advances in Neural Information Processing Systems, pp. 529–536 (2005)
9. Gulrajani, I., Ahmed, F., Arjovsky, M., Dumoulin, V., Courville, A.C.: Improved training of wasserstein gans. In: Advances in Neural Information Processing Systems, pp. 5767–5777 (2017)
10. Kingma, D.P., Ba, J.: Adam: A method for stochastic optimization. In: Proceeding of the 3rd International Conference for Learning Representations (ICLR) (2014)
11. Kipf, T.N., Welling, M.: Semi-supervised classification with graph convolutional networks. In: Proceedings of the 2017 International Conference on Learning Representations (ICLR) (2017)
12. Lee, D.H.: Pseudo-label: The simple and efficient semi-supervised learning method for deep neural networks. In: Workshop on Challenges in Representation Learning, ICML. vol. 3, p. 2 (2013)
13. Rasmus, A., Berglund, M., Honkala, M., Valpola, H., Raiko, T.: Semi-supervised learning with ladder networks. In: Advances in Neural Information Processing Systems, pp. 3546–3554 (2015)
14. Spinelli, I., Scardapane, S., Uncini, A.: Missing Data Imputation with Adversarially-trained Graph Convolutional Networks. arXiv preprint arXiv:1905.01907 (2019)
15. Xie, Q., Dai, Z., Hovy, E., Luong, M.T., Le, Q.V.: Unsupervised Data Augmentation. arXiv preprint arXiv:1904.12848 (Apr 2019)
16. Yoon, J., Jordon, J., van der Schaar, M.: Gain: Missing data imputation using generative adversarial nets. In: Proceedings of the 35th International Conference of Machine Learning (ICML), pp. 1–10 (2018)
17. Zhang, H., Cisse, M., Dauphin, Y.N., Lopez-Paz, D.: mixup: Beyond empirical risk minimization. arXiv preprint arXiv:1710.09412 (2017)

Flexible Generative Adversarial Networks with Non-parametric Activation Functions

Eleonora Grassucci, Simone Scardapane, Danilo Comminiello, and Aurelio Uncini

Abstract Generative adversarial networks (GANs) have become widespread models for complex density estimation tasks such as image generation or image-to-image synthesis. At the same time, training of GANs can suffer from several problems, either of stability or convergence, sometimes hindering their effective deployment. In this paper we investigate whether we can improve GAN training by endowing the neural network models with more flexible activation functions compared to the commonly used rectified linear unit (or its variants). In particular, we evaluate training a deep convolutional GAN wherein all hidden activation functions are replaced with a version of the kernel activation function (KAF), a recently proposed technique for learning non-parametric nonlinearities during the optimization process. On a thorough empirical evaluation on multiple image generation benchmarks, we show that the resulting architectures learn to generate visually pleasing images in a fraction of the number of the epochs, eventually converging to a better solution, even when we equalize (or even lower) the number of free parameters. Overall, this points to the importance of investigating better and more flexible architectures in the context of GANs.

E. Grassucci · S. Scardapane (✉) · D. Comminiello · A. Uncini
Department of Information Engineering, Electronics and Telecommunications (DIET), "Sapienza" University of Rome, Via Eudossiana 18, 00184 Rome, Italy
e-mail: simone.scardapane@uniroma1.it

D. Comminiello
e-mail: danilo.comminiello@uniroma1.it

A. Uncini
e-mail: aurelio.uncini@uniroma1.it

A. Esposito et al. (eds.), *Progresses in Artificial Intelligence and Neural Systems*, Smart Innovation, Systems and Technologies 184,
https://doi.org/10.1007/978-981-15-5093-5_7

1 Introduction

Recently, density estimation in high-dimensional spaces (e.g., pixels of an image or frames of a video) has seen rapid advancements thanks to the use of deep convolutional networks. Among the different techniques, generative adversarial networks (GANs) [6, 12] have become one of the most popular methods to this end, because of their capacity to scale up to large resolutions with no apparent loss in perceptual quality [3].

A GAN is a neural network model that is trained to perform a deterministic transformation between a sample of a very simple probability distribution (generally uniform) and a sample from the distribution of interest (e.g., a distribution over all possible photos of chairs). Because of the complexity of directly training such a network, a second neural network, originally called a discriminator [6], is trained at the same time to guide the learning process in some way. In the original implementation of GANs [6], the discriminator is trained to classify between original samples and generated samples, although other techniques have been designed recently to make the optimization smoother [2] (see also the discussion in Sect. 2). Apart from high-quality image synthesis [3], GANs have been successfully applied to a variety of contexts ranging from noise reduction in the medical domain [18] to image retrieval [16], image super resolution [17], and many others, for which we refer interested readers to the survey in [4].

The two networks are trained on competitive objectives, making GAN training connected to the search for a proper game-theoretic equilibrium, and a significantly complex problem in practical terms. A main contribution in this sense was the introduction of the Wasserstein GAN (WGAN) [2, 7], defining a stabler metric to optimize, leading to faster convergence and less dependence on hyper-parameters. Both GANs and WGANs are nonetheless still prone to a variety of issues, ranging from mode collapse of the underlying distribution to training instability and loss of perceptual quality. Because of this, multiple techniques and empirical solutions are being routinely evaluated in the research community, ranging from feature matching and gradient averaging [13], to least-squares training [10]. Generally speaking, finding ways to make GAN training faster and less reliant on hyperparameters remains an open research problem.

1.1 Contributions of the Paper

In this paper we start a preliminary investigation on the influence of the activation functions on the stability and convergence of GANs. In recent years, multiple authors have investigated the use of more advanced activation functions in neural networks compared to the standard squashing functions and rectifier ones, ranging from parametric solutions such as the leaky ReLU [9], to more advanded techniques such as maxout neurons [19], piecewise linear units [1], and kernel activation functions

(KAFs) [14, 15]. A shared conclusion of many of these papers is that the choice of nonlinearity strongly affects what a network can easily learn, and even small changes in their design can have wide ranging impacts on the number of layers and on the feasibility of the optimization task. However, no paper to the best of our knowledge investigated this in the context of GAN training, where this added flexibility could be even more beneficial.

Summarizing the contributions of this paper:

1. Focusing on a state-of-the-art version of the WGAN, we propose a more flexible architecture wherein all activation functions in the two neural networks are replaced with extensions of the KAF. The KAF defines each activation function as a small one-dimensional kernel model, making the overall function flexible while remaining easy to optimize.
2. We evaluate empirically the results on a number of image synthesis problems involving multiple datasets, ranging from the standard MNIST to a dataset of Latin characters extracted from historical manuscripts [5].
3. We show that GANs endowed with KAFs can converge significantly faster to a solution of higher perceptual quality, even with a much reduced architecture.

1.2 Structure of the Paper

Section 2 describes the WGAN we consider in the paper. In Sect. 3 we propose our modified WGAN with the use of KAFs. Next, we evaluate the proposal in Sect. 4 before concluding in Sect. 5.

2 Wasserstein Generative Adversarial Networks

Suppose we are given a set of n elements $\{x_i\}_{i=1}^n$ sampled from some probability distribution $p(x)$. We are interested in being able to approximate p to a high degree in the case where x represents a high-dimensional, complex object (e.g., an image). A GAN solves the problem by simultaneously training two models, a generator $G(z)$ translating a latent vector z drawn from a uniform distribution $u(z)$ into a sample from the desired probability distribution p, and a discriminator D (equivalently called a critic in WGAN) to help guide the optimization process.

In the original GAN implementation [6], the discriminator is trained to classify whether a sample w comes from the original distribution p or from the generator G, leading to the following minimax optimization problem:

$$\min_g \max_d \left\{ \mathop{\mathbb{E}}_{x\sim p(x)}\Big[\log\left(D(x)\right)\Big] + \mathop{\mathbb{E}}_{z\sim u(z)}\Big[\log\left(1 - D(G(z))\right)\Big]\right\} \tag{1}$$

Problem (1) can be solved by interleaving optimization steps over G and over D using mini-batches of samples from our dataset. At optimality, one can discard the discriminator D and use G to obtain new samples from p.

The authors in [2] showed that training the GAN is this fashion is equivalent to minimizing the Jensen-Shannon divergence between the real distribution and the learned one. In the same paper, they also argue that minimizing instead the 1-Wasserstein distance (also called the Earth Mover distance) is a more sensible choice and empirically leads to a more stable optimization behaviour. After some mathematical manipulation, the resulting optimization problem becomes:

$$\min_{g} \max_{d \in \mathcal{L}} \left\{ \mathbb{E}_{x \sim p(x)} \Big[D(x) \Big] - \mathbb{E}_{z \sim u(z)} \Big[D(G(z)) \Big] \right\} \tag{2}$$

where the discriminator (critic) is now restricted to the family $\mathcal{L}$ of 1-Lipschitz functions, and it is not interpretable anymore as a classifier. While the resulting Wasserstein GAN (WGAN) is more stable, enforcing the property of Lipschitz continuity is not trivial if one uses standard unconstrained stochastic gradient descent algorithms. While the original implementation [2] used weights clipping to enforce the constraint, most recent implementations consider instead a gradient penalty term added to the original loss function L:

$$\widetilde{L} = L + \lambda \mathbb{E}_{\hat{x} \sim \hat{p}(\hat{x})} \left[\left(\|\nabla_{\hat{x}} D(\hat{x})\| - 1 \right)^2 \right], \tag{3}$$

where λ is a weighting hyper-parameter, and $\hat{p}(\hat{x})$ is defined as sampling uniformly from either p or the generator G. For our experiments we use a WGAN with gradient penalty as in (3). However, we note that a large amount of variations are possible, as described more in depth in Sect. 1.

3 Proposed WGAN with Kernel Activation Functions

We now focus on the design of the generator G and the critic D. When x is a vector, these can be implemented as standard NNs built by stacking classical fully connected layers:

$$l(\mathbf{h}) = \phi\left(\mathbf{W}\mathbf{h} + \mathbf{b}\right), \tag{4}$$

where $\mathbf{h}$ is the input to the layer, $\mathbf{W}$ and $\mathbf{b}$ are trainable parameters, and $\phi(\cdot)$ is any nonlinear activation function. When working with images, one can replace linear projections with convolutive layers [12], and interleave these with up-sampling/down-sampling operations when required.

Concerning the choice of $\phi(\cdot)$, most works either consider ReLUs $\phi(s) = \max(0, s)$ or some variants of it, such as the Leaky ReLU [9]:

$$\phi(s) = \begin{cases} s & \text{if } s \geq 0, \\ \alpha s & \text{otherwise,} \end{cases} \tag{5}$$

where α is a small positive parameter (e.g., $\alpha = 0.01$). In this paper, we instead propose to replace all ϕs in a standard WGAN implementation by KAFs, a more flexible technique for jointly learning the shape of the activation functions together with the parameters of the linear projection in (4). The assumption is that increasing the flexibility in this fashion can significantly improve the optimization process, possibly leading to higher-quality samples.

Practically, we model each ϕ neuron-wise as a one-dimensional kernel model with a fixed number of coefficients as [15]:

$$\phi(s) = \sum_{k=1}^{K} \alpha_k \kappa\left(s, s_k\right) , \tag{6}$$

where the number of coefficients K and the scalar kernel function κ are hyper-parameters of the model. For simplicity, the elements $s_1, \ldots, s_K$ where the kernel is evaluated are fixed at the beginning by sampling uniformly the real line around zero, while the coefficients $\alpha_1, \ldots, \alpha_K$ are optimized during training by performing back-propagation on them. We use the Gaussian kernel in the experiments:

$$\kappa(a, b) = \exp\left\{-\gamma(a - b)^2\right\} , \tag{7}$$

where the scalar γ is selected using the rule-of-thumb also proposed in [15], to which we refer the interested readers for additional details on the practical implementation of KAFs. We also note that additional insights can be obtained from [11], where a similar model was independently proposed starting from a functional optimization perspective.

4 Experimental Evaluation

4.1 Experimental Setup

The baseline network we consider is a conditional deep convolutional WGAN with gradient penalty, inspired from [12]. The generator is composed by an initial linear layer, four convolutive layers with 128, 64, 32 and 1 filters respectively, and a tanh activation function in output to map the output to the interval $[-1, 1]$.

The critic has a similar stack of four convolutive layers of dimension 64, 128, 256 and 512, followed by a linear projection in the output. As proposed in [12], instead of using pooling layers we have applied fractional-strided convolutions in the generator and strided convolutions in the critic. In addition, batch normalization is used in the generator only, whereas layer normalization is preferred in the critic as suggested in [7]. Afterwards, the conditioning one-hot encoded labels are concatenated with the noise vector in the generator, whereas in the critic we have created a one-hot encoded mask to be chained in the channels dimension. Namely, there will be as many masks as classes where the mask corresponding to the conditioning label will have all $1s$ instead of all $0s$.

Two versions of this network are evaluated: according to [12], the baseline one has ReLUs activation functions in the generator and Leaky ReLUs in the critic. In comparison, the second one uses KAFs in both networks as described in Sect. 3. KAFs are initialized to approximate an exponential linear unit function as described in [15].

We also test a reduced version of this architecture, maintaining the number of layers but decreasing the number of filters by approximately 60%. More in depth, the above described standard WGAN has 10593858 parameters, whereas the smaller version just 3750914. Similarly, the reduced KAF architecture has 3757954 parameters to train (that is, in this setting KAFs add approximately 7000 parameters).

To test the performance of our method against the classical conditional WGAN, three different datasets are used: MNIST, Fashion MNIST and a subset of the dataset coming from the In Codice Ratio (ICR) research project [5]. The last one is a collection of 23 different latin characters coming from the Vatican Registers of the Vatican Secret Archives. From this set we have selected 10 classes.

4.2 Experimental Results

The following figures show some samples of generated images from the ICR dataset during training (respectively after 1, 20, and 40 epochs) for three different configurations: in Fig. 1 the reduced GAN with KAFs (KAF-WGAN), in Fig. 2 the same small architecture but having ReLUs—Leaky ReLUs activation functions. Finally, Fig. 3 shows the results of the larger model with ReLUs—Leaky ReLUs.

The remarkable result of this experiment is the capacity of the KAF-WGAN to learn to distinguish between the background and the overall character shape in one epoch. While the other models still generate mostly noise. Moreover, focusing on the small architectures, the KAF network achieves an appreciable image quality with just 20 epochs of training, while the standard WGAN does not generate characters yet (neither in 40 epochs). Furthermore, the proposed method can also compete with the heavier architecture, generating better images with less trainable parameters. These results are evident in Fig. 4 also where samples from the original ICR dataset are compared with images generated by the KAF-WGAN network and with samples of

Fig. 1 Samples of generated images on a subset coming from the ICR dataset with the reduced KAF-WGAN after 1, 20, and 40 epochs of training

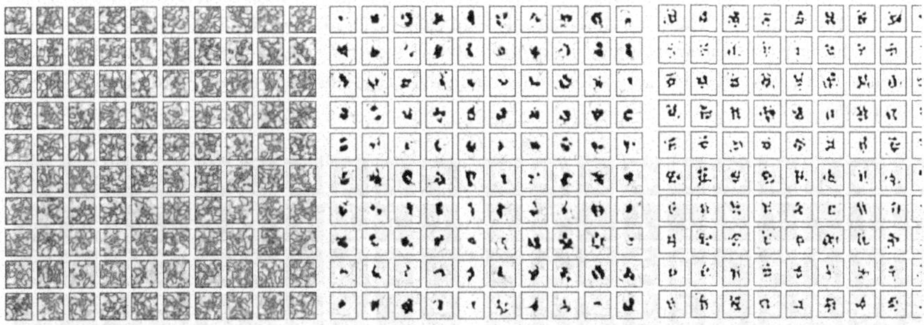

Fig. 2 Samples of generated images on a subset coming from the ICR dataset with the reduced standard WGAN having ReLUs activation function in the generator and Leaky ReLUs activation function in the critic after 1, 20, and 40 epochs of training

Fig. 3 Samples of generated images on a subset coming from the ICR dataset with ReLUs activation function in the generator and Leaky ReLUs activation function in the critic after 1, 20, and 40 epochs of training. These results are obtained by the larger baseline WGAN having 10593858 parameters

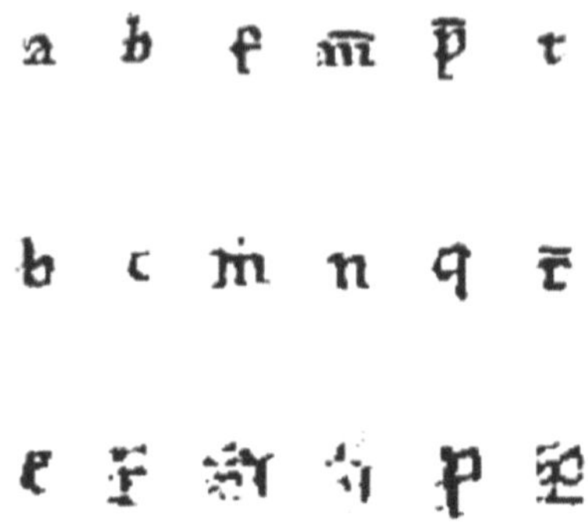

Fig. 4 Examples on the ICR dataset. On top, characters randomly sampled from the original ICR dataset (ground truth). Then, random sampled images generated by the KAF-WGAN network. Instead, the last line shows samples from the network having ReLUs activation function in the generator and Leaky ReLUs activation function in the critic and the full architecture (10593858 parameters)

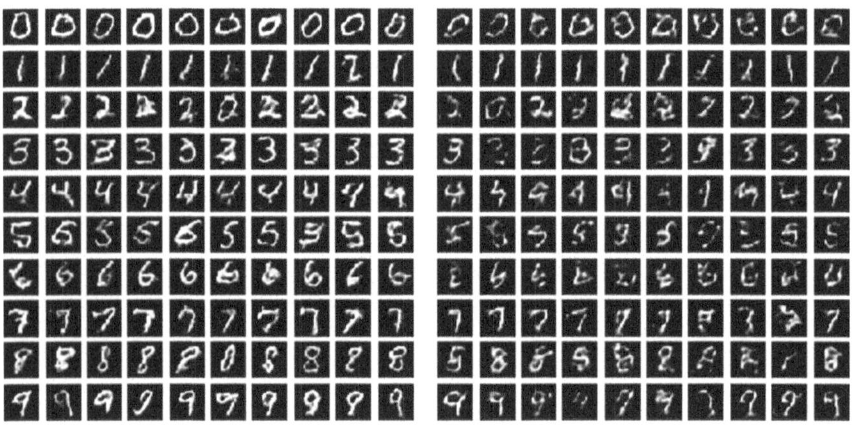

Fig. 5 Samples of generated images from MNIST after 20 epochs of training. On the left, the results of our proposed KAF-WGAN. On the right, the product of the standard WGAN with ReLUs activation function in the generator and Leaky ReLUs in the critic

standard WGAN with full architecture. As it is clear, the proposed method is able to generate samples much closer to the ground truth with respect to the standard less flexible network.

Additionally, samples of generated images after 20 epochs are reported both for MNIST in Fig. 5 and for Fashion MNIST in Fig. 6. Since the task was easier than for ICR, the last generator layer was removed for these tests. Also in this case, the proposed KAF-WGAN produces samples with better quality with respect to the standard WGAN.

Finally, to supplement these results we show the evolution of the Wasserstein loss in Fig. 7 for the proposed architecture on the left and the baseline WGAN on the

Fig. 6 Samples of generated images from Fashion MNIST after 20 epochs of training. On the left, the results of our proposed KAF-WGAN. On the right, the product of the standard WGAN with ReLUs activation function in the generator and Leaky ReLUs in the critic

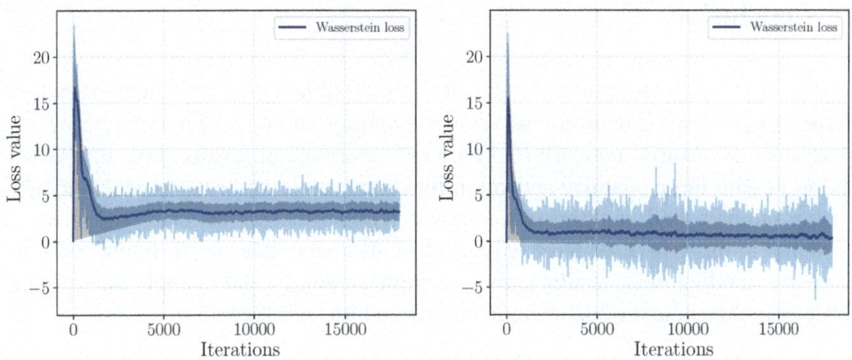

Fig. 7 Critic losses on Fashion MNIST. The proposed KAF-WGAN with ELU initialization (left) is a little bit stabler than standard WGAN with Leaky ReLU activation functions (right)

right, and the Fréchet Inception distance (FID) [8] on the ICR dataset in Sect. 8. The FID score is evaluated by using a pretrained Inception model on CIFAR. This may produce higher scores than usual values when the dataset is completely different from the baseline as in our case. Nevertheless, the plot shows the capacity of our method to learn faster to generate high quality images with respect to the standard ones.

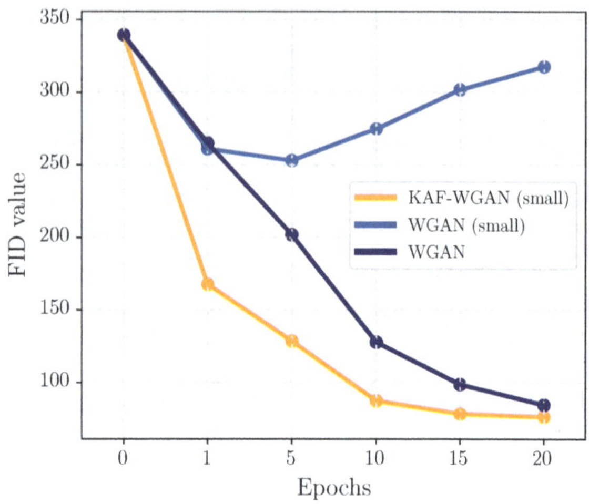

Fig. 8 Fréchet Inception Distance (FID) on the ICR dataset for three models: standard WGAN, reduced standard WGAN and our reduced KAF-WGAN. For a definition of the FID score see [8] and the details in the text

5 Conclusions

In this paper we investigated the use of more flexible activation functions, whose shape can be adapted neuron-wise from the dataset, in the design and training of a generative adversarial network (GAN). Our experimental results show that GANs endowed with these adaptive activation functions can learn faster to generate high-quality samples, both in terms of loss and in terms of perceptual quality (Fig. 8).

While in this paper we have only used small image datasets, in future work we plan on extending these results to more complex datasets, and to apply the resulting strategy in the context of data augmentation and image restoration.

References

1. Agostinelli, F., Hoffman, M., Sadowski, P., Baldi, P.: Learning activation functions to improve deep neural networks. arXiv preprint arXiv:1412.6830 (2014)
2. Arjovsky, M., Chintala, S., Bottou, L.: Wasserstein generative adversarial networks. In: Proceedings of the 2017 International Conference on Machine Learning (ICML), pp. 214–223 (2017)
3. Brock, A., Donahue, J., Simonyan, K.: Large scale GAN training for high fidelity natural image synthesis. In: Proceedings of the 2019 International Conference on Learning Representations (ICLR) (2019)
4. Creswell, A., White, T., Dumoulin, V., Arulkumaran, K., Sengupta, B., Bharath, A.A.: Generative adversarial networks: an overview. IEEE Signal Process. Mag. **35**(1), 53–65 (2018)
5. Firmani, D., Merialdo, P., Nieddu, E., Scardapane, S.: In codice ratio: OCR of handwritten Latin documents using deep convolutional networks. In: CEUR Workshop Proceedings, pp. 9–16 (2017)

6. Goodfellow, I., Pouget-Abadie, J., Mirza, M., Xu, B., Warde-Farley, D., Ozair, S., Courville, A., Bengio, Y.: Generative adversarial nets. In: Advances in Neural Information Processing Systems, pp. 2672–2680 (2014)
7. Gulrajani, I., Ahmed, F., Arjovsky, M., Dumoulin, V., Courville, A.C.: Improved training of wasserstein GANs. In: Advances in Neural Information Processing Systems, pp. 5767–5777 (2017)
8. Heusel, M., Ramsauer, H., Unterthiner, T., Nessler, B., Hochreiter, S.: GANs trained by a two time-scale update rule converge to a local Nash equilibrium. In: Advances in Neural Information Processing Systems, pp. 6626–6637 (2017)
9. Maas, A.L., Hannun, A.Y., Ng, A.Y.: Rectifier nonlinearities improve neural network acoustic models. In: Proceedings of the 2013 International Conference on Machine Learning (ICML), vol. 30, p. 3 (2013)
10. Mao, X., Li, Q., Xie, H., Lau, R.Y., Wang, Z., Paul Smolley, S.: Least squares generative adversarial networks. In: Proceedings of the 2017 IEEE International Conference on Computer Vision (ICCV), pp. 2794–2802 (2017)
11. Marra, G., Zanca, D., Betti, A., Gori, M.: Learning neuron non-linearities with kernel-based deep neural networks. arXiv preprint arXiv:1807.06302 (2018)
12. Radford, A., Metz, L., Chintala, S.: Unsupervised representation learning with deep convolutional generative adversarial networks. arXiv preprint arXiv:1511.06434 (2015)
13. Salimans, T., Goodfellow, I., Zaremba, W., Cheung, V., Radford, A., Chen, X.: Improved techniques for training GANs. In: Advances in Neural Information Processing Systems, pp. 2234–2242 (2016)
14. Scardapane, S., Van Vaerenbergh, S., Comminiello, D., Totaro, S., Uncini, A.: Recurrent neural networks with flexible gates using kernel activation functions. In: Proceedings of the 2018 IEEE 28th International Workshop on Machine Learning for Signal Processing (MLSP). pp. 1–6. IEEE (2018)
15. Scardapane, S., Van Vaerenbergh, S., Totaro, S., Uncini, A.: Kafnets: kernel-based non-parametric activation functions for neural networks. Neural Networks **110**, 19–32 (2019)
16. Song, J., He, T., Gao, L., Xu, X., Hanjalic, A., Shen, H.T.: Binary generative adversarial networks for image retrieval. In: Proceedings of the Thirty-Second AAAI Conference on Artificial Intelligence (2018)
17. Wang, X., Yu, K., Wu, S., Gu, J., Liu, Y., Dong, C., Qiao, Y., Change Loy, C.: ESRGAN: Enhanced super-resolution generative adversarial networks. In: Proceedings of the 2018 European Conference on Computer Vision (ECCV) (2018)
18. Wolterink, J.M., Leiner, T., Viergever, M.A., Išgum, I.: Generative adversarial networks for noise reduction in low-dose ct. IEEE Trans. Med. Imaging **36**(12), 2536–2545 (2017)
19. Zhang, X., Trmal, J., Povey, D., Khudanpur, S.: Improving deep neural network acoustic models using generalized maxout networks. In: Proceedings of the 2014 IEEE International Conference on Acoustics, Speech and Signal Processing (ICASSP), pp. 215–219. IEEE (2014)

Low-Power Hardware Accelerator for Sparse Matrix Convolution in Deep Neural Network

Erik Anzalone, Maurizio Capra, Riccardo Peloso, Maurizio Martina, and Guido Masera

Abstract Deep Neural Networks (DNN) have reached an outstanding accuracy in the past years, often going beyond human abilities. Nowadays, DNNs are widely used in many Artificial Intelligence (AI) applications such as computer vision, natural language processing and autonomous driving. However, these incredible performance come at a high computational cost, requiring complex hardware platforms. Therefore, the need for dedicated hardware accelerators able to drastically speed up the execution by preserving a low-power attitude arise. This paper presents innovative techniques able to tackle matrix sparsity in convolutional DNNs due to non-linear activation functions. Developed architectures allow to skip unnecessary operations, like zero multiplications, without sacrificing accuracy or throughput and improving the energy efficiency. Such improvement could enhance the performance of embedded limited-budget battery applications, where cost-effective hardware, accuracy and duration are critical to expanding the deployment of AI.

E. Anzalone · M. Capra (✉) · R. Peloso · M. Martina · G. Masera
Politecnico di Torino, 10129 Torino, Italy
e-mail: MaurizioCapra@polito.it

E. Anzalone
e-mail: ErikAnzalone@polito.it

R. Peloso
e-mail: RiccardoPeloso@polito.it

M. Martina
e-mail: MaurizioMartina@polito.it

G. Masera
e-mail: GuidoMasera@polito.it

A. Esposito et al. (eds.), *Progresses in Artificial Intelligence and Neural Systems*, Smart Innovation, Systems and Technologies 184,
https://doi.org/10.1007/978-981-15-5093-5_8

1 Introduction

Artificial Intelligence (AI) has definitely become an undeniable part of human life. The increasing interest the research world is addressing towards this topic is the result of the astonishing performance of the AI approach. Scientists look at the AI as a playground, where they get to reproduce human brain behaviors such as reasoning, learning and problem-solving. AI will soon permeate every aspect of our lives, changing the way we perceive the world, offering new tools able to assist people both during work time and in their everyday life. In fact, many are the rising applications aimed at improving the quality of the work such as the medical environment [1], where, for example, aiding software have been designed to refine specialist X-rays scans interpretation.

Even though the state of the art is far from producing devices with intrinsic man-like properties of reasoning and creativity, recent developments in brain-inspired algorithms have enormously enhanced computers ability in classification tasks [2–6], going beyond human accuracy, leading the emergence of the Machine Learning (ML) and its narrower area, named Deep Learning (DL) [7].

In recent years, DL gained visibility due to its potential in speech recognition, computer vision and robotics. Such models rely on Neural Network (NN), or more specifically on Deep Neural Network (DNN) [8], brain-inspired structures composed of several layers of artificial neurons, that represent the fundamental building blocks. These blocks are a mathematical representation of the physic principle of how a neuron is supposed to work. Basically, a multiply and accumulate (MAC) operation, followed by a non-linear activation function.

Thanks to subsequent layers of neurons, DNNs are able to extract high-level features from raw data, making them accuracy superior to any other known approach. The most effective DNNs used today, especially in computer vision, are the convolutional ones, where the basic task performed by each layer is the convolution between a feature map and a filter, also called kernel. Even though such operations are not complex, they come in a huge amount, making the above structures very computation-hungry and, subsequently, power-hungry.

DNNs can work in two separated phases, namely training and inference. During the former, the network tries to learn and generalize a given labeled dataset by tuning its weights (kernels), while in the latter it exploits the trained weights to predict and classify the input.

In the above-mentioned scenario, where DL energy greedy applications are becoming more and more pervasive, the need for tailored low-power hardware platforms arises [8]. Whilst during the training stage, parallel computation and high precision (floating point architecture) are required, making the GPU (Graphics Processing Unit) the only possible choice in order to speed up the process, such constraints are not imperative for inference, allowing for a different approach. CPU represents a reasonable choice, but its general purpose design appears not optimized enough to reach high-efficiency levels. Therefore, two hardware solutions exist able to lead DNNs towards a low-power implementation such as FPGAs and ASICs. Although

FPGAs allow for greater flexibility by means of their built-in re-programmable property, their power consumption is still too high compared to what ad-hoc integrated solutions are able to achieve. This is the reason why many hardware accelerators for DNNs have been designed as ASIC to be integrated into more complex architectures [8].

From the hardware point of view, two are the main sources of power consumption, namely memory access and multiplication operations. The former is the most energy-expensive, generally tackled at algorithm level by exploiting reutilization policies (introduced in Sect. 2) in order to ward off memory readings. The latter, instead, can be handled by approximate computing, or more effectively, by preventing useless operations like multiplications by zero. This current work is based on the design of a hardware accelerator for convolutional DNN that adopts a rescheduled data flow in order to fulfill the maximum weights reuse. Moreover, algorithmic low-power techniques are applied in order to avoid unnecessary or negligible multiplication between activations and weights.

The paper is organized as follows. Section 2 describes the basic architecture of the hardware accelerator for the convolutional task, including reutilization policies and rescheduled data flow. Section 3 introduces the low-power techniques developed in this work, with emphasis on the architectural implementations and driving motivations. Section 4 illustrates the results obtained, focusing not only on the outcome of hardware synthesis and validation, but also on comparisons with other known architectures. Section 5 draws the conclusions and possible future works.

2 Basic Hardware Accelerator

The core structure of a typical convolution accelerator is the MAC (Multiply and Accumulate) unit, as the convolution operation is usually performed as the sum of the products between the convolution kernel coefficients and the corresponding input samples. Equation 1 refers to a typical discrete convolution adopted in deep learning, with dimensionality equal to 2.

$$\text{output}(x, y) = \sum_{\substack{\text{row}\\ \text{column}}} \text{kernel}[r, c] \cdot \text{input}[x + r, y + c] \tag{1}$$

where r and c represent the dimension of the kernel, while x and y are the dimensions of output feature map. There are various possible implementations of this equation in hardware, by interchanging the order of summations and by deciding the flow of data. In particular, there are three possible architectural implementations for fetching and transferring input and weight data from memories to the processing elements (PEs) [9]:

- **Broadcast**, where data are fetched one by one and they are sent to multiple PEs every clock cycle. Therefore, in the same clock cycle each PE receives the same

data. Despite one input memory is required, some registers can be used to separate the data flow between memory and PEs;
- **Forwarding**, where data are fetched one by one from the input memory but each PE receives them in different clock cycles, i.e. each PE contains the logic to process the data and some registers to store them. Thus, the data flowing from one PE to the next one are delayed by one clock cycle;
- **Stay**, where data, once loaded inside a PE, are kept fixed for the entire convolution. This is a good way to reduce the number of memory accesses as data are reused by the same PE.

There are also three architectural possibilities to classify the process of partial-sum accumulation:

- **Aggregation**, where all partial-sums are added at the same time by means of a tree adder. Nowadays this solution is the least used one due to the relevant complexity which would be required a parallel structure;
- **Migration**, where the partial sum is transferred to a neighboring PE or to the same PE that generated it;
- **Sedimentation**, where each partial sum is stored into an on-chip memory for each PE.

In this work, a FSM (*Forwarding-Stay-Migration*) approach is used because it is well suited to support low power techniques, as described in Sect. 3. The *Forwarding* approach, applied to inputs loading, allows a delayed parallelization of all the computations. This ensures that input is fetched only once from the input memory. Since the kernel does not change during a convolutional operation, the *Stay* approach is an optimum choice, as kernel weights are loaded just once and remain inside architecture during the entire operation. Eventually, *Migration* has been chosen for the output process, since it allows to use internal registers to transfer the partial sum between two PEs, thus reducing memory operations.

The data input memory (herein called "activation memory"), the weight memory, the output memory and the memory controller are assumed to be off-chip (as in [10, 11]) to focus on the hardware accelerator itself, composed by a 3-channels datapath and its control unit. Differently from *Eyeriss* [10], where each PE receives inputs from the same row, and from *Zena* [11], where a zig-zag access pattern is required, in this work a column-by-column access is required for the *Forwarding* input processing. Input data is represented on 9 bits per color, hence 27 bits per pixel, with a resolution of 128×128 pixels, leading to an activation memory of 16384 rows and 27 columns. Due to the *Stay* approach, all the kernel is needed at the same clock cycle. Weights are quantized on 6 bits per color, leading to 162 bits of memory for a $3 \times 3 \times 3$ convolution kernel. The output memory is composed by 126×126 rows of 3×16 bits each.

The Processing Elements, the internal registers and the On-chip memories are embedded in the structure, as shown in Fig. 1. Using the rescheduled dataflow from [9], a series of registers is inserted between the inputs of the Datapath and the various PEs accordingly to the kernel size, in this case 2 registers to cope with a 3×3 kernel.

Fig. 1 Basic architecture for *FSM* approach

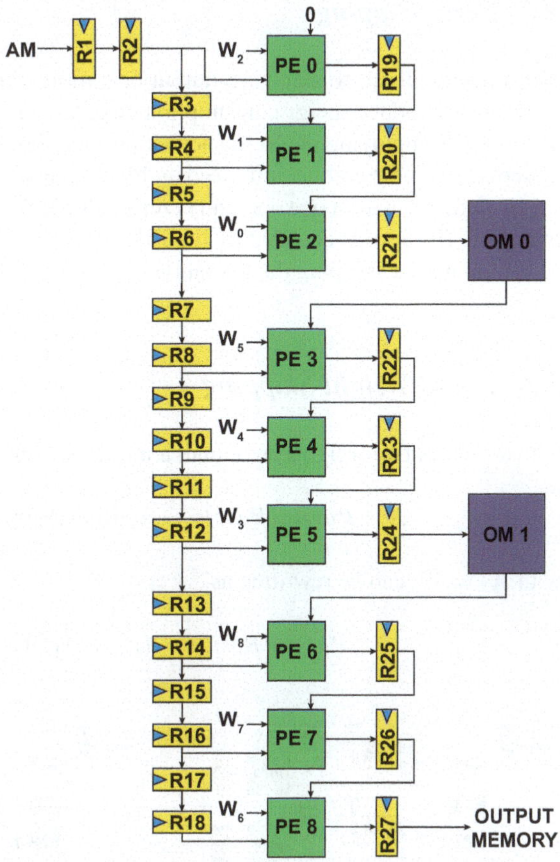

3 Power Consumption Reduction

Rescheduled dataflow is an effective technique to reduce power consumption due to on-chip memory access. Besides, the multiplication is known to be another important source of power consumption. As a consequence, techniques aimed at reducing the power consumption of multipliers are worth studying. For instance, activations and weights can be analyzed to understand which operations can be skipped to create low power architectures. For this reason, in the next sections, four different approaches applied to the starting architecture are presented: Zero Skipping (ZS) Architecture, Equal Weights Skipping (EWS) Architecture, Approximation Skipping (AS) Architecture and Hybrid EWS and AS Architecture.

3.1 Zero Skipping

This method allows to skip zero-output operations caused by zero weights or zero activations to reduce energy consumption. In particular this technique is very effective in networks employing the ReLU activation function, which outputs a large number of zeros. The new architecture needs a block named *Zero Input Recognizer* (ZIR) able to detect if at least one operand is zero-valued, skipping the multiplier, as shown in Fig. 2a. If a zero is detected by the ZIR, a zero-flag is sent to a modified version of PE capable of skipping the operation.

3.2 Equal Weight Skipping

If two consecutive weights are equal, a multiplication can be traded for an addition as

$$PartialResult = A_2W_2 + A_1W_1 + A_0W_0 \tag{2}$$

with $W_2 = W_1$ can be rewritten as

$$PartialResult = (A_2 + A_1)W_1 + A_0W_0 \tag{3}$$

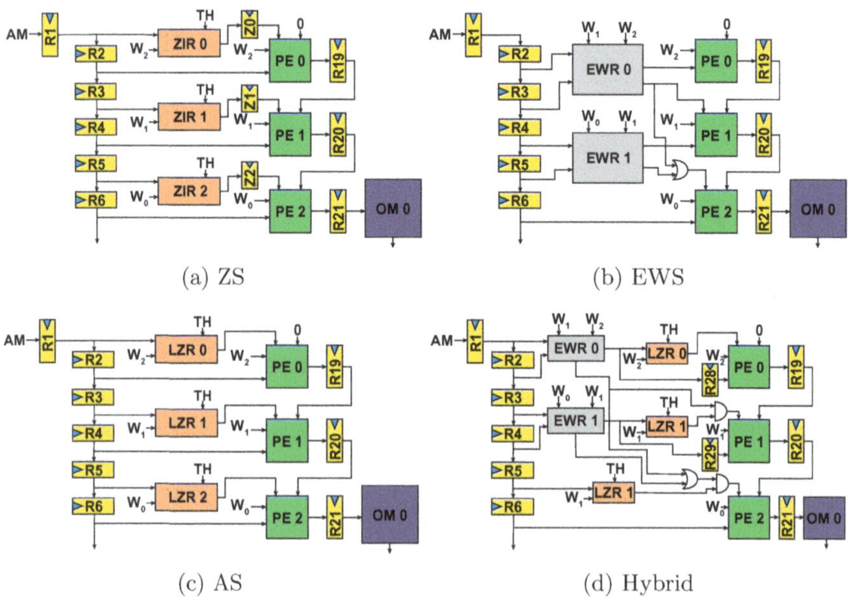

(a) ZS (b) EWS

(c) AS (d) Hybrid

Fig. 2 Low power optimizations (partial view)

which requires one less multiplication. This optimization requires a new block called *Equal Weights Recognizer* (EWR) which looks for equal weights, as shown in Fig. 2b. This results to be a low-power optimization thanks to the *Stay* approach.

3.3 Approximation Skipping

The third proposed low power technique consists of an approximation of the convolution operation. It is possible to skip multiplications which results are known to be negligible by counting the number of operands *leading zeros*. The multiplication returns a number having as much *leading zeros* as the sum of the ones in the input. A threshold can be set to avoid multiplications having too small results. This is accomplished as in Fig. 2c, resorting to a *Leading Zeros Recognizer* (LZR) based on a carry-lookahead Leading Zero Counting structure proposed in [12].

3.4 Hybrid Equal Weights and Approximation Skipping

The last proposed architecture is a combination of the Equal Weights Skipping and the Approximation Skipping techniques and it is depicted in Fig. 2. Thanks to this configuration, even more multiplications can be skipped.

4 Experimental Results

The following section presents the results of simulation, validation and synthesis. Indeed, the architecture has been simulated in order to verify the correct behavior, then validated on the well known AlexNet NN [3]. Finally, the architecture has been synthesized and simulated, thus the occupied area and power consumption, which will be discussed later.

4.1 Validation

The proposed architecture has been validated on AlexNet [3], with the aim of testing the performance of the developed techniques on an existing test case. Such NN is composed of a total of 8 layers, but only the first five are convolutional, thus they represent where the proposed techniques are applied. The mentioned NN, already trained on ImageNet [13], has been described both in Matlab and VHDL. In such way, every time a new technique is applied to the VHDL code, the performance is evaluated by comparing the results with the exact model in Matlab.

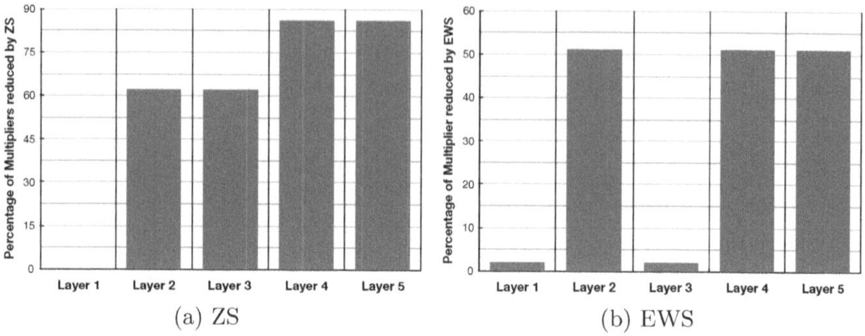

Fig. 3 Multiplications reduction across AlexNet layers

The first analysis is performed on the Zero Skipping Architecture. As it is possible to notice from Fig. 3a, multiplications are significantly reduced. Since in the first layer no zero-padding is applied, no reduction is noticed, but in the successive layers, both for zero-padding and ReLU, the amount of skipped operations increases until the 85%.

The next approach to be analyzed is the Equal Weights Skipping. Such technique is not as efficient as the previous one (see Fig. 3b), in fact, this method is strictly coupled with kernels and so with the NN model. In this case, layers 2, 4 and 5 allow for about 50% reduction of multiplications.

For what concerns the approximated architectures, a study on the range of approximation has been conducted with several thresholds going from $[-2^{-5}, 2^{-5}]$ to $[-2 : 2]$. The idea is to measure the impact of the approximation on the accuracy of the network and to evaluate the complexity reduction, i.e. the number of multiplications, not only with various threshold values but also with uniform and non-uniform thresholds for different layers.

Considering uniform boundaries, simulations showed that the NN is able to produce acceptable results with an accuracy around the 90% until $[-2^0, 2^0]$. Since the sparsity of the features maps increases going deeper into the NN, the amount of saved multiplications increases as well in the last layers. In Fig. 4a it is possible to notice the behavior of the Approximation Skipping architecture respect to the thresholds in all the five different layers. Obviously, the error (Fig. 4b) intensifies moving the threshold towards the integer part of the binary representation. Such an error is calculated as the normalized average of the absolute difference between the pixel values of the ideal and the approximate feature map.

The Hybrid solution (Equal Weights and Approximation Skipping Architecture) applied to AlexNet [3], does not provide any clear improvements with respect to the previous architectures.

Regarding the non-uniform thresholds, no significant improvements have been noticed. This is mainly due to the simplicity of the NN and the fact that the layers are not independent as previous approximations affect next layers. So, in the first layers small-magnitude thresholds are needed, as such layers are trying to extract as

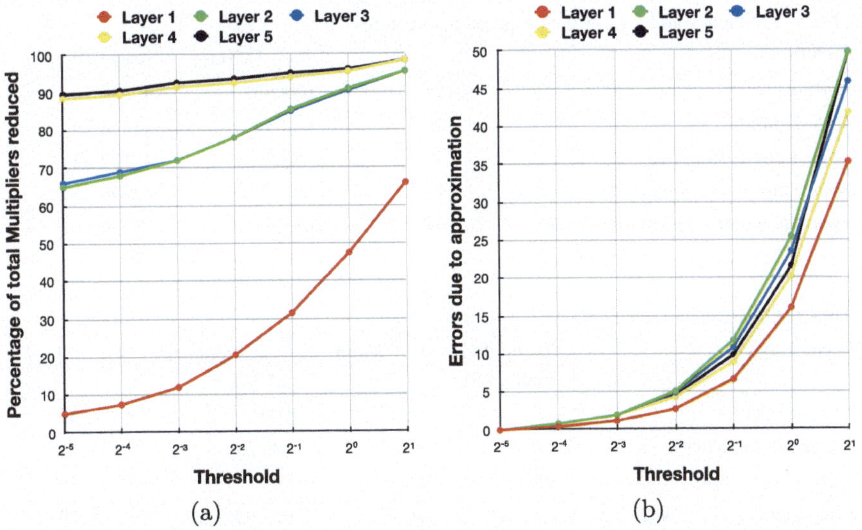

Fig. 4 Left: multiplication reduction across AlexNet layers due to Approximation Skipping Architecture. Right: normalized average error

many features as possible. So the last layers could have higher thresholds, but this approach is not very useful because in such layers the sparsity is very high and a Zero Skipping architecture is already enough.

4.2 *Synthesis*

The synthesis of the architecture has been performed by using *Synopsys Design Compiler* with the UMC 65 nm technology. After the logic synthesis, a simulation was run on the obtained netlist in order to verify the correct behavior.

As it is possible to notice from Table 1, the low-power approaches present a lower maximum frequency compared to the starting architecture and an increased area. The size of the area increases going towards more complex solutions such as the hybrid one.

Here, a simulation is run again exploiting *Moldelsim*, recording the switching activity, which is fundamental to obtain an accurate estimation of the power consumption. In the following Table 2, a comparison between developed architectures and other works is presented.

From Table 2 is clear how the ASIC solutions lead to a reduced power consumption with respect to CPUs and GPUs, even though the frequency is about the half. Presented architectures perform slightly better than those in [14], even though the number of multipliers is 9 and 256, respectively.

Table 1 Frequency and area of the proposed achitectures

Architecture	T_{clock} [ns]	F_{max} [MHz]	Area [μm^2]
Starting	1.45	689.65	40948.20
Zero skipping	1.53	653.59	41575.76
Equal weights skipping	1.53	653.59	42182.64
Approximation skipping	1.53	653.59	47296.08
Hybrid EW and APP skipping	1.53	653.59	51039.68

Table 2 Comparison with existing platforms

Accelerators	Technology node [nm]	Bid Width	CLK Frequency [MHz]	Power [mW]
Starting [14]	65	16	500	59
Starting [**This work**]	65	16	689.65	**21**
Zero skipping [14]	65	16	500	38
Zero skipping [**This work**]	65	16	653.39	**20**
Approximation skipping [14]	65	16	500	31
Approximation skipping [**This work**]	65	16	653.59	**19**
CPU Core-i7 5930k [15]	22	Not given	3500	73000
GPU GeFore Titan X [15]	28	Not given	1075	159000
mGPU Tegra K1 [15]	28	Not given	852	5100

5 Conclusion

This work proposes several hardware architectures for convolutional neural networks able to address the problem of the matrix sparsity caused by non-linear activation function like ReLU. Such architectures are capable of avoiding unnecessary operations like zero multiplications. Moreover, this paper presents an approximate technique based on leading-zeros able to skip operations involving parameters lower than a certain preset threshold.

After a detailed introduction of the different approaches, a validation, conducted on a well know network like AlexNet, is examined. Performance are compared doing a deep analysis of the NN layers, ranging through several thresholds. Synthesis results present similar performance with respect to other existing architecture, reaching even better low-power levels.

The percentage of skipped multiplications is encouraging in all the simulations, reaching up to the 85%. Such results demonstrate that the matrix sparsity in NNs could be exploited to further reduce the power consumption thus enabling new low-power frontiers.

However, the proposed architectures could be still improved by applying such techniques to a scheduler able to skip multiplication completely, reducing the latency besides the power. Anytime an operation must be skipped, a different one is scheduled at its place increasing the execution speed. Such scheduler would require major modification of the initial structure.

References

1. Rajabi Shishvan, O., Zois, D., Soyata, T.: Machine intelligence in healthcare and medical cyber physical systems: A survey. IEEE Access **6**, 46419–46494 (2018)
2. Le Cun, Y., Jackel, L.D., Boser, B., Denker, J.S., Graf, H.P., Guyon, I., Henderson, D., Howard, R.E., Hubbard, W.: Handwritten digit recognition: applications of neural network chips and automatic learning. IEEE Commun. Mag. **27**(11), 41–46 (1989)
3. Krizhevsky, Alex, Sutskever, Ilya, Hinton, Geoffrey E.: Imagenet classification with deep convolutional neural networks. Neural Inf. Proc. Syst. **25**, 01 (2012)
4. Simonyan, K., Zisserman, A.: Very deep convolutional networks for large-scale image recognition. arXiv:1409.1556 (2014)
5. Szegedy, C., Liu, W., Jia, Y., Sermanet, P., Reed, S., Anguelov, D., Erhan, D., Vanhoucke, V., Rabinovich, A.: Going deeper with convolutions. In: 2015 IEEE Conference on Computer Vision and Pattern Recognition (CVPR), pp. 1–9 (June 2015)
6. He, K., Zhang, X., Ren, S., Sun, J.: Deep residual learning for image recognition. In: 2016 IEEE Conference on Computer Vision and Pattern Recognition (CVPR), pp. 770–778 (June 2016)
7. Goodfellow, I., Bengio, Y., Courville, A.: Deep Learning. The MIT Press (2016)
8. Sze, V., Chen, Y., Yang, T., Emer, J.S.: Efficient processing of deep neural networks: a tutorial and survey. Proc. IEEE **105**(12), 2295–2329 (2017)
9. Jo, J., Kim, S., Park, I.: Energy-efficient convolution architecture based on rescheduled dataflow. IEEE Trans. Circuits Syst. I Regul. Pap. **65**(12), 4196–4207 (2018)
10. Chen, Y., Krishna, T., Emer, J.S., Sze, V.: Eyeriss: an energy-efficient reconfigurable accelerator for deep convolutional neural networks. IEEE J. Solid-State Circuits **52**(1), 127–138 (2017)
11. Kim, D., Ahn, J., Yoo, S.: Zena: zero-aware neural network accelerator. IEEE Des. Test **35**(1), 39–46 (2018)
12. Dimitrakopoulos, G., Galanopoulos, K., Mavrokefalidis, C., Nikolos, D.: Low-power leading-zero counting and anticipation logic for high-speed floating point units. IEEE Trans. Very Large Scale Integr. VLSI Syst. (VLSI) **16**(7), 837–850 (2008)
13. Deng, J., Dong, W., Socher, R., Li, L.-J., Li, K., Fei-Fei, L.: ImageNet: A Large-Scale Hierarchical Image Database. In: CVPR09 (2009)
14. Huan, Y., Qin, Y., You, Y., Zheng, L., Zou, Z.: A low-power accelerator for deep neural networks with enlarged near-zero sparsity (2017)
15. Han, S., Liu, X., Mao, H., Pu, J., Pedram, A., Horowitz, M.A., Dally, W.J.: EIE: efficient inference engine on compressed deep neural network. arXiv e-prints, arXiv:1602.01528 (Feb 2016)

Use of Deep Learning for Automatic Detection of Cracks in Tunnels

Vittorio Mazzia, Fred Daneshgaran, and Marina Mondin

Abstract Tunnel cracks on concrete surfaces are one of the earliest indicators of degradation, and if not promptly treated, they could result in full closure of an entire infrastructure or even worse in a structural failure of it. Visual inspection, carried out by trained operators, is still the most commonly used technique, and according to the literature, automatic assessment systems are arguably expensive and still rely on old image processing techniques, precluding the possibility to afford a large quantity of them for a high-frequency monitoring. So, this article proposes a low cost, automatic detection system that exploits deep convolutional neural network (CNN) architectures for identifying cracks in tunnels relying only on low-resolution images. The trained model is obtained with two different methods: a custom CNN trained from scratch and a retrained 48-layer network, using supervised learning and transfer learning, respectively. Both architectures have been trained and tested with an image database acquired with the first prototype of the video acquisition system.

1 Introduction

The increased urbanization and population density in major urban centers of the world have led to a greater demand for new underground transportation infrastructures. Indeed, a well-organized network of tunnels in a densely built-up urban area can solve problems such as traffic congestion, noise, and air pollution. Neverthe-

V. Mazzia
Department of Electronics and Telecommunications Engineering,
Politecnico di Torino, Turin, Italy

PIC4SeR - Politecnico Interdepartmental Centre for Service Robotic, Turin, Italy

Big Data and Data Science Laboratory, SmartData@PoliTo, Turin, Italy

F. Daneshgaran · M. Mondin (✉)
California State University, 5151 State University Drive, Los Angeles, CA 90037, USA
e-mail: marina.mondin@calstatela.edu

A. Esposito et al. (eds.), *Progresses in Artificial Intelligence and Neural Systems*,
Smart Innovation, Systems and Technologies 184,
https://doi.org/10.1007/978-981-15-5093-5_9

less, the design, construction, and maintenance of underground infrastructure still rely on empirical methodologies that, most of the time, result in a cost increase and extended handling times. In this context, where automatic assessments could be an important aid for the maintenance of tunnels, the presented article proposes an affordable and reliable low cost, cracks detection system as a replacement to traditional visual inspection techniques. Underground infrastructures are characterized by poor lighting conditions and filthy surfaces covered by dust and mold. Moreover, most of the time, huge scuffs on the walls, due to the aging of the lining, could easily resemble shapes of cracks. Due to these main reasons, automated inspection of tunnels is still an open challenge with not so many research studies and few enterprises that offer solutions. In the last few years, Deep Learning has demonstrated promising results in image detection and recognition tasks. This intrinsic capability has been taken into evidence by the annual ImageNet Large Scale Visual Recognition Challenge that has been showing the potentiality of this technique since 2012 [1]. For that reason, a convolutional neural network has been used as a key element to obtain an automatic system able to rely only on low resolution, consumer-grade equipment. Several image processing techniques (IPTs) have been proposed for identifying civil infrastructure defects, but they all require high-resolution images, drastically increasing the cost of the overall system. Indeed, only a low-cost device that allows for a high-frequency monitoring could be a valuable substitution for the actual manual assessment procedure. An image database has been obtained using the first prototype of the video acquisition system developed by our team, and it has been exploited to train the two different solutions of the model that, considering the highly challenging task, achieved scores of test accuracy above 90%. Finally, artificially expanding the training data has been exploited in order to increase the dimension of the available dataset [2]. Rectified linear units as neurons of the network, GPUs in order to speed up the training process and Dropout technique in order to prevent the overfitting problem have been used extensively [3].

2 Related Works

As previously introduced, visual inspection of underground infrastructure is by far the most used technique for cracks detection and tunnels maintenance. Several studies and solutions have been presented to automate and simplify this process, but currently, these systems are known to be expensive, and in many cases, not reliable.

2.1 Cracks Detection Using Image Processing

Before 2010, Japan and South Korea were at the leading of research for crack detection in tunnels. Researchers in Japan proposed an automatic monitoring system to detect cracks in tunnels based on a mobile robot [4]. They used edge enhancement and graph searching technology to extract the cracks on images. More recently, due to the high expenses, which imply visual inspections, engineers were driven to develop completer and more reliable solutions. This is true especially for all those countries with a spread network of underground infrastructures. That is the case of China, where, for example, the network of a subway in Beijing has been rapidly developed and should be maintained. At present, urban rail transportation in China is still mainly dominated by manual checking, missing in efficiency, and security. In this context, the Beijing Municipal Commission of Education Beijing Jiaotong University developed an algorithm to detect cracks in tunnels based on image processing [5]. The acquisition system is based on CCD cameras with high-frequency data collection. A Laser is used as a light source, and a custom IPT is used as the core of the module. Toshiba Research Europe in 2014 tried to investigate a low-cost system using a technique known as Structure from Motion (SfM) in order to recover a 3D version of the tunnel surface and find defects with comparisons [6]. A different research project has been developed by Pavemetrics Systems Inc. making use of top-grade and expensive equipment to acquire both 2D images, and high-resolution 3D profiles, of infrastructure surfaces at speeds up to 30 Km/h (tests have been carried out at 20 Km/h) [7]. More recently, in 2016, a research has been carried out in order to automatically collect and organize images taken from tunnels carrying high-voltage cables [8]. In order to improve image quality, the project is developed by making use of consumer DSLR cameras and high-power polarized LEDs mounted on a lightweight aluminum frame, which is studied to dampen vibrations during data capture.

2.2 Cracks Detection Using Machine Learning

As the first attempt in 2002, during the golden age of Support Vector Machines, four researchers proposed an efficient tunnel crack detection and recognition method [9]. No further recorded attempts have been made using machine learning algorithms until 2016, when, for the first time, a deep learning model has been used within the detection algorithm [10]. The system exploits the algorithm to devise an automatic robotic inspector for tunnel assessment. The robotic platform consists of an autonomous mobile vehicle with all sensors attached to a crane arm. Two sets of stereo-cameras are used for taking the necessary images, and an Arduino Uno board is used as a pulse generator synchronizing the two cameras. The first stereo pair is responsible for the crack and the other defects detection that lay on the tunnel lining. The second stereo pair is used for the full 3D reconstruction of high fidelity models of the areas of cracks. A FARO 3D laser scanner is deployed when a crack is detected for a precise calculation of any tunnel deformation that could be present.

3 Overview of the Devised System

The basic architecture of the system is primarily composed of four core blocks. The first one is a video acquisition block. It is composed of three consumer-grade cameras with their related LEDs array as lighting sources. This composition ensures coverage of about 180° of the tunnel lining with the minimum equipment cost. All the devices are mounted on a steel framework supported by a vibration isolation pad that increases the performances of the image stabilization system of the cameras. Finally, a series of sensors like accelerometer, encoder, help in tracking, inside a tunnel, a possible detected crack. The second block is part of the software and divides the input videos into their frames, which feed the model block. That is a pre-trained CNN that gives a prediction through the output layer. It is not needed in a real-time evaluation, and so, the computational power required is very low. Finally, the resulting classification and all data coming from the sensors are stored and organized by the last block of the system.

4 Video Acquisition Block

The low equipment cost, which is more than an order of magnitude cheaper than the industrial one, the great portability and modularity are the key points of this module of the devised system. The video acquisition block has a huge impact on the price, and its quality greatly affects the accuracy of the model block. The capability of the neural network to identify cracks of small dimensions is highly determined by the quality of the input images and the dataset for the training session. The camera to be chosen, as well as the light source and the mobile base, should be carefully weighed, trying to have a compromise between quality and price. Naturally, the presented prototype, in the following sub-section, is one of several possible options, and only further field experimentation can determine what is the best one.

4.1 Acquisition System's First Prototype

Two major reasons has led to the choice of the design of the prototype-1. First, it had to be easy to install on a vehicle, remotely controlled within it, and not annoying for other drivers. Secondly, it has been used to test and compare the acquisition system with an unusual source of light. The prototype-1 was made of four components: a consumer-grade camcorder, an infrared light torch, a remotely controlled pan tilt, and a bicycle rack as a base. Infrared was selected in order to try a different spectrum of light, looking for advantages and disadvantages of it. The chosen infrared flashlight was the Evolva Future Technology T20 with a 38 mm lens, emitting infrared light at 850 nm and powered by one 18,650 Li-Lon rechargeable battery (3.7–4.2V). Due to the chosen light source, a consumer-grade camera with a CCD sensor has been chosen. CCD sensors are very sensitive to infrared light but with the drawback to present a lower frame rate than CMOS sensors. Consumer-grade cameras all have a hot filter for infrared light, which must be removed to detect light in this spectral region. It is not simple to manually remove this filter, but some cameras have a night vision mode that simply removes it mechanically, turning the RGB image into a gray-scale one. The total cost of the prototype, shown in Fig. 1, was less than $200. In order to have an order of magnitude, a leading manufacturer of industrial cameras, Imaging Source, has been contacted for a quotation. The cheapest camera without lens was around $850, four times more expensive than the entire prototype.

4.2 Dataset

The prototype, shown in Fig. 1, has been mounted on the roof of a minivan and carried through the major tunnels (300 m on average) of Los Angeles, maintaining a speed of approximately 10 Km/h. With the first data acquisition attempt, it has been possible to achieve a collection of 1920×1800 videos lasting approximately 40 min (total time: 40.09, total number of frames: $2405.4 \times 30 = 72162$ frames). After a proper selection, it has been possible to obtain a dataset of 6494 images.

- Crack detection folder: 4094
- No-Cracks detection folder: 2400
- Total dataset images: 6494.

Fig. 1 Prototype-1 mounted on the top of the vehicle

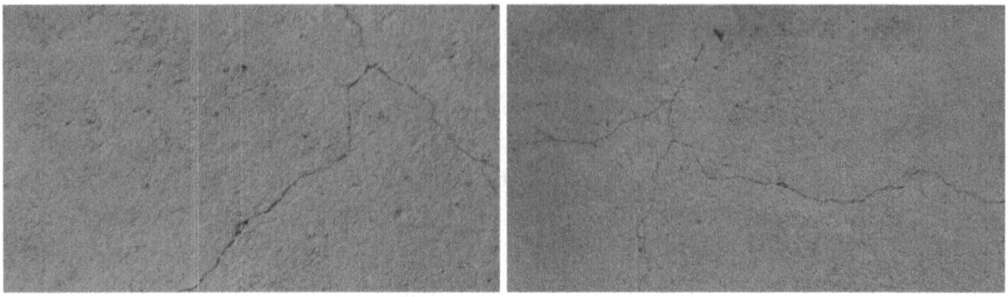

Fig. 2 Practical example of artificially expanding the training on a crack image of the dataset. The picture on the left is the one on the right randomly rotated of 209° counterclockwise

Finally, using the already introduced technique, artificially expanding the training data, through simple image modifications like cropping, rotating, and scaling, it has been possible to increase the number of the available images drastically. An example of the potentiality of this technique is presented in Fig. 2.

5 The Model

The model algorithm is the third block of the system and differs from almost all related works present in literature. It exploits recent achievements in computer vision, brought by the application of Deep Learning. Indeed, this groundbreaking technique, all of a sudden, has exceeded all old IPTs and traditional machine learning algorithms, allowing, for this project, to use only images taken by low-cost consumer-grade equipment. That has been made possible by the capabilities of convolutional neural networks to divide the problem in simpler ones and to identify more complex patterns increasingly from the input images. After a model is trained on the desired features, it is simple to embed it in an actual working system. It will analyze input images coming from the pre-processing block giving a probability prediction for each class set. In the next sections, we present two methodologies with related results that have been followed to produce models trained on the specific task of crack detection. The results of the two approaches are proposed with different values of hyperparameters, such as learning rate, η, number of training epochs, regularization parameter, λ, and mini-batch size.

5.1 Re-trained CNN

Transfer Learning is becoming a prevalent topic in the machine learning community. Moreover, experimental evidence with Deep Learning techniques has demonstrated the possibility to successfully re-purpose an already trained deep convolutional net-

Table 1 Transfer learning results applying different hyperparameters

Experiment	Iterations	Learning rate	Train b. size	Cross entropy	Test accuracy (%)
1	4000	0.01	100	0.143	95
2	9000	0.01	100	0.109	97.5
3	9000	0.001	100	0.132	96.4
4	15,000	0.001	100	0.147	95.1
5	15,000	0.01	100	0.103	97.6
6	9000	0.01	300	0.102	97.6
7	9000	0.01	1000	0.106	97.5
8	9000	0.01	1000	0.094	98.1

work with new generic tasks (11). Indeed, all convolution and pooling layers extract increasingly abstract features that can be used to classify different types of objects. Instead, the fully connected layers and the classifier need to be re-trained on the new task, using supervised learning with the proper image database. In light of this, as the first step of the model generation, we have taken a state of the art CNN and retrained it to detect the presence of cracks. In this way, it has been possible to train a large model in less than one hour (a model that usually required 2–3 weeks to be fully trained). The selected, trained model is the Inception-v4 model, a slimmed-down version of the relative Inception-v3 model (12). Unlike the other, this novel architecture has been designed specifically for the TensorFlow library. It has shown excellent performance at relatively low computational cost, and in order to achieve these performances, it exploits inception blocks, batch normalization, and residual connections in its sibling version Inception-ResNet-v2 (13). The model has been retrained with input images of 299 × 299 pixels using different settings and hyper-parameters. The final results are presented in (Table 1). The last training session has been performed using the artificially expanded dataset. The choice of hyperparameters has been made taking into account common heuristic rules, and the final accuracy has been computed with the test set, only presented at the last epoch. The test set was tuned at 10% of the dataset with stratified sampling and containing 649 different pictures. The ninth test that achieved 98.1% of test accuracy, classified in the correct class 637 images with only 12 misplaced. This result demonstrates the huge potential of transfer learning that already with the first simulations has proven to be able to reach high values of accuracy

5.2 *Custom CNN*

A custom convolutional neural network, despite the lack of a large dataset, has been trained using supervised learning. Unlike Inception-v4, a CNN trained from scratch on a small number of classes is more suitable for the specific assigned task. Indeed, all weights and biases are specifically calibrated to recognize features of the selected classes. The definition of network architecture does not have a closed solution. Many different approaches can be applied, and usually, only the actual simulation can discern what the most suitable framework is. For this project, all decisions have been made trying to keep the number of parameters low. A high number of parameters not only increases the training process time but also makes the network more inclined to overfit input data. Figure 3 gives a detailed overview of the designed architecture. It has two similar stages of convolution and pooling and a final soft-max classifier that has as output a certain prediction over two different classes. Going deeper with the network, the number of parameters of a single feature decreases, and after the second pooling, the output matrix takes the shape of an array in order to be suitable for the fully connected layer. The input layer has output with dimension 128 × 128 pixels. Indeed, images camera acquisitions in the infrared band are represented as a single channel grayscale images. The first convolutional layer (CL), like the second one, has a 5 × 5 filter with zero padding in order to analyze the presence of cracks in the entire picture. Then, if a crack is detected, it is not essential anymore its position within the image so that a max pooling (PL) is exploited to decrease the number of parameters. Finally, fully connected layers (FCL) analyze the extracted features, and the softmax output layer generates a prediction. All neurons, except the one of the last layer (sigmoid neuron), are rectified linear units (ReLu). That type of neurons has proved to be simplier and faster to train [11]. In Table 2 presents an overlook of the designed convolutional neural network. This is only one possible architecture. Many other frameworks can be devised, and as it is for hyperparameters, only empirical and not deterministic rules are available for the designing process. In order to compare the results achieved with this model with the ones reached with the Inception-v4 model, we selected an equal size of the test set (10%) and the same percentage of validation images (20%). In every simulation, the validation occurs every 150 steps. Unfortunately, it was not possible to maintain the same input dimension of the pictures due to a lack of available HW. Surprisingly, with only input images of 128 px per side, it was possible to achieve 93.1% of test accuracy. So, it is highly probable that higher dimension of the input images, carrying more

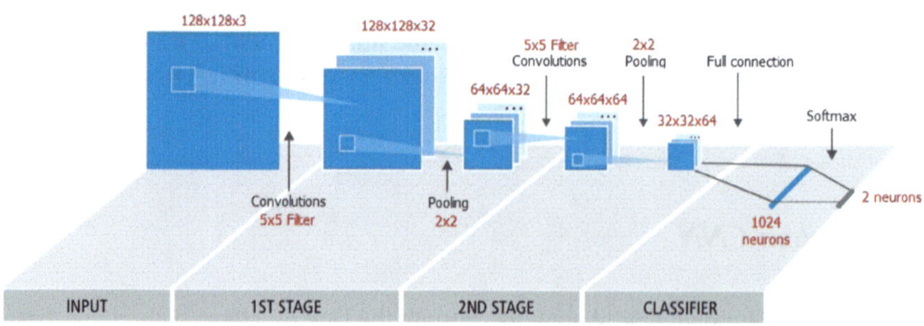

Fig. 3 Architecture layout of the convolutional neural network

Table 2 Summary table of the network specifications

Section	Input	Output	Kernel	Stride	Filters
Input layer	1920 × 1080	128 × 128	N.A.	N.A.	3
1st CL	128 × 128	128 × 128	5 × 5	1	32
1st PL	128 × 128 × 32	64 × 64	2 × 2	1	32
2nd CL	64 × 64 × 32	64 × 64	5 × 5	1	64
2nd PL	64 × 64 × 64	32 × 32	2 × 2	1	64
FCL	64 × 64 × 32	1024	N.A.	N.A.	N.A.
Sofmax layer	1024	2	N.A.	N.A.	N.A.

Table 3 Summary of hyper-parameters selected for the last simulation

Image size	Hyper-parameters				
	Train b.	Validation b.	Dropout	L. rate	L2 Reg.
128 px	150	200	0.5	1e–5	10

information, could significantly help the network learning more robust and useful features with a resulting higher accuracy. Again, the training procedures have been carried out with a different combination of hyperparameters, and only the ones of the last attempt are reported in Table 3. The last simulation 2has been performed with 10,000 steps and Dropout, L2 regularization, and artificially expanding the training data in order to tackle the overfitting problem. Finally, the model has been trained with GPUs in order to speed up the entire process.

5.3 *Comparison of the Two Models*

Both networks are easily implementable. After the training session, it is possible to store the model in a binary format file. This file can be loaded and used in the software of the device in order to make predictions about new input images. Further works and improvements are needed, but the early results have been auspicious. As expected, transfer learning, which needs less data for its learning process, has overtaken the custom neural network in terms of learning time and test accuracy. Considering the significant number of parameters that had to be trained, the result of 93.1% achieved by the custom CNN has to be considered as high. Little adjustments of the network, but especially a larger dataset and higher dimension of input images, should increase the accuracy level around the value achieved by Inception-v4. Indeed, the graph presented in Fig. 4 shows that, due to the dimension of the image database, the custom deep neural network, after 7000 iterations, starts to overfit the data.

In conclusion, both methodologies have successfully proven that convolutional neural networks, with a small effort, can overcome most of the traditionally complex

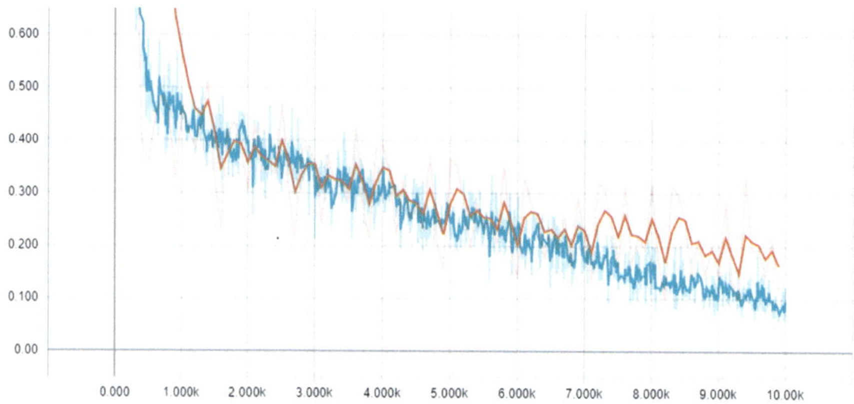

Fig. 4 Cross-entropy functions. Blueline is the training function, and red line is the validation function, smoothing = 0.4

and accurately calibrated image processing techniques. Moreover, libraries like TensorFlow have drastically simplified network design and implementation and, their tools and features have gradually opened the possibility to train deeper and more precise neural networks.

6 Conclusion

A low cost, easily enforceable automatic cracks detection system, based on recent developments in image recognition and computer vision, has been proposed as a replacement to the commonly used manual inspection. Results have pointed out the suitability of deep learning architectures for the tunnel defect inspection problem relying only on low range, consumer-grade equipment. Further works and researches may automate on a large scale the current tedious and unreliable work of underground infrastructures assessment.

References

1. Deng, J., Dong, W., Socher, R., Li, L. J., Li, K., Fei-Fei, L.: ImageNet: a large-scale hierarchical image database. In: 2009 IEEE Conference on Computer Vision and Pattern Recognition, pp. 248–255. IEEE, Miami (2009)
2. Simard, P.Y., Steinkraus, D., Platt, J.C.: Best practices for convolutional neural networks applied to visual document analysis. In: ICDAR, vol. 3, pp. 958–962. IEEE Computer Society (2003)
3. Srivastava, N., Hinton, G., Krizhevsky, A., Sutskever, I., Salakhutdinov, R.: Dropout: a simple way to prevent neural networks from overfitting. J. Mach. Learn. Res. **15**(1), 1929–1958 (2014)

4. Yamaguchi, T., Nakamura, S., Hashimoto, S.: An efficient crack detection method using percolation-based image processing. In: 2008 3rd IEEE Conference on Industrial Electronics and Applications, pp. 1875–1880. IEEE (2008)
5. Qi, D., Liu, Y., Wu, X., Zhang, Z.: An algorithm to detect the crack in the tunnel based on the image processing. In: 2014 Tenth International Conference on Intelligent Information Hiding and Multimedia Signal Processing, pp. 860–863. IEEE (2014)
6. Stent, S., Gherardi, R., Stenger, B., Soga, K., Cipolla, R.: Visual change detection on tunnel linings. Mach. Vis. Appl. **27**(3), 319–330 (2014)
7. Laurent, J., Fox-Ivey, R., Dominguez, F.S., Garcia, J.A.R.: Use of 3D scanning technology for automated inspection of tunnels. In: Proceedings of the World Tunnel Congress, pp. 1–10 (2014)
8. Liu, Z., Suandi, S.A., Ohashi, T., Ejima, T.: Tunnel crack detection and classification system based on image processing. In: Machine Vision Applications in Industrial Inspection X. International Society for Optics and Photonics, vol. 4664, pp. 145–152 (2002)
9. Stent, S.A.I., Girerd, C., Long, P.J.G., Cipolla, R.: A low-cost robotic system for the efficient visual inspection of tunnels. In: ISARC Proceedings of the International Symposium on Automation and Robotics in Construction, vol. 32, p. 1. IAARC Publications (2015)
10. Protopapadakis, E., Makantasis, K., Kopsiaftis, G., Doulamis, N., Amditis, A.: Crack identification via user feedback, convolutional neural networks and laser scanners for tunnel infrastructures. VISIGRAPP/VISAPP **4**, 725–734 (2016)
11. Nair, V., Hinton, G.E.: Rectified linear units improve restricted Boltzmann machines. In: Proceedings of the 27th International Conference on Machine Learning, pp. 807–814 (2010)

SoCNNet: An Optimized Sobel Filter Based Convolutional Neural Network for SEM Images Classification of Nanomaterials

Cosimo Ieracitano, Annunziata Paviglianiti, Nadia Mammone, Mario Versaci, Eros Pasero, and Francesco Carlo Morabito

Abstract In this paper an optimized deep Convolutional Neural Network (CNN) for the automatic classification of Scanning Electron Microscope (SEM) images of homogeneous (HNF) and nonhomogeneous nanofibers (NHNF) produced by electrospinnig process is presented. Specifically, SEM images are used as input of a Deep Learning (DL) framework consisting of: a Sobel filter based pre-processing stage followed by a CNN classifier. Here, such DL architecture is denoted as *SoCNNet*. The Polyvinylacetate (PVAc) SEM image of NHNF and HNF dataset collected at the *Materials for Environmental and Energy Sustainability Laboratory* of the University Mediterranea of Reggio Calabria (Italy) is used to evaluate the performance of the developed system. Experimental results (average accuracy rate up to 80.27% $\pm$ 0.0048) demonstrate the potential effectiveness of the proposed SoCNNet in the industrial chain of nanofibers production.

1 Introduction

Nanofibers (NF) produced by electrospinning process have gained a great deal of interest due to their unique mechanical properties and the wide range of potential real-world applications, including electronics [1], medicine [2], tissue engineering [3], drug delivery [4] and so on. NF are very thin fibers and exhibit diameters less

C. Ieracitano (✉) · M. Versaci · F. C. Morabito
DICEAM Department, University Mediterranea of Reggio Calabria,
89124 Reggio Calabria, Italy
e-mail: cosimo.ieracitano@unirc.it

A. Paviglianiti · E. Pasero
Department of Electronic Engineering, Polytechnic of Turin, C.so Duca Degli Abruzzi 24,
10137 Turin, Italy

N. Mammone
IRCCS Centro Neurolesi Bonino-Pulejo, Via Palermo c/da Casazza, SS. 113,
98124 Messina, Italy

A. Esposito et al. (eds.), *Progresses in Artificial Intelligence and Neural Systems*,
Smart Innovation, Systems and Technologies 184,
https://doi.org/10.1007/978-981-15-5093-5_10

then 100 nm. However, the fabrication of NF is very difficult to control. Indeed, electrospun fibers may be affected by manufacturing faults due to the instability of the polymeric jet attributable to undesirable processing parameters such viscosity, surface tension or applied voltage [5, 6]. The result is an array of nonhomogeneous nanofibers (NHNF) where the most common problem is the presence of *beads*. Notably, beads are micro/nano aggregates that alter the morphology and properties of the material, observed especially with low values of polymeric concentration. One of the most effective method to monitor the quality and morphology of electrospun fibers is to examine Scanning Electron Microscope (SEM) images obtained from the NF sample under analysis. However, visual examination of SEM images is time consuming and is not the most efficient practice to detect and analyze potential defects in NF. In this context, intelligent systems able to discriminate automatically SEM images of homogeneous nanofibers (HNF, anomalies-free) and NHNF via advanced machine learning techniques (i.e. *deep learning*, DL [7]) have been emerging. DL has been successfully employed in several applications [8–12] but only a few works on anomaly detection in NF are reported in the literature (Table 1). In [13], Carrera et al. developed an one-class classification approach based on a dictionary of patches of HNF proposed in [14]. The dictionary was applied to detect defects in a patch-wise fashion, reporting very good performance in identifying also small defects. Specifically, the area under the curve (AUC) was employed as unique quantitative performance indicator, achieving AUC = 93%, over a dataset of 45 SEM images (including 5 images without anomalies and 40 with defects). In a second work [15], the authors implemented a CNN based system for detecting and localizing defects in SEM images. Anomaly patches were identified via similarity among test-patches under analysis and HNF patch of the dictionary. Similarly, only the AUC index was used to evaluate the effectiveness of the proposed method. In particular, the authors claimed AUC = 97%. Recently, Ieracitano et al. [16] proposed a DL based anomaly detection system for classifying NHNF and HNF of PVAc nanfibers. The authors developed a deep CNN and used a dataset of 160 SEM images (75 anomaly-free and 85 with defects), reporting accuracy rate up to 80%. However, no data-preprocessing or validation techniques were applied. In contrast, here, motivated by the promising results achieved in [16], we propose an optimized DL system for discriminating SEM images of NHNF and HNF. Specifically, the proposed DL framework consists of three main modules: *electrospinning process, SEM image pre-processing, SEM-image classification*. Electrospinning process module includes electrospun NF production and NHNF/HNF SEM images collection [16]. SEM image pre-processing module includes the application of three different set of filters (i.e. Sobel, Laplacian, Fuzzy) in order to detect only the edge of each image and consequently make the classification task easier. In the SEM-image classification module, instead, pre-processed SEM images are used as input of the deep CNN for performing the NHNF vs HNF classification task. Experimental results showed that the Sobel filtering was able to improve the CNN discrimination performance (accuracy of 80.27% $\pm$ 0.0048, Table 2). Here, such optimized DL based anomaly detection system (consisting of Sobel + CNN) is denoted as *SoCNNet* (Fig. 4).

Table 1 State-of-the-art of anomaly detection systems in nanofibrous materials

Authors	Dataset	Results
Carrera et al. [13]	45 SEM images: 5 anomaly-free 40 with anomalies	AUC = 93%
Napoletano et al. [15]	45 SEM images: 5 anomaly-free 40 with anomalies	AUC = 97%
Ieracitano et al. [16]	160 SEM images: 75 anomaly-free 85 with anomalies	Accuracy = 80%

The rest of this work is organized as follows. Section 2 introduces the proposed method, including the electrospinning process, SEM-image pre-processing and CNN based SEM-image classification. Section 3 reports the achieved results. Section 4 concludes the paper.

2 Methodology

Figure 1 reports the proposed framework. It includes three main processing units:

1. *Electrospinnig process*. PVAc nanofibers are produced through electrospinning process by dissolving PVAC in Ethanol (EtOH) solvent and SEM images of HNF and NHNF are stored on a computer according to the procedure described in [16].
2. *SEM image pre-processing*. Each HNF/NHNF SEM image is pre-processed by using Sobel, Laplacian and Fuzzy filtering, capable of providing information on the object contours of an image under analysis.
3. *SEM image classification*. Pre-processed SEM images are used as input of a CNN based classifier able to discriminate HNF and NHNF images automatically.

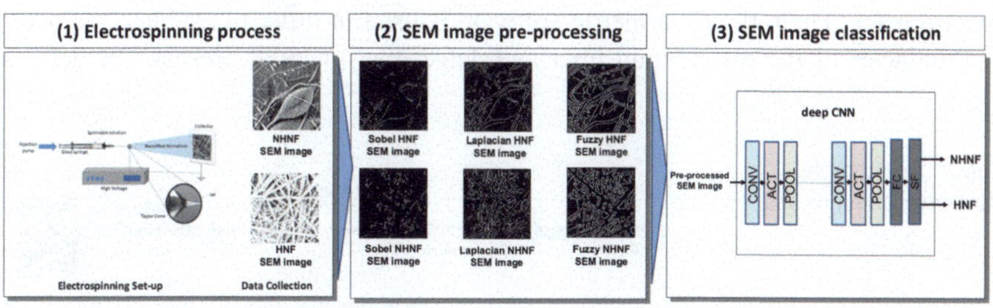

Fig. 1 Procedure of the proposed method

2.1 *Electrospinning Process*

The basic set-up of the electrospinning (ES) process is schematically shown in Fig. 2. It consists of a high voltage generator, a syringe pump and a grounded collector plane. Firstly, the polymer fluid is introduced into a glass syringe and extruded through the spinneret by external pumping (at a constant and controllable flow rate) until a small droplet is formed. Then, a high voltage is applied between the collecting (i.e. collector surface, cathode) and spinning (i.e. needle, anode) electrode. As the electric field increases, the droplet deforms into a conical shape, known as *Taylor cone* [17]. Specifically, when the electrostatic repulsion is greater than the surface tension of the droplet, a charged jet is ejected from the tip of the cone towards the collector plane. During the jet emission, the solvent evaporates and the solidified fibers are collected on the target.

It has been proven that viscosity and concentration of the polymeric solution mainly affect the diameter and morphology of nanofibers. For example, low values of concentration cause the production of micro-particles (i.e. *beads*) due to the electrospray phenomenon [18]. Other important electrospinning parameters are applied voltage, tip-collector distance and flow-rate [19].

2.1.1 Materials

Here, Polyvinylacetate (PVAc) with average molecular weight (M_w) of $170 * 10^3$ g/mol and Ethanol (EtOH) are employed as polymer and solvent, respectively. Specifically, PVAc is dissolved in EtOH solvent by using a magnetic stirrer until a clear fluid is achieved. The CH-01 Electrospinner 2.0 (Linari Engineering s.r.l.) with a 20 mL glass syringe with a stainless steel needle of 40 mm length and 0.8 mm thick is used for the nanofiber production. Moreover, it is to be noted that the spinning process is carried out at a temperature and air humidity of 20 ± 1 C and 40% respectively. The morphology of the produced PVAc nanofiber is analyzed through the Phenom Pro-X scanning electron microscope (SEM) that included an energy-dispersive x-ray spectrometer. Then, the Fibermetric software is used in order to evaluate the average diameter of the electrospun fiber and detect the potential presence of defects

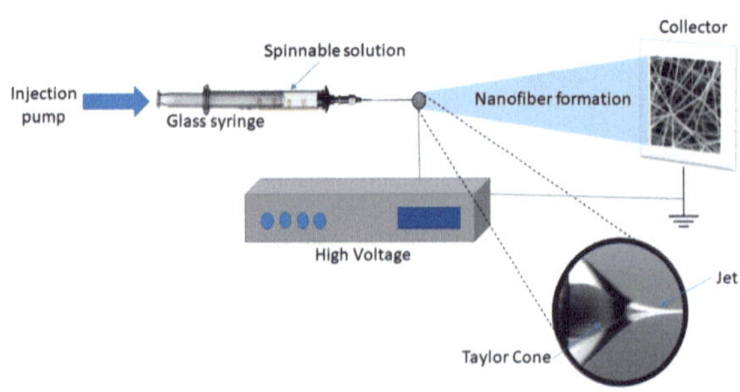

Fig. 2 Set-up of the electrospinning process

(i.e. beads). The experiments included 16 different setup, where concentration (e_1), voltage e_2, flow rate e_3 and tip-collector distance e_4 parameter were changed one at time in well-known working conditions: 10–25 wt.% e_1; 10–17.5 kV e_2; 100–300 μL/min e_3; 10–15 cm e_4. Further details of the ES experiments can be found in [16].

2.1.2 Dataset Description

The SEM images dataset proposed in [16] was used. Specifically, it consists of 160 SEM images labeled by an expert as image of homogeneous nanofibers (HNF) or nonhomogeneous nanofibers (NHNF). Notably, the dataset includes 75 HNF and 85 NHNF sized 128 × 128 [16]. It is worth mentioning that the production HNF is typically observed with high values of voltages and concentrations; whereas, NHNF are affected by the presence of micro or nano structural anomalies (i.e. beads) that can occur when the polymeric solution is made up of low values of concentrations or when the TCD is too high. As an example, Fig. 3 reports a NHNF and HNF SEM image of PVAc electrospun nanofiber.

2.2 SEM Image Pre-processing

In order to make the classification task easier for the proposed classier (Sect. 2.3.1), each NHNF/HNF SEM image $I(x, y)$, has been pre-processed by reducing the number of gray-scale levels but, simultaneously, maintaining the texture of the individual image as much as possible [20]. With this goal in mind, edge detection techniques are excellent candidates as they segment images based on information on the edges providing information on the object contours using some edge-detection operators finding discontinuity in the gray levels, color, texture, etc. the edge pixel (x, y) are pixel in which the intensity of brightness, $f(x, y)$, of the image changes abruptly and the edges (or segments of edge) are sets of connected pixels. By means of Sobel technique [20], edge detection is achieved by of a differential operator consisting of two

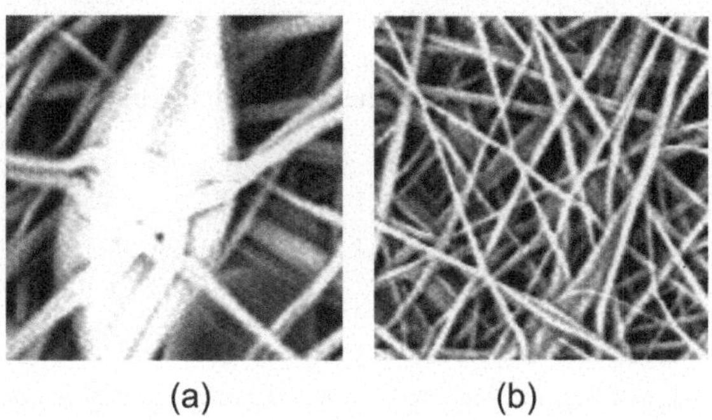

Fig. 3 **a** SEM image of nonhomogeneous nanofibers (NHNF) due to *beads*. **b** SEM image of homogeneous nanofibers (HNF)

convolution matrices 3×3 with integer values, $G_x = [0\ 1\ 2;\ -1\ 0\ 1;\ -2\ -1\ 0]$ and $G_y = [-2\ -1\ 0;\ -1\ 0\ 1;\ 0\ 1\ 2]$, which convoluted with the image I calculate an approximate value of $\nabla f(x, y) = [f_x(x, y),\ f_y(x, y)] = [G_x * I,\ G_y * I]$ identifying the direction of greater variation of $f(x, y)$, $\theta = \tan^{-1}(f_y(x, y)/f_x(x, y))$, together with its speed in the same direction identified by its magnitude $|\nabla f(x, y)| = \sqrt{f_x(x, y)^2 + f_y(x, y)^2}$. According to Marr and Hildred, instead, edge detection can be implemented using the filter $\nabla^2 G$, with $G(x, y) = e^{-\frac{x^2+y^2}{2\sigma^2}}$ obtaining $\nabla^2 G = ((x^2 + y^2 - 2\sigma^2)/\sigma^4)e^{-\frac{x^2+y^2}{2\sigma^2}}$ which represents the LoG filter (Laplacian of the Gaussian) [20]. However, to reduce the computational complexity of LoG, usually a convolution matrix 5×5, such as $[0\ 0\ -1\ 0\ 0;\ 0\ -1\ -2\ -1\ 0;\ -1\ -2\ 16\ -2\ -1;\ 0\ -1\ -2\ -1\ 0;\ 0\ 0\ -1\ 0\ 0]$, is used that approximates $\nabla^2 G$. Fuzzy edge detection is an alternative approach to edge detection which considers the image to be fuzzy because, often, in most of the images the edge are not clearly defined, so that detection can becomes very difficult. In this paper, a modified Chaira and Ray approach [21] exploiting the fuzzy divergence between the image window and each of a set of 16 convolution matrices (3×3, whose elements belong to the set $\{0.3,\ 0.8\}$ to ensure good edge detection) which represent the edge profile of different types is presented. Specifically, after normalizing the image I, the center of each convolution matrix is place on each pixel (x, y) of I. Then, fuzzy divergence measure, $Div(x, y)$, between each of the elements of the image window and the template is calculated and the minimum value is selected. This procedure is repeated for all of 16 convolution matrices selecting the maximum value among the 16 divergence values obtained. Then, we obtain a divergence matrix on which a threshold technique must be applied. For this purpose, in this paper a new entropic $2D$ fuzzy thresholding method based on minimization of fuzzy entropy is proposed. In particular, for each threshold T, set a square matrix W of size r centered on (x, y) and considered another window W' of the same dimensions centerd on another pixel (x', y'), their distance is first calculated by the fuzzy divergence.[1] The average value of all the fuzzy divergences obtained by moving (x', y') in all possible positions is then calculated. Moreover, we calculate the further average value obtained by moving (x, y) in all possible position. We indicate with $Mean_r$ the latter average value obtained. We repeat the procedure for square windows of size $r + 1$, obtaining $Mean_{r+1}$. Then, Fuzzy Entropy depending on T, $FE(T)$ can be computed as $FE(T) = \ln(Mean_r/Mean_{r+1})$ so that the optimum threshold, $T_{optimum}$ can be computed by means of $T_{optimum} = \arg;\min_T|FE(T)|$. Obviously, if necessary, a pre-treatment such as contrast enhancement could be implemented to improve the image quality globally [22–24].

[1]Fuzzy divergence can be considered as a distance because it satisfies all the axioms of the metric spaces.

2.3 SEM Image Classification

2.3.1 CNN Classifier

Convolutional Neural Networks (CNN) is a DL technique capable of learning the most relevant features from the input representations through an architecture organized hierarchically. A standard CNN includes the following processing modules:

1. *convolutional layer (CONV)*: where, K_j filters (sized $k_1 \times k_2$) convolve with the ith input image I (sized $h \times w$), producing j *features maps* (A) of size $a_1 \times a_2$. Notably:
$$A_j = \sum I_i * K_j + B_j \tag{1}$$
where B_j represents the bias and * the convolution operation;
$$a_1 = \frac{h - k_1 + 2p}{s} + 1 \tag{2}$$
$$a_2 = \frac{w - k_2 + 2p}{s} + 1 \tag{3}$$
where s and p are the stride (or shift) and zero padding parameters, respectively. Specifically, the jth filter convolves with a sub-region of the ith input and shift over the whole input map with a stride s; whereas, p is typically used to control the output matrix dimension by padding the input edges with null values.
2. *activation layer (ACT)*: it includes a nonlinear transfer function. Specifically, "*Rectified Linear Unit*" (ReLu, $l(x) = \max(0, x)$) activation function is typically used in CNN architecture (ACT_{ReLU}). Indeed, it achieves very good performance in terms of generalization and learning time [25].
3. *pooling layer (POOL)*: it reduces the input spatial size by evaluating the average (*average pooling*, $POOL_{\text{avg}}$) or maximum (*max pooling*, $POOL_{\text{max}}$) value of a sub-matrix conveniently selected by a filter sized $\tilde{f}_1 \times \tilde{f}_2$. Here, the $POOL_{\text{max}}$ is employed. Notably, the filter slides over the input map with stride $\tilde{s}$ and takes the maximum of the sub-matrix under analysis. The results is a downsampled representation of A_j sized $\tilde{a}_1 \times \tilde{a}_2$ with
$$\tilde{a}_1 = \frac{a_1 - \tilde{f}_1}{\tilde{s}} + 1 \tag{4}$$
and
$$\tilde{a}_2 = \frac{a_2 - \tilde{f}_2}{\tilde{s}} + 1 \tag{5}$$

The CNN typically ends with a standard fully connected (FC) neural network for classification purposes.

Here, the pre-processed (and raw) SEM images were used as input of the deep CNN previously proposed in [16]. Specifically, it included five modules of $CONV$ + ACT_{ReLU} + $POOL_{\max}$ and one fully connected layer (FC) with 40 hidden units followed by a softmax output layer to perform the NHNF versus HNF classification task. Each $CONV$ layer had filters size $k_1 \times k_2 = 3 \times 3$, whereas shift and padding values of $s = 1$ and $p = 1$, respectively. Each $POOL_{\max}$ layer had filters size $\tilde{f}_1 \times \tilde{f}_2 = 2 \times 2$ and stride $\tilde{s} = 2$. All the learning parameters were set up by following the recommendations reported in [26]. The network was initialized through a Gaussian distribution having mean 0 and standard deviation 0.01. Moreover, the stochastic gradient descent technique with momentum $= 9 * 10^{-1}$, weight decay $= 10^{-4}$, learning parameter $= 10^{-2}$, mini-batch $= 32$, was used. Further details can be found in [16].

3 Results

The evaluation performances were quantified in terms of precision (PC), recall (RC), F-measure (FM) and accuracy (ACC):

$$PR = \frac{TP}{TP + FP} \tag{6}$$

$$RC = \frac{TP}{TP + FN} \tag{7}$$

$$FM = 2 * \frac{PR * RC}{PR + RC} \tag{8}$$

$$ACC = \frac{TP + TN}{TP + TN + FP + FN} \tag{9}$$

where TP are the true positives: number of NHNF SEM images properly classified as NHNF; TN are the true negatives: number of HNF SEM images properly classified as HNF; FP are the false positives: number of HNF SEM images erroneously classified as NHNF; FN are the false negatives: number of NHNF SEM images erroneously classified as HNF. Moreover, the k-fold cross validation procedure (with $k = 15$) was used. Notably, train and test sets included 70% and 30% of images (in each fold), respectively. Thus, all the outcomes are reported as average value $\pm$ standard deviation. Table 2 reports the NHNF versus HNF classification performance when the CNN receives as input: raw SEM images (RaCNNet), pre-processed SEM images with Sobel filter (SoCNNet), pre-processed SEM images with Laplacian filter (LaCNNet) and pre-processed SEM images with Fuzzy based filter (FuCNNet). As can be seen, the Sobel approach, SoCNNet (Fig. 4), outperforms all the others, achieving accuracy rate up to 80.27% $\pm$ 0.048 and F-measure of 82.81% $\pm$ 0.046. To the best

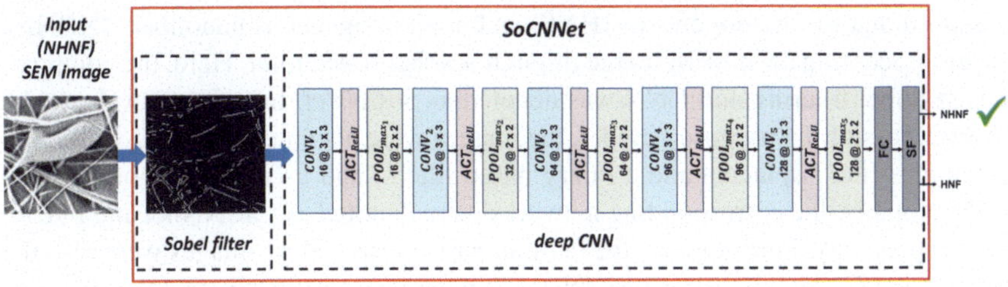

Fig. 4 SoCNNet: optimized CNN, consisting of Sobel filer and the deep CNN proposed in [16]

Table 2 Classification performance of the CNN when raw SEM images (RaCNNet), pre-processed SEM images with Sobel filter (SoCNNet), pre-processed SEM images with Laplacian filter (LaCNNet) and pre-processed SEM images with Fuzzy based filter (FuCNNet) are used as input. All the results are reported as average value ± standard deviation

Method	Precision	Recall	F-measure	Accuracy
RaCNNet	$84.51\% \pm 7.7 \times 10^{-2}$	$71.46\% \pm 6.1 \times 10^{-2}$	$77.01\% \pm 3.3 \times 10^{-2}$	$74.69\% \pm 4.5 \times 10^{-2}$
SoCNNet	$83.62\% \pm 5.7 \times 10^{-2}$	$82.76\% \pm 8.7 \times 10^{-2}$	$82.81\% \pm 4.6 \times 10^{-2}$	$80.27\% \pm 4.8 \times 10^{-2}$
LaCNNet	$73.52\% \pm 7.6 \times 10^{-2}$	$68.00\% \pm 7.4 \times 10^{-2}$	$70.07\% \pm 3.5 \times 10^{-2}$	$66.54\% \pm 4.7 \times 10^{-2}$
FuCNNet	$74.20\% \pm 6.6 \times 10^{-2}$	$66.46\% \pm 9.5 \times 10^{-2}$	$69.70\% \pm 6.5 \times 10^{-2}$	$65.06\% \pm 6.6 \times 10^{-2}$

of our knowledge, this is the first work on SEM images classification of HNF and NHNF of PVAc electrospun nanofibers by using a Sobel filter as pre-processor of a CNN architecture. There are only a few works that used DL for the automatic anomalies detection of SEM images. Notably, for fair comparison, we compared the results here presented with a recent work [16], where the same CNN structure and dataset was employed, reporting a classification accuracy of 80%. However, in [16], raw SEM images were used as input and no validation technique was applied. In contrast, here, we observed that the performance decreased to 74.69% with raw SEM images (RaCNNet, Table 2) using the 15-fold cross validation technique and most important the Sobel approach allowed to improve the classification performance of about 6% (SoCNNet, Table 2).

4 Conclusion

In this research, we presented an optimized DL system for the automatic anomaly detection in SEM images of nanofibers produced by electrospinning process. Specifically, we improved the performance of the CNN architecture proposed in [16], used to classify images of homogeneous (HNF) and nonhomogeneous nanofibers (NHNF), by pre-processing each SEM image through a Sobel based filter. Here, the combination of Sobel filtering and CNN was denoted as *SoCNNet*. In order to evaluate the effectiveness of the proposed model, the images were also pre-processed with other techniques (i.e. Laplacian and a Fuzzy based filters) and used as input of the deep CNN. Notably, the corresponding networks were denoted as LaCNNnet and FuCNNet, respectively. Furthermore, for fair comparison, raw SEM images were also the input of the CNN classifier (RaCNNet). Comparative results showed that the proposed SoCNNet (Fig. 4) outperformed all the other systems LaCNNnet, FuCNNet and RaCNNet achieving accuracy rate up to $80.27\% \pm 0.048$. However, it is worth mentioning that this is a preliminary study for a more accurate and versatile system. In the future, a more accurate investigation of the applied filters will be addressed. In addition, in order to estimate the feasibility of the proposed SoCNNet a lager image dataset produced by electrospinning process of PVAc and others polymers will be taken into account.

Acknowledgments This work is supported by the project code: GR-2011-02351397. The authors would also like to thank the research group of the *Materials for Environmental and Energy Sustainability Laboratory* from the University Mediterranea of Reggio Calabria (Italy) for providing the SEM image dataset used in this work.

References

1. Wu, Y., Qu, J., Daoud, W.A., Wang, L., Qi, T.: Flexible composite-nanofiber based piezo-triboelectric nanogenerators for wearable electronics. J. Mater. Chem. A (2019)
2. Yang, Y., Chawla, A., Zhang, J., Esa, A., Jang, H.L., Khademhosseini, A.: Applications of nanotechnology for regenerative medicine; healing tissues at the nanoscale. In: Principles of Regenerative Medicine, pp. 485–504. Elsevier (2019)
3. Mo, X., Sun, B., Wu, T., Li, D.: Electrospun nanofibers for tissue engineering. In: Electrospinning: Nanofabrication and Applications, pp. 719–734. Elsevier (2019)
4. Topuz, F., Uyar, T.: Electrospinning of cyclodextrin functional nanofibers for drug delivery applications. Pharmaceutics **11**(1), 6 (2019)
5. Entov, V., Shmaryan, L.: Numerical modeling of the capillary breakup of jets of polymeric liquids. Fluid Dyn. **32**(5), 696–703 (1997)
6. Yarin, A.L.: Free liquid jets and films: hydrodynamics and rheology. Longman Publishing Group (1993)
7. LeCun, Y., Bengio, Y., Hinton, G.: Deep learning. Nature **521**(7553), 436–444 (2015)
8. Ieracitano, C., Adeel, A., Gogate, M., Dashtipour, K., Morabito, F.C., Larijani, H., Raza, A., Hussain, A.: Statistical analysis driven optimized deep learning system for intrusion detection. In: International Conference on Brain Inspired Cognitive Systems, pp. 759–769. Springer (2018)
9. Ieracitano, C., Adeel, A., Morabito, F.C., Hussain, A.: A novel statistical analysis and autoencoder driven intelligent intrusion detection approach. Neurocomputing **387**, 51–62. Elsevier (2020)
10. Ieracitano, C., Mammone, N., Bramanti, A., Hussain, A., Morabito, F.C.: A convolutional neural network approach for classification of dementia stages based on 2d-spectral representation of EEG recordings. Neurocomputing **323**, 96–107 (2019)

11. Dashtipour, K., Gogate, M., Adeel, A., Ieracitano, C., Larijani, H., Hussain, A.: Exploiting deep learning for Persian sentiment analysis. In: International Conference on Brain Inspired Cognitive Systems, pp. 597–604. Springer (2018)
12. Ieracitano, C., Mammone, N., Hussain, A., Morabito, F.C.: A novel multi-modal machine learning based approach for automatic classification of EEG recordings in dementia. Neural Netw. **123**, 176–190. Elsevier (2020)
13. Carrera, D., Manganini, F., Boracchi, G., Lanzarone, E.: Defect detection in sem images of nanofibrous materials. IEEE Trans. Industr. Inf. **13**(2), 551–561 (2017)
14. Boracchi, G., Carrera, D., Wohlberg, B.: Novelty detection in images by sparse representations. In: 2014 IEEE Symposium on Intelligent Embedded Systems (IES), pp. 47–54. IEEE (2014)
15. Napoletano, P., Piccoli, F., Schettini, R.: Anomaly detection in nanofibrous materials by CNN-based self-similarity. Sensors **18**(1), 209 (2018)
16. Ieracitano, C., Pantó, F., Mammone, N., Paviglianiti, A., Frontera, P., Morabito, F.C.: Towards an automatic classification of SEM images of nanomaterial via a deep learning approach. In: Neural Approaches to Dynamics of Signal Exchanges. pp. 61–72. Springer (2020)
17. Doshi, J., Reneker, D.H.: Electrospinning process and applications of electrospun fibers. J. Electrostat. **35**(2–3), 151–160 (1995)
18. Fenn, J.B., Mann, M., Meng, C.K., Wong, S.F., Whitehouse, C.M.: Electrospray ionization for mass spectrometry of large biomolecules. Science **246**(4926), 64–71 (1989)
19. Theron, S., Zussman, E., Yarin, A.: Experimental investigation of the governing parameters in the electrospinning of polymer solutions. Polymer **45**(6), 2017–2030 (2004)
20. Gonzales, R., Woods, R.: Digital Image Processing. Pearson-Prentice Hall (2018)
21. Chaira, T., Ray, A.K.: Fuzzy Image Processing and Applications with MATLAB. CRC Press (2009)
22. Versaci, M., Morabito, F.C., Angiulli, G.: Adaptive image contrast enhancement by computing distances into a 4-dimensional fuzzy unit hypercube. IEEE Access **5**, 26922–26931 (2017)
23. Versaci, M., Calcagno, S., Morabito, F.C.: Fuzzy geometrical approach based on unit hyper-cubes for image contrast enhancement. In: IEEE International Conference on Signal and Image Processing Applications (ICSIPA 2015), pp. 488–493. IEEE (2015)
24. Versaci, M., Calcagno, S., Morabito, F.C.: Image contrast enhancement by distances among points in fuzzy hyper-cubes. In: IEEE International Conference, CAIP 2015, pp. 494–505. IEEE (2015)
25. Nair, V., Hinton, G.E.: Rectified linear units improve restricted Boltzmann machines. In: Proceedings of the 27th International Conference on Machine Learning (ICML-10), pp. 807–814 (2010)
26. Bengio, Y.: Practical recommendations for gradient-based training of deep architectures. In: Neural Networks: Tricks of the Trade, pp. 437–478. Springer (2012)

Intent Classification in Question-Answering Using LSTM Architectures

Giovanni Di Gennaro, Amedeo Buonanno, Antonio Di Girolamo, Armando Ospedale, and Francesco A. N. Palmieri

Abstract Question-answering (QA) is certainly the best known and probably also one of the most complex problem within Natural Language Processing (NLP) and artificial intelligence (AI). Since the complete solution to the problem of finding a generic answer still seems far away, the wisest thing to do is to break down the problem by solving single simpler parts. Assuming a modular approach to the problem, we confine our research to intent classification for an answer, given a question. Through the use of an LSTM network, we show how this type of classification can be approached effectively and efficiently, and how it can be properly used within a basic prototype responder.

1 Introduction

Despite the remarkable results obtained in the different areas of Natural Language Processing, the solution to the Question-Answering problem, in its general sense, still seems far away [1]. This lies in the fact that the search for an answer to a specific

G. Di Gennaro (✉) · A. Di Girolamo · A. Ospedale · F. A. N. Palmieri
Dipartimento di Ingegneria, Universitá degli Studi della Campania "Luigi Vanvitelli", via Roma 29, Aversa, CE, Italy
e-mail: giovanni.digennaro@unicampania.it

A. Di Girolamo
e-mail: antonio.digirolamo@studenti.unicampania.it

A. Ospedale
e-mail: armando.ospedale@studenti.unicampania.it

F. A. N. Palmieri
e-mail: francesco.palmieri@unicampania.it

A. Buonanno
ENEA, Energy Technologies Department, Portici Research Centre, P. E. Fermi, 1, Portici, NA, Italy
e-mail: amedeo.buonanno@enea.it

A. Esposito et al. (eds.), *Progresses in Artificial Intelligence and Neural Systems*, Smart Innovation, Systems and Technologies 184,
https://doi.org/10.1007/978-981-15-5093-5_11

question requires many different phases, each of which representative of a separate problem. For this reason, in this work, we have approached the problem confining our attention to the classification of the intent of a response, given a specific question. In other words, our objective is not to classify the incoming questions according to their meaning, but rather referring them to the type of response they may require.

The interest in this case study is not limited to the achievement of the aforementioned objective, but also to the building of a processing block that could be inserted in a larger architecture of an autonomous responder. Furthermore, current chatbot-based dialogue interfaces can also already take advantage of these type of structures [2, 3].

By acting just as an aid in the separation of intents, it is possible to assess incoming questions in a more targeted manner. The objective is to conceive simple systems, such as those based on AIML (Artificial Intelligence Markup Language), to circumvent the extremely complex problem of evaluating all possible variations of the input.

In the following, after an introduction to the LSTM and its main features, the predictive model is presented in two successive steps. Finally, we report an example of a network trained to respond after it has been trained on previous dialogues.

2 A Particular RNN: The LSTM

The understanding of natural language is directly linked to human thought, and as such is persistent. During reading, the comprehension of a text does not happen simply through single word understanding, but mostly through the way in which they are arranged. In other words, there is the need to model the dynamics with which individual words arise.

Unfortunately, traditional feedforward neural networks are not directly fit to extrapolate information from the temporal order in which the inputs occur, as they are limited to considering only the current input for their processing. The idea of considering blocks of input words as independent from each other is too limited within the NLP context. Indeed, just as the process of reading for a human being does not lead to a memorization of all the words of the text, but to the extraction of the fundamental concepts expressed, we need a "memory" that is not limited only to explicit consideration of the previous inputs, but compresses all the relevant information acquired at each step into state variables.

This has naturally led to the so-called Recurrent Neural Networks (RNN) [4] that, by introducing the concept of "internal state" (obtained on the basis of the previous entries), have already shown great promise, especially in NLP tasks. The RNNs contain loops inside them, which allow the previous information to pass through the various steps of the analysis. An explicit view of the recurrent system can be obtained by unrolling the network activations considering the individual subnets as copies (updated at each step) of the same network, linked together to form a chain in which each one sends a message to his successor.

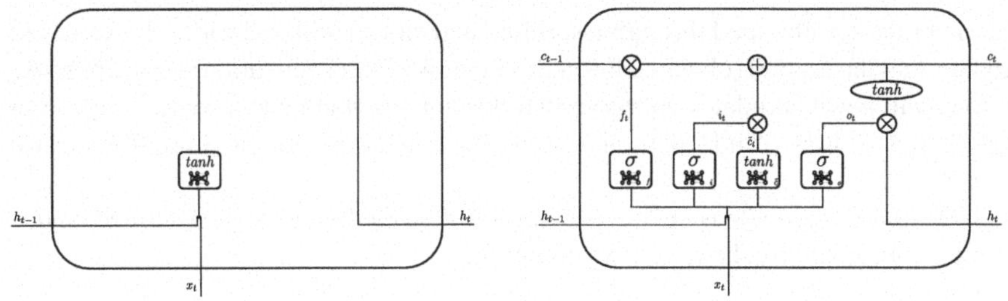

(a) RNN (b) LSTM

Fig. 1 Schematic representation of a classic RNN and an LSTM network

In theory, RNNs should be able to retain in their states the information that the learning algorithm finds in a training sequence. Unfortunately, in the propagation of the gradient backwards through time, the effect of desired output values that contribute to the cost function can become so small that, after only a few steps, there is no longer sufficient contribution to parameter learning (vanishing gradient problem). This problem, explored by [5], means that the RNNs can only retain a short-term memory, because the parts of the sequence further away in time is gradually less important. This makes the RNNs useful only for very short sequences.

To overcome (at least in part) the problem of short-term memory, the Long short-term memory (LSTM) [6] architectures were introduced. Unlike a classic RNN network (Fig. 1a), the LSTM has a much more complex single-cell structure (Fig. 1b), in which a single neural network is replaced by four that interact with each other. However, the distinctive element of LSTM is the cell state c, which allows information to flow along the chain through simple linear operations. The addition or removal of information from the state is regulated by three structures, called "gates", each one with specific objectives.

The first, called "forget gate," has the purpose of deciding the information that must be eliminated from the state. To reach this goal, the state is pointwise multiplied with:

$$f_t = \sigma(W_f \cdot [h_{t-1}, x_t] + b_f), \tag{1}$$

that is obtained from a linear (actually affine because of the biases) block that combines the joint vector of the input with the previous output ($[h_{t-1}, x_t]$), followed by a standard logistic sigmoidal activation function ($\sigma(x) = \frac{1}{1+e^{-x}}$). The forget gate makes possible to delete (values close to zero), or to maintain (values close to one), individual state vector components.

The second gate, called "input gate", has instead the purpose of conditioning the addition of new information to the state. This operation is obtained through pointwise multiplication between two vectors:

$$i_t = \sigma(W_i \cdot [h_{t-1}, x_t] + b_i), \tag{2}$$

$$\tilde{c}_t = \tanh(W_c \cdot [h_{t-1}, x_t] + b_c) \tag{3}$$

the first (always obtained through a sigmoid activation) which decides the values to update, and the second (obtained through a layer with tanh activation) whose purpose is to create new candidates. Observe that the function tanh has also the purpose of regulating the flow of information, forcing the activations to remain in the interval $[-1; 1]$.

Note that the cell status update depends only on the two gates just defined, and is in fact represented by the following equation:

$$c_t = f_t * c_{t-1} + i_t * \tilde{c}_t, \tag{4}$$

Finally, there is the "output gate", that controls the generation of the new output h_t for this cell in relation to the current input and previous visible state:

$$o_t = \sigma(W_o \cdot [h_{t-1}, x_t] + b_o), \tag{5}$$

$$h_t = o_t * \tanh(c_t). \tag{6}$$

Also, in this case there is a sigmoid activation layer that determines which part of the state to send out, multiplying them by values between zero and one. Note that, again to limit output values, the tanh function is applied to each element of the state vector (this is a simple function without any neural layer) before it is multiplied by the vector determined by the gate.

3 Implementation

Through the recurrent neural networks, with LSTM type architecture, the models used to achieve the intended objective are analysed and described below. Despite the relative simplicity of these architectures (from a general point of view in the dynamic text processing), they prove extremely efficient in being able to catalogue the intent of the answer; demonstrating how the decomposition of the general complex problem can also be tackled simply in its individual parts.

3.1 Embedding

Obviously, and regardless of the type of neural network used, having to deal with text in natural language it is essential to define the type of embedding used. In fact, unlike formal languages, which are completely specified, natural language emerges from the simple need for communication, and is therefore the bearer of a large number of ambiguities. To be able to at least try to understand it, it is therefore necessary to specify a sort of "semantic closeness" between the various terms, transforming the single words into vectors with real values within an appropriate space [7]. The resulting embedding is therefore able to map the single words into a numerical representation that "preserves their meaning", making it, in a certain sense, "understandable" even to the computer.

Nowadays there are various ways to obtain this semantic space, generally known as Word Embeddings techniques, each with its own peculiarities. For the prefixed purpose it was decided to use a pre-trained embedding known as GloVe [8], based on a vocabulary of 400,000 words mapped in a 300-dimensional space.

3.2 Dataset

A heterogeneous dataset, consisting of questions both manually constructed and published by TREC and USC [9], was used to train and test the various models. This dataset contains 5500 questions in English in the training set and another 500 in the test set. Each question has a label that classifies the answer in a two-level hierarchy. The highest level contains six main classes (abbreviation, entity, description, human, location, numeric), each of which is in turn specialized in distinct subclasses, which together represent the second level of the hierarchy (e.g. human/group, human/ind, location/city, location/country, etc). In total, from the combination of all the categories and the sub-categories we get 50 different labels with which the answer to the supplied question is classified. An example extracted from the dataset is the following (label is marked with bold and the question in italics):

HUMAN:ind *Who was the 16th President of the United States?*

As you can see, the label is composed of two parts separated by the symbol ':', representing the main class and the sub class respectively. In this example, the main class indicates that the answer must communicate a person or a group, while the sub class informs that we want to identify a specific individual (obviously through his name).

It should be noted that the representation provided by GloVe covers the totality of the words (9123 words) present in the Dataset, without therefore the need for further work in this sense. However, once the data has been extracted from the dataset, a first manipulation is carried out, which consists in cleaning the strings from any special characters and punctuation symbols (excluding the question mark). Moreover, all the words contained in the strings are transformed into lowercase, thus avoid multiple codings for the same word.

3.3 Models Analysis and Results

The search for the problem solution has been divided in two steps. It was in fact preferred to create a basic model first, which aimed to classify only the main class, and then continue with a second model that also used the secondary class.

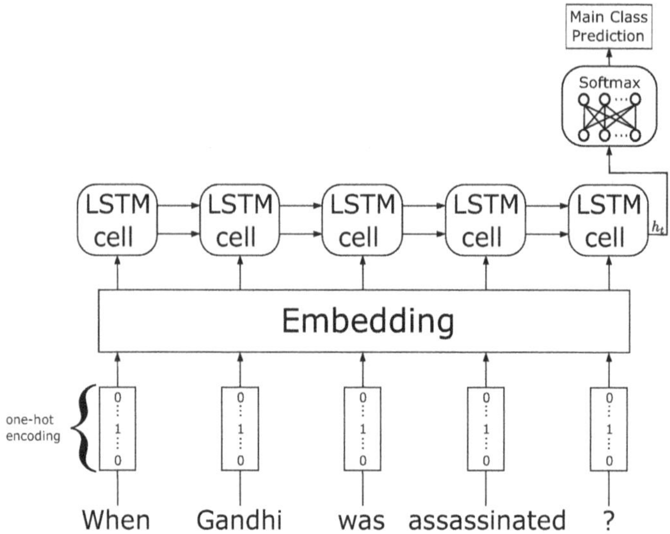

Fig. 2 Representation of the first classification model

3.3.1 First Model

To achieve the first objective, the strategy pursued was to map the entire sequence, entering the network, into a vector of fixed dimensions, equal to the number of main categories. In other words, the procedure foresees that every single word constituting the question (according to the temporal order and after having been mapped through the level of embedding) is placed in input to the LSTM, whose final state is mapped through a classical neural network with Softmax activation in one of the six main prediction classes (Fig. 2).

The Table 1 shows the results of the accuracy obtained, both on the training set and on the test set, varying the size of the h state of the LSTM.

The supervised learning of the model was carried out using the Backpropagation through time (BPTT) [10] algorithm with the classical Categorical Cross-Entropy cost function:

$$L(t, p) = -\sum_{n}\sum_{j} t_{n,j} \log(p_{n,j}) \tag{7}$$

where the first summation is extended on the number of evaluated samples (examples) while the second on the number of classes owned; $t_{n,j}$ is the target of the example n represented as one-hot vector where only the j-th entry is 1 and $p_{n,j}$ is the predicted probability that example n belong to class j.

Table 1 Accuracy relative to the size of the state for the first model

h Dimension	Training set (%)	Test set (%)
25	99.26	87.80
50	99.94	89.80
75	99.94	90.60
100	99.98	90.20

3.3.2 Second Model

The good results obtained in the prediction of the main class, have further encouraged the progress in the development of the model, maintaining the first part unchanged. In this second part we have included the subclass prediction, that having to represent a specialization of the main classes, it must be influenced in some way by the prediction of the first.

The basic idea was therefore simply to add a further element to the end of the question. This added padding element does not carry any kind of information, but it is necessary only to evaluate the output following the one corresponding to the last element of the question. The second last exit, linked to the last element of the question, can be considered just like in the previous case, while the latter depends only on the previous state (represented both by h_{t-1} and c_{t-1}) since the embedding representative of the padding element is previously set to a vector of all zeros and hence doesn't contribute to computation of h_t (look at Sect. 2).

By training the network, in a supervised manner, to associate the sub-category to this output, a dependence will therefore be created on both the main classification and the question itself. The two information coming out of the recursive part are subsequently discriminated by two distinct fully-connected layers with Softmax, thus mapping the input sequence into two fixed size vectors representing the main class and the associated sub class. The schematic representation of the second model just described can be observed in Fig. 3.

The BPTT algorithm is still used for the training phase of the model, but with a subtle difference compared to the previous case: in fact, there will not be a single (Categorical Cross-Entropy) cost function but two, since our goals have doubled.

Table 2 shows the results obtained, with an accuracy of around 80% for the prediction of the sub-category of the samples belonging to the test set. Furthermore, as was to be expected, there are almost identical performances on the main category prediction.

It should be noted that, despite the excellent performance of the model on the training set, it does not seem to have gone into overfitting. This affirmation can be confirmed by observing through the accuracy trend for the model with $H = 100$ (Fig. 4) both on the training and on the test set at different epochs, which shows how the greater accuracy on the training set does not become pejorative for the test set. In fact, both trends seem to stabilize around a regime value, with slight fluctuations that do not seem to affect the generalization characteristics of the network.

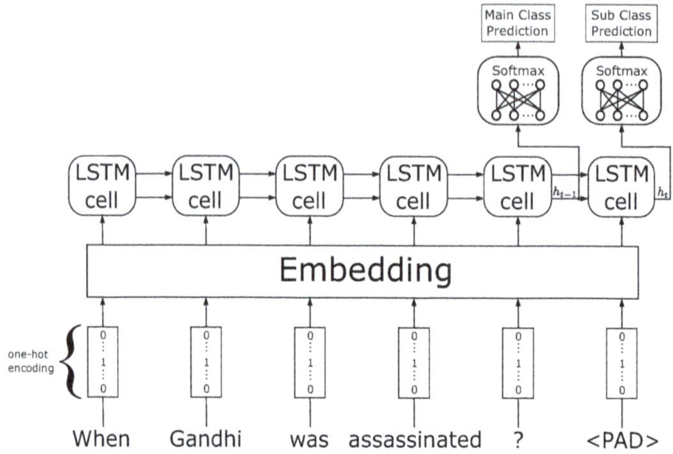

Fig. 3 Representation of the double classification model

Table 2 Accuracy relative to the size of the state for the second model

h Dimension	Training set		Test set	
	Main class (%)	Sub class (%)	Main class (%)	Sub class (%)
25	99.24	96.86	86.20	74.40
50	99.94	99.83	90.00	80.00
75	99.94	99.72	91.00	78.60
100	99.82	99.67	91.20	82.20

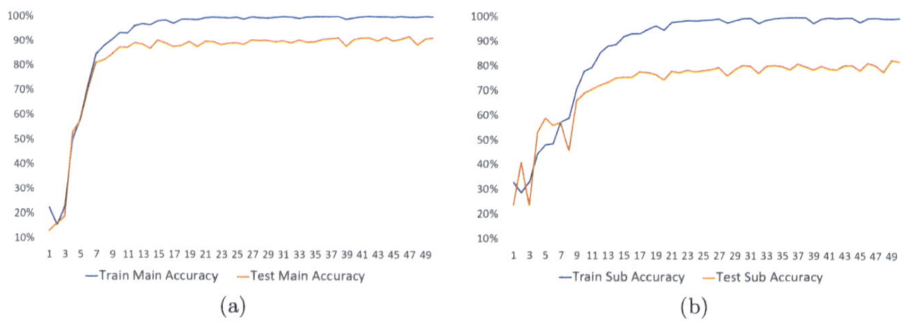

Fig. 4 Accuracy trend for the Main (**a**) and Sub (**b**) classes at different epochs

3.4 Prototype Responder

Finally, in order to test (at least briefly) the importance of what was achieved in the classification of intents on the ultimate purpose of creating a responder, it was decided to create a prototype that could exploit the categorization obtained to generate a response. This prototype (Fig. 5) uses a bidirectional LSTM (BLSTM) network [11] to review the inbound application, so that it can generate an answer only after acquiring the entire question.

The status of this BLSTM network is conditioned by the prediction returned by the previous model. The two vectors representing the main and the sub class are in fact linked together, forming a single vector that will represent the initial state of the BLSTM network. The exits of the network, which constitute the words forming the answer, are thus generated by the analysis of both contexts (future and past), and strongly influenced by the categorization provided.

The supervised training of the network was performed on a set of 500 question-answer samples, and in the Table 3 are shown some of the network outputs relating

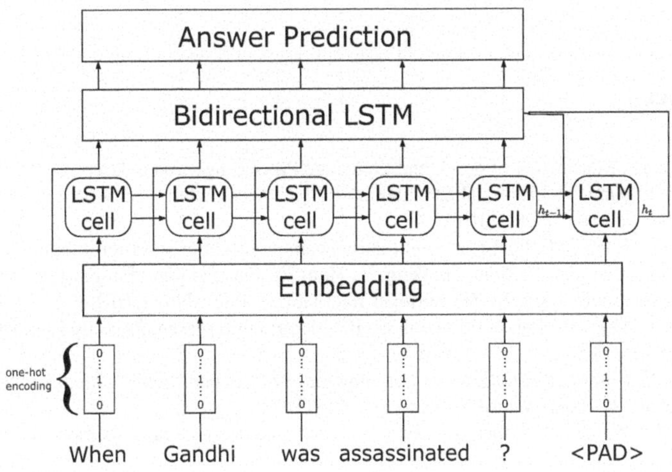

Fig. 5 Representation of the responder prototype

Table 3 Examples of answers for questions not belonging to the training set

Question	Answer
How many people speak French?	13
What day is today?	The first day may nights and the
Who will win the war?	North
Who is Italian first minister?	Francisco vasquez
When World War II ended?	March
When Gandhi was assassinated?	1976

to questions not present in the training set. It should be noted that the purpose of this prototype is not to provide a correct answer (which is quite impossible given the limited dataset and since no knowledge of the answer not relating to it is never provided to the network) but to show how the simple excellent categorization of the intent allows, already alone, to get consistent answers to the context of the question.

4 Conclusion

The results obtained from the models presented show very high accuracy values, both for the training set and for the test set. Networks of this type are actually very effective in this type of classification, perhaps because of their simplicity. The example of the responder prototype then confirms how the choice to classify the questions based on the intent of the answer is extremely effective in limiting and contextualizing the outgoing answer. Consequently, this approach can be of help for the development of more complex systems, being already able to compensate for the deficiencies of "traditional" systems that can benefit from the classification of intent provided.

References

1. Jurafsky, D., Martin, J.H.: Speech and Language Processing, 3rd ed. Draft (2019)
2. Shen, L., Zhang, J.: Empirical Evaluation of RNN Architectures on Sentence Classification Task. arXiv e-prints, Sept 2016
3. Meng, L., Huang, M.: Dialogue intent classification with long short-term memory networks. In: Huang, X., Jiang, J., Zhao, D., Feng, Y., Hong, Y. (eds.) Natural Language Processing and Chinese Computing, pp. 42–50, Springer International Publishing (2018)
4. Rumelhart, D.E., Hinton, G.E., Williams, R.J.: Learning representations by back-propagating errors. Nature **323**, 533–536 (1986)
5. Bengio, Y., Simard, P., Frasconi, P.: Learning long-term dependencies with gradient descent is difficult. IEEE Trans. Neural Netw. **5**, 157–166 (1994)
6. Hochreiter, S., Schmidhuber, J.: Long short-term memory. Neural Comput. **9**, 1735–1780 (1997)
7. Bengio, Y., Ducharme, R., Vincent, P.: A neural probabilistic language model. In: Leen, T.K., Dietterich, T.G., Tresp, V. (eds.) Advances in Neural Information Processing Systems, vol. 13, pp. 932–938. MIT Press (2001)
8. Pennington, J., Socher, R., Manning, C.: Glove: Global Vectors for Word Representation, vol. 14, pp. 1532–1543 (2014)
9. Li, X., Roth, D.: Learning question classifiers. In: Proceedings of the 19th International Conference on Computational Linguistics, COLING '02, Stroudsburg, PA, USA, vol. 1, pp. 1–7. Association for Computational Linguistics (2002)
10. Mozer, M.: A focused backpropagation algorithm for temporal pattern recognition. Complex Syst. **3**, 01 (1995)
11. Schuster, M., Paliwal, K.K.: Bidirectional recurrent neural networks. IEEE Trans. Signal Process. **45**, 2673–2681 (1997)

A Novel Proof-of-concept Framework for the Exploitation of ConvNets on Whole Slide Images

A. Mascolini, S. Puzzo, G. Incatasciato, F. Ponzio, E. Ficarra, and S. Di Cataldo

Abstract Traditionally, the analysis of histological samples is visually performed by a pathologist, who inspects under the microscope the tissue samples, looking for malignancies and anomalies. This visual assessment is both time consuming and highly unreliable due to the subjectivity of the evaluation. Hence, there are growing efforts towards the automatisation of such analysis, oriented to the development of computer-aided diagnostic tools, with a ever-growing role of techniques based on deep learning. In this work, we analyze some of the issues commonly associated with providing deep learning based techniques to medical professionals. We thus introduce a tool, aimed at both researchers and medical professionals, which simplifies and accelerates the training and exploitation of such models. The outcome of the tool is an attention map representing cancer probability distribution on top of the Whole Slide Image, driving the pathologist through a faster and more accurate diagnostic procedure.

1 Introduction

The field of pathology often relies on the analysis of microscopic images to perform a diagnosis. Whole Slide Imaging represents a technology, through which glass slides are digitalized in the form of minimally compressed images, featuring a pyramid structure with various levels of magnification. This process enables microscopic images of tissues to be analyzed by advanced digital tools [1].

The Whole Slides Images (WSIs) have been used for a wide variety of both educational and clinical purposes, and several authors have reported good diagnostic concordance between the analysis of WSIs and glass slides [2]. It can be thus reasonably assumed that digital image classification, as well as deep learning techniques, can play a key role in the delicate and time consuming process of diagnosis. On

A. Mascolini · S. Puzzo · G. Incatasciato · F. Ponzio (✉) · E. Ficarra · S. Di Cataldo
Politecnico di Torino, Corso Duca Degli Abruzzi 24, 10129 Turin, Italy
e-mail: francesco.ponzio@polito.it

A. Esposito et al. (eds.), *Progresses in Artificial Intelligence and Neural Systems*, Smart Innovation, Systems and Technologies 184,
https://doi.org/10.1007/978-981-15-5093-5_12

one hand, they can serve as a way to double check and mitigate the extreme inter-operators variability. On the other hand, they can significantly reduce the evaluation time spent by the clinicians, by providing accurate and automatic information in a reasonably short time.

While literature features an extensive amount of papers in which machine and deep learning techniques are successfully applied to the field of Computer Aided Diagnosis (CAD) [3–5], there are very few instances of these techniques being used in a daily medical practice. This is mainly due to:

- the very long processing times and specialized hardware required to implement most of the newly developed architectures. For instance, ScanNet, a framework to analyze WSIs in a fully connected fashion, takes 15 min on a Titan X GPU to analyze a single Whole Slide Image [6], which is totally unfeasible in a everyday clinical scenario.
- The lack of easy-to-use interfaces for non-technical users, such as the medical staff.

With the aim of facing the above-mentioned challenges, we have thus developed a standalone, cross-platform, CPU based tool, featuring an easy-to-use Graphical User Interface to facilitate the user experience. Our tool features a staining normalization phase and an asynchronous sample pre-fetching to optimize computational time, and applies a dynamic resolution approach. Beside being a CAD oriented tool, our apparatus also allows researchers to train and prototype new WSIs segmentation models, without having to worry about tedious and extensive pre-processing, being the model easily embeddable in the framework.

The tool has been evaluated on a highly challenging dataset consisting of histological images of a specific type of tumor known as Colorectal Carcinoma (CRC). Nowadays, CRC is the third most frequent cancer that afflicts mankind with 1.8 million new cases in 2018 [7]. CRC is a type of epithelial cancer, coming from the colon or the rectum, which provokes the uncontrolled proliferation of mucosal cells covering the last part of the intestine. The initial diagnosis of CRC is performed by means of colonoscopy, i.e. the endoscopic inspection of the large and the distal part of the small intestine, during which the surgeon may perform a biopsy on the suspicious colorectal lesions. This surgical step is then usually followed by a diagnostic procedure carried out by the pathologist to determine the nature of lesions, studying the tissue sample under the microscope or through an analysis of the corresponding WSI. The importance of the early diagnosis of the tumour, crucial for the survival of a large number of patients, makes the CRC an interesting case study to test the feasibility of our method.

The rest of the paper is organized as follows. In Sect. 2 we describe our dataset and then introduce the design characteristics of our proposed approach. In Sect. 3, we report our experimental results and we discuss our findings. Finally, Sect. 4 concludes the paper.

2 Materials and Methods

2.1 Dataset

Our case study dataset was extracted from a public repository of H&E stained whole-slide images (WSIs) of colorectal tissues, available on line at http://www.virtualpathology.leeds.ac.uk/. In order to obtain a statistically significant dataset, in terms of inter-subjects and inter-class variability, 18 WSIs were selected among univocal subjects (i.e. one WSI per patient) and then split into regions containing either exclusively cancer or exclusively healthy tissue via a sliding windows approach, as shown in Fig. 1.

The cropped patches were separated into a training and a testing set with a 75–25% split, ensuring that regions coming from a single patient always belong to the same set. These sets were fed to the network first to train it and then to evaluate the patch-wise predictions. A second independent cohort of 11 patients, never fed to the network model during training, were randomly selected to serve as the validation set for performance evaluation in terms of WSIs attention maps, which is ultimately a tissue segmentation task.

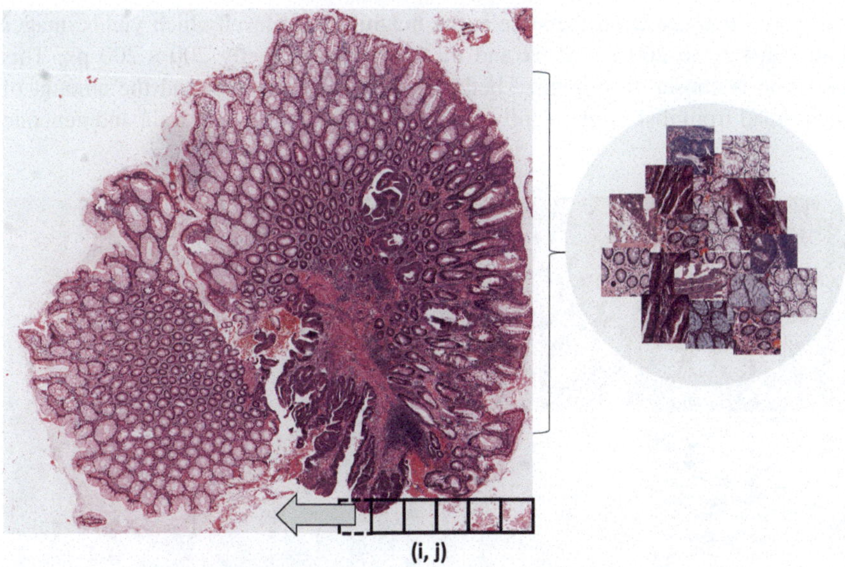

Fig. 1 A WSI example depicting the sliding window cropping technique

2.2 Slide Analysis

As initial step, aimed at reducing the large inter-patients variability in terms of slide color (see Fig. 2), we perform staining normalization on the input H&E stained WSIs, using the well-known stain vector variation and correction method [8], providing the network with consistent colors. The slide is cropped using a sliding window, and references to every crop are stored in memory, while the crop itself is not loaded until necessary.

At runtime, two heuristics identify and remove crops which are either white (due to saturation of colors or absence of tissues) or contaminated by blood.

- To identify blood contaminated crops, the following value is calculated

$$\delta = 2 \cdot \frac{\overline{Red}}{\overline{Green} + \overline{Blue}}$$

  where $\overline{Channel}$ represents the mean value of the channel in the image. If $\delta > 1.5$ the slide is discarded due to too much blood.
- To identify white crops, the mean luminance and standard deviation are checked to be respectively higher and lower than two fixed thresholds.

Every crop is taken from the slide at the magnification level which yields images closest in size to 200 × 200 px and then resized to exactly 200 × 200 px. This resolution is chosen to minimize both the amount of neurons and the amount of data to read from disk, significantly reducing the time it takes to train and generate

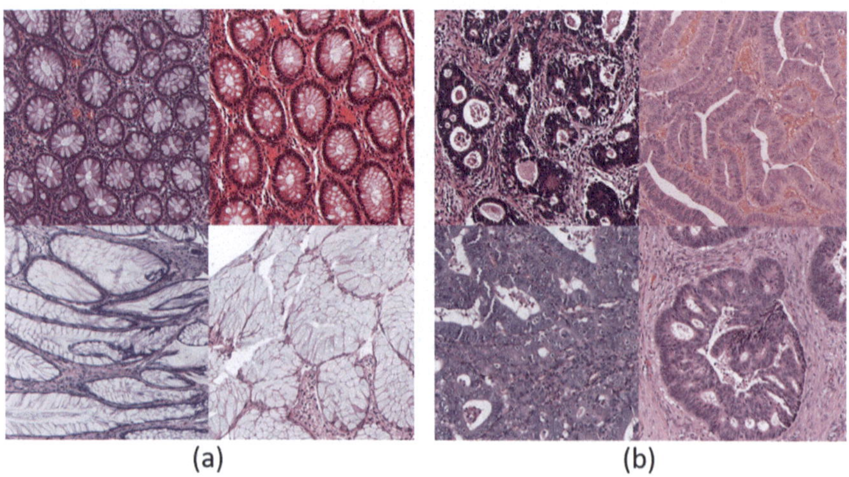

Fig. 2 Histological images of colorectal tissues (cropped patches) presenting a very different staining effect. **a** Healthy tissue; **b** Cancer

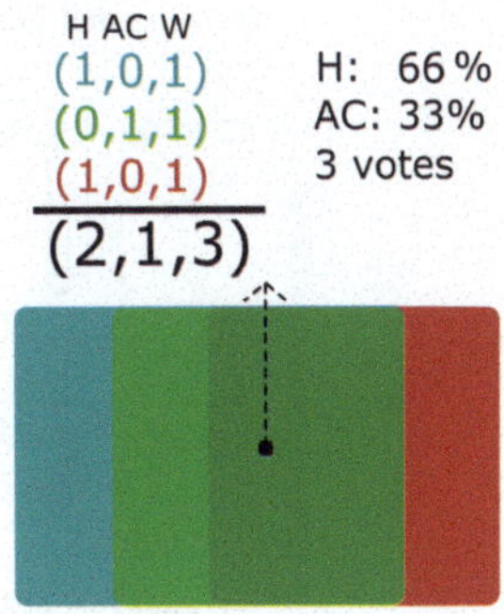

Fig. 3 Simplified example of probability image generation

predictions with the net, while providing a high enough resolution for the classifier to obtain good results.

When two neighbouring crops are assigned to different classes, they are recognized as a border. Additional windows are then added between border crops to increase the resolution of the classification where needed, in an iterative process which can be repeated as many times as the user requires. The crop loading process is asynchronous, allowing the CPU to process the previous sample while the next is fetched from the disk.

The tool stores the location and classification of every crop. When an area of the slide needs to be shown with its corresponding attention map(s), the tool averages the votes received by every pixel in the required area of the slide to create a series of attention maps, one for each class the network is capable of identifying. In other words, every crop containing that pixel votes for a specific class and the results are averaged, as show in Fig. 3.

Our tool offers a specific module to manage and crop WSIs, automatically generating patches, removing invalid areas, performing H&E stain normalization and producing data which is ready to be fed into a neural model. The implementation relies on CropList objects which can be joined and split at will to freely create datasets. The crops generated by our tool are asyncronously loaded in memory at runtime, allowing the employment of large datasets while making efficient use of the system resources.

2.3 Neural Network Architecture

Our tool provides an architecture-agnostic way to perform segmentation over a WSI; in order to test it, we created and trained a simple supervised neural network for patch classification of CRC, featuring AlexNet style CNN based features extraction (see Fig. 4). The architecture is made up of a base unit of 2 convolutional layers, with kernel size 3×3, stride 1 and no padding, which start from a 200×200 input

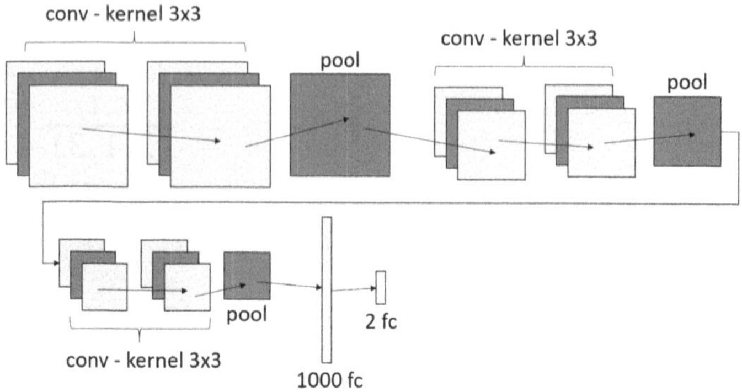

Fig. 4 Overview of the CNN architecture

size and progressively shrink, followed by batch normalization, ReLU activation and max pooling.

This unit is repeated 3 times, followed by a fully connected layer of 1000 neurons and an output layer of 2 neurons with softmax activation. Between the two fully connected layers, there is a dropout layer which randomly drops 40% of its inputs.

The network uses categorical Cross-Entropy as its loss function and is optimized using Adam optimizer with alpha = 0.001, as suggested by the original paper [9]. Our accuracy showed little variability between epochs, we thus believe the learning rate does not need adjusting.

2.4 Graphical User Interface

The python back-end communicates using the Eel library with the javascript code in the front-end, providing a seamless user experience featuring (see Fig. 5):

- a file browser;
- the possibility of choosing the stride of the sliding window during the first step of the classification problem;
- a choice of which classes to show;
- an estimate of the time necessary to analyze the entire WSI;
- an area in which to annotate information about the slide.

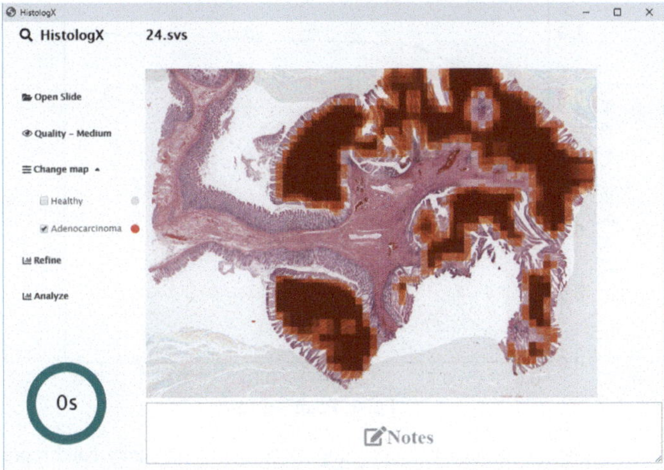

Fig. 5 Example of WSI analysis using the presented tool

3 Classification Accuracy

3.1 Performance Metrics

To evaluate the quality of the predictions yielded by the architecture coupled with our tool, we had a group of professional pathologists annotating various WSIs containing cancer which the network had never seen before. These will be referred as the validation set.

3.2 Results and Discussion

The Dice coefficient is a metric used to evaluate the overlap between two discrete sets (X and Y), and is calculated as:

$$DSC = \frac{2|X \cap Y|}{|X| + |Y|}$$

It is commonly employed to evaluate the similarity between segmentation masks [10]. The results obtained by our approach in terms of Dice coefficient are shown in Table 1, alongside with the true positive (i.e. Sensitivity) and true negative rate (Specificity), as well as a the pixel-wise accuracy (Ac), defined as follows:

$$Ac = \frac{N_c}{N}$$

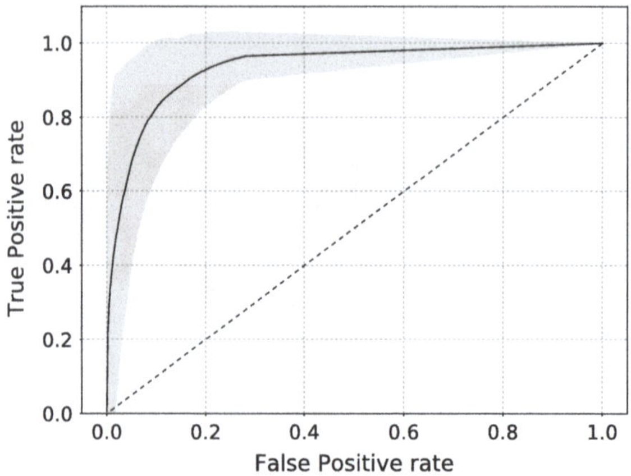

Fig. 6 Mean ROC curve over all validation subjects ($n = 11$). The grey band represents 95% confidence intervals

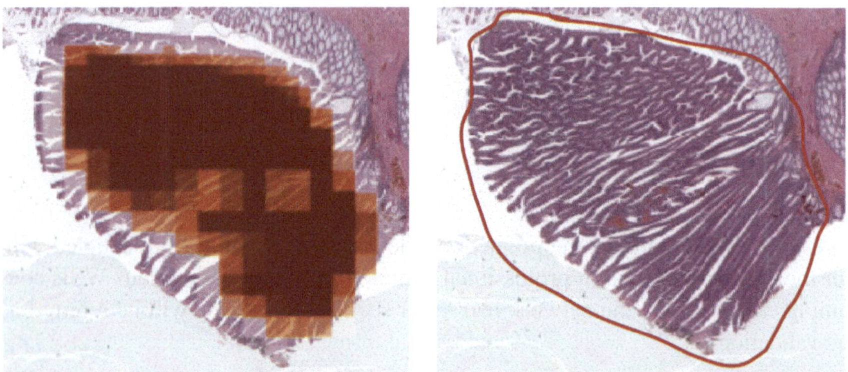

Fig. 7 Example showing the tendency of the pathologist to over-segment the lesion (right) with respect to out tool (left)

where N_c is the number of pixels which were correctly classified and N is the total number of pixels in an image. The two rows of the table respectively show mean and standard deviations of the figures of merit in the validation set.

Since our classifier is binary, we had to choose a threshold to indicate a pixel as belonging to the positive class. We did so by using the Receiver Operating Characteristics curve (see Fig. 6), which yield to a threshold of 27/255.

As reported in Table 1, we obtained a mean Dice coefficient of 0.80 and an average per-pixel accuracy of 87%, with no bias towards any of the two classes. It must be noticed that a loss of performance in terms of Dice coefficient is possibly due to the tendency of the human pathologist to over-segment tumor-containing areas compared to our automatic algorithm, as it can be gathered from Fig. 7.

Table 1 Performance of our test architecture on the validation set

	Dice	Ac	Sensitivity	Specificity
Mean	0.80	0.87	0.87	0.86
Std	0.07	0.06	0.10	0.08

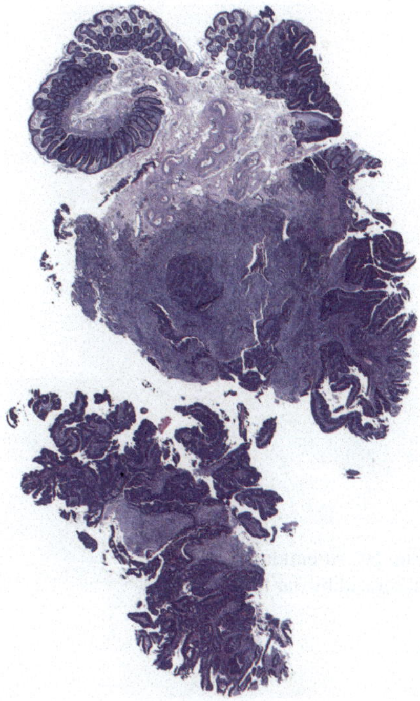

Fig. 8 Patient 1 original WSI

To fully assess our tool, we compared it to the most similar framework we found in literature to evaluate WSIs, which is described in [6]. To compare the segmentation performance, we used AUC (Area under ROC curve) on the entire validation set as the figure of merit, while the processing time was compared on a sample $200 \times 100 \times 10^3$ WSI, as reported in Table 2. We tested our framework on the same hardware as the one used in [6], i.e. the Nvidia Titan X GPU, and on a widely available hardware consisting in a standard Intel i5 7200U CPU for personal computers.

As it can be gathered from Table 2, despite the slightly lower accuracy in terms of pixel-wise classification, our tool allows a quicker classification on cheaper, less powerful and widely available hardware (Figs. 8, 9 and 10).

The trade-off between accuracy and efficiency can be explained as follows. The proof-of-concept of our work was built on the idea of exploiting the classification potentials of some existing deep learning models, with the aim of achieving sufficiently accurate characterization of tissue regions with little or no extra effort in re-design of the model. The patch-based approach we used, which significantly dif-

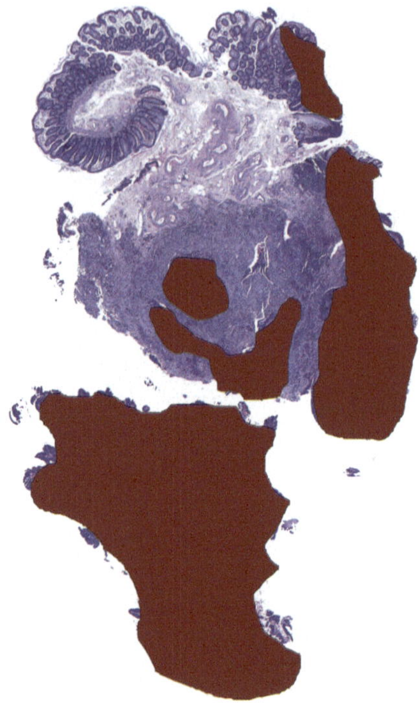

Fig. 9 Manual segmentation by pathologist

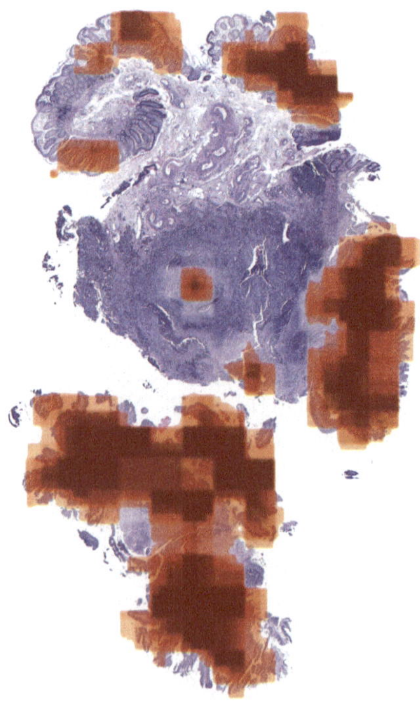

Fig. 10 Attention map generated by our tool

Table 2 Comparison between our framework and the one described in [6]

	ScanNet	Ours	
Hardware	Nvidia Titan X	Nvidia Titan X	Intel i5 7200U
Type	Pixel-based	Patch-based	Patch-based
AUC (mean ± std.)	98.75 ± n.a.	93.44 ± 0.05	93.44 ± 0.05
Time (single WSI)	15 min	12 min	8 min
Asyncronous	Yes	Yes	Yes

fers from the dense per-pixel prediction implemented by the state-of-the-art tool, seems to be a promising choice, showing a good accuracy coupled with much lower processing time on a less powerful hardware.

While ScanNet features a highly parallelizable Fully Convolutional architecture which runs significantly faster on the GPU, our Patch Based approach proves to be I/O bound and as such more suitable to low hardware specifications, such as standard personal computers. The test slide is processed significantly faster on the CPU by saving on the overhead required to constantly move data between the system and the graphics processor. Our architecture is thus best optimized for consumer devices such as laptops and tablets.

4 Conclusions and Future Work

In this work, we built and tested a novel framework which we believe could be useful in accelerating the development and adoption of deep learning techniques in the every-day digital pathology. We demonstrated our approach on WSI segmentation, showing that our easy-to-use framework can be run on cheap and widely available hardware with limited amount of processing time. Overall, the network we trained using our framework obtained results which agree with the manual segmentation performed by human pathologists.

References

1. Zarella, M., et al.: A practical guide to whole slide imaging: a white paper from the digital pathology association. Arch. Pathol. Lab. Med. **143** (2018). https://doi.org/10.5858/arpa.2018-0343-RA
2. Pantanowitz, L., Farahani, N., Parwani, A.: Whole side imaging in pathology: advantage, limitations, and emerging perspectives. Pathol. Lab. Med. Int. **2015** (2015). https://doi.org/10.2147/PLMI.S59826
3. Xu, Y., et al.: Large scale tissue histopathology image classification, segmentation, and visualization via deep convolutional activation features. In: BMC Bioinformatics (2017)

4. Ponzio, F., et al.: Dealing with lack of training data for convolutional neural networks: the case of digital pathology. Electronics **8**, 256 (2019). https://doi.org/10.3390/electronics8030256
5. Xing, F., et al.: Deep learning in microscopy image analysis: a survey. IEEE Trans. Neural Netw. Learn. Syst. **99**, 1–19 (2017)
6. Lin, H., et al.: ScanNet: a fast and dense scanning framework for metastatic breast cancer detection from whole-slide images (2017). arXiv:1707.09597
7. Bray, F., et al.: Global cancer statistics 2018: GLOBOCAN estimates of incidence and mortality worldwide for 36 cancers in 185 countries. CA Canc. J. Clin. **68**(6), 394–424 (2018). https://doi.org/10.3322/caac.21492. https://onlinelibrary.wiley.com
8. Macenko, M., et al.: A method for normalizing histology slides for quantitative analysis. **9**, 1107–1110 (2009). https://doi.org/10.1109/ISBI.2009.5193250
9. Kingma, D.P., Ba, J.: Adam: a method for stochastic optimization (2014). arXiv:1412.6980 [cs.LG]
10. Tustison, N.J., Gee, J.C.: Introducing Dice, Jaccard, and other label overlap measures to ITK. Insight. J. **2** (2009)

An Analysis of Word2Vec for the Italian Language

Giovanni Di Gennaro, Amedeo Buonanno, Antonio Di Girolamo, Armando Ospedale, Francesco A. N. Palmieri, and Gianfranco Fedele

Abstract Word representation is fundamental in NLP tasks, because it is precisely from the coding of semantic closeness between words that it is possible to think of teaching a machine to understand text. Despite the spread of word embedding concepts, still few are the achievements in linguistic contexts other than English. In this work, analysing the semantic capacity of the Word2Vec algorithm, an embedding for the Italian language is produced. Parameter setting such as the number of epochs, the size of the context window and the number of negatively backpropagated samples is explored.

G. Di Gennaro (✉) · A. Di Girolamo · A. Ospedale · F. A. N. Palmieri
Dipartimento di Ingegneria, Universitá degli Studi della Campania "Luigi Vanvitelli", via Roma 29, Aversa, CE, Italy
e-mail: giovanni.digennaro@unicampania.it

A. Di Girolamo
e-mail: antonio.digirolamo@studenti.unicampania.it

A. Ospedale
e-mail: armando.ospedale@studenti.unicampania.it

F. A. N. Palmieri
e-mail: francesco.palmieri@unicampania.it

A. Buonanno
ENEA, Energy Technologies Department, Portici Research Centre, P. E. Fermi, 1, Portici, NA, Italy
e-mail: amedeo.buonanno@enea.it

G. Fedele
MAZER s.r.l, Marcianise, CE, Italy
e-mail: gianfranco.fedele@laila.tech

A. Esposito et al. (eds.), *Progresses in Artificial Intelligence and Neural Systems*, Smart Innovation, Systems and Technologies 184,
https://doi.org/10.1007/978-981-15-5093-5_13

1 Introduction

In order to make human language comprehensible to a computer, it is obviously essential to provide some word encoding. The simplest approach is the one-hot encoding, where each word is represented by a sparse vector with dimension equal to the vocabulary size. In addition to the storage need, the main problem of this representation is that any concept of word similarity is completely ignored (each vector is orthogonal and equidistant from each other). On the contrary, the understanding of natural language cannot be separated from the semantic knowledge of words, which conditions a different closeness between them. Indeed, the semantic representation of words is the basic problem of Natural Language Processing (NLP). Therefore, there is a necessary need to code words in a space that is linked to their meaning, in order to facilitate a machine in potential task of "understanding" it. In particular, starting from the seminal work [1], words are usually represented as dense distributed vectors that preserve their uniqueness but, at the same time, are able to encode the similarities.

These word representations are called Word Embeddings since the words (points in a space of vocabulary size) are mapped in an embedding space of lower dimension. Supported by the distributional hypothesis [2, 3], which states that a word can be semantically characterized based on its context (i.e. the words that surround it in the sentence), in recent years many word embedding representations have been proposed (a fairly complete and updated review can be found in [4, 5]). These methods can be roughly categorized into two main classes: *prediction-based models* and *count-based models*. The former is generally linked to work on Neural Network Language Models (NNLM) and use a training algorithm that predicts the word given its local context, the latter leverage word-context statistics and co-occurrence counts in an entire corpus. The main prediction-based and count-based models are respectively Word2Vec [6] (W2V) and GloVe [7].

Despite the widespread use of these concepts [8, 9], few contributions exist regarding the development of a W2V that is not in English. In particular, no detailed analysis on an Italian W2V seems to be present in the literature, except for [10, 11]. However, both seem to leave out some elements of fundamental interest in the learning of the neural network, in particular relating to the number of epochs performed during learning, reducing the importance that it may have on the final result. In [10], this for example leads to the simplistic conclusion that (being able to organize with more freedom in space) the more space is given to the vectors, the better the results may be. However, the problem in complex structures is that large embedding spaces can make training too difficult.

In this work, by setting the size of the embedding to a commonly used average value, various parameters are analysed as the number of learning epochs changes, depending on the window sizes and the negatively backpropagated samples.

2 Word2Vec

The W2V structure consists of a simple two-level neural network (Fig. 1) with one-hot vectors representing words at the input. It can be trained in two different modes, algorithmically similar, but different in concept: Continuous Bag-of-Words (CBOW) model and Skip-Gram model. While CBOW tries to predict the target words from the context, Skip-Gram instead aims to determine the context for a given target word. The two different approaches therefore modify only the way in which the inputs and outputs are to be managed, but in any case, the network does not change, and the training always takes place between single pairs of words (placed as one-hot in input and output).

The text is in fact divided into sentences, and for each word of a given sentence a window of words is taken from the right and from the left to define the context. The central word is coupled with each of the words forming the set of pairs for training. Depending on the fact that the central word represents the output or the input in training pairs, the CBOW and Skip-gram models are obtained respectively.

Regardless of whether W2V is trained to predict the context or the target word, it is used as a word embedding in a substantially different manner from the one for which it has been trained. In particular, the second matrix is totally discarded during use, since the only thing relevant to the representation is the space of the vectors generated in the intermediate level (embedding space).

2.1 *Sampling Rate*

The common words (such as “the”, “of”, etc.) carry very little information on the target word with which they are coupled, and through backpropagation they tend

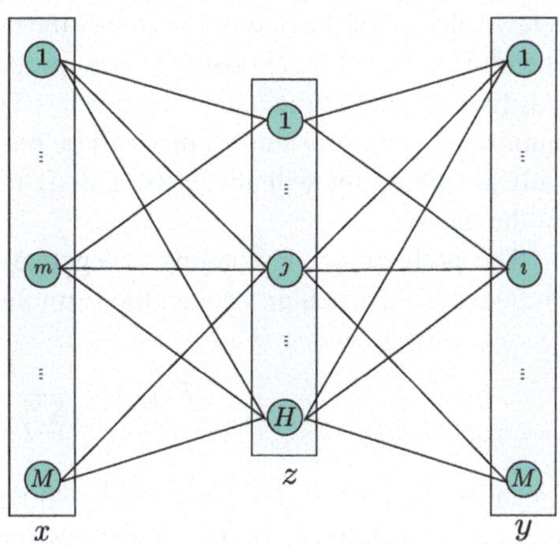

Fig. 1 Representation of Word2Vec model

to have extremely small representative vectors in the embedding space. To solve both these problems the W2V algorithm implements a particular "subsampling" [12], which acts by eliminating some words from certain sentences. Note that the elimination of a word directly from the text means that it no longer appears in the context of any of the words of the sentence and, at the same time, a number of pairs equal to (at most) twice the size of the window relating to the deleted word will also disappear from the training set.

In practice, each word is associated with a sort of "keeping probability" and, when you meet that word, if this value is greater than a randomly generated value then the word will not be discarded from the text. The W2V implementation assigns this "probability" to the generic word w_i through the formula:

$$P(w_i) = \left(\sqrt{\frac{f(w_i)}{s}} + 1\right)\frac{s}{f(w_i)}, \qquad (1)$$

where $f(w_i)$ is the relative frequency of the word w_i (namely $count(w_i)/total$), while s is a sample value, typically set between 10^{-3} and 10^{-5}.

2.2 Negative Sampling

Working with one-hot pairs of words means that the size of the network must be the same at input and output, and must be equal to the size of the vocabulary. So, although very simple, the network has a considerable number of parameters to train, which lead to an excessive computational cost if we are supposed to backpropagate all the elements of the one-hot vector in output.

The "negative sampling" technique [12] tries to solve this problem by modifying only a small percentage of the net weights every time. In practice, for each pair of words in the training set, the loss function is calculated only for the value 1 and for a few values 0 of the one-hot vector of the desired output. The computational cost is therefore reduced by choosing to backpropagate only K words "negative" and one positive, instead of the entire vocabulary. Typical values for negative sampling (the number of negative samples that will be backpropagated and to which therefore the only positive value will always be added), range from 2 to 20, depending on the size of the dataset.

The probability of selecting a negative word to backpropagate depends on its frequency, in particular through the formula:

$$P(w_i) = \frac{f(w_i)^{3/4}}{\sum_{j=0}^{n} f(w_j)^{3/4}} \qquad (2)$$

Negative samples are then selected by choosing a sort of "unigram distribution", so that the most frequent words are also the most often backpropagated ones.

3 Implementation Details

The dataset needed to train the W2V was obtained using the information extracted from a dump of the Italian Wikipedia (dated 2019.04.01), from the main categories of Italian Google News (WORLD, NATION, BUSINESS, TECHNOLOGY, ENTERTAINMENT, SPORTS, SCIENCE, HEALTH) and from some anonymized chats between users and a customer care chatbot (Laila[1]). The dataset (composed of 2.6 GB of raw text) includes 421,829,960 words divided into 17,305,401 sentences.

The text was previously preprocessed by removing the words whose absolute frequency was less than 5 and eliminating all special characters. Since it is impossible to represent every imaginable numerical value, but not wanting to eliminate the concept of "numerical representation" linked to certain words, it was also decided to replace every number present in the text with the particular $\langle NUM \rangle$ token; which probably also assumes a better representation in the embedding space (not separating into the various possible values). All the words were then transformed to lowercase (to avoid a double presence) finally producing a vocabulary of 618,224 words.

Note that among the special characters are also included punctuation marks, which therefore do not appear within the vocabulary. However, some of them ('.', '?' and '!') are later removed, as they are used to separate the sentences.

The Python implementation provided by Gensim was used for training the various embeddings all with size 300 and sampling parameter (s in Eq. 1) set at 0.001.

4 Results

To analyse the results we chose to use the test provided by [11], which consists of 19,791 analogies divided into 19 different categories: 6 related to the "semantic" macro-area (8915 analogies) and 13 to the "syntactic" one (10,876 analogies). All the analogies are composed by two pairs of words that share a relation, schematized with the equation: $a : a^* = b : b^*$ (e.g. "man : woman = king : queen"); where b^* is the word to be guessed ("queen"), b is the word coupled to it ("king"), a is the word for the components to be eliminated ("man"), and a^* is the word for the components to be added ("woman").

The determination of the correct response was obtained both through the classical additive cosine distance (3COSADD) [6]:

$$\arg\max_{b^* \in V} \cos(b^*, b - a + a^*) \tag{3}$$

[1]https://laila.tech/.

and through the multiplicative cosine distance (3COSMUL) [13]:

$$\arg\max_{b^* \in V} \frac{\cos(b^*, b)\cos(b^*, a^*)}{\cos(b^*, a) + \epsilon} \tag{4}$$

where $\epsilon = 10^{-6}$ and $\cos(x, y) = \frac{x \cdot y}{\|x\|\|y\|}$. The extremely low value chosen for the ϵ is due to the desire to minimize as much as possible its impact on performance, as during the various testing phases we noticed a strange bound that is still being investigated. As usual, moreover, the representative vectors of the embedding space are previously normalized for the execution of the various tests.

4.1 Analysis of the Various Models

We first analysed 6 different implementations of the Skip-gram model each one trained for 20 epochs. Table 1 shows the accuracy values (only on possible analogies) at the 20th epoch for the six models both using 3COSADD and 3COSMUL. It is interesting to note that the 3COSADD total metric, respect to 3COSMUL, seems to have slightly better results in the two extreme cases of limited learning (W5N5 and W10N20) and under the semantic profile. However, we should keep in mind that the semantic profile is the one best captured by the network in both cases, which is probably due to the nature of the database (mainly composed of articles and news that principally use an impersonal language). In any case, the improvements that are obtained under the syntactic profile lead to the 3COSMUL metric obtaining better overall results.

Figure 2 shows the trends of the total accuracy at different epochs for the various models using 3COSMUL (the trend obtained with 3COSADD is very similar). Here we can see how the use of high negative sampling can worsen performance, even causing the network to oscillate (W5N20) in order to better adapt to all the data. The choice of the negative sampling to be used should therefore be strongly linked to the choice of the window size as well as to the number of training epochs.

Continuing the training of the two worst models up to the 50th epoch, it is observed (Table 2) that they are still able to reach the performances of the other models. The W10N20 model at the 50th epoch even proves to be better than all the other previous models, becoming the reference model for subsequent comparisons. As the various epochs change (Fig. 3a) it appears to have the same oscillatory pattern observed previously, albeit with only one oscillation given the greater window size. This model is available at: https://mlunicampania.gitlab.io/italian-word2vec/.

Various tests were also conducted on CBOW models, which however proved to be in general significantly lower than Skip-gram models. Figure 3b shows, for example, the accuracy trend for a CBOW model with a window equal to 10 and negative sampling equal to 20, which on 50 epochs reaches only 37.20% of total accuracy (with 3COSMUL metric).

Table 1 Accuracy at the 20th epoch for the 6 Skip-gram models analysed when the W dimension of the window and the N value of negative sampling change

		3COSADD			3COSMUL		
		Semantic (%)	Syntactic (%)	Total (%)	Semantic (%)	Syntactic (%)	Total (%)
W = 5	N = 5	**40.93**	**38.85**	**39.85**	39.62	37.78	38.67
	N = 10	52.99	45.57	49.14	**53.35**	**46.71**	**49.91**
	N = 20	**53.66**	44.06	48.68	53.56	**45.80**	**49.53**
W = 10	N = 5	**53.63**	45.55	49.44	53.39	**46.79**	**49.97**
	N = 10	**55.76**	45.92	50.66	55.56	**47.54**	**51.40**
	N = 20	**45.56**	**34.95**	**40.06**	44.35	33.52	38.74

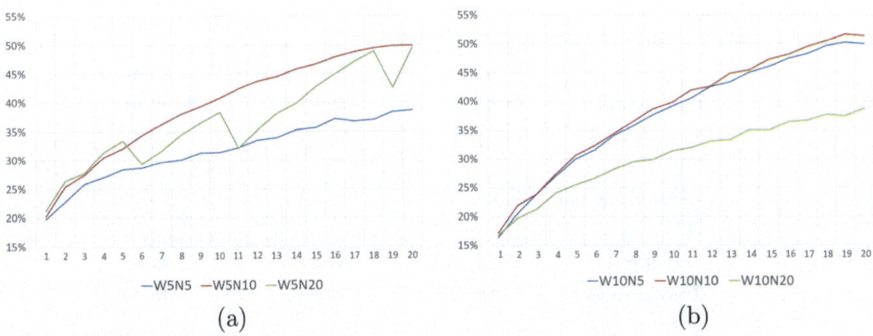

Fig. 2 Total accuracy using 3COSMUL at different epochs with negative sampling equal to 5, 10 and 20, where: **a** window is 5 and **b** window is 10

Table 2 Accuracy at the 50th epoch for the two worst Skip-gram models

	3COSADD			3COSMUL		
	Semantic (%)	Syntactic (%)	Total (%)	Semantic (%)	Syntactic (%)	Total (%)
W5N5	49.59	45.25	47.34	**49.78**	**46.84**	**48.26**
W10N20	**59.20**	46.98	52.86	59.07	**48.80**	**53.74**

4.2 Comparison with Other Models

Finally, a comparison was made between the Skip-gram model W10N20 obtained at the 50th epoch and the other two W2V in Italian present in the literature [10, 11]. The first test (Table 3) was performed considering all the analogies present, and therefore evaluating as an error any analogy that was not executable (as it related to one or more words absent from the vocabulary).

As it can be seen, regardless of the metric used, our model has significantly better results than the other two models, both overall and within the two macro-areas.

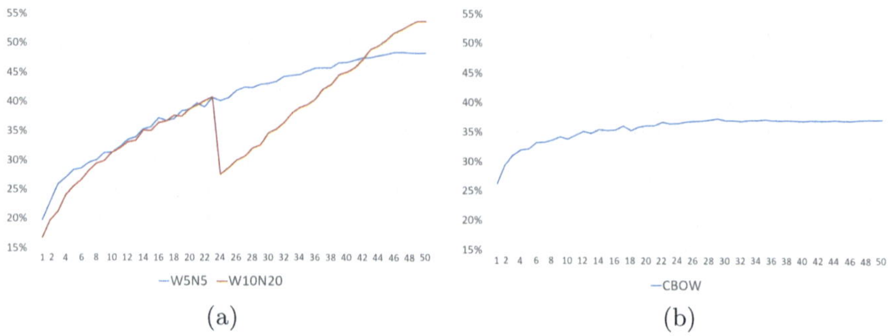

Fig. 3 Total accuracy using 3COSMUL up to the 50th epoch for: **a** the two worst Skip-gram models and **b** CBOW model with $W = 10$ and $N = 20$

Table 3 Accuracy evaluated on the total of all the analogies

		Semantic (%)	Syntactic (%)	Total (%)
3COSADD	**Our model**	**58.42**	**40.92**	**48.81**
	Tipodis model [10]	53.21	37.37	44.51
	Berardi's model [11]	48.81	32.62	39.91
3COSMUL	**Our model**	**58.31**	**42.51**	**49.62**
	Tipodis model [10]	55.56	39.60	46.79
	Berardi's model [11]	49.59	33.70	40.86

Table 4 Accuracy evaluated only on the analogies common to both vocabularies

		Semantic (%)	Syntactic (%)	Total (%)
3COSADD	**Our model**	**59.20**	**47.95**	**53.43**
	Tipodi's model [10]	53.92	43.94	48.81
3COSMUL	**Our model**	**59.08**	**49.75**	**54.30**
	Tipodi's model [10]	56.30	46.57	51.31

Furthermore, the other two models seem to be more subject to the metric used, perhaps due to a stabilization not yet reached for the few training epochs.

For a complete comparison, both models were also tested considering only the subset of the analogies in common with our model (i.e. eliminating from the test all those analogies that were not executable by one or the other model). Tables 4 and 5 again highlight the marked increase in performance of our model compared to both.

Table 5 Accuracy evaluated only on the analogies common to both vocabularies

		Semantic (%)	Syntactic (%)	Total (%)
3COSADD	**Our model**	**59.20**	**48.48**	**53.73**
	Berardi's model [11]	49.45	38.73	43.98
3COSMUL	**Our model**	**59.08**	**50.35**	**54.63**
	Berardi's model [11]	50.25	40.00	45.02

5 Conclusion

In this work we have analysed the Word2Vec model for Italian Language obtaining a substantial increase in performance respect to other two models in the literature (and despite the fixed size of the embedding). These results, in addition to the number of learning epochs, are probably also due to the different phase of data pre-processing, very carefully executed in performing a complete cleaning of the text and above all in substituting the numerical values with a single particular token. We have observed that the number of epochs is an important parameter and its increase leads to results that rank our two worst models almost equal, or even better than others.

Changing the number of epochs, in some configurations, creates an oscillatory trend, which seems to be linked to a particular interaction between the window size and the negative sampling value. In the future, thanks to the collaboration in the Laila project, we intend to expand the dataset by adding more user chats. The objective will be to verify if the use of a less formal language can improves accuracy in the syntactic macro-area.

References

1. Bengio, Y., Ducharme, R., Vincent, P.: A neural probabilistic language model. In: Leen, T.K., Dietterich, T.G., Tresp, V. (eds.) Advances in Neural Information Processing Systems, vol. 13, pp. 932–938. MIT Press (2001)
2. Harris, Z.: Distributional structure. Word **10**(23), 146–162 (1954)
3. Firth, J.R.: A synopsis of linguistic theory 1930–55, pp. 1–32 (1952–59) (1957)
4. Almeida, F., Xexéo, G.: Word Embeddings: A Survey (2019)
5. Zhang, Y., Rahman, M.M., Braylan, A., Dang, B., Chang, H.-L., Kim, H., McNamara, Q., Angert, A., Banner, E., Khetan, V., McDonnell, T., Nguyen, A.T., Xu, D., Wallace, B.C., Lease, M.: Neural information retrieval: a literature review. In: CoRR (2016). arXiv:1611.06792
6. Mikolov, T., Chen, K., Corrado, G., Dean, J.: Efficient estimation of word representations in vector space. In: CoRR (2013). arXiv:1301.3781
7. Pennington, J., Socher, R., Manning, C.: Glove: Global Vectors for Word Representation, vol. 14, pp. 1532–1543 (2014)
8. Schnabel, T., Labutov, I., Mimno, D., Joachims, T.: Evaluation methods for unsupervised word embeddings. In: Proceedings of the 2015 Conference on Empirical Methods in Natural Language Processing, Lisbon, Portugal, pp. 298–307. Association for Computational Linguistics (2015)

9. Bakarov, A.: A survey of word embeddings evaluation methods (2018)
10. Tripodi, R., Li Pira, S.: Analysis of Italian word embeddings (2017)
11. Berardi, G., Esuli, A., Marcheggiani, D.: Word embeddings go to Italy: a comparison of models and training datasets. In: CEUR Workshop Proceedings, vol. 1404 (2015)
12. Mikolov, T., Sutskever, I., Chen, K., Corrado, G., Dean, J.: Distributed representations of words and phrases and their compositionality. Adv. Neural Inf. Process. Syst. **26**, 10 (2013)
13. Levy, O., Goldberg, Y.: Linguistic regularities in sparse and explicit word representations. In: Proceedings of the Eighteenth Conference on Computational Natural Language Learning, pp. 171–180, June 2014

On the Role of Time in Learning

Alessandro Betti and Marco Gori

Abstract By and large the process of learning concepts that are embedded in time is regarded as quite a mature research topic. Hidden Markov models, recurrent neural networks are, amongst others, successful approaches to learning from temporal data. In this paper, we claim that the dominant approach minimizing appropriate risk functions defined over time by classic stochastic gradient might miss the deep interpretation of time given in other fields like physics. We show that a recent reformulation of learning according to the principle of Least Cognitive Action is better suited whenever time is involved in learning. The principle gives rise to a learning process that is driven by differential equations, that can somehow describe the process within the same framework as other laws of nature.

1 Introduction

The process of learning has been recently formulated under the framework of laws of nature derived from variational principle [1]. While the paper addresses some fundamental issues on the links with mechanics, a major open problem is the one connected with the satisfaction of the boundary conditions of the Euler-Lagrange equations of learning.

This paper springs out from recent studies especially on the problem of learning visual features [2–4] and it is also stimulated by a nice analysis on the interpretation of Newtonian mechanics equations in the variational framework [5]. It is pointed out that the formulation of learning as Euler-Lagrange (EL) differential equation is

A. Betti
University of Florence, Florence, Italy
e-mail: alessandro.betti@unifi.it

A. Betti · M. Gori (✉)
SAILab, University of Siena, Siena, Italy
e-mail: marco@diism.unisi.it
URL: http://sailab.diism.unisi.it

A. Esposito et al. (eds.), *Progresses in Artificial Intelligence and Neural Systems*, Smart Innovation, Systems and Technologies 184,
https://doi.org/10.1007/978-981-15-5093-5_14

remarkably different with respect to classic gradient flow. The difference is mostly originated from the continuous nature of time; while gradient flow has a truly algorithmic flavor, the EL-equations of learning, which are the outcome of imposing a null variation of the action, can be interpreted as laws of nature.

The paper shows that learning is driven by fourth-order differential equations that collapses to second-order under an intriguing interpretation connected with the mentioned result given in [5] concerning the arising of Newtonian laws.

2 Euler-Lagrange Equations

Consider an integral functional $F\colon X \to \overline{\mathbb{R}} := \mathbb{R} \cup \{-\infty, +\infty\}$ of the following form

$$\mathcal{A}(q) := \int_{t_1}^{t_N} L(t, q(t), \dot{q}(t))\, dt \tag{1}$$

where $L \in C^1(\mathbb{R} \times \mathbb{R}^n \times \mathbb{R}^n)$ maps a point (t, q, p) into the real number $L(t, q, p)$ and $t \mapsto q(t) \in \mathbb{R}^n$ is a map of X. Consider a partition $t_1 < t_2 < \cdots < t_N$ of the interval $[t_1, t_N]$ into $N-1$ subintervals of length ε. Given a function q one can identify the point $(q(t_1), q(t_2), \ldots q(t_N)) \in \mathbb{R}^N$, and in general one can define the subset of $\mathbb{R}^N$

$$X_\varepsilon := \{(q(t_1), q(t_2), \ldots q(t_N)) \in \mathbb{R}^N : q \in X\}.$$

Now consider the and consider the following "approximation" $\mathcal{A}_\varepsilon\colon X_\varepsilon \to \mathbb{R}$ of the functional integral F:

$$\mathcal{A}_\varepsilon(x_1, \ldots, x_N) := \varepsilon \sum_{k=1}^{N-1} L(k, x_k, \Delta_\varepsilon x_k),$$

where $\Delta_\varepsilon x_k = (x_{k+1} - x_k)/\varepsilon$. The stationarity condition on $\mathcal{A}_\varepsilon$ is $\nabla \mathcal{A}_\varepsilon(x) = 0$, thus we have

$$\nabla_i \mathcal{A}_\varepsilon(x) = \varepsilon \nabla_i [L(i-1, x_{i-1}, \Delta_\varepsilon x_{i-1}) + L(i, x_i, \Delta_\varepsilon x_i)].$$

Using the fact that $\partial(\Delta_\varepsilon x_i)/\partial x_i = -1/\varepsilon$ and $\partial(\Delta_\varepsilon x_{i-1})/\partial x_i = 1/\varepsilon$ we get

$$\begin{aligned}\nabla_i \mathcal{A}_\varepsilon(x) &= \varepsilon[L_p(i-1, x_{i-1}, \Delta_\varepsilon x_{i-1})\varepsilon^{-1} + L_q(i, x_i, \Delta_\varepsilon x_i) - L_p(i, x_i, \Delta_\varepsilon x_i)\varepsilon^{-1}] \\ &= \varepsilon L_q(i, x_i, \Delta_\varepsilon x_i) - \varepsilon \frac{L_p(i, x_i, \Delta_\varepsilon x_i) - L_p(i-1, x_{i-1}, \Delta_\varepsilon x_{i-1})}{\varepsilon}.\end{aligned} \tag{2}$$

This means that the condition $\nabla \mathcal{A}_\varepsilon(x) = 0$ implies

$$L_q(i, x_i, \Delta_\varepsilon x_i) - \Delta_\varepsilon L_p(i-1, x_{i-1}, \Delta_\varepsilon x_{i-1}) = 0, \qquad i = 2, \ldots, N-1, \tag{3}$$

where, consistently with our previous definition we are assuming that $\Delta_\varepsilon L_p(i - 1, x_{i-1}, \Delta_\varepsilon x_{i-1}) = [L_p(i, x_i, \Delta_\varepsilon x_i) - L_p(i-1, x_{i-1}, \Delta_\varepsilon x_{i-1})]/\varepsilon$.

This last equation is indeed the discrete counterpart of the Euler-Lagrange equations in the continuum:

$$L_q(t, u(t), \dot{u}(t)) - \frac{d}{dt} L_p(t, u(t), \dot{u}(t)) = 0, \qquad t \in [t_1, t_N]. \tag{4}$$

The discovery of stationary points of the cognitive action defined by Eq. 1 is somewhat related with the gradient flow that one might activate to optimize $\mathcal{A}$, namely by the classic updating rule

$$X_\epsilon \leftarrow X_\epsilon - \eta \nabla \mathcal{A}_\epsilon. \tag{5}$$

This flow is clearly different with respect to Eq. 3 (see also its continuous counterpart (4)). Basically, while the Euler-Lagrange equations yield an updating computation model of x_i, the gradient flow moves X_ϵ.

3 A Surprising Link with Mechanics

Let us consider the action

$$\mathcal{A} = \int_0^T dt\, h(t) \bar{L}(x(t), q(t), \dot{q}(t)). \tag{6}$$

The Euler-Lagrange equations are

$$h\bar{L}_q - \dot{h}\bar{L}_{\dot{q}} - h\frac{d}{dt}\bar{L}_{\dot{q}} = 0. \tag{7}$$

Since $h > 0$ we have

$$\frac{d}{dt}\bar{L}_{\dot{q}} + \frac{\dot{h}}{h}\bar{L}_{\dot{q}} - \bar{L}_q = 0. \tag{8}$$

In case we make no assumption on the variation then these equations must be joined with the boundary condition $\left[hL_{\dot{q}}\right]_0^T = 0$. Now suppose $\bar{L} = T + \gamma V$, with $\gamma \in \mathbb{R}$. Then Eq. 8 becomes

$$\frac{d}{dt}T_{\dot{q}} + \frac{\dot{h}}{h}T_{\dot{q}} - \gamma V_q = 0. \tag{9}$$

The Lagrangian $\bar{L} = T + \gamma V$, with $T = \frac{1}{2}m\dot{q}^2$ and $\gamma = -1$, and $h(t) = e^{\theta t}$, is the one used in mechanics, which returns the Newtonian equations

$$m\ddot{q} + \theta\dot{q} + V_q = 0$$

of the damping oscillator. We notice in passing that this equation arises when choosing the classic action from mechanics, which does not seem to be adequate for machine learning since the potential (analogous to the loss function) and the kinetic energy (analogous to the regularization term) come with different sign. It is also worth mentioning that the trivial choice $h = 1$ yields a pure oscillation with no dissipation, which is on the opposite the fundamental ingredient of learning.

This Lagrangian, however, does not convey a reasonable interpretation for a learning theory, since one very much would like $\gamma > 0$, so as γ^{-1} could be nicely interpreted as a temporal regularization parameter. Before exploring a different interpretation, we notice in passing that large values of θ/m, which corresponds with strong dissipation on small masses yields the gradient flow

$$\dot{q} = -\frac{1}{\theta}V_q$$

4 Laws of Learning and Gradient Flow

While the discussion in the previous section provides a somewhat surprising links with mechanics, the interpretation of the learning as a problem of least actions is not very satisfactory since, just like in mechanics, we only end up into stationary points of the actions that are typically saddle points.

We will see that an appropriate choice of the Lagrangian function yields truly laws of nature where Euler-Lagrange equations turns out to minimize corresponding actions that are appropriate to capture learning tasks. We consider kinetic energies that also involve the acceleration and two different cases which depend on the choice of h. The new action is

$$\mathcal{A}_2 = \int_0^T dt\ L(t, q(t), \dot{q}(t), \ddot{q}(t)), \tag{10}$$

where $L = h\bar{L}$. In the continuum setting, the corresponding Euler-Lagrange equations can be determined by considering the variation associated with $q \rightsquigarrow q + sv$, where v is a variation and $s \in \mathbb{R}$. We have

$$\delta\mathcal{A}_2 = s \int_0^T dt\ (L_q v + L_p \dot{v} + L_a \ddot{v}). \tag{11}$$

If we integrate by parts, we get

$$\int_0^T dt\ L_p \dot{v} = -\int_0^T dt\ v \frac{d}{dt} L_p + [v L_p]_0^T$$
$$\int_0^T dt\ L_a \ddot{v} = -\int_0^T dt\ \dot{v} \frac{d}{dt} L_a + [\dot{v} L_a]_0^T = \int_0^T dt\ v \frac{d^2}{dt^2} L_a - \left[v \frac{d}{dt} L_a\right]_0^T + [\dot{v} L_a]_0^T,$$

and, therefore, the variation becomes

$$\delta \mathcal{A}_2 = s \int_0^T dt\ v\left(\frac{d^2}{dt^2} L_a - \frac{d}{dt} L_p + L_q\right) + \left[v\left(L_p - \frac{d}{dt} L_a\right)\right]_0^T + [\dot{v} L_a]_0^T = 0.$$

Now, suppose we give the initial conditions. In that case we can promptly see that this is equivalent with posing $v(0) = 0$ and $\dot{v}(0) = 0$. Hence, we get the Euler-Lagrange equation when posing

$$v(T)\left(L_p\Big|_{t=T} - \frac{d}{dt} L_a\Big|_{t=T}\right) + \dot{v}(T) L_a\Big|_{t=T} = 0.$$

Now if we choose $v(t)$ as a constant we immediately get

$$L_p\Big|_{t=T} - \frac{d}{dt} L_a\Big|_{t=T} = 0, \tag{12}$$

while if we choose v as an affine function, when considering the above condition we get

$$L_a\Big|_{t=T} = 0. \tag{13}$$

Finally, the stationary point of the action corresponds with the Euler-Lagrange equations

$$\frac{d^2}{dt^2} L_a - \frac{d}{dt} L_p + L_q = 0, \tag{14}$$

that holds along with Cauchy initial conditions on $q(0)$, $\dot{q}(0)$ and boundary conditions (12) and (13).

Now, let us consider the case in which $L = h\bar{L}$. The Euler-Lagrange equations become

$$\frac{d^2}{dt^2} \bar{L}_a + 2\frac{\dot{h}}{h} \frac{d}{dt} \bar{L}_a + \frac{\ddot{h}}{h} \bar{L}_a - \frac{\dot{h}}{h} \bar{L}_p - \frac{d}{dt} \bar{L}_p + L_q = 0. \tag{15}$$

If we consider again the case $\bar{L} = T + \gamma V$ we get

$$\frac{d^2}{dt^2}T_a + 2\frac{\dot{h}}{h}\frac{d}{dt}T_a + \frac{\ddot{h}}{h}T_a - \frac{\dot{h}}{h}T_p - \frac{d}{dt}T_p + \gamma V_q = 0. \tag{16}$$

Now we consider the kinetic energy associated with the differential operator $P = \alpha_1 \frac{d}{dt} + \alpha_2 \frac{d^2}{dt^2}$

$$T = \frac{1}{2\theta^2}(Pq)^2 = \frac{1}{2\theta^2}(\alpha_1\dot{q} + \alpha_2\ddot{q})^2 = \frac{1}{2}\frac{\alpha_1^2}{\theta^2}\dot{q}^2 + \frac{\alpha_1\alpha_2}{\theta^2}\dot{q}\ddot{q} + \frac{1}{2}\frac{\alpha_2^2}{\theta^2}\ddot{q}^2 \tag{17}$$

Let us consider the following two different cases of $h(t)$. In both cases, they convey the unidirectional structure of time.

i. $h(t) = e^{\theta t}$
In this case, when plugging the kinetic energy in Eq. 17 into Eq. 16 we get

$$\frac{1}{\theta^2}q^{(4)} + \frac{2}{\theta}q^{(3)} + \frac{\alpha_1\alpha_2\theta + \alpha_2^2\theta^2 - \alpha_1^2}{\alpha_2^2\theta^2}\ddot{q} + \frac{\alpha_1\alpha_2\theta^2 - \alpha_1^2\theta}{\alpha_2^2\theta^2}\dot{q} + \frac{\gamma}{\alpha_2^2}V_q = 0. \tag{18}$$

These equations hold along with Cauchy conditions and boundary conditions given by Eq. 12 and 13, that turn out to be

$$\frac{\alpha_1^2}{\theta^2}\dot{q}(T) - \frac{\alpha_2^2}{\theta^2}q^{(3)}(T) = 0 \tag{19}$$

$$\frac{\alpha_1\alpha_2}{\theta^2}\dot{q}(T) + \frac{\alpha_2^2}{\theta^2}\ddot{q}(T) = 0. \tag{20}$$

A possible satisfaction is $\dot{q}(0) = \ddot{q}(0) = q^{(3)}(0) = 0$. Notice that as $\theta \to \infty$ the Euler-Lagrange Eq. 18 reduces to

$$\ddot{q} + \frac{\alpha_1}{\alpha_2}\dot{q} + \frac{\gamma}{\alpha_2^2}V_q = 0. \tag{21}$$

and the corresponding boundary conditions are always verified.

ii. $h(t) = e^{-t/\epsilon}$
Let us assume that $\beta = 0$ in the kinetic energy (17) and $h(t) = e^{-t/\epsilon}$. In particular we consider the action

$$\mathcal{A} = \int_0^T dt\, e^{-t/\epsilon}\left(\frac{1}{2}\epsilon^2\rho\ddot{q}^2 + \frac{1}{2}\epsilon\nu\dot{q}^2 + V(q,t)\right) \tag{22}$$

In this case the Lagrange equations turn out to be

$$\epsilon^2\rho q^{(4)} - 2\epsilon\rho q^{(3)} + (\rho - \epsilon\nu)\ddot{q} + \nu\dot{q} + \gamma V_q = 0, \tag{23}$$

along with the boundary conditions

$$\epsilon^2 \rho \dddot{q}(T) = 0 \tag{24}$$

$$\epsilon \nu \ddot{q}(T) - \rho \epsilon^2 q^3(T) = 0. \tag{25}$$

Interesting, as $\epsilon \to 0$ the Euler-Lagrange equations become:

$$\rho \ddot{q} + \nu \dot{q} + \gamma V_q = 0, \tag{26}$$

where the boundary conditions are always satisfied.

Remark 1 Notice that while we can choose the parameters in such a way that Eq. 18 is stable, the same does not hold for Eq. 23. Interestingly, stability can be gained for $\epsilon = 0$, which is corresponds with a singular solution. Basically if we denote by q_ϵ the solution associated with $\epsilon \in \mathbb{R}$, we have that q_ϵ does not approximate q corresponding at $\epsilon = 0$ in case in which we can choose arbitrarily large domains $[0, T]$.

5 Conclusions

While machine learning is typically framed in the statistical setting, in this case time is exploited in such a way that one relies on a sort of underlying ergodic principle according to which statistical regularities can be captured in time. This paper shows that the continuous nature of time gives rise to computational models of learning that can be interpreted as laws of nature. Unlike traditional stochastic gradient, the theory suggests that, just like in mechanics, learning is driven by the Euler-Lagrange equations that minimize a sort of functional risk. The collapsing from forth- to second-order differential equations opens the doors to an in-depth theoretical and experimental investigation.

Acknowledgments We thank Giovanni Bellettini for insightful discussions.

References

1. Betti, A., Gori, M.: The principle of least cognitive action. Theor. Comput. Sci. **633**, 83–99 (2016)
2. Betti, A., Gori, M.: Convolutional networks in visual environments. In: CoRR (2018). arXiv:1801.07110
3. Betti, A., Gori, M., Melacci, S.: Cognitive action laws: the case of visual features. In: CoRR (2018). arXiv:1808.09162
4. Betti, A., Gori, M., Melacci, S.: Motion invariance in visual environments. In: CoRR (2018). arXiv:1807.06450
5. Liero, M., Stefanelli, U.: A new minimum principle for Lagrangian mechanics. J. Nonlinear Sci. **23**, 179–204 (2013)

Preliminary Experiments on Thermal Emissivity Adjustment for Face Images

Marcos Faundez-Zanuy, Xavier Font-Aragones, and Jiri Mekyska

Abstract In this paper we summarize several applications based on thermal imaging. We emphasize the importance of emissivity adjustment for a proper temperature measurement. A new set of face images acquired at different emissivity values with steps of 0.01 is also presented and will be distributed for free for research purposes. Among the utilities, we can mention: (a) the possibility to apply corrections once an image is acquired with a wrong emissivity value and it is not possible to acquire a new one; (b) privacy protection in thermal images, which can be obtained with a low emissivity factor, which is still suitable for several applications, but hides the identity of a user; (c) image processing for improving temperature detection in scenes containing objects of different emissivity.

1 Introduction

In the past we have used thermal images for a wide range of applications, including face biometric recognition, providing a new database freely available to the scientific community [1–3], hand morphology biometric recognition, including a hand image database distributed for free too [4–6], biomedical application for tuberculosis detection using tuberculine test and thermal imaging [7], and facial emotion recognition using thermal imaging in an induced emotion database [8]. We have also performed studies about the focusing of thermal images [9] and the fusion of different images containing objects at different focal distances [10, 11].

In this paper we deal with a new research topic, which is the emissivity configuration in a thermal camera, and the possibility of several applications.

M. Faundez-Zanuy (✉) · X. Font-Aragones
ESUP Tecnocampus (Pompeu Fabra University), Av. Ernest Lluch 32, 08302 Mataró, Spain
e-mail: faundez@tecnocampus.cat

J. Mekyska
Department of Telecommunications, Brno University of Technology, Brno, Czech Republic

A. Esposito et al. (eds.), *Progresses in Artificial Intelligence and Neural Systems*, Smart Innovation, Systems and Technologies 184,
https://doi.org/10.1007/978-981-15-5093-5_15

In order to obtain a correct temperature measurement, the emissivity adjustment of the thermal camera must be properly setup. Otherwise, wrong measurements are obtained. Table 1 represents some emissivity values [12, 13]. The problem appears when several objects of interest with different emissivity are inside the scene. In this case, only the temperature of one of them can be directly estimated with a single thermal camera image.

In this paper we present a set of images acquired from a fixed scene varying the emissivity configuration of the camera. With a set of images acquired with different emissivity configuration, several research possibilities exist:

(a) A combined image can be obtained with correct temperatures for all the objects inside the scene despites their different emissivity values. This can be achieved by means of image processing in a procedure similar to our previous work where we obtained a focused image containing objects at very different focal distance, being each object focused properly in a different image [10].
(b) When dealing with human images privacy problems could appear [14]. However, in some applications such as activity surveillance of elder people inside their home, fall down detection, etc., a very precise detail is not necessary. A coarse description would be correct, for instance, to detect fall downs in the

Table 1 Emissivity values for several materials

Material	Emissivity
Aluminum foil	0.03
Aluminum, anodized	0.9
Asphalt	0.88
Brick	0.90
Concrete, rough	0.91
Copper, polished	0.04
Copper, oxidized	0.87
Glass, smooth (uncoated)	0.95
Human skin	0.98
Ice	0.97
Limestone	0.92
Marble (polished)	0.89 to 0.92
Paint (including white)	0.9
Paper, roofing or white	0.88 to 0.86
Plaster, rough	0.89
Silver, polished	0.02
Silver, oxidized	0.04
Snow	0.8 to 0.9
Transition metal Disilicides (e.g. MoSi2 or WSi2)	0.86 to 0.93
Water, pure	0.96

shower/bathroom using thermal images. In this scenario, bad emissivity adjustment could be an alternative to hide the details of the face/body in order to protect the user's privacy. Especially in a situation where the user is typically naked.

(c) When an image is acquired at wrong emissivity value and there is no possibility to acquire another one with proper settings, some kind of compensation/transformation can be done in order to correct the wrong adjustment effect. For this purpose, a set of images of the same object acquired at different emissivity values can help to establish the mathematical relation to perform this compensation.

In this paper we have worked with object and face images, and special effort is put on this last research possibility.

The paper is organized as follows: Sect. 2 describes the experiments and databases acquired. Section 3 summarizes the main conclusions.

2 Expertiments

We have performed a first experiment with human faces, where we have acquired a static individual (his head is resting on the wall in order to make it easy to stay in a fix position for the whole set of acquisitions).

Figure 1 shows the relation between temperatures acquired in the middle of the eyebrows of a face image for different emissivity values. The images have been acquired with a TESTO 880 camera, which provides a sensibility of 100 mK and a

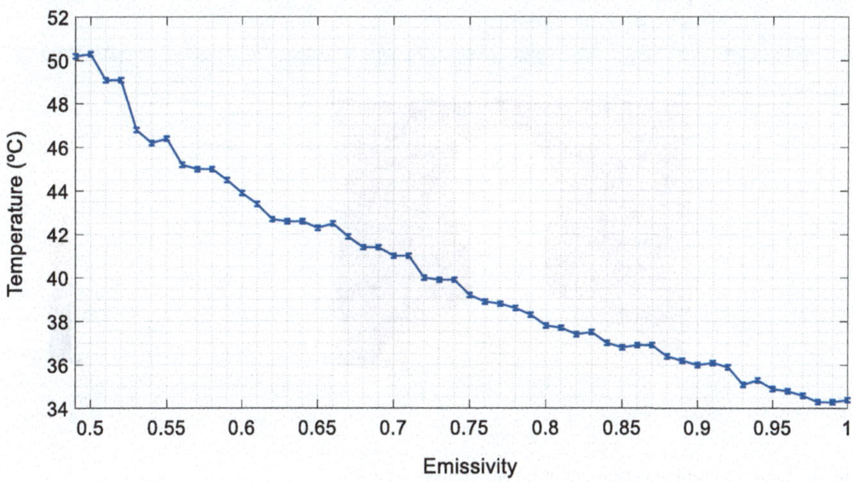

Fig. 1 Relation between temperatures acquired in the middle of the eyebrows of a face image for different emissivity values

resolution of 160 × 120 pixels without interpolation. This has been done by manual measurements using IRsoft from TESTO freely available for download from TESTO. Figure 2 shows the face image acquired at the correct emissivity (0.98) while Fig. 3 shows a wrong set up for human skin (emissivity = 0.62).

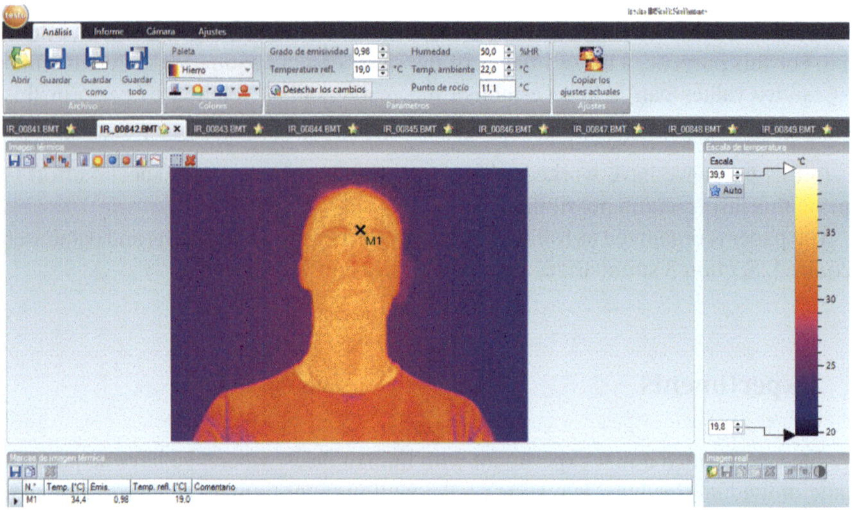

Fig. 2 Temperature measurement at correct emissivity value

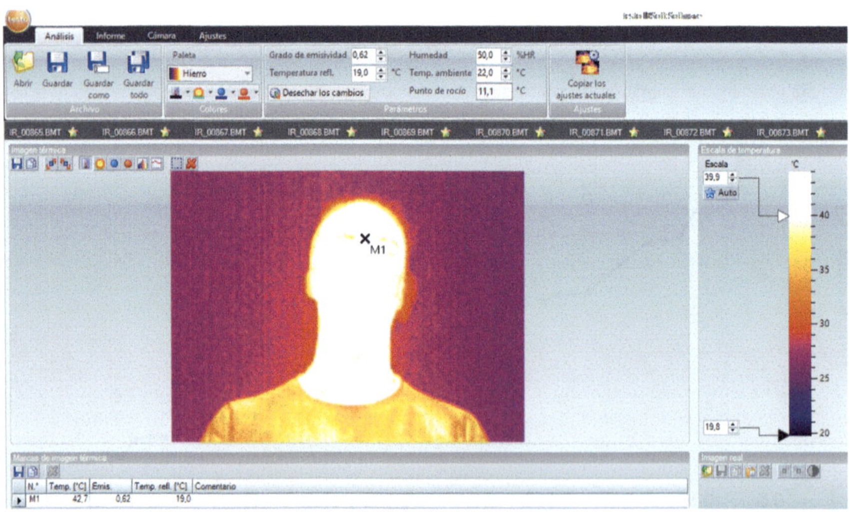

Fig. 3 Temperature measurement at wrong emissivity value

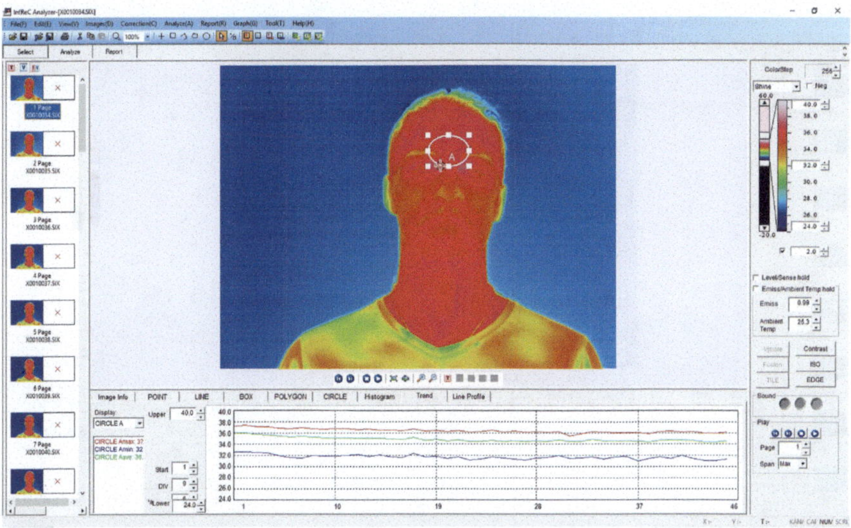

Fig. 4 Temperature measurement with NEC software in a sequence of 46 snapshots acquired at different emissivity values

We also acquired a set of images of the same subject with a NEC H2640 camera. NEC H2640 provides a 640 × 480 pixels resolution without interpolation and resolution 0.06 °C or better (at 30 °C, 30 Hz). In this case, NEC provides proprietary software that requires a license key. This software permits an automatic analysis of the whole set of images. Figure 4 shows a screen shot of NEC software. On the bottom there is the automatic result of the analysis of the temperature in the ellipsoid in the centre of the eyebrows. From this analysis it is evident that fixing the emissivity to a value to 0.99 we always obtain the same result (almost flat plot for average, low and high temperature values in each acquisition). Thus, one important conclusion is that emissivity is important during visualization but not during acquisition.

In a second experiment we acquired a set of objects of different material, which has different emissivity (see Table 1). Figure 5 shows the acquired scene. Different objects should be at same temperature (the room temperature) but they are visualized with different colour because of the different emissivity (wood, plastic, wax, chalk, metal). In this case the temperature analysis is performed inside the large square that covers almost the whole scene. This set of images can be used for future research analysis on measurements of objects with different emissivity at the same temperature. While a single image will be enough for the purpose, the existence of several consecutive acquisitions with small variations can be useful for algorithm development, which will not have to be based on a single image.

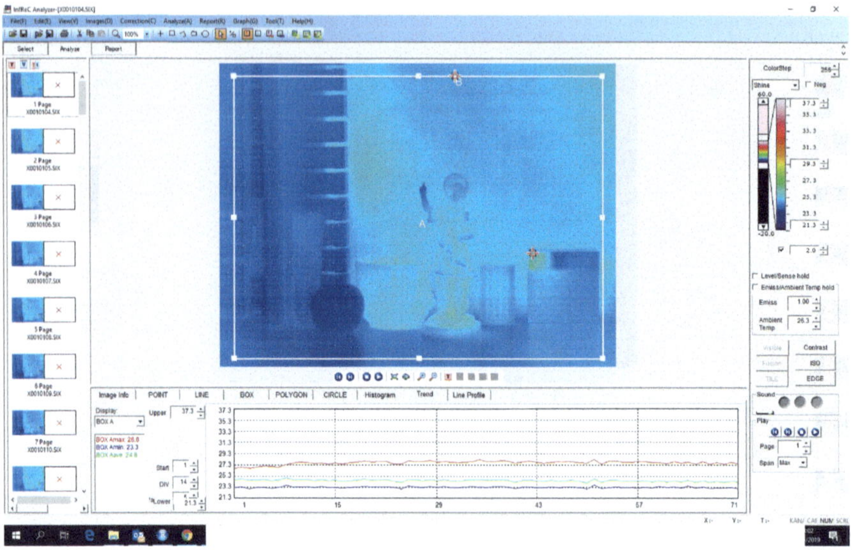

Fig. 5 An image of a set of 71 images acquired of a fix scene modifying the emissivity configuration in each image

3 Conclusions

In this paper we dealt with two thermal cameras and their own infrared image processing software. These cameras have been a TESTO 880 and NEC H2640. With these cameras we acquired three sets of images where each image has a different emissivity configuration. Based on these sequences we can establish:

(a) The emissivity value has influence during visualization of the image but not during acquisition. During the acquisition the image file stores the values provided by the sensor and analog to digital converter. During visualization these raw values are converted into pseudocolors using the emissivity information provided by the analysis software, which can be the same fix during acquisition or different.
(b) User privacy can be protected in some way visualizing the image with a wrong emissivity configuration. However, privacy can be compromised once the adjustment returns to correct value. In this context privacy is considered as the possibility of indentifying the user inside the image and/or the possibility "to see" the shape and lengths of parts of his body.
(c) Measurement of different objects, each one with different emissivity, is a problem that cannot be addressed with current thermal cameras. This is due to the fact that emissivity is adjusted for the whole image. Some algorithm and software has to be developed for segmenting objects, assigning emissivities to each of them, and measuring proper values. This remains an open research issue.

Acknowledgements This work has been supported by FEDER and MEC, TEC2016-77791-C4-2-R, PID2019-109099RB-C41 and LO1401.

References

1. Espinosa, V., Faundez-Zanuy, M., Mekyska, J.: Beyond cognitive signals. Cogn. Comput. Springer. **3**, 374–381, June (2011)
2. Espinosa, V., Faundez-Zanuy, M., Mekyskya, J., Monte, E.: A criterion for analysis of different sensor combinations with an application to face biometrics. Cogn. Comp. **2**(3), 135–141. September (2010)
3. Mekyska, J., Espinosa-Duró V., Faundez-Zanuy, M.: Face segmentation: a comparison between visible and thermal images. IEEE 44th International Carnahan Conference on Security Technology ICCST 2010, San José, USA. 5–8 October (2010)
4. Faundez-Zanuy, M., Mekyska, J., Font-Aragonès, X.: A new hand image database simultaneously acquired in visible, near-infrared and thermal spectrums. Cogn. Comput. **6**(2), 230–240 (2014)
5. Font-Aragones, X., Faundez-Zanuy, M., Mekysk, J.: Thermal hand image segmentation for biometric recognition. IEEE Aerospace Elect. Sys. Mag. **28**(6), 4–14, June (2013)
6. Mekyska, J., Font, X., Faundez-Zanuy, M., Hernández-Mingorance, R., Morales, A., Ángel Ferrer-Ballester, M.: Thermal hand image segmentation for biometric recognition (pp. 26–30), 45th IEEE Carnahan Conference on Security Technology ICCST'2011, 18-21, Mataró October (2011)
7. Antonio Fiz, J., Lozano, M., Monte-Moreno, E., González-Martínez, A., Faundez-Zanuy, M., Becker, C., Rodriguez-Pons, L., Ruiz Manzano, J.: Tuberculine reaction measured by infrared thermography. Comp. Methods Prog. Biomed. **122**(2), 199–206, November (2015)
8. Esposito, A., Capuano, V., Mekyska, J., Faundez-Zanuy, M.: A naturalistic database of thermal emotional facial expressions and effects of induced emotions on memory (pp. 158–173), Proceedings of the 2011 International Conference on Cognitive Behavioural Systems. LNCS Springer-Verlag Berlin, Heidelberg ©2012 ISBN: 978-3-642-34583-8 Dresden, February (2011)
9. Faundez-Zanuy, M., Mekyska, J., Espinosa, V.: On the focusing of thermal images. Patt. Recog. Lett. **32**, 1548–1557, Elsevier, August (2011)
10. Benes, R., Dvorak, P., Faundez-Zanuy, M., Espinosa-Duro, V., Mekyska, J.: Multi-focus thermal image fusión. Patt. Rec. Lett. **34**(5), 536–544, 1 April (2013)
11. Espinosa-Duró, V., Faundez-Zanuy, M., Mekyska, J.: Contribution of the temperature of the objects to the problem of thermal imaging focusing. 46 ICCST'2012 (pp. 363–366). Boston, USA IEEE Catalog Number: CFP12ICR-USB ISBN: 978-1-4673-2449-6
12. Brewster, M. Quinn (1992). Thermal Radiative Transfer and Properties. John Wiley & Sons. p. 56. ISBN 9780471539827
13. Ashrae Handbook: Fundamentals - IP Edition. Atlanta: American Society of Heating, Refrigerating and Air-Conditioning Engineers (2009) ISBN 978–1-933742-56-4
14. Faundez-Zanuy, M.: Privacy issues on biometric systems. IEEE Aero. Elect. Sys. Mag **20**(2), 13–15, ISSN: 0885-8985. February (2005)

Psychological Stress Detection by 2D and 3D Facial Image Processing

Livia Lombardi and Federica Marcolin

Abstract This work aims to identify people psychological stress through the capture of micro modifications and motions within their facial expression. Exogenous and endogenous causes of stress, from environment and/or psychological conditions that could induce stress, have been reproduced in the experimental test involving real subjects, and their face expressions have been recorded by 2D and 3D image capturing tools to create a sample of emotional database. Successively, 2D and 3D analyses have been performed on recorded data according to the respective protocols, by deep learning and machine learning techniques, and a data driven model of the databases has been developed by neural network approach, to classify the psycho-behavioral answers to the different kinds of stress conditions induced on tested people. The ultimate aim of the study is to demonstrate the possibility to analyze data collected on participants from 2D shooting and 3D scans in a consistent way by means of deep learning and machine learning techniques, so that to provide a methodology to identify and classify some of the subtle facial micro-expressions of people involved in stressing activities.

1 Introduction

Emotions are an active part of humans being. Many philosophers of the ancient world undertook studies about physiological relation among emotional states, brain activities, body reactions. Aristotle himself established a classification of emotions showing that body behavior can be considered as a reaction to external impulses.

L. Lombardi (✉)
Università di Roma La Sapienza, Rome, Italy
e-mail: livia.n.lombardi@gmail.com

F. Marcolin
Politecnico di Torino, Turin, Italy
e-mail: federica.marcolin@polito.it

A. Esposito et al. (eds.), *Progresses in Artificial Intelligence and Neural Systems*, Smart Innovation, Systems and Technologies 184,
https://doi.org/10.1007/978-981-15-5093-5_16

"Emotions are pathos in which the body processes involvement is of common evidence" [1]. In 1884, William James presented his emotion theory rooted in the bodily experience. Thanks to Schachter et alii [2] and Ekman et alii [3] facial expressions have been recognized to be among the most evident body reactions that are able to catch the major part of emotions. These studies encoded emotions as a combination of facial muscle movements, no matter the gender, the culture and the age, giving rise to the FACS (Facial Action Coding System) [4]. The importance of emotion recognition systems pertains to a lot of application fields, from health care and safe protection, up to economic and advertisement improvement. However, the implementation is still a complex task: many researches have been performed [5] and several approaches have been developed but most of them relies on high quality, specialized datasets and a generalized approach is still envisaged to investigate on different sets of classified emotions. This work focuses on emotions that involve anxiety and stress states. Different approaches have been already proposed in this field, as well. However, all these works refer to specific targets: in [6] a workplace experiment is set to recognize the psychological chronic stress; in [7, 8] stress detection applications range from driver tiredness to astronaut performance. All the mentioned studies aim to extract emotions from biological signals (electromyography, skin temperature, etc.) and face expressions (FACS). In spite of the captured symptoms, the output of the investigations is highly affected by participant intrinsic features, rather than the emerging emotion itself.

The present study is seeking for those imperceptible nuances that emotions imprint on the face of every humans, aiming to discriminate among those reactions that mostly refer to psychological stress states. Computation techniques are based on Neural Networks tools for the 2D input signals, and Machine Learning tools for the 3D ones. Two different experimental trials have been performed. The former has been run on a first group of subjects and has led to preliminary analyses on 2D images with just two emotional states or set of classes. The latter—including a richer number of experimental tests—has been run on a second group of subjects, and has led to both, 2D and 3D analyses, with multiple emotional states, or set of classes. The first experimental trial has been based on a binary classification of stressed and non-stressed status recognition. Starting from these results, the second experimental trial introduces another emotional state, the happiness, to observe the behavior of subjects under different kind of classes, thus enabling a sort of judgement on system consistency. A further fallout of the second experimental trial is obtained by comparing the two observation runs, namely the 2D and 3D methods, which offer the opportunity to control the robustness of the experiment. In the following sections the study is deepened from the experimental set-up and samples collection to the data processing methods. In Sects. 3 and 4 are presented the overall findings and conclusions.

2 Method

2.1 Bi-dimensional Binary Classification

A small dataset based on 13 Caucasian voluntary subjects has been created, including both males and females ranging between 18 and 59 years old. Each subject has been required to undergo different tasks able to stimulate different emotions, giving rise to different kind of expressions, neutral and stressed. To record the experiments a smartphone camera has been used, set on a tripod located at 60 cm distance from the face of the person. To reduce the influence of exogenous variables, experiments have been conducted in a room with white walls and artificial light.

Because stress is a complex psychological state and can occur in a variety of ways, different stress-inducing experiments have been created. For instance, in the first test, named 'The Impossible game', subjects have been asked to play a seemingly simple game, featured with exponential increments in difficulty. In fact, failing in what seems a simple objective to the player's eyes, quite often it can induce stress on him/her. Every time the user fails, the game quickly restarts, leading the subject to make several mistakes in a short span of time. In most cases, such mistakes result in stressed emotions reflected on the subject face motions.

The second test stress has been induced thanks to the empathy with someone else who is undergoing a frightening experience. The subject has been asked to watch an aspiring 'Pilot training' video, where the pilot undergoes a series of physical and psychological tests, such as dealing with a high increase in acceleration, which ultimately results in the pilot to increase the heart rate and show scared facial expressions.

In the third test, subjects have been unexpectedly 'interviewed' and required to talk about themselves in a foreign language. In general, speaking about yourself when you are not ready, or you do not expect to do it, may induce stress. The stress increases if you are required to speak about yourself in a language that is different from your native one; in most cases, subjects think about what to say in their native language and translate it in another language. This generates stress throughout the extra work of the brain required to satisfy the expected performance.

For each of the different tasks a couple of frames from the video have been extracted, which display the most amount of stress according to literature and objective moments (e.g. game failures). Additionally, a picture of the subject in a neutral state (not stressed) has been captured, relaxing in front of the camera while waiting for the beginning of the different tasks.

2.2 Bi-dimensional Multi-class Classification

The experiment is composed of five tasks. The first three of them are the same of the 2D binary case, namely the 'The Impossible game', the 'Pilot training' video and the 'Interview' test. During the experimental setups, more 'happiness' displays have been introduced and neutral expressions captured in a better way. Indeed, this trial has been expanded to induce and capture happiness emotional states with the aim to develop a multi-classifier neural network implementation.

In the fourth test subjects have been asked to watch a theatre show, held by a famous Italian comedian. It has been observed that the video aroused amusement in almost all subjects.

Additionally, neutral expressions have been captured again by asking participants to explore a virtual reality environment, hosted on a website, composed of several rooms with access doors and short environment descriptions. At the end, participants have been asked to answer a questionnaire about the virtual environment experience and the self-assessed emotional state has been used to compare and verify the pictures that have been taken during the experiment, when the subject has been supposed to be in neutral state.

2.3 Bi-dimensional Data Processing

A suitable data cleaning has been adopted and performed on frames in two steps, using OpenCv Python interface version. At first, original images have been converted from RGB to grayscale ones. Then, face edges have been refined by the use of a sharpener kernel. The second step involved the use of the Dlib library to detect face windows and crop the relative face square frame to get a uniform dataset and display the highest number of facial features with low background noise. Subsequent operations are then performed on the cropped images stored as a new gallery. Through the previously mentioned pre-trained database, Dlib also provides a landmark predictor that enables the detection of 68 coordinates on the image, able to identify the fundamental face traits (lips, eye, brows etc.) pointed out in Fig. 1a and b. Face landmarks on the gallery have been stored in a json file (one for each subject). The landmark coordinates played a fundamental role to properly train the Neural Network for emotion detection and classification. Because of dataset characteristics, it was chosen to anchor all the landmarks to the central 'nose' one (landmark 34).

2.4 Three-dimensonal Multi-class Classification

Simultaneously to the previous 2D data logs, within the same experiments, every subject has also been recorded with a 3D scanner, set up on the laptop adopted for subjects to play games and watch videos. The sensor adopted is the *Intel Real Sense Camera SR300 module* with front facing orientation, coded light fast VGA 60fps technology and infrared laser projector system class.

The experimental trial is intrinsically made of the five tasks described above.

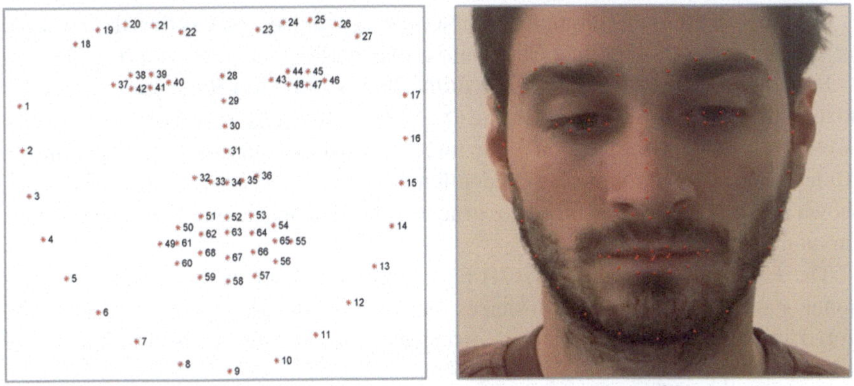

(a) (b)

Fig. 1 Representation of the landmark: **a** Dlib pattern; **b** landmark prediction implemented on the subject

2.5 Three-dimensional Data Processing

After registering the participants with the depth scanners, facial activities videos were analyzed with the Intel Real Sense Viewer Software. Depth video-frames obtained by the scanner were exported in both standards: ".raw" and ".png" files.

To avoid divergencies in the color maps (Fig. 2a) and the consequent loss of information in the grayscale conversion (Fig. 2b), a look-up table was applied to the image by the HSV (Hue Saturation Brightness) function.

In the first experimental trial, the analysis focused on depth information. However, the scanner output does not represent the effective 3D depth map of the subject face. For this reason, ".raw" files generated by the scanner were furtherly processed. Raw data were organized in 480×640 matrices with *UInt16* value and rotated 90 degrees.

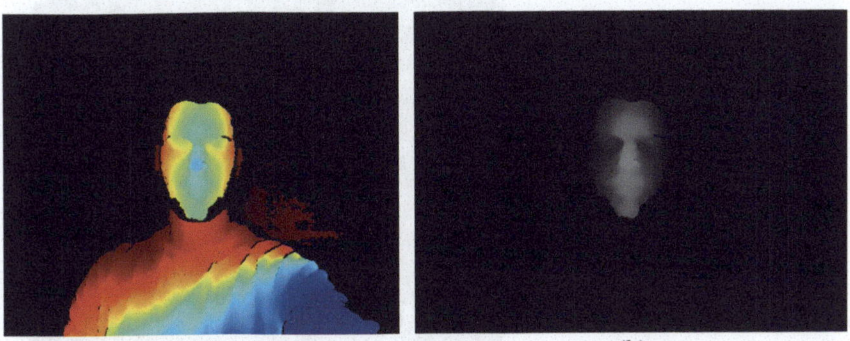

(a) (b)

Fig. 2 Example of the raw output: **a** divergent color scale; **b** grayscale conversion

Finally, only depth-data belonging to subject face (within the face surface level range) were extracted and stored in a NaN matrix, disregarding all other data (Fig. 3).

Face depth values, or masks, were furtherly processed to get the face boundaries boxes and create a reference 3D frame (X, Y, Z), so that a X, Y grid of the depth size (Z) is created. A revised version of the three-dimensional landmarking algorithm [9, 10] has been adapted to process the depth matrix above. Seventeen reference points, shown in Fig. 4, were automatically extracted in correspondence of significant surface curvatures.

The landmark vectors (V [1,17]) have been stored for each subject and emotional state. The Euclidean distances between landmarks were computed and stored in a matrix, containing all ‘distances’ clustered for each emotional state (‘neutral’,’happiness’, ‘stress’).

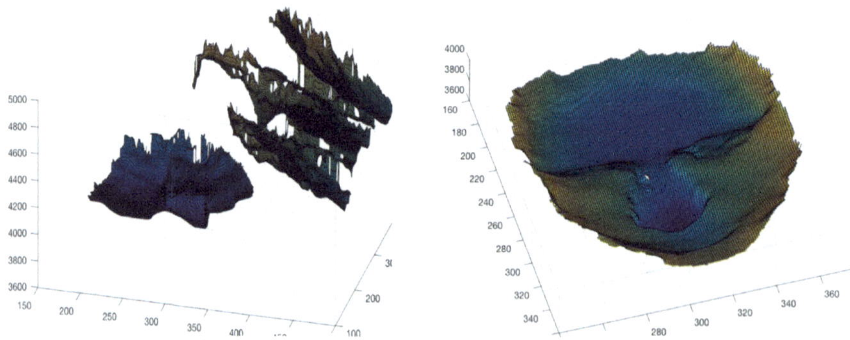

(a) (b)

Fig. 3 Example of Mask reconstruction: **a** 3D data extraction; **b** face boundaries

Fig. 4 The set of landmarks localized in this work

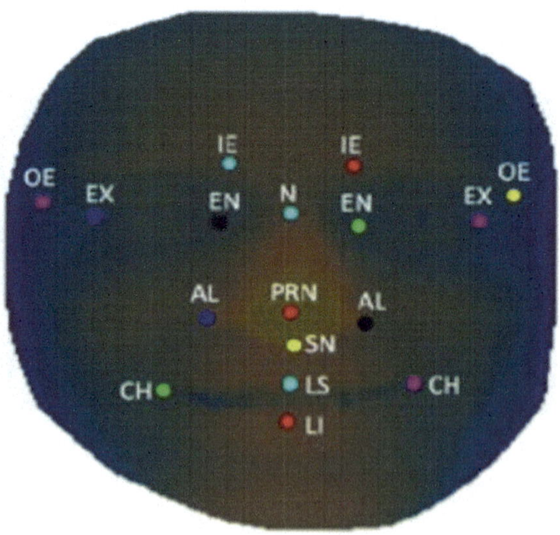

Snapshots of micro-expression, extremely fleeting, have been saved, according to the sample selection criterion described in Sect. 2.6.

Because of the 3D scanner position (on top of the laptop screen), sometimes screen vibrations blurred the captured data, resulting in the presence of holes in some of the depth scanned frames. Frames with holes have been discarded from the sample, causing some reduction in the training samples. For this reason, authors didn't adopt a deep neural network, but preferred to develop a machine learning supervised model. Data have been normalized to compare different models: logistic regression, decision trees, k-nearest neighbors and SVM.

2.6 Sample Selection

For each subject task a 2–3 min video of the requested performance has been captured. In the 2D analysis, the facial expressions in each frame have been analyzed to identify facial landmarks, i.e. typical points of the face. Starting from the video baseline, crucial frames have been selected that are consequent to the objective critical conditions. For instance, during the 'Impossible game' task, video frames of the facial expressions have been chosen in coincidence with the failure events in the game. In addition to objective selection methods, other symptomatic frames have been extracted when noteworthy micro-expression—Action Units—appeared, in accordance with characteristic traits identified by the literature [4] (Fig. 5)

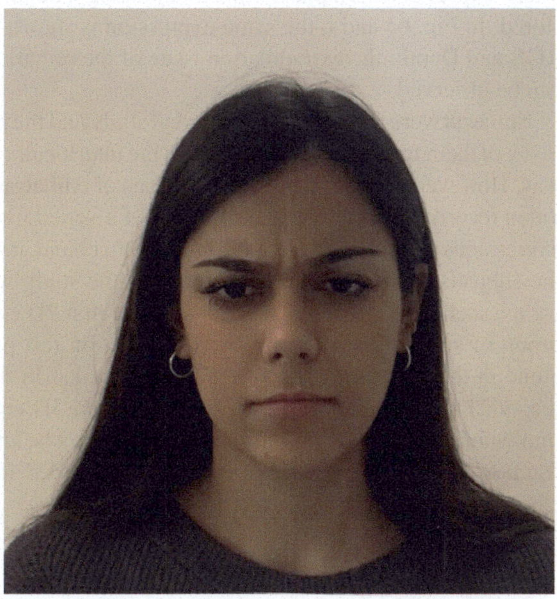

Fig. 5 Example of Action Unit 4 micro-expression from FACS system

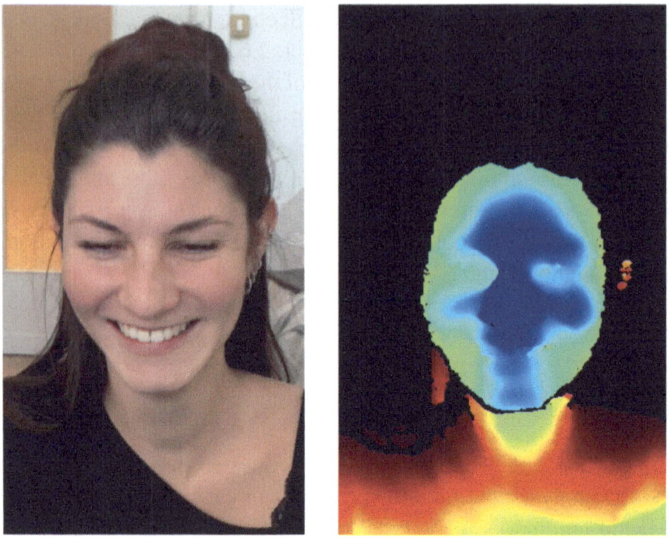

(a) (b)

Fig. 6 Micro-expression captures: **a** RGB images; **b** Depth Map

In the 3D analysis, a different method has been adopted to select frames because of the coarse depth resolution in 3D maps. Depth maps scansions has been synchronized and associated to RGB images, to select the proper frames and capture the 3D facial micro-expressions. Thanks to this methodology, depth maps have been selected and stored. In Fig. 6a and b the same expression is shown in the two different versions, RGB and Depth; the extrapolation issue of the remarkable facial expression frames can be observed.

Subjects were not aware of the kind of study and the purpose of the experiment. So, many of them tried to be compliant with the interlocutor and behaved in a quite natural way. However, some of them showed signs of embarrassment in being observed and video recorded. Therefore, the presence of a veiled filter in the naturalness of facial expressions should be considered. On the other hand, it was considered that informing the subjects a priori about the purpose of the study would have led to higher bias forms, such as ostentatious reactions. In the first 2D experimental trial with the first group of subjects, after some short briefing on test procedures, subjects were left alone in the room to perform the required tasks. In the second experimental trial, the synchronized management of both 2D and 3D equipment made impossible to implement such a condition and subjects could not be left alone. For this reason, many frames of embarrassed smiles have been discarded for the purpose of multi-class classification.

3 Results

Several analyses have been conducted before coming up with the best input to feed the Neural Network, or the Machine Learning model. Summaries of the analyses results are reported in the following sections for each methodology.

3.1 Bi-dimensional Analysis

In the binary model, the input layer consisted in arrays made up of 91 instances and 67 features; for each participant, an array containing the distances of all the landmarks from landmark number 34 was collected. Then a fully connected hidden layer comprised of 67 neurons was exploited to shrink and extract important structures. A rectifier activation function (*Relu*) was adopted and data were weighted with Gaussian initialization. The output layer is composed of a single neuron initialized with a sigmoid function to produce a probabilistic function and the binary cross entropy loss function, adopted for the classification. Adam optimizer algorithm has been used to calculate gradient descent as well as the accuracy matrix. Cross-validation with 7-folds has been performed using an initialized random seed in order to obtain a deterministic result. This analysis gave a good accuracy ranging between 75–90%.

In the multi-classification model, the input layer consisted in arrays made up of 91 instances and 67 features; for each participant, the distance arrays from landmark number 34 were collected as well. Then a fully connected hidden layer comprised of 67 neurons was exploited to shrink and extract important structures. A tangent activation function (Tanh) has been adopted. The output layer was composed of three neurons as the number of label classes, initialized with *softmax* function to produce a probabilistic output over the set of label possibilities and the categorical cross entropy loss function utilized for the multiclass classification problem. Adam optimizer algorithm was used to calculate gradient descent as well as the accuracy matrix. Cross-validation with 11-fold has been performed using an initialized random seed in order to obtain a deterministic result. This analysis gave an accuracy ranging between 65–75%.

3.2 Three-dimensional Analysis

The different architectures—logistic regression, decision trees, k-nearest neighbors and SVM—have been performed under several parameters tuning trials. Finally, the most performing k-neighbors model has been selected. In the analysis of the 3D dataset the KNN algorithm has been tuned with $k = 5$ and similarity Euclidean matrix has shown the highest performance, with 69% of accuracy. Figure 7 shows the confusion matrix of the KNN classification algorithm with 5 nearest neighbors.

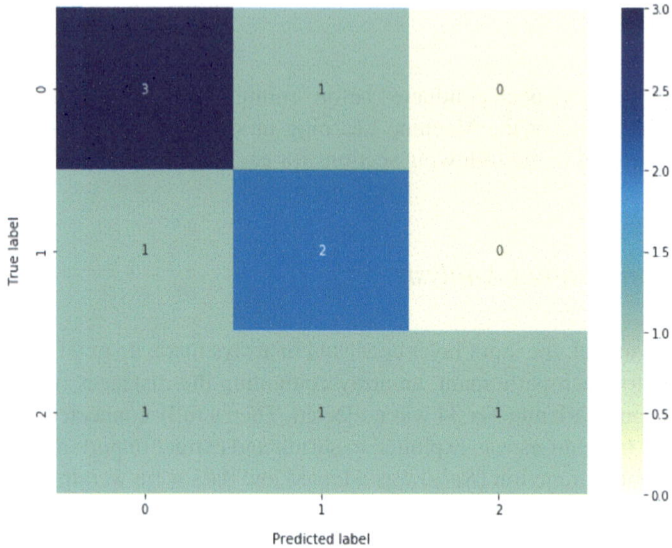

Fig. 7 KNN Confusion Matrix

In a minority of stressed cases, the training sample appeared to be unbalanced

4 Conclusions

This work proposes a novel preliminary methodology for detecting stress with facial expression recognition techniques on both 2D and 3D data. Instead of feeding NN with images as input data, which would request more training samples than the present, a face landmark distances approach is proposed in this work, so that less training samples than the former method are needed. In future work, different landmark selection and distance metrics will be implemented, in order to improve the method accuracy and efficiency.

In spite of the small size in the training sample, a good accuracy level was reached in all the three-different setups.

Three-dimensional model especially showed the potential to perform despite the loss and the discard of multiple depth maps.

In addition, for the purpose of this demo as mentioned above, the gallery was always composed of only 11–13 subjects; a more reliable model can be obtained with a much larger gallery. Artificial Neural Network played a key role for the 2D model development and showed the potential for discriminating the emotions through face regions maps. Results showed a first evidence of the research consistency. Three-dimensional extension with more training smoothed cases could be the natural continuation of the project. For a more depth application, a wider variety of emotional displays should be added to the gallery to discourage false positives: for example, a smile which implies face muscles stretching could be misidentified as different emotional states. Also, the missing value reconstruction could be performed in a more accurate way (thanks to the symmetry property of the face landmarking) by mirroring X, Y and Z coordinates over the plane.

References

1. Quarantotto, D.: Aristotele la psicofisiologia delle emozioni e l'ilemorfismo, De Anima I.1, Aristotele. Bruniana & Campanelliana XXIII **1**, 183–200 (2017)
2. Schachter, S., Singer, J.: Cognitive, social, and physiological determinants of emotional state. Psychol. Rev. **69**(5), 379–399 (1962)
3. Ekman, P., Friesen, W.: Unmasking the Face: A Guide to Recognizing Emotions From Facial Clues (2003)
4. Li Tian, Y., Kanade, T., Cohen, J.F.: Facial action coding system (FACS), recognizing facial action units for facial expression analysis. IEEE Trans. Pattern Anal. Mach. Intell. **23**(2), 97–115 (2001)
5. Gasparre, A.: Application field of facial action coding system: psychopathology and psychotherapy. Psychology Department University of Bari, Bari (2010)
6. Viegas, C., Lau, S., Maxion, R., Hauptmann, A.: Towards independent stress detection: a dependent model using facial action units. In: 2018 International Conference on Content-Based Multimedia Indexing (CBMI), La Rochelle, pp. 1–6 (2018)
7. Gao, H., Yuce, A., Thiran, J.P.: Detecting emotional stress from facial expressions for driving safety. IEEE Int. Conf. Image Proc. (2014). https://doi.org/10.1109/ICIP.2014.7026203
8. Dingers, D.F., Rider, R.L., Dorrian, J., Mc Glinchery, E.L., Rogers, N.L., Cizman, Z., Goldenstein, S.K., Vogler, C., Venkataraman, S., Metaxan, D.N.: Optical computer recognition of facial expression associated with stress inducedby performance demands. Art. Aviation Space Env. Med. **76**(6, Suppl.), B172–182 (2005)
9. Vezzetti, E., Marcolin, F., Tornincasa, S., Ulrich, L., Dagnes, N.: 3D geometry-based automatic landmark localization in presence of facial occlusions. Mult. Tools Appl. 1–29 (2018)
10. Dagnes, N., Marcolin, F., Nonis, F., Tornincasa, S., Vezzetti, E.: 3D geometry-based face recognition in presence of eye and mouth occlusions. Int. J. Interact. Des. Manuf. (IJIDeM), 1–19 (2019)

Unsupervised Geochemical Analysis of the Eruptive Products of Ischia, Vesuvius and Campi Flegrei

Antonietta M. Esposito, Giorgio Alaia, Flora Giudicepietro, Lucia Pappalardo, and Massimo D'Antonio

Abstract This work aimed to study geochemical data, composed of major and trace elements describing volcanic rocks collected from the Campanian active volcanoes of Vesuvius, Campi Flegrei and Ischia Island. The data were analyzed through the Self-Organizing Map (SOM) unsupervised neural net. SOM is able to group the input data into clusters according to their intrinsic similarities without using any information derived from previous geochemical-petrological considerations. The net was trained on a dataset of 276 geochemical patterns of which 96 belonged to Ischia, 94 to Vesuvius and 86 to Campi Flegrei volcanoes. Two investigations were carried out. The first one aimed to cluster geochemical data mainly characterizing the type of volcanic rocks of the three volcanic areas. The SOM clustering well grouped the oldest volcanic products of Ischia, Vesuvius and Campi Flegrei identifying a similar behaviour for the rocks emplaced in the oldest activity periods (>19 ka), and showing their different evolution over time. In the second test, devoted to inferring information on the magmatic source, the ratios of significant trace elements and K_2O/Na_2O have been used as input data. The SOM results highlighted a high degree of affinity between the geochemical element ratios of Vesuvius and Campi Flegrei that were separated from the products of Ischia. This result was also evidenced through isotope ratios by using traditional two-dimensional diagrams.

A. M. Esposito (✉) · G. Alaia · F. Giudicepietro · L. Pappalardo
Istituto Nazionale di Geofisica e Vulcanologia, Sezione di Napoli Osservatorio Vesuviano, Naples, Italy
e-mail: antonietta.esposito@ingv.it

G. Alaia
e-mail: giorg.alaia@studenti.unina.it

F. Giudicepietro
e-mail: flora.giudicepietro@ingv.it

L. Pappalardo
e-mail: lucia.pappalardo@ingv.it

G. Alaia · M. D'Antonio
Università degli Studi di Napoli Federico II, Naples, Italy
e-mail: masdanto@unina.it

A. Esposito et al. (eds.), *Progresses in Artificial Intelligence and Neural Systems*, Smart Innovation, Systems and Technologies 184,
https://doi.org/10.1007/978-981-15-5093-5_17

1 Introduction

The proposed analysis aims to discriminate the origins and the evolutionary processes of the magmas of distinct volcanic systems, searching for possible similarities or differences among their volcanic products by using neural networks. In particular, the Neapolitan volcanoes (Vesuvius, Campi Flegrei and Ischia) were chosen as a case study, for which different hypotheses on the source and evolution have been proposed in the literature [6, 22]. The generation of the wide alkaline magmatic systems that feed these volcanic districts, their mechanisms of magma differentiation and the timescales of magma recharge are crucial issues for volcanic risk evaluation, particularly in these densely populated areas.

This research was realized with the application of the unsupervised neural network called SOM [18]. Compared to the two-dimensional (2D) diagrams, commonly used in petrology, reporting projections of observed variables (such as elemental concentration, elemental ratio, isotopic ratio, and their combinations), the main advantage of using SOM consists in allowing the use of large databases with data described by a large number of variables. Indeed, two-dimensional diagrams are usually used to compare/distinguish rock suites. They trace the concentrations of one element or oxide with respect to those of another element (generally SiO_2). However, their limitation is that similar trends can result from more than one geochemical process. In order to fully extract information from multivariate geochemical data sets, we can then consider groups of parameters, such as major and trace elements, to diversify the magmatic processes.

To perform the analysis with the SOM algorithm, a dataset of 276 geochemical patterns was selected from the 2005 and 2017 databases by Peccerillo [23, 24], with 96 patterns originated from Ischia, 94 from Vesuvius and 86 from Campi Flegrei.

Two investigations were carried out on these samples. The first one was devoted to the geochemical patterns (described through their major and trace elements) related to the type of rocks of the three volcanoes. In the second examination, significant trace element ratios and K_2O/Na_2O ratio were taken into account in order to discriminate among the three magmatic sources. The results obtained with the SOM algorithm, particularly in the second test, were compared with those obtained with standard 2D diagrams. From the comparison, it emerges that SOM performed a good clustering, well separating the volcanic products of the three volcanoes.

In the following, we introduce the geological context, describe the proposed dataset, illustrate the neural methodology selected for the analysis and finally present the obtained results and our conclusions.

2 The Geological Context

The Neapolitan area shallow crust is characterized by a mainly Mesozoic-Tertiary sedimentary bedrock (251.90–5.30 Ma). The Mesozoic-Tertiary bedrock is covered by the sequences of the Campania plain (Fig. 1), which formed during the Plio-Pleistocene epoch due to the extensional tectonics related to the opening of the southern Tyrrhenian Sea. Starting from the Quaternary (2.58 Ma–to date), a new extensional stress regime allowed the genesis and the rise of magma that fed the volcanism of Vesuvius, Campi Flegrei and Ischia. These volcanoes are still active and have produced eruptions even in recent times [21]. In particular, Vesuvius had its last eruption in 1944, the Campi Flegrei caldera in 1538 and Ischia in 1302.

The Somma-Vesuvio volcanic complex consists of an old volcanic edifice, the Somma, truncated by a summit caldera in which Mt. Vesuvius is located. The Vesuvius cone was formed during the recent period (post 79 AD) characterized by very frequent eruptions.

Campi Flegrei and Ischia are large caldera systems, which have produced some of the largest explosive eruptions in the world.

The Campi Flegrei are characterized by a caldera with a geometry strictly influenced by volcano-tectonic collapses related to two events of great intensity: the eruption of the Campanian Ignimbrite, which occurred around 39000 years ago [12], and the Neapolitan Yellow Tuff eruption, which dates back to about 15000 years [7]. The eruption of the Neapolitan Yellow Tuff also marks the beginning of an intense long-term deformation phase of the caldera. In fact, the base of the caldera was subject to structural resurgence and fractured into blocks with relative movements. In more recent times, these vertical movements, which are called bradyseism, produced two main crises, occurred respectively from 1969 to 1972 and from 1982 to 1984, that generated an overall uplift of 170 and 180 cm respectively.

Fig. 1 The Campanian volcanic district of Ischia, Campi Flegrei and Vesuvius, with typical examples of their eruptive products

The Ischia Island is an active volcanic field that extends for about 46 km^2 and is dominated, in the central area, by Mount Epomeo (787 m a.s.l.). The main eruption is that of the Green Tuff of Mt. Epomeo, which occurred about 55 thousand years ago [1] and generated a large caldera. This depression was filled by pyroclastic flows generated by the Green Tuff eruption, which produced a huge volume of trachytic ignimbrites. Ischia is characterized by a deep magmatic chamber with poorly evolved magma, connected to shallow and evolved magmatic bodies that fed the recent activity [1].

3 Data Description

The considered dataset contains the major and trace elements contents of 276 volcanic products collected from the three abovementioned volcanoes. The major elements define the type of rock and its degree of evolution (i.e. basalt, trachyte, latite, etc.), while the trace elements are indicative of the magma source or the geodynamic environment in which magma is formed. Thus, the combination of these elements allows both the classification of the rocks and the identification of the magmatic sources.

The dataset includes:

- 96 patterns from Ischia (labelled as **IS**);
- 94 patterns from Vesuvius (labelled as **VE**);
- 86 patterns form Campi Flegrei (labelled as **CF**).

Two tests were performed: in the first one, the input to the SOM was a vector of 26 components of which 10 were major elements and 16 trace elements. This information is useful to classify the type of rock. An example of this input sample is illustrated in the Fig. 2.

10 major elements

SiO2	TiO2	Al2O3	Fe2O3	MnO	MgO	CaO	Na2O	K2O	P2O5
59,63981	0,405996	19,05059	4,320217	0,124922	0,874453	3,133458	3,476994	8,775765	0,197793
59,95419	0,406081	19,02332	4,227405	0,124948	0,843399	3,009163	3,425656	8,80883	0,1770096
59,40021	0,403309	19,06929	4,364012	0,124095	0,879007	3,216132	3,505688	8,831437	0,2068252
59,83385	0,404984	19,11734	4,23676	0,134995	0,809969	3,05296	3,489097	8,74351	0,1765317
60,21243	0,371747	18,87414	3,834307	0,127456	0,817844	3,30324	3,526288	8,773234	0,1593202
60,37796	0,360973	19,01476	3,768978	0,127402	0,764412	3,153201	3,471706	8,833209	0,1274021
57,8753	0,474734	19,47463	5,317017	0,137145	1,244857	3,903365	3,354784	7,975525	0,2426416
57,73961	0,445293	19,96395	4,56955	0,148431	0,922392	3,392706	3,94402	8,693808	0,1802375
57,72942	0,460978	20,48671	5,017153	0,139365	0,943396	3,11964	3,569897	8,361921	0,1715266
58,61775	0,469283	20,21118	4,276877	0,127986	1,023891	3,743601	3,412969	7,903157	0,2133106

16 trace elements

Y	Zn	Zr	La	Ce	Pr	Nd	Sm	Eu	Gd	Tb	Dy	Ho	Er	Tm	Yb
27,4	84,63	256	69,4	130	14	50	8,7	2,3	6,7	0,96	5,3	0,91	2,6	0,37	2,5
34,7	105,4	341	89	165	16,8	57,9	10,1	2,6	7,9	1,12	6,1	1,1	3	0,44	3
31,1	104,8	286	79,7	148	15,4	53,7	9,4	2,6	7,5	1,06	5,6	1	2,7	0,4	2,7
29,4	88,49	280	74,5	140	14,3	49,4	8,7	2,3	6,7	0,97	5,3	0,92	2,6	0,38	2,6
32,7	91,21	327	85,5	159	17	57	9,7	2,3	7,7	1,12	6	1,11	3,1	0,46	3,1
34,4	98,05	348	85,6	167	17,8	60,3	10,3	2,4	8,2	1,17	6,3	1,15	3,2	0,48	3,2
35,1	107,8	349	87,8	167	17,6	60,4	10,5	2,3	8,3	1,23	6,5	1,19	3,3	0,47	3,3
34,5	104,6	347	92,6	171	18,1	61,2	10,5	2,3	7,9	1,18	6,1	1,12	3,2	0,45	3,1
34,3	111,8	382	92,2	172	17,9	61,5	10,5	2,2	8,1	1,17	6,2	1,13	3,2	0,45	3,1
35	97,9	355	88,2	165	17,2	60,3	10,4	2,3	8	1,19	6,3	1,19	3,3	0,47	3,3

Fig. 2 An example of input vector related to the first test, composed by 10 major elements (blue rectangle) and 16 trace elements (green rectangle). Major elements are reported as weight percentages (wt%), trace elements as parts per million (ppm)

In the second test, each input sample to the SOM was composed by a vector of four geochemical ratios, i.e. Zr/Y, La/Yb, K_2O/Na_2O and Ti/Y, defined on both major and trace elements and chosen to better discriminate the geochemical data of the three volcanoes.

4 Methodology and Results

Neural Network [4, 16] based methods have had a strong growth in recent years, also in the volcanological field, both for analysis and monitoring purposes [8–10, 13–15, 17] and for petrographic investigations [2, 3, 5, 11]. In particular, the Self-Organizing Map algorithm offers several advantages such as the possibility of performing the data clustering considering large datasets of high-dimensional patterns, without any prior information about their distribution (i.e. it is unsupervised) and visualizing the results on an easily understandable bi-dimensional map.

In our tests, the SOM algorithm parameters have been settled in line with [19] and the SOM Toolbox for Matlab (http://www.cis.hut.fi/somtoolbox/). The results for the first and second test are visualized in Fig. 3 and Fig. 4, respectively. In both cases, a SOM map of 15 nodes (5 × 3 size) was selected. All the colored hexagons

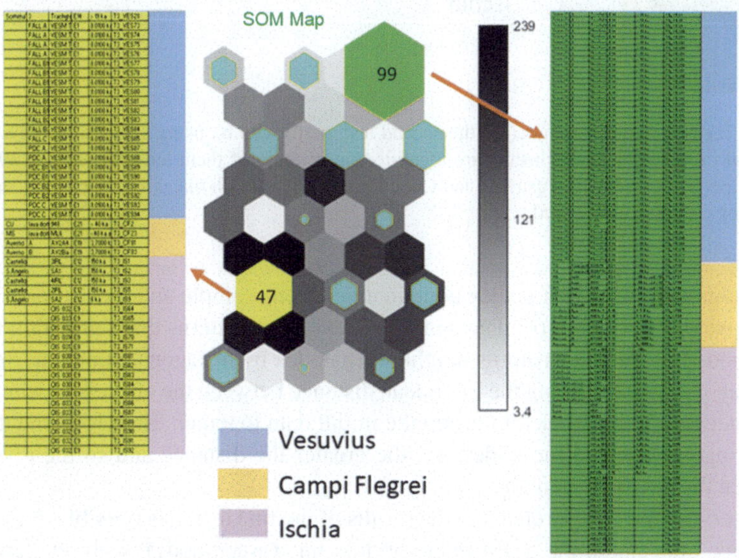

Fig. 3 The SOM clustering related to first test performed by using 10 major elements and 16 trace elements. Two principal clusters are recognized on the SOM map: the first one (green node) includes mainly the oldest volcanic samples of Vesuvius, Ischia and Campi Flegrei, while the second one (yellow node) includes mainly the youngest rocks of Vesuvius, Ischia, and few ones of Campi Flegrei

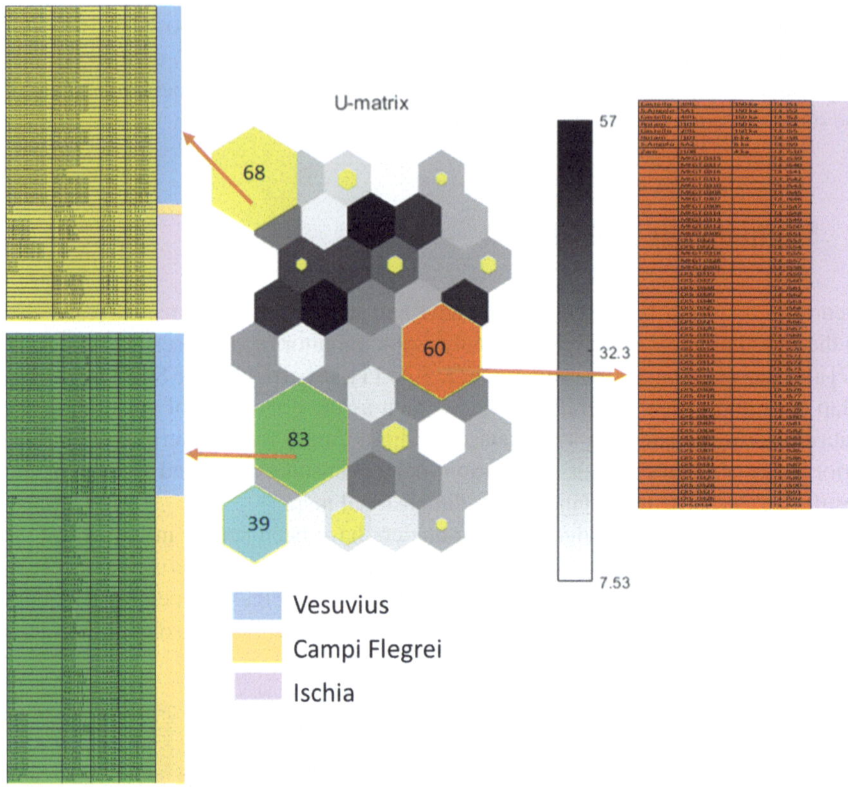

Fig. 4 The SOM map associated to the second test, performed by using a vector of only 4 geochemical ratios. Four main clusters are identified on it: three of them have 68, 83 and 39 input data, respectively, from the Vesuvius and Campi Flegrei; the fourth one contains 60 rock samples belonging only to Ischia Island

are non-empty nodes that is they contain at least one sample and their dimension is indicative of the number (or data density) of volcanic patterns that fall within them. Each node is separated from its neighboring nodes by hexagons painted according to a gray scale representing the Euclidean distance between the clusters in which the SOM network has provided to group the initial data together. Therefore, the darker the hexagons between the nodes are, the greater the distance and so the diversity between the data they contain.

Observing the Fig. 3, related to the results of the first test, it is possible to identify two main clusters on the SOM map: the first one (green node), with 99 elements, includes mainly the oldest products of Vesuvius, Ischia and Campi Flegrei volcanic activity; the second one (yellow node), with 47 patterns, contains mainly the youngest products of Vesuvius, Ischia, and few ones of Campi Flegrei. Thus, the SOM clustering shows a similarity for the rocks erupted by Ischia, Vesuvius and Campi Flegrei mainly in the oldest volcanic periods.

Looking at the SOM map obtained in the second experiment (Fig. 4), four main clusters are detected: three of them, with 68, 83 and 39 samples respectively, group data mostly from Vesuvius and Campi Flegrei; the fourth one, with 60 samples, includes instead rocks exclusively from Ischia. About the remaining nodes, three are empty and eight have in total 26 patterns, which compared to the previous four cover less than 10% of the total dataset (276 in total).

5 Conclusions

This work has proposed a neural network-based application for the unsupervised clustering of geochemical patterns. In particular, the Neapolitan active volcanoes (Vesuvius, Campi Flegrei and Ischia Island) were selected as case study. For these volcanoes, even though the petrological and geophysical studies of the last decades have allowed to put constrains on the evolution and in part the magma feeding system structure in terms of number and size of magma reservoirs, some questions remain unsolved (for example the exact extension and the depth of the sources/magma chambers). To apply the SOM technique, a geochemical database was prepared based on the datasets present in Peccerillo 2005 and 2017 [23, 24]. Two tests were accomplished, choosing in the first case vectors with major and trace elements to find similarities on rock evolution processes on the basis of geochemical patterns; while in the second one, to investigate aspects related to the magmatic source, vectors of the ratio between significant major and trace elements were selected (Zr/Y; La/Yb; K_2O/Na_2O; Ti/Y). In this last case, the SOM grouped most of the Ischia patterns (60 data on a total of 96, i.e. about 70%) in a single cluster, separating it from the clusters containing the Vesuvius and Campi Flegrei products, which were more similar.

When the same datasets are organized in conventional binary diagrams by using major and trace elements (Fig. 5), they are generally able to highlight the differences observed with the SOM network. To obtain consistent results by using two-dimensional diagrams it is necessary the application of radiogenic isotopes data

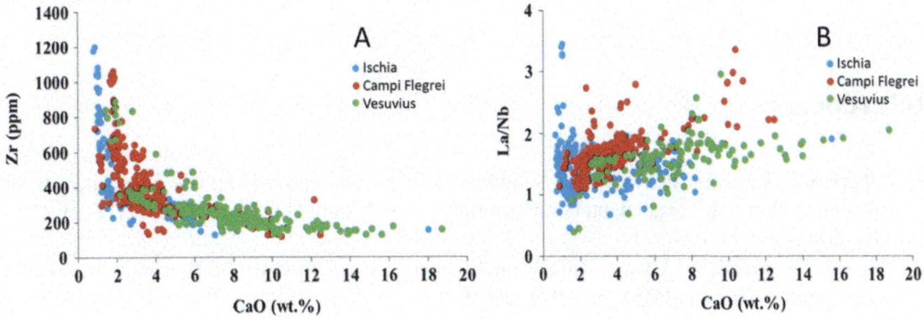

Fig. 5 **a** Binary diagram CaO (wt%) versus Zr (ppm); **b** binary diagram CaO (wt%) versus La/Nb

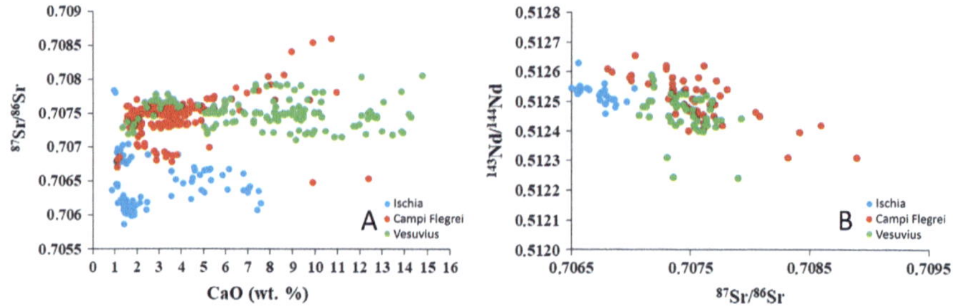

Fig. 6 **a** Binary diagram CaO (wt%) versus $^{87}Sr/^{86}Sr$; **b** Binary diagram $^{87}Sr/^{86}Sr$ versus $^{143}Nd/^{144}Nd$

(Fig. 6) which are more discriminating but more difficult and expensive to measure especially on a large number of samples.

Finally, these results suggest, as already hypothesized in the literature [20], that the magmas of Ischia eruptions evolved in different chemical-physical conditions compared to those of Vesuvius and Campi Flegrei which appear more similar to each other [22]. The dataset used in this work (consisting mainly of non-primitive rocks' compositions) does not allow to obtain information on the mantle source of the Neapolitan volcanoes' magmatism.

It is worth to note that in the future, by using more detailed data, neural networks can contribute to the solution of petrological questions, such as the possible existence of distinct sources/magma chambers for different volcanoes. In particular, as regards the Campanian volcanism, the use of datasets containing geochemical patterns of primitive rocks can give information on the type of source(s) from which the parental magmas derive and then differentiate them in single or multiple shallower magma chambers.

In conclusion, the tests showed that the SOM neural network allowed clustering and therefore discriminating the considered volcanic samples, even by using only some major and trace elements. The results indicate a geochemical affinity between the oldest products of the three Neapolitan active volcanoes (Vesuvius, Campi Flegrei and Ischia) and suggest a relationship also between the evolution processes.

References

1. Alberico, I., Lirer, L., Petrosino, P., Scandone, R.: Volcanic hazard and risk assessment from pyroclastic flows at Ischia island (southern Italy). J. Volcanol. Geotherm. Res. **171**(1–2), 118–136 (2008)
2. Ali, M., Chawathé, A.: Using artificial intelligence to predict permeability from petrographic data. Comput. Geosci. **26**(8), 915–925 (2000)
3. Aminian, K., Ameri, S.: Application of artificial neural networks for reservoir characterization with limited data. J. Petrol. Sci. Eng. **49**(3–4), 212–222 (2005)
4. Bishop, C.: Neural Networks for Pattern Recognition, 500 pp. Oxford University Press (1995)

5. Corsaro, R.A., Falsaperla, S., Langer, H.: Geochemical pattern classification of recent volcanic products from Mt. Etna, Italy, based on Kohonen maps and fuzzy clustering. Int. J. Earth Sci. **102**(4), 1151–1164 (2013)
6. D'Antonio, M., Tonarini, S., Arienzo, I., Civetta, L., Dallai, L., Moretti, R., Orsi, G., Andria, M., Trecalli, A.: Mantle and crustal processes in the magmatism of the Campania region: inferences from mineralogy, geochemistry, and Sr–Nd–O isotopes of young hybrid volcanics of the Ischia island (South Italy). Contrib. Mineral. Petrol. **165**(6), 1173–1194 (2013)
7. Deino, A.L., Orsi, G., de Vita, S., Piochi, M.: The age of the Neapolitan Yellow Tuff caldera-forming eruption (Campi Flegrei caldera–Italy) assessed by 40Ar/39Ar dating method. J. Volcanol. Geotherm. Res. **133**(1–4), 157–170 (2004)
8. Esposito, A.M., D'Auria, L., Giudicepietro, F., Peluso, R., Martini, M.:. Automatic recognition of landslide seismic signals based on neural network analysis of seismic signals: an application to the monitoring of Stromboli volcano (Southern Italy). In: Pure and Applied Geophysics PAGEOPH, vol. 170, pp. 1821–1832, © Springer Basel 2012 (2013). https://doi.org/10.1007/s00024-012-0614-1. http://link.springer.com/article/10.1007/s00024-012-0614-1
9. Esposito, A.M., D'Auria, L., Giudicepietro, F., Caputo, T., Martini, M.:. Neural analysis of seismic data: applications to the monitoring of Mt. Vesuvius, in special issue "Mt. Vesuvius monitoring: the state of the art and perspectives". Ann Geophys 56(4), S0446 (2013). ISSN: 1593-5213. https://doi.org/10.4401/ag-6452. http://www.annalsofgeophysics.eu/index.php/annals/article/view/6452
10. Esposito, A.M., D'Auria, L., Giudicepietro, F., Martini, M.: Waveform variation of the explosion-quakes as a function of the eruptive activity at Stromboli Volcano, in neural nets and surroundings. In: Apolloni, B., Bassis, S., Esposito, A., Morabito, F.C. (eds.) 22nd Italian Workshop on Neural Nets, WIRN 2012, May 17–19, Vietri sul Mare, Salerno, Italy. Smart Innovation, Systems and Technologies (SIST), vol. 19, pp. 111–119 (2013). ISBN 978-3-642-35466-3. https://doi.org/10.1007/978-3-642-35467-0_12. http://www.earth-prints.org/handle/2122/8692
11. Esposito, A.M., De Bernardo, A., Ferrara, S., Giudicepietro, F., Pappalardo, L.:SOM-based analysis of volcanic rocks: an application to Somma–Vesuvius and Campi Flegrei volcanoes (Italy). In: Esposito, A., Faundez-Zanuy, M., Morabito, F., Pasero, E. (eds.) Neural Approaches to Dynamics of Signal Exchanges. Smart Innovation, Systems and Technologies, vol. 151, pp. 55–60, Springer, Singapore (2020) . https://doi.org/10.1007/978-981-13-8950-4_6
12. Gebauer, S.K., Schmitt, A.K., Pappalardo, L., Stockli, D.F., Lovera, O.M.: Crystallization and eruption ages of Breccia Museo (Campi Flegrei caldera, Italy) plutonic clasts and their relation to the Campanian ignimbrite. Contrib. Mineral. Petrol. **167**, 953 (2014). https://doi.org/10.1007/s00410-013-0953-7
13. Giudicepietro, F., Esposito, A., D'Auria, L., Martini, M., Scarpetta, S.: Automatic analysis of seismic data by using neural networks: applications to Italian volcanoes. In: Marzocchi, W., Zollo, A. (eds.) Conception, Verification, and Application of Innovative Techniques to Study Active Volcanoes, pp. 399–415. Copyright © 2008 Istituto Nazionale di Geofisica e Vulcanologia (2008)
14. Giudicepietro, F., Esposito, A.M., Ricciolino, P.: Fast discrimination of local earthquakes using a neural approach. Seismol. Res. Lett. **88**(4), 1089–1096 (2017)
15. Ham, F.M., Iyengar, I., Hambebo, B.M., Garces, M., Deaton, J., Perttu, A., Williams, B.: A neurocomputing approach for monitoring plinian volcanic eruptions using infrasound. Procedia Comput. Sci. **13**, 7–17 (2012)
16. Haykin, S.: Neural Networks: A Comprehensive Foundation, 2nd edn. Prentice-Hall, Englewood Cliffs, NJ (1999)
17. Ibs-von Seht, M.: Detection and identification of seismic signals recorded at Krakatau volcano (Indonesia) using artificial neural networks. J. Volcanol. Geotherm. Res. **176**(4), 448–456 (2008)
18. Kohonen, T., Hynninen, J., Kangas, J., Laaksonen, J.: SOM_PAK: the self-organizing map program package. Report A31, Helsinki University of Technology, Laboratory of Computer

and Information Science, Espoo, Finland (1996). http://www.cis.hut.fi/research/som_lvq_pak.shtml
19. Kohonen, T.: Self-Organizing Maps, 2nd edn. Series in Information Sciences, vol. 30. Springer, New York (1997)
20. Melluso, L., Morra, V., Guarino, V., De'Gennaro, R., Franciosi, L., Grifa, C.: The crystallization of shoshonitic to peralkaline trachyphonolitic magmas in a H_2O–Cl–F-rich environment at Ischia (Italy), with implications for the feeder system of the Campania Plain volcanoes. Lithos **210**, 242–259 (2014)
21. Pappalardo, L., Piochi, M., D'Antonio, M., Civetta, L., Petrini, R.: Evidence for multi-stage magmatic evolution during the past 60 kyr at Campi Flegrei (Italy) deduced from Sr, Nd and Pb isotope data. J. Petrol. **43**(8), 1415–1434 (2002)
22. Pappalardo, L., Mastrolorenzo, G.: Rapid differentiation in a sill-like magma reservoir: a case study from the Campi Flegrei caldera. Sci. Rep. **2**, 712 (2012)
23. Peccerillo, A.: Plio-Quaternary Volcanism in Italy, vol. 365. Springer, Berlin, Heidelberg, New York (2005)
24. Peccerillo, A.: Cenozoic Volcanism in the Tyrrhenian Sea Region, p. 399. Springer (2017)

A Novel System for Multi-level Crohn's Disease Classification and Grading Based on a Multiclass Support Vector Machine

S. Franchini, M. C. Terranova, G. Lo Re, M. Galia, S. Salerno, M. Midiri, and S. Vitabile

Abstract Crohn's disease (CD) is a chronic inflammatory condition of the gastrointestinal tract that can highly alter patient's quality of life. Diagnostic imaging, such as Enterography Magnetic Resonance Imaging (E-MRI), provides crucial information for CD activity assessment. Automatic learning methods play a fundamental role in the classification of CD and allow to avoid the long and expensive manual classification process by radiologists. This paper presents a novel classification method that uses a multiclass Support Vector Machine (SVM) based on a Radial Basis Function (RBF) kernel for the grading of CD inflammatory activity. To validate the system, we have used a dataset composed of 800 E-MRI examinations of 800 patients from the University of Palermo Policlinico Hospital. For each E-MRI image, a team of radiologists has extracted 20 features associated with CD, calculated a disease activity index and classified patients into three classes (no activity, mild activity and severe activity). The 20 features have been used as the input variables to the SVM classifier, while the activity index has been adopted as the response variable. Different feature reduction techniques have been applied to improve the classifier performance, while a Bayesian optimization technique has been used to find the optimal hyperparameters of the RBF kernel. K-fold cross-validation has been used to enhance the evaluation reliability. The proposed SVM classifier achieved a better performance when compared with other standard classification methods. Experimental results show an accuracy index of 91.45% with an error of 8.55% that outperform the operator-based reference values reported in literature.

S. Franchini (✉) · M. C. Terranova · G. Lo Re · M. Galia · S. Salerno · M. Midiri · S. Vitabile
Dipartimento di Biomedicina, Neuroscienze e Diagnostica Avanzata, University of Palermo, Via del Vespro, 129, 90127 Palermo, Italy
e-mail: silvia.franchini@unipa.it

A. Esposito et al. (eds.), *Progresses in Artificial Intelligence and Neural Systems*, Smart Innovation, Systems and Technologies 184,
https://doi.org/10.1007/978-981-15-5093-5_18

1 Introduction

Crohn's disease (CD) is a chronic inflammatory disorder that affects mainly the gastrointestinal tract with extra intestinal manifestations [1, 2]. This disease presents a high tendency to relapse and often leads to severe disability. Although its exact etiology remains uncertain, results from several researches indicate that CD is a heterogeneous disease arising from an interaction between various genetic alterations and environmental factors. The onset and reactivation of disease are triggered by environmental factors in genetically susceptible individuals.

CD diagnosis often involves a combination of different clinical investigations, including radiological, endoscopic, and histological examinations [3]. However, colonoscopy is a burdensome monitoring tool for CD patients [4], while alternative non-invasive approaches, such as Enterography Magnetic Resonance Imaging (E-MRI), offer several advantages to monitor CD patients compared to colonoscopy [5] such as better acceptability, concomitant evaluation of the small bowel and the colon, and detection of extra-enteric CD complications [6]. Therefore, in the last decade, the use of MRI in clinical practice for CD diagnosis and monitoring has been increasing [7, 8]. Sensitivity and specificity indexes of 93% and 90%, respectively, have been achieved in MRI-based diagnosis [9], while different MRI-based disease activity indexes and scores have been proposed [6, 10, 11]. The E-MRI-based diagnosis is performed by radiology experts starting from the extraction and evaluation of a set of specific features that have been proved to be associated with CD [12, 13]. However, the manual classification process of CD affected patients, performed by expert radiologists starting from the E-MRI exam, is time-consuming and expensive, while machine learning methods allow for automatic classification and can aid physicians to speed up the CD diagnosis and monitoring process.

1.1 Related Works

Machine learning includes both supervised learning methods, which train a model on known input data and output responses so that they can generate reasonable predictions for the response to new data, and unsupervised learning methods, which find hidden patterns, intrinsic structures or groupings in input data. Supervised learning uses classification techniques to develop predictive models. Common classification algorithms include support vector machines (SVM), decision trees, k-nearest neighbor, Naïve Bayes, discriminant analysis, and neural networks. Clustering is the most common unsupervised learning method. Common clustering algorithms include k-means, hierarchical clustering, Gaussian mixture models, hidden Markov models, self-organizing maps, and fuzzy c-means clustering.

Different machine learning approaches have been proposed to classify patients into positive or negative with respect to some kind of disease starting from MRI images. In [14], the authors have used an unsupervised method, namely fuzzy c-means, to classify MR brain images, while a supervised learning method, namely k-nearest neighbor, has been used in [15] to classify MR brain tissues. In [16], the authors have classified MR brain images using both an unsupervised technique based on self-organizing maps and a supervised method based on a support vector machine. An SVM-based classifier with leave-one-out cross-validation has been also used to predict medication adherence in Heart Failure (HF) patients [17]. Another classification method based on SVM is presented in [18] to classify MR brain images. In this work, the digital wavelet transform is used to extract features from MR images, while a Principal Component Analysis (PCA) technique is applied to reduce the feature space dimension. Finally, the classification process is performed by using an SVM classifier with gaussian kernel and 5-fold cross-validation. In general, SVM-based classifiers show usually a better performance in terms of accuracy than other supervised learning methods; furthermore, these classification methods present other advantages as elegant mathematical formulation and intuitive geometric interpretation.

Different machine learning techniques have been used and compared in [19] to classify pediatric patients affected by inflammatory bowel diseases, comprising Crohn's disease and ulcerative colitis. As reported in the paper, the SVM classifier with linear kernel achieved a classification accuracy of 82.7% using combined endoscopic and histological data as inputs. In [20], we have presented the evaluation of an SVM-based method for Crohn's disease classification. The SVM uses as input variables 20 parameters extracted by expert radiologists starting from the E-MRI images, while the histological specimen is used as the Ground-Truth for CD diagnosis. The classifier was trained and tested using an E-MRI dataset composed of 800 patients. Preliminary results of this approach had been presented in [21] for a reduced dataset of 300 patients. Experimental results presented in [20] show an accuracy of 96.54% and an error of 3.46%. However, the binary classification technique proposed in [20] can only detect the presence/absence of CD and categorize patients into two classes (positive/negative with respect to CD), but it is not able to grade the disease inflammatory activity.

1.2 Our Contribution

This paper proposes a novel classification method based on a multiclass SVM with Radial Basis Function (RBF) kernel that allows for grading CD inflammatory activity into three classes: no activity, mild-level activity, and severe-level activity. The proposed classifier has been trained, tested and validated by using a real dataset composed of 800 E-MRI exams of 800 patients of the University of Palermo Policlinico Hospital. Starting from the E-MRI examination of each patient, a team of expert radiologists have extracted 20 features usually associated with CD and calculated an activity index. The activity index is defined on the basis of the total summa of the following features: wall thickening greater than 4 mm, intramural and mesenteric edema, mucosal hyperemia, wall enhancement (and enhancement pattern), transmural ulceration and fistula formation, vascular engorgement, and inflammatory mesenteric lymph nodes [13]. Based on these parameters, each patient has been classified into one of three classes according to the value of the activity index calculated by the radiologists (class 0 for no activity, class 1 for mild-level activity and class 2 for severe-level activity). The vectors composed of the 20 input parameters and the activity index have been used to train and evaluate different multiclass classifiers. The SVM classifier achieved the better results with respect to the other standard multiclass classification methods. Two different feature reduction techniques, namely Sequential Forward Feature Selection (SFFS) and Recursive Feature Elimination (RFE) [22, 23], have been used to reduce the feature space dimensionality with the aim to both exclude redundant or noisy features and reduce the computational cost of the SVM algorithm. Furthermore, different SVM models with different kernels have been compared, while a Bayesian optimization algorithm has been used to find the optimal hyperparameters of the RBF kernel [24]. To improve the reliability of the SVM evaluation results, a 15-fold cross-validation technique has been used to train and test the different SVM classifiers. Experimental results have shown that the classifier with the best performance is the SVM classifier with RBF kernel that uses a reduced set of 12 features. This classifier allowed us to achieve an accuracy of 91.45% and an error of 8.55%.

The rest of the paper is organized as follows: Sect. 2 describes the E-MRI dataset used to test and validate the proposed classification system. Section 3 presents the classification method as well as the algorithms used to optimize its performance and compares the proposed system with other multiclass classifiers. Experimental results are summarized in Sect. 4, while Sect. 5 contains the conclusions of the paper.

2 Materials

The dataset used to test and validate the proposed classification system is composed of 800 E-MRI images of 800 patients (427 females, 373 males, mean age 30.1 years) from the University of Palermo Policlinico Hospital (400 healthy individuals and 400 individuals with histologically proved CD). The following sequences have been used according to the MRI protocol for image acquisition: Steady State Free Precession (BFFE)—axial and coronal planes, Contrast 3D Spoiled Gradient Echo (T1 e-Thrive)—axial and coronal planes, Single Shot Fast Spin Echo (T2 SPAIR)—axial and coronal planes, HASTE thick slab—coronal plane, and DWI—axial plane.

A team of expert radiologists have extracted for each patient a vector composed of 20 features, which, as reported in literature [7, 8, 13], are associated with Crohn's disease. The same 20 features have been used in the SVM-based binary classification method of CD presented in [20].

Table 1 reports the 20 features that have been used as the input variables to the classifier and describes the disease activity index calculated by radiologists and used as the classifier target response variable. The team of radiologists classified

Table 1 Features extracted by radiologists and used to train and test the classifier

Classifier input variables 20 features extracted by radiologists	*Classifier response variable* Disease activity index calculated by radiologists
Bowel cleaning protocol, bowel distention protocol, terminal ileum thickening, length, pseudopolyps, single lesion/skip lesions, lumen, fat wrapping, lymph nodes, sinus, fistulas, surgery, mucosal layer, free fluid, complications, intestinal obstruction, breathing/peristalsis artifacts, DWI, T2W imaging, post contrast T1 imaging	The activity index is derived as the total summa of the following features [13]: • wall thickening greater than 4 mm • intramural and mesenteric edema • mucosal hyperemia • wall enhancement (and enhancement pattern) • transmural ulceration and fistula formation • vascular engorgement • inflammatory mesenteric lymph nodes

each patient into one of three classes corresponding to three different values of the activity index: class 0 (no activity), class 1 (mild activity), class 2 (severe activity).

3 Methods

The proposed classification method for CD activity grading is based on a multiclass Support Vector Machine with Radial Basis Function kernel.

3.1 Comparison of Different Classification Methods

Using the same dataset presented in Sect. 2, which consists of 800 observations (each composed of the 20 extracted features and the calculated activity index), we have trained and evaluated three different multiclass classifiers: Feed Forward Neural Network (FFNN), K-Nearest Neighbor (KNN) and Support Vector Machine (SVM). Table 2 lists the accuracy and error values measured for the different classifiers. It can be observed that the multiclass SVM classifier achieves the better performance

Table 2 Comparison of three different multiclass classification methods

	Multiclass classification method		
	Feed forward neural network (%)	K-nearest neighbor (K = 10) (%)	Support vector machine (%)
Accuracy (as defined in Table 5)	85.48	83.85	91.45
Error (as defined in Table 5)	14.52	16.15	8.55

with respect to the other classifiers. Therefore, we have chosen multiclass SVM for the multilevel classification and grading of CD affected patients.

3.2 Feature Reduction Methods

The features extracted from the E-MRI exams are used as the predictive variables to be provided as inputs to the classifier. Some of these features may be noisy, highly correlated and redundant. Therefore, it is useful to apply proper algorithms to automatically reduce the number of features. Reducing the feature space dimensionality is important to reduce the computational costs as well as the memory resource utilization. Furthermore, a predictive model based on a reduced number of predictors is simpler and can be more easily interpreted and generalized.

Automated feature reduction methods include feature transformation methods, such as Principal Component Analysis (PCA) [25], and feature selection methods, including Sequential Forward Feature Selection (SFFS) and Recursive Feature Elimination (RFE) [22, 23]. PCA transforms the feature space in a new orthogonal coordinate system thus obtaining the so-called principal components that are ordered by the variation explained in the data. The components beyond a given threshold of explained variance are then discarded leading to a dimensionality reduction. Feature selection methods select a subset of the most relevant features to be included in the predictive model. The SFFS technique adds new features to the model as long as the error is decreasing, while the RFE algorithm is a backward elimination procedure that recursively fits the model and removes the weakest feature (or features) until the specified number of features is eventually reached. The final output of the RFE algorithm is a ranked list with variables ordered according to their relevance. In this work, we have applied both a sequential forward feature selection method and an RFE algorithm and compared the results obtained by the two techniques. Results are reported in Sect. 4.

3.3 Multiclass Support Vector Machines

Support Vector Machines were originally designed for binary classification [26]. To extend this method for multiclass classification, different approaches have been proposed in literature. Most of these solutions construct the multiclass SVM classifier by combining several binary SVM classifiers. These methods include one-vs-all and one-vs-one approaches [27]. In the one-vs-all approach, one classifier per class is trained. For a given class C_i, the classifier assumes samples with C_i as positives and the rest as negatives. To solve a multiclass classification problem with N classes, the one-vs-all approach requires N binary classifiers. In the one-vs-one approach, a separate binary classifier is trained for each different pair of classes. Therefore, considering N classes, the one-vs-one approach requires $N(N-1)/2$ binary classifiers.

The one-vs-one method has a higher computational cost. Other solutions have been proposed that consider all classes at once in a larger one-step optimization problem [28]. However, the latter methods are more computationally expensive. To solve our CD classification and grading problem, we have implemented and compared both one-vs-one and one-vs-all multiclass SVM classifiers. Finally, we have chosen the one-vs-one approach since it gave the best performance in terms of classification accuracy.

3.4 K-Fold Cross-Validation

K-fold cross-validation is a statistical method used to obtain a reliable estimation of the skill of a machine learning model on new unseen data. In general, in order to train and test a classification model, the original dataset is split into a training set and a test set. The model is fit to the training set and evaluated using the test set. However, the model accuracy estimation is obtained for that particular test set. It is not possible to say if the model would generalize well to new test data. Conversely, the K-fold cross-validation procedure splits the dataset into K groups or folds of equal size. The first fold is used as a validation set, and the method is fit on the remaining K − 1 folds. This procedure is repeated for all the K folds and finally the overall K-fold error is calculated as the average error from all the folds. When K is equal to the number of observations, then a single observation is used each time for validation. This approach is called leave-one-out cross-validation. In our classification system, we have used a 15-fold cross-validation technique in order to obtain a more reliable evaluation of the classifier accuracy.

3.5 Bayesian Optimization

Bayesian optimization is an optimization algorithm used to optimize a given objective or cost function. This algorithm can be used to find the optimal hyperparameters of the SVM model capable of minimizing the cross-validation loss of the classifier. We have used this optimization technique to find the optimal hyperparameters of the Radial Basis Function kernel of the SVM model. The RBF kernel is defined as

$$K(x, y) = e^{-\frac{\|x-y\|^2}{2\sigma^2}} \tag{1}$$

where x and y are two input feature vectors, while σ is the scaling factor. The Bayesian optimization algorithm searches the best values of the parameters σ and C that minimize the cross-validation loss, where C is a parameter, called box constraint, which, in the SVM formulation, is used when input data is not linearly separable. In the non-separable case, often called soft-margin SVM, C is the cost or penalty term assigned to misclassifications. The parameter C controls the trade-off between the penalty term and the size of the margin.

Results deriving from the application of the Bayesian optimization technique are reported in Sect. 4.

4 Experimental Results

This section describes the results of the experimental tests performed to evaluate the performance of the proposed multiclass SVM classifier. The dataset presented in Sect. 2 has been used for experimental tests.

First, the two feature reduction methods SFFS and RFE described in Sect. 3.2 have been applied to the original dataset composed of 20 features. The 12 features selected by the SFFS algorithm are listed in Table 3 along with the related classification error of the reduced SVM model, while Fig. 1 reports the predictor importance estimates calculated by the RFE algorithm. It can be observed that the first 12 features with the most relevant scores obtained by the RFE method coincide with the 12 features selected by the SFFS method. Based on these results, we have trained and evaluated the 3 following multiclass SVM models based on different feature sets:

full model (20 features listed in Table 1)
reduced model 1 (17 features with an importance estimate higher than 10^{-4}): Bowel cleaning protocol, Terminal ileum thickening, Length, Pseudopolyps, Single lesion/skip lesions, Lumen, Fat wrapping, Lymph nodes, Sinus, Fistulas, Surgery, Mucosal Layer, Free Fluid, Complications, Water diffusion restriction (DWI), T2W imaging, Post contrast T1 imaging)
reduced model 2 (12 features with an importance estimate higher than 10^{-3}): Terminal ileum thickening, Length, Single lesion/skip lesions, Lumen, Fat wrapping, Lymph nodes, Fistulas, Mucosal Layer, Free fluid, Water diffusion restriction (DWI), T2W imaging, Post contrast T1 imaging).

The evaluation results of these 3 SVM classifiers, derived by using a 15-fold cross-validation scheme, are reported in Table 4 for two different kernels (linear and radial basis function).

The multiclass SVM classifier with RBF kernel optimized by the Bayesian optimization technique that uses the reduced set composed of 12 features shows the

Table 3 Sequential Forward Feature Selection (SFFS) results

12 selected features	SVM reduced model error (%)
Terminal ileum thickening, length, single lesion/skip lesions, lumen, fat wrapping, lymph nodes, fistulas, mucosal layer, free fluid, DWI, T2W imaging, post contrast T1 imaging	8.55

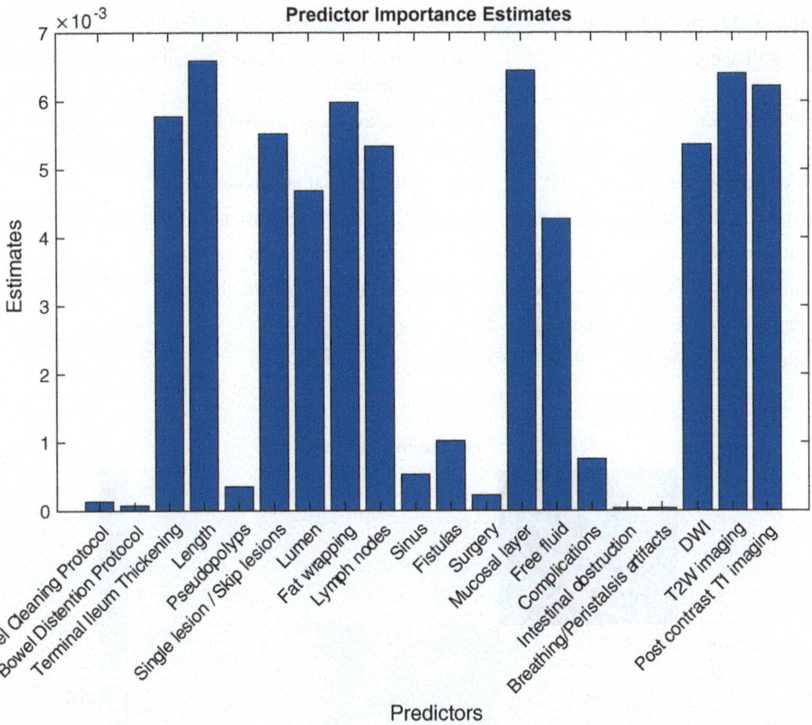

Fig. 1 Predictor importance estimates calculated by the Recursive Feature Elimination (RFE) algorithm

Table 4 Measured errors for 3 multiclass SVM classifier models: full model (20 features), reduced model 1 (17 features), and reduced model 2 (12 features). The errors have been measured using two different kernels (linear and RBF) and a 15-fold cross-validation scheme

	Full model (20 features) (%)	Reduced model 1 (17 features) (%)	Reduced model 2 (12 features) (%)
Linear kernel	14.76	14.12	13.64
Radial Basis Function (RBF) kernel (Bayesian optimization)	10.33	9.75	**8.55**

best performance in terms of overall classification error. This classifier has been further validated by calculating the metrics reported in Table 5 conventionally used to measure the performance of a multiclass classification system.

Starting from the confusion matrix calculated for our classifier and reported in Fig. 2, we have derived the metrics reported in Table 6. As it can be observed, the

Table 5 Metrics conventionally used to measure the performance of a multiclass classifier

Overall accuracy	Sum of correct classifications divided by the total number of classifications
Overall error	Sum of misclassifications divided by the total number of classifications
True positives of class C_i (TP_i)	All C_i instances that are classified as C_i
True negatives of class C_i (TN_i)	All non-C_i instances that are not classified as C_i
False positives of class C_i (FP_i)	All non-C_i instances that are classified as C_i
False negatives of class C_i (FN_i)	All C_i instances that are not classified as C_i
Sensitivity or Recall of class C_i	$\text{Sensitivity}_i = \frac{TP_i}{TP_i+FN_i}$
Specificity of class C_i	$\text{Specificity}_i = \frac{TN_i}{TN_i+FP_i}$

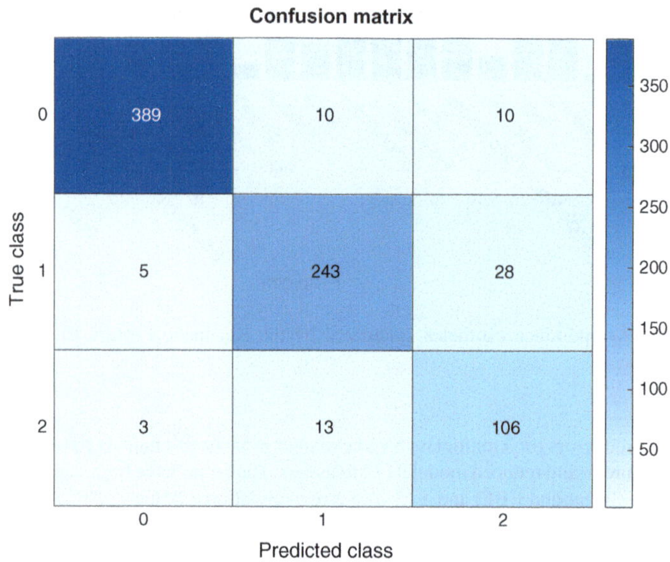

Fig. 2 Confusion matrix of the multiclass SVM classifier with RBF kernel optimized by the Bayesian optimization technique that uses the reduced set composed of 12 features

proposed classifier achieves an overall accuracy of 91.45%, an overall error of 8.55%, sensitivity values of 95.11% for class 0, 88.04% for class 1, and 86.89% for class 2, and specificity values of 97.99%, 95.67%, and 94.45% for class 0, class 1, and class 2, respectively, while the average values of sensitivity and specificity are 90.01% and 96.04%, respectively.

Table 6 Measured metrics for the multiclass SVM classifier with RBF kernel that uses a reduced set composed of 12 features. 15-fold cross-validation has been used to train and validate the classifier

Overall accuracy	91.45%			
Overall error	8.55%			
	Class 0 (no disease activity) (%)	Class 1 (mild activity) (%)	Class 2 (severe activity) (%)	Average values (%)
True positives (TP)	48.20	30.11	13.14	30.48
True negatives (TN)	48.33	62.95	80.17	63.82
False positives (FP)	1.00	2.85	4.71	2.85
False negatives (FN)	2.48	4.09	1.98	2.85
Sensitivity or recall	95.11	88.04	86.89	90.01
Specificity	97.99	95.67	94.45	96.04

5 Conclusions

A novel classification method based on a multiclass Support Vector Machine with RBF kernel has been proposed for the multi-level grading of Crohn's disease (CD) activity. To evaluate the classifier performance, we have used a dataset composed of 800 Enterography Magnetic Resonance Imaging (E-MRI) exams of 800 patients from the University of Palermo Policlinico Hospital. Starting from the E-MRI examination, a team of radiologists have extracted a set of 20 features usually associated with CD. The radiologists have also calculated a proper activity index that classifies and grades the disease inflammatory activity into three different classes, namely class 0 (no disease activity), class 1 (mild disease activity), and class 2 (severe disease activity). These 800 observations, each composed of the 20 features and the activity index, have been used to train and test the multiclass SVM classifier. The 20 features represent the input predictors for the classifier, while the activity index is provided as the target response variable. A 15-fold cross-validation scheme has been used to obtain a more robust estimation of the classifier performance. Two different feature reduction techniques, namely the Sequential Forward Feature Selection (SFFS) and the Recursive Feature Elimination (RFE), have been applied and their results have been compared. Furthermore, a Bayesian optimization algorithm has been applied to find the optimal hyperparameters of the RBF kernel that allow to minimize the 15-fold loss of the classifier. Different SVM classifiers based on different kernels and different numbers of predictors have been evaluated and compared in terms of accuracy and error. Experimental results show that the best classifier is the SVM with RBF kernel that uses a reduced set of predictors composed of 12 features. For this classifier, we have measured an overall accuracy of 91.45%, an overall error of 8.55%, sensitivity values of 95.11% for class 0, 88.04% for class 1, and 86.89% for class 2, and specificity values of 97.99%, 95.67%, and 94.45% for class 0, class 1, and class 2, respectively, while average values of sensitivity and specificity are 90.01% and 96.04%, respectively. To the best of our knowledge, this is the first work that presents a classification method capable not only of classifying patients into positive/negative with respect to CD, but also of grading the disease inflammatory activity. The availability of such an automatic tool for CD diagnosis and activity grading is very important since it allows for early and correct CD evaluation without requiring the expensive and time-consuming manual classification process by radiologists with specific gastrointestinal expertise. Early diagnosis is fundamental since it has a positive impact on disease course and prognosis, while it has been extensively reported that late diagnosis may lead to severe complications, as fistulizing events and strictures, which require more aggressive treatments. Our future work will be aimed at developing a novel multi-level classification method capable of classifying CD patients into four or more classes with respect to the disease activity grade.

References

1. Peyrin-Biroulet, L., Loftus Jr., E.V., Colombel, J.F., Sandborn, W.J.: The natural history of adult Crohn's disease in population-based cohorts. Am. J. Gastroenterol. **105**(2), 289–297 (2010)
2. Peyrin-Biroulet, L., Cieza, A., Sandborn, W.J., Coenen, M., Chowers, Y., Hibi, T., et al.: Development of the first disability index for inflammatory bowel disease based on the international classification of functioning, disability and health. Gut **61**(2), 241–247 (2012)
3. Gomollón, F., Dignass, A., Annese, V., Tilg, H., Van Assche, G., Lindsay, J.O., Peyrin-Biroulet, L., Cullen, G.J., Daperno, M., Kucharzik, T., et al.: 3rd European evidence-based consensus on the diagnosis and management of Crohn's disease 2016: part 1: diagnosis and medical management. J. Crohns Colitis **11**, 3–25 (2016)
4. Buisson, A., Gonzalez, F., Poullenot, F., Nancey, S., Sollellis, E., Fumery, M., et al.: Comparative acceptability and perceived clinical utility of monitoring tools: a nationwide survey of patients with inflammatory bowel disease. Inflamm. Bowel Dis. **23**(8), 1425–1433 (2017)
5. Taylor, S.A., Avni, F., Cronin, C.G., Hoeffel, C., Kim, S.H., Laghi, A., et al.: The first joint ESGAR/ESPR consensus statement on the technical performance of cross-sectional small bowel and colonic imaging. Eur. Radiol. **27**(6), 2570–2582 (2016)
6. Buisson, A., Pereira, B., Goutte, M., Reymond, M., Allimant, C., Obritin-Guilhen, H., Bommelaer, G., Hordonneau, C.: Magnetic resonance index of activity (MaRIA) and Clermont score are highly and equally effective MRI indices in detecting mucosal healing in Crohn's disease. Dig Liver Dis **49**(11), 1211–1217 (2017)
7. Sinha, R., Verma, R., Verma, S., Rajesh, A.: Mr enterography of Crohn disease: part 1, rationale, technique, and pitfalls. Am. J. Roentgenol. **197**(1), 76–79 (2011)
8. Lo Re, G., Midiri, M.: Crohn's Disease: Radiological Features and Clinical-Surgical Correlations. Springer, Heidelberg (2016)
9. Panes, J., Bouzas, R., Chaparro, M., García-Sánchez, V., Gisbert, J., Martinez de Guereñu, B., Mendoza, J.L., Paredes, J.M., Quiroga, S., Ripollés, T., et al.: Systematic review: the use of ultrasonography, computed tomography and magnetic resonance imaging for the diagnosis, assessment of activity and abdominal complications of Crohn's disease. Aliment. Pharmacol. Ther. **34**(2), 125–145 (2011)

10. Kitazume, Y., Fujioka, T., Takenaka, K., Oyama, J., Ohtsuka, K., Fujii, T., Tateisi, U.: Crohn disease: a 5-point MR enterocolonography classification using enteroscopic findings. Am. J. Roentgenol. **212**(1), 67–76 (2019)
11. Puylaert, C.A.J., et al.: Comparison of MRI activity scoring systems and features for the terminal ileum in patients with Crohn disease. Am. J. Roentgenol. **212**(2), W25–W31 (2019)
12. Sinha, R., Verma, R., Verma, S., Rajesh, A.: MR enterography of Crohn disease: part 2, imaging and pathologic findings. Am. J. Roentgenol. **197**(1), 80–85 (2011)
13. Tolan, D.J., Greenhalgh, R., Zealley, I.A., Halligan, S., Taylor, S.A.: MR enterographic manifestations of small bowel Crohn disease 1. Radiographics **30**(2), 367–384 (2010)
14. Agnello, L., Comelli, A., Ardizzone, E., Vitabile, S.: Unsupervised tissue classification of brain MR images for voxel-based morphometry analysis. Int. J. Imaging Syst. Technol. **26**(2), 136–150 (2016)
15. Cocosco, C.A., Zijdenbos, A.P., Evans, A.C.: A fully automatic and robust brain MRI tissue classification method. Med. Image Anal. **7**(4), 513–527 (2003)
16. Chaplot, S., Patnaik, L., Jagannathan, N.: Classification of magnetic resonance brain images using wavelets as input to support vector machine and neural network. Biomed. Signal Process. Control **1**(1), 86–92 (2006)
17. Son, Y.J., Kim, H.G., Kim, E.H., Choi, S., Lee, S.K.: Application of support vector machine for prediction of medication adherence in heart failure patients. Healthc. Inform. Res. **16**(4), 253–259 (2010)
18. Zhang, Y., Wang, S., Ji, G., Dong, Z.: An MR brain images classifier system via particle swarm optimization and Kernel support vector machine. Sci. World J. **2013**, 9 (2013)
19. Mossotto, E., Ashton, J.J., Coelho, T., Beattie, R.M., MacArthur, B.D., Ennis, S.: Classification of paediatric inflammatory bowel disease using machine learning. Sci. Rep. **7**(1), 1–10 (2017)
20. Franchini, S., Terranova, M.C., Lo Re, G., Salerno, S., Midiri, M., Vitabile, S.: Evaluation of a support vector machine based method for Crohn's disease classification. In: Esposito, A., Faundez-Zanuy, M., Morabito, F.C., Pasero, E. (eds.) Neural Approaches to Dynamics of Signal Exchanges. Smart Innovation, Systems and Technologies, vol. 151, Chapter 29, pp. 313–327. Springer, Singapore. ISSN: 2190-3018, Print ISBN: 978-981-13-8949-8, Online ISBN: 978-981-13-8950-4. https://doi.org/10.1007/978-981-13-8950-4_29
21. Comelli, A., Terranova, M.C., Scopelliti, L., Salerno, S., Midiri, F., Lo Re, G., Petrucci, G., Vitabile, S.: A kernel support vector machine based technique for Crohn's disease classification in human patients. In: Barolli, L., Terzo, O. (eds.) Complex, Intelligent, and Software Intensive Systems. CISIS 2017. Advances in Intelligent Systems and Computing, vol 611. Springer, Cham (2018)
22. Guyon, I., Weston, J., Barnhill, S., Vapnik, V.: Gene selection for cancer classification using support vector machines. Mach. Learn. **46**(1–3), 389–422 (2002)
23. Sanz, H., Valim, C., Vegas, E., Oller, J.M., Reverter, F.: SVM-RFE: selection and visualization of the most relevant features through non-linear kernels. BMC Bioinform. **19**(1), 432 (2018)
24. Hastie, T., Tibshirani, R., Friedman, J.: The Elements of Statistical Learning—Data Mining, Inference, and Prediction, 2nd edn. Springer (2008)
25. Jolliffe, I.T:. Principal Component Analysis, 2nd edn. Springer (2002)
26. Christianini, N., Shawe-Taylor, J.C.: An Introduction to Support Vector Machines and Other Kernel-Based Learning Methods. Cambridge University Press, Cambridge, UK (2000)
27. Deng, N., et al.: Support Vector Machines: Optimization Based Theory, Algorithms and Extensions. Data mining and Knowledge Discovery Series. Chapman & Hall /CRC (2012)
28. Weston, J., Watkins, C.: Multi-class support vector machines. In: Verleysen, M. (ed.) Proceedings of ESANN99, pp. 219–224. D. Facto Press, Brussels (1999)

Preliminary Study on the Behavioral Traits Obtained from Signatures and Writing Using Deep Learning Algorithms

Xavier Font, Angel Delgado, and Marcos Faundez-Zanuy

Abstract This work describes how a new multimodal data base is constructed to try to solve the link between signature/handwriting and personality. With the help of two devices, one of them responsible to report the mechanical writing process (using a tablet) and the other one to acquire brain activity (with an EEG headset) we will be able to carry out different sessions through a set of experiments. Because the data base is not completed yet, and it is well known that deep learning requires larges amount of data, the main results about signature and personality factors were not good enough. The different deep convolutional neural networks (DCNN) tested does not obtain a reasonable minimum threshold. However the same incomplete data base gives promising results when solving a completely different problem such as signature recognition (where a performance of 80% was reached) using the same DCNN architecture.

Keywords Injury detection · Thermal images · Inertial data · EEG data · Penalized logistic regression · Classification

1 Introduction

This is an ongoing study that has its main focus on the possible link between a person's signature or handwriting with its behavioral traits. Few results are available showing acceptable results (see for example [1] where they use the structure of signature and handwriting). We have heard many accounts about how an expert or specialist in

X. Font · A. Delgado · M. Faundez-Zanuy (✉)
Tecnocampus (UPF), Mataró, Barcelona, Spain
e-mail: faundez@tecnocampus.cat

X. Font
e-mail: font@tecnocampus.cat

A. Delgado
e-mail: adelgadog@edu.tecnocampus.cat

A. Esposito et al. (eds.), *Progresses in Artificial Intelligence and Neural Systems*, Smart Innovation, Systems and Technologies 184,
https://doi.org/10.1007/978-981-15-5093-5_19

graphology assess analysis over the handwriting patterns that can distill personality evaluations on us. Obviously the claims are far from being accepted and evidence is not clear so far.

With the help of different sensors and the use of a well-established personality test (big five used in many context i.e academic achievement [2]), a set of experiments will be carried out to acquire data in two directions. First, a set of data generated when a user is actively writing. And second, a set of data coming from a survey. Although the former data set will have a major impact because it will be repeated many times, it is no less true that data coming from the survey play a role when associations want to be found in the ground of personal behavioral patterns.

Keep in mind that the output of the personality test will be treated as described from OCEAN factors:

- Openness
- Conscientiousness
- Extraversion
- Agreeableness
- Neuroticism

Some advice given by calligraphic experts to explain what your writing is saying about ourselves, like letter angle, size, volume or legibility will not be taken into account. The reason for this relays on the approach: let the data speak for themselves. Once the data will be correctly pre-processed a Deep Convolutional Neural Network (for general review see [3, 4]) will be used to figure out the links between these two sides: writing and personality.

It is worth mentioning that the data will let us check additional problems usually found in current journals like user identification or verification through signature or writing (see for instance [5] or [6]), gender identification [7, 8] or user handedness as well. Although several online signature handwritten databases exist and some of them are multimodal (see [9, 10] or [11]) to the best of our knowledge none of them acquired EEG during the signature process.

2 Experimentation and Sensors

2.1 Survey Data

The questionnaire was divided into four parts: the first one with the aim of acquiring possible key segmentation variables such as age, gender, etc. The second one was prepared to gather personality traits and fulfill some previously defined standards in the field (big five personality test [12, 13]). The last two parts were used as completely sided information where users pick a tree-drawing from a list (see [14]) and then, they select an animal as the one they would like to become. Once they have selected an animal, they list three descriptive adjectives that best describes it.

Fig. 1 Wacom Intuos Pro Large PTH-851 tablet

The survey can be reached at:
https://forms.gle/UbeFkatYt8Kr9cbt7

2.2 Sensors

Two sensors were used extensively in these preliminary study. The first sensor allow users to write sentence and perform signatures while all data were recorded.

Wacom Intuos Pro Large PTH-851 Tablet
The main characteristic of this tablet is that allows to record writing pressures with a press-sensitive pen. One of the main advantage of this tablet is its size making possible to use in different writing settings (see Fig. 1).

Information provided by the device:

- x and y positions
- pressure, azimut and angle
- time stamp

Cognionics Quick-30 Headset
This equipment for acquiring EEG signals has the following main characteristics:

- Combination of active electrodes and active shielding (30 eeg sensors)
- Resistant against electrical and movement artifacts
- Real-time measurement of sensor impedance
- 24-bits ADC low noise and high dynamic range inputs

As it is shown in Fig. 2 the headset can work with additional sensors such as ECG/EMG, Respiration, GSR and more. Given its ability to work in a wireless environment (see [15]) makes it a plus when designing the experiments.

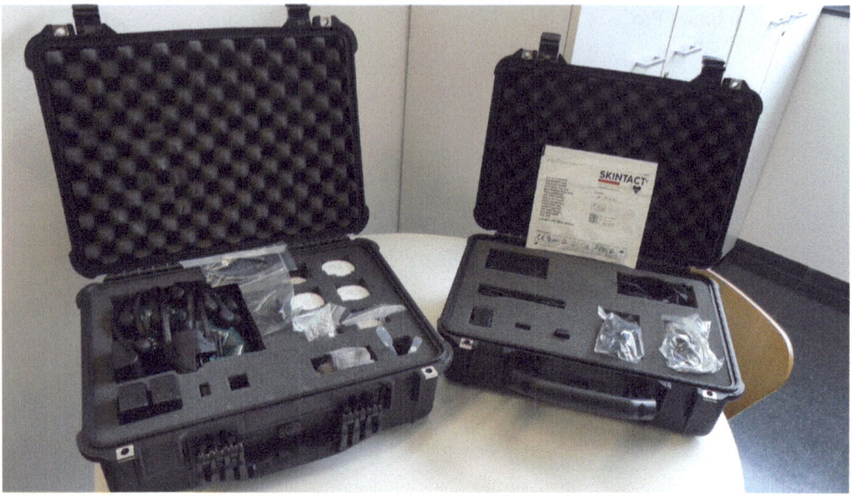

Fig. 2 Quick-30 headset

Table 1 Predesigned schedule for user acquisition in the first three sessions in 2019

Volunteer	Right/left-hand	Male/Female	1st ses.	2nd ses.	3rd ses.
User 3	R	M	22-May	29-May	05-June
User 10	R	M	24-May	30-May	05-June
User 27	R	M	30-May	06-June	16-June
User …	…	…	…	…	…

2.3 *Experimentation*

With the goal to reach 100 user and five experimentation sessions, the first steep acquire almost half of them (43 users) and obtain three sessions.

Table 1 shown the proposed schedule for different users through the first three experimentation session. Each session last almost 20 min and the user start following the protocol for correct data acquisition. Each user involved in the experimentation had to sign the data transfer—data protection agreement and fill out the survey. Once each user completed these two simple steps they could start the acquisition process.

The preparation of the tablet and the eeg headset takes no more than seven minutes. Thanks to the dry sensors the eeg usually report no further complications and all signals reach green lights. The acquisition (see Fig. 3) of the data starts with the user signature and then it follows the phrase writing according to the randomized experiment (see Table 2). The type of sentence was either a positive one: *I'm going to make my life a masterpiece*, or a negative one: *I am wasting my time, I'm useless*. The process of signature and writing was repeated six times. Thus, each session generate 6 signatures, 6 sentence writing, three of them positive and the remaining three negative ones.

Fig. 3 Experimentation process with the tablet and the EEG headset working alongside to acquire multimodal data

Table 2 Writing sentence order. The experiment was randomized, so user could not expect a predefined sentence (either positive or negative)

Session	Sent1	Sent2	Sent3	Sent4	Sent5	Sent6	Imitation
1a	Neg	Pos	Neg	Pos	Pos	Neg	1
2a	Neg	Pos	Pos	Neg	Pos	Neg	1
3a	Pos	Neg	Pos	Pos	Neg	Neg	1
…	…	…	…	…	…	…	1

3 Methods and Data Insights

Given the target variable as one of the possible personality traits (Openness, Conscientiousness, Extraversion, Agreeableness and Neuroticism) and some of the acquired writing data (see Fig. 4) a set of deep learning methods have been tested. From the first works on deep learning (see [16, 17]) one of the main advantages reported has been its ability to enhance image recognition with the well known Convolutional Neural Network architecture (see [18] or [19]).

The basic idea is to take a single vector as our target Y_i, where the sub index can be any of the desirable personality factor and the set of images coming from the writing process X. As a preliminary study only signatures have been used, and all of them have been represented as a black and white static images.

The approach taken is through an optimization process using a loss function (see Eq. 1). The Stochastic Gradient Descendent Algorithm (SGD) implemented in google TensorFlow API simplify the overall process.

$$L_n(\theta) = \sum_{i=1}^{n} l(f(X_i, \theta), Y_i) + \lambda\Omega(\theta) \quad (1)$$

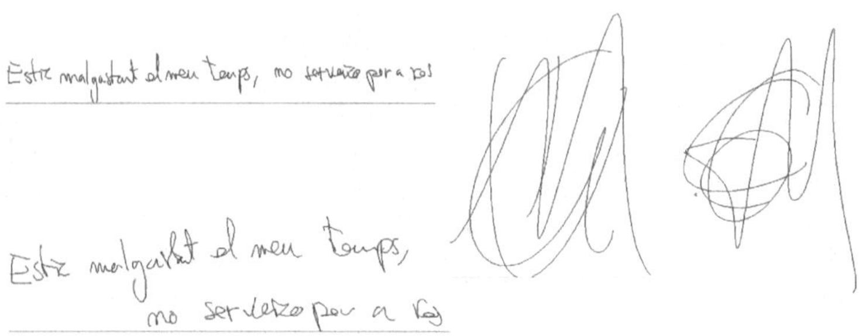

Fig. 4 Writing and signature example

4 Results and Discussion

In this paper we have proposed an experimentation protocol to validate the relation between online handwriting and personality. To the best of our knowledge this is the first paper devoted to this topic, which should be deeper analyzed. The problem to asses a correct prediction for the set of factors was not good enough. In fact each of the models tested failed, even with a completely different set of network algorithms parameters such as: learning rate ϵ, the batch size, and the number of epochs.

Why the results were not satisfactory? There are at least two clear reasons behind the performance of the network:

- First: The number of users and sessions are not good enough to make CNN models work properly.
- Second: The diversity of the people involved in the experimentation were not good enough as well. In fact, most of the users share the same socio-economical status and many common behavior can be found among them. Most of the users were from engineering studies with similar ages, interests and hobbies.

Although a number of variations are ready to use to enhance the model, like data augmentation (see [20] or [21]) or dropout (see [22]) none of them work.

4.1 Signature Identification

One problem that arises after the previous results was to check if a problem involving user identification through signature will give a kind of similar results.

Taking the same set of signature images, the problem of user identification was convincing. The performance of the model with the current signature data base reach a 80% accuracy (compare with signature recognition [23]). Again, larger database size is required to improve the results when dealing with deep learning. For small databases more simple algorithms outperform deep learning.

5 Conclusions

Although results are far from being considered good enough to be reported, the construction of a multimodal data base using two sensors add a clear advantage for future research on the topic. Additional deep learning models can take advantage of the two sensors to assess better predictions about the personality factors.

Because it may be very helpful to enroll users from different backgrounds, the future line of this work will be to add additional data to complete this preliminary data set, so that users with a more diversity background will be present in the upcoming studies.

Acknowledgements This work has been supported by FEDER and MEC, TEC2016-77791-C4-2-R and PID2019-109099RB-C42.

References

1. Djamal, E.C., Darmawati, R., Ramdlan, S.N.: Application image processing to predict personality based on structure of handwriting and signature. In: 2013 International Conference on Computer, Control, Informatics and Its Applications (IC3INA), pp. 163–168. IEEE, Nov 2013
2. Komarraju, M., Karau, S.J., Schmeck, R.R., Avdic, A.: The Big Five personality traits, learning styles, and academic achievement. Personal. Individ. Differ. **51**, 472–477 (2011)
3. Schmidhuber, J.: Deep Learning in Neural Networks: An Overview, Apr 2014
4. Voulodimos, A., Doulamis, N., Doulamis, A., Protopapadakis, E.: Deep learning for computer vision: a brief review. Comput. Intell. Neurosci. **2018**, 1–13 (2018)
5. Pascual-Gaspar, J.M., Faundez-Zanuy, M., Vivaracho, C.: Fast on-line signature recognition based on VQ with time modeling. Eng. Appl. Artif. Intell. **24**, 368–377 (2011)
6. Faundez-Zanuy, M., Pascual-Gaspar, J.M.: Efficient on-line signature recognition based on multi-section vector quantization. Pattern Anal. Appl. **14**, 37–45 (2011)
7. Sesa-Nogueras, E., Faundez-Zanuy, M., Roure-Alcobé, J.: Gender classification by means of online uppercase handwriting: a text-dependent allographic approach. Cogn. Comput. **8**, 15–29 (2016)
8. Faundez-Zanuy, M., Sesa-Nogueras, E.: Preliminary Experiments on Automatic Gender Recognition Based on Online Capital Letters, pp. 363–370 (2014)
9. Faundez-Zanuy, M., Fierrez-Aguilar, J., Ortega-Garcia, J., Gonzalez-Rodriguez, J.: Multimodal biometric databases: an overview. IEEE Aerosp. Electron. Syst. Mag. **21**(8), 29–37 (2006)
10. Fierrez, J., Galbally, J., Ortega-Garcia, J., Freire, M.R., Alonso-Fernandez, F., Ramos, D., Toledano, D.T., Gonzalez-Rodriguez, J., Siguenza, J.A., Garrido-Salas, J., Anguiano, E., Gonzalez-de Rivera, G., Ribalda, R., Faundez-Zanuy, M., Ortega, J.A., Cardeñoso-Payo, V., Viloria, A., Vivaracho, C.E., Moro, Q.I., Igarza, J.J., Sanchez, J., Hernaez, I., Orrite-Uruñuela, C., Martinez-Contreras, F., Gracia-Roche, J.J.: BiosecurID: a multimodal biometric database. Pattern Anal. Appl. **13**, 235–246 (2010)
11. Ortega-Garcia, J., Fierrez-Aguilar, J., Simon, D., Gonzalez, J., Faundez-Zanuy, M., Espinosa, V., Satue, A., Hernaez, I., Igarza, J.-J., Vivaracho, C., Escudero, D., Moro, Q.-I.: MCYT baseline corpus: a bimodal biometric database. IEE Proc. Vis. Image Signal Process. **150**(6), 395 (2003)
12. Cobb-Clark, D.A., Schurer, S.: The stability of big-five personality traits. Econ. Lett. **115**, 11–15 (2012)
13. Furnham, A., Cheng, H.: The Big-Five personality factors, mental health, and social-demographic indicators as independent predictors of gratification delay. Personal. Individ. Differ. **150**, 109533 (2019)

14. Stanzani Maserati, M., Matacena, C., Sambati, L., Oppi, F., Poda, R., De Matteis, M., Gallassi, R.: The tree-drawing test (Koch's Baum test): a useful aid to diagnose cognitive impairment. Behav. Neurol. **2015**, 1–6 (2015)
15. Chi, Y.M., Wang, Y., Wang, Y.-T., Jung, T.-P., Kerth, T., Cao, Y.: A Practical Mobile Dry EEG System for Human Computer Interfaces, pp. 649–655 (2013)
16. Lecun, Y.: Generalization and network design strategies. Technical report, University of Toronto (1989)
17. LeCun, Y., Bengio, Y., Hinton, G.: Deep learning. Nature **521**, 436–444 (2015)
18. Chollet, F.: Xception: Deep Learning with Depthwise Separable Convolutions, Oct 2016
19. Krizhevsky, A., Sutskever, I., Hinton, G.E.: ImageNet classification with deep convolutional neural networks. Commun. ACM **60**, 84–90 (2017)
20. Perez, L., Wang, J.: The Effectiveness of Data Augmentation in Image Classification Using Deep Learning, Dec 2017
21. Shima, Y.: Image augmentation for object image classification based on combination of pre-trained CNN and SVM. J. Phys. Conf. Ser. **1004**, 012001 (2018)
22. Salakhutdinov, N.S., Hinton, G., Krizhevsky, A., Sutskever, I., Ruslan, S.: Dropout: a simple way to prevent neural networks from overfitting. J. Mach. Learn. Res. **15**, 1929–1958 (2014)
23. Han, K., Sethi, I.K.: Handwritten signature retrieval and identification. Pattern Recognit. Lett. **17**, 83–90 (1996)

An Ensemble Based Classification Approach for Persian Sentiment Analysis

Kia Dashtipour, Cosimo Ieracitano, Francesco Carlo Morabito, Ali Raza, and Amir Hussain

Abstract In recent years, sentiment analysis received a great deal of attention due to the accelerated evolution of the Internet, by which people all around the world share their opinions and comments on different topics such as sport, politics, movies, music and so on. The result is a huge amount of available unstructured information. In order to detect positive or negative subject's sentiment from this kind of data, sentiment analysis technique is widely used. In this context, here, we introduce an ensemble classifier for Persian sentiment analysis using shallow and deep learning algorithms to improve the performance of the state-of-art approaches. Specifically, experimental results show that the proposed ensemble classifier achieved accuracy rate up to 79.68%.

Keywords Persian sentiment analysis · Natural language processing · Deep learning · Ensemble classifier

K. Dashtipour (✉)
Department of Computing Science and Mathematics, University of Stirling, Stirling, UK
e-mail: kd28@cs.stir.ac.uk

C. Ieracitano · F. Carlo Morabito
DICEAM, University Mediterranea of Reggio Calabria, Via Graziella, Feo di Vito, 89060 Reggio Calabria, Italy

A. Raza
Department of Electrical Engineering and Computing, Rochester Institute of Technology, Dubai, UAE

A. Hussain
School of Computing, Edinburgh Napier University, Merchiston Campus, Edinburgh EH10 5DT, UK

A. Esposito et al. (eds.), *Progresses in Artificial Intelligence and Neural Systems*, Smart Innovation, Systems and Technologies 184,
https://doi.org/10.1007/978-981-15-5093-5_20

1 Introduction

Sentiment analysis (SA) is the computational study of peoples behaviours, sentiments, emotions and attitudes through the extraction of significant information from unstructured big data (such as products and services). The rise of social media, blogs and forums allow people to share continuously comments and ideas on different topics. For example, in buying a consumer product, there is no longer any need to ask friends or families, as many reviews and discussions are available online on that specific product. Similarly, for a company or organization, there is not anymore need to conduct surveys to understand public opinions about their services as social media help to reorganize their business. SA uses natural language processing (NLP) to extract useful information such as identifying the polarity of the text from online data [9, 10].
There are different types of opinions available online:

- Explicit opinion: expresses opinion about an entity directly. For example, "I do not like Spider-man movie" .من فیلم مرد عنکبوتی دوست ندارم.
- Comparative opinions: expresses by comparing one entity by another entity. For example, "The spider-man movie is better than Avengers" .بهتر از انتقام جویّان است فیلم مرد عنکبوتی.

However, most of the current approaches are available in English and only a few works regard other languages such as Persian (the official language of Iran and Afghanistan that includes more than 80 million speakers). Hence, in order to fill the gap, we propose a novel framework for Persian sentiment analysis. Specifically, as deep learning (DL, [15, 16]) has been achieving impressive results in several real-world problems ([14, 18–21]), we propose an ensemble classifier based on shallow (Support Vector Machines (SVM), Multilayer Perceptron (MLP)) and DL (Convolution Neural Network (CNN)) techniques to detect polarity in Persian sentiment analysis. The hotel reviews dataset gathered from [3] is used to evaluate the proposed approach.

The rest of the paper is organized as follows: Sect. 2 presents related work. Section 3 presents methodology and experimental results. Finally, Sect. 4 concludes this paper.

2 Related Work

In the literature, extensive research was carried out to model novel sentiment analysis models using both shallow and deep learning algorithms. For example, García-Pablos et al. [13] proposed a supervised approach for aspect-based sentiment analysis. The approach was evaluated using different languages such as English, Spanish, French and Dutch. The hotel, restaurants and electronic devices datasets are used to evaluate the performance of the approach. Experimental results showed that the proposed approach (89%) obtained better accuracy as compared with CNN (82.31%) and LSTM (79.25%). Dashtipour et al. [8] proposed a framework for using deep learning classifiers to detect polarity in Persian movie reviews. The proposed deep learning classifiers were compared with shallow MLP. Simulation results reported that DL classifiers such as 1D-CNN (82.86%) showed better performance as compared with MLP (78.49%). Dos Santos et al. [11] proposed an approach to detect polarity in short text using deep learning classifiers. The character embedding used to convert the short text. Experimental results showed the CNN (86.4%) achieved better accuracy as compared with LSTM (85.7%). Poria et al. [30] proposed a novel framework for detecting concept sentiment analysis which merges linguistic patterns and machine learning to identify the polarity of the sentence. Experimental results showed the proposed approach (76.33%) achieved better performance as compared with sentic pattern (70.84%). However, the linguistic patterns are used for English sentiment analysis and it cannot detect polarity in any other language such as Persian. Sohangir et al. [33] proposed a model using deep learning classifiers to detect sentiment for StockTwits. Experimental results demonstrated that the CNN outperformed LSTM, doc2vec and logistic regression, achieving accuracy rate up to 90.93%. Al-Smadi et al. [2] proposed an aspect-based sentiment analysis for Arabic hotel reviews using LSTM and neural network. The character level bidirectional LSTM used for aspect extraction in Arabic hotel reviews. Experimental results demonstrated that the proposed approach (82.7%) achieved better performance as compared with the traditional approach with lexicon (76.4%). Nakayama et al. [25] proposed a method to detect polarity in Japanese hotel reviews. The Yelp hotel reviews used which consist of 157 million reviews. The results are compared with English reviews. The experimental results indicated the English reviews explicitly express the polarity in their comments as compared with Japanese reviews. Poria et al. [29] proposed a parser to breaks texts into words and extract meaningful concept from sentences. There are different rules are developed to identify the polarity of the sentence. The experimental results showed that the proposed approach (86.10%) received better performance as compared with the part-of-speech tag (92.21%). Ozturk et al. [27] proposed a novel method to detect polarity in public opinions towards the Syrian refugee crisis. There are 2,381,297 million tweets are collected in English and Turkish languages. The results indicate the Turkish tweets express more positive opinions towards Syrian refugees than English tweets. On the other hand, most of the English tweets expressed neutral sentiments. However, the proposed method did not use any machine learning classifiers. Li et al. [23] proposed the architecture for sentiment recognizer on the call centre. The proposed architecture used openSMILE to identify the polarity of the sentences. The experimental results indicated the architecture is successful to identify the polarity of the sentence. Minaee et al. [24] presents a model based on ensemble classification of deep learning classifiers to capture the temporal information of the data (English dataset). The experimental results indicated the ensemble classification (90%) achieved better accuracy as compared with CNN (89.3%) and LSTM (89%).

Table 1 Summarized results for different approaches

References	Purpose	Language	Approach	Accuracy
Chen et al. [6]	Detect polarity in Chinese reviews	Chinese	LSTM	78%
Shuai et al. [32]	Detect polarity in Hotel reviews	Chinese	SVM	81.16%
Kirilenko et al. [22]	Detect polarity for Tourism	English	SVM	75.23%
Al-Smadi et al. [1]	Detect sentiment in Arabic reviews	Arabic	Deep Recurrent neural network SVM	78%
Dashtipour et al. [7]	Feature combination for Persian	Persian	SVM	81.24%
Cambria et al. [5]	Develop lexicon for English	English	CNN	94.6%
Dragoni et al. [12]	Detect polarity in multidomain	English	SVM	88.81

Rogers et al. [31] developed a new corpus for Russian language. The data is collected from social media and it was labelled into positive, negative and neutral. After annotation, the data converted using word2vec (fastText) and TF-IDF. The experimental results demonstrated neural net classifier (71.7%) achieved better F-measure as compared with Linear SVM (62.6%) and logistic regression (63.2%). Hazarika et al. [17] proposed a model to detect sarcasm expressed in the text. The hybrid model employed content and context-driven for detecting sarcasm in social media discussions. CNN is trained to detect sarcasm in the sentence. Experimental results indicated discourse features and embedding of the words play important roles to detect sarcasm in the sentence. Peng et al. [28] proposed an approach to detect sentiment in Chinese reviews. The SemEval dataset used to evaluate the performance of the approach. Experimental results showed the proposed approach (75.59%) achieved better accuracy as compared with SVM (66.92%), LSTM (74.63%), BiLSTM (74.15%). In Table 1 some of the sentiment analysis approaches are depicted. However, the none of the aforementioned studies explored the ensemble classifier to detect polarity for Persian sentences. Most of the current approaches use instead ensemble classifier to identify the polarity for English sentences. In contrast, in the present research, we propose a novel ensemble based classification framework for Persian sentiment analysis.

3 Methodology

The proposed methodology includes three main stages: *data pre-processing, feature extraction* and *classification*. Each processing module is described as follows.

3.1 Data Description and Pre-processing

The hotel reviews dataset gathered from [3] is used in this work. consists of 3000 reviews: 1500 positive and 1500 negative. For classification purpose, 60% of data is used as a train set, 30% as a test set and finally and 10% as a validation set. The corpus was pre-processed using the following techniques: *tokenisation*, *normalisation* and *stemming*.

- Tokenisation is used to convert the sentences into words or tokens. For example, فِلم خوب بود (Movie is great), is divided into بود, خوب, فیلم ; or, مر۳۰ (Thanks30) is converted into مرسي.
- Normalisation technique is used to convert these words into their normal forms;
- Stemming is the process of converting words into their roots.

3.2 Feature Extraction

After the data pre-processing, the N-gram (unigram, bigram and trigram) features are extracted. N-grams represent continuous sequences of *n* items in the text. When n = 1 the ngram is called unigram, when n = 2 bigram, when n = 3 trigram and so on. For example, “I like this movie” من این فیلم دوست دَارم . The unigram is من, این, فیلم, دوست, دَارم, “I”, “like”, “this”, “movie”. The bigram features are من فیلم, فیلم دوست, دوست دَارم and trigram features are من فیلم دوست and فیلم دوست دَارم.

3.3 Classification

In order to classify negative and positive reviewers, standard (i.e. SVM, MLP) and deep (CNN) machine learning classifiers are employed. In order to train the classifiers, we used word2vec. Specifically, the sentences are converted into 300-dimensional vectors by using fastText (a Python package) [4].

Convolutional neural network (CNN). The proposed CNN consists of 11 layers (4 convolution layers, 4 max pooling and 3 fully connected layers). Convolution layers have 15 filters sized 1×2 with stride size 1. Each convolution layer is followed by a max polling layer with window size 1×2. The last max pooling layer is followed by a standard MLP with hidden layers size 5000, 500 and 4. In the final layer, softmax activation function is used for classification purpose.

Support Vector Machine. The support vector machines (SVM) is used to find the decision boundary to separate different classes. The Sklearn Python package is used to train the proposed SVM classifier. In addition, the linear kernel is used.

Multilayer Perceptron (MLP). The MLP is a supervised machine learning technique. It typically consists of an input, hidden and output layer. Here, a single hidden layer MLP with 50 hidden units was developed and trained for 100 iterations. A softmax output layer was then used for positive vs negative classification task. In addition, the alpha 0.5 and adaptive learning rate is used.

Ensemble Classifier. Ensemble method consists of employing different classifiers and combining their predictions to train a meta-learning model. The ensemble is typically used to enhance the accuracy of a specific system [26]. In this study, the predictions of the aforementioned classifiers (SVM, MLP, 1D-CNN) are used as input of a linear-SVM based architecture. Figure 1 shows the proposed ensemble classification system. It is to be noted that the parameters and topology of each classifier has been set-up empirically after several simulation experiments.

3.4 Experimental Results

In order to evaluate the performance of the proposed approach, precision, recall, F-measure and accuracy metrics were used:

$$Precision = \frac{TP}{TP + FP} \tag{1}$$

$$Recall = \frac{TP}{TP + FN} \tag{2}$$

$$F\_measure = 2 * \frac{Precision * Recall}{Precision + Recall} \tag{3}$$

$$Accuracy = \frac{TP + TN}{TP + TN + FP + FN} \tag{4}$$

where TP denotes true positive, TN is true negative, FP is a false positive, and FN is false negative. The experimental results are shown in Table 2. As can be seen, MLP achieves accuracy of 74.26%, 72.22%, 63.23% when unigram, bigram, trigram were

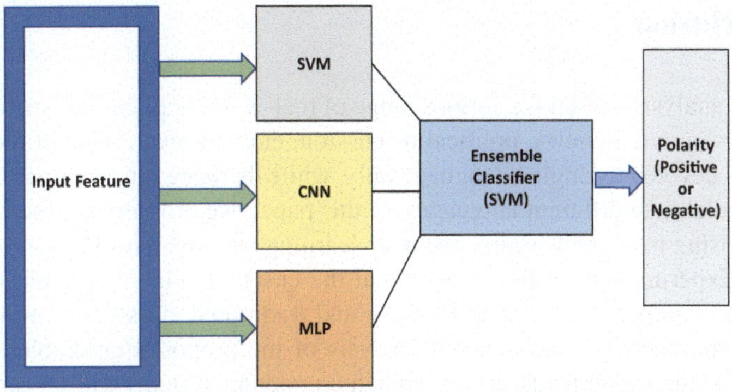

Fig. 1 Proposed ensemble classifier

Table 2 Results of the proposed SVM, MLP, 1D-CNN and ensemble classifier

Method	Input	Precision	Recall	F-measure	Accuracy
SVM	Unigram	0.68	0.64	0.64	68.25
	Bigram	0.71	0.65	0.59	64.92
	Trigram	0.69	0.59	0.45	58.80
MLP	Unigram	0.76	0.74	0.74	74.26
	Bigram	0.74	0.72	0.71	72.22
	Trigram	0.69	0.63	0.56	63.23
1D-CNN	Unigram	0.73	0.80	0.76	78.02
	Bigram	0.71	0.76	0.72	74.25
	Trigram	0.70	0.73	0.71	72.45
SVM (ensemble classifier)	Unigram	0.79	0.78	0.76	79.68
	Bigram	0.80	0.79	0.75	78.18
	Trigram	0.73	0.80	0.76	78.02

used as input respectively. As regards SVM the optimal result was observed with unigram features input (accuracy of 68.25%). The 1D-CNN, instead achieved better classification accuracy, reporting 78.02%. However, the proposed ensemble classifier (based on SVM) outperforms all the other classifiers using unigram features achieving an accuracy rate up to 79.68%. The ensemble classifier successfully identify the overall polarity of the sentence. For example, the polarity of the sentence "I really liked comedy movies" (من فیلم کمدي دوست دَارم) was correctly detect as positive.

4 Conclusion

Sentiment analysis is used for various range of real-world applications such as product reviews, movie reviews, political discussion, etc. However, most of the current research is devoted to English language only, while there are lots of important information available in different languages. In this paper, we propose an ensemble classification using machine learning and deep learning classifiers for Persian sentiment analysis. Experimental results showed that the ensemble classifier achieved better accuracy as compared with deep learning and traditional classifiers. In the future, a more comprehensive and detailed analysis of the proposed ensemble approach (including statistical considerations of each competing system) will be carried out. In addition, we intend to build a novel approach to detect polarity in multilingual sentiment analysis using ensemble classifier.

References

1. Al-Smadi, M., Qawasmeh, O., Al-Ayyoub, M., Jararweh, Y., Gupta, B.: Deep recurrent neural network vs. support vector machine for aspect-based sentiment analysis of Arabic hotels reviews. J. Comput. Sci. **27**, 386–393 (2018)
2. Al-Smadi, M., Talafha, B., Al-Ayyoub, M., Jararweh, Y.: Using long short-term memory deep neural networks for aspect-based sentiment analysis of Arabic reviews. Int. J. Mach. Learn. Cybern. 1–13 (2018)
3. Alimardani, S., Aghaie, A.: Opinion mining in Persian language using supervised algorithms (2015)
4. Bojanowski, P., Grave, E., Joulin, A., Mikolov, T.: Enriching word vectors with subword information. arXiv:1607.04606 (2016)
5. Cambria, E., Poria, S., Hazarika, D., Kwok, K.: SenticNet 5: discovering conceptual primitives for sentiment analysis by means of context embeddings. In: Thirty-Second AAAI Conference on Artificial Intelligence (2018)
6. Chen, H., Li, S., Wu, P., Yi, N., Li, S., Huang, X.: Fine-grained sentiment analysis of Chinese reviews using LSTM network. J. Eng. Sci. Technol. Rev. **11**(1) (2018)
7. Dashtipour, K., Gogate, M., Adeel, A., Hussain, A., Alqarafi, A., Durrani, T.: A comparative study of Persian sentiment analysis based on different feature combinations. In: International Conference in Communications, Signal Processing, and Systems, pp. 2288–2294. Springer (2017)
8. Dashtipour, K., Gogate, M., Adeel, A., Ieracitano, C., Larijani, H., Hussain, A.: Exploiting deep learning for Persian sentiment analysis. In: International Conference on Brain Inspired Cognitive Systems, pp. 597–604. Springer (2018)
9. Dashtipour, K., Hussain, A., Zhou, Q., Gelbukh, A., Hawalah, A.Y., Cambria, E.: PerSent: a freely available Persian sentiment lexicon. In: International Conference on Brain Inspired Cognitive Systems, pp. 310–320. Springer (2016)
10. Dashtipour, K., Poria, S., Hussain, A., Cambria, E., Hawalah, A.Y., Gelbukh, A., Zhou, Q.: Multilingual sentiment analysis: state of the art and independent comparison of techniques. Cogn. Comput. **8**(4), 757–771 (2016)
11. Dos Santos, C., Gatti, M.: Deep convolutional neural networks for sentiment analysis of short texts. In: Proceedings of COLING 2014, the 25th International Conference on Computational Linguistics: Technical Papers, pp. 69–78 (2014)

12. Dragoni, M., Petrucci, G.: A fuzzy-based strategy for multi-domain sentiment analysis. Int. J. Approx. Reason. **93**, 59–73 (2018)
13. García-Pablos, A., Cuadros, M., Rigau, G.: W2VLDA: almost unsupervised system for aspect based sentiment analysis. Expert Syst. Appl. **91**, 127–137 (2018)
14. Gasparini, S., Campolo, M., Ieracitano, C., Mammone, N., Ferlazzo, E., Sueri, C., Tripodi, G., Aguglia, U., Morabito, F.: Information theoretic-based interpretation of a deep neural network approach in diagnosing psychogenic non-epileptic seizures. Entropy **20**(2), 43 (2018)
15. Gogate, M., Adeel, A., Hussain, A.: Deep learning driven multimodal fusion for automated deception detection. In: 2017 IEEE Symposium Series on Computational Intelligence (SSCI), pp. 1–6. IEEE (2017)
16. Gogate, M., Adeel, A., Marxer, R., Barker, J., Hussain, A.: DNN driven speaker independent audio-visual mask estimation for speech separation. arXiv:1808.00060 (2018)
17. Hazarika, D., Poria, S., Gorantla, S., Cambria, E., Zimmermann, R., Mihalcea, R.: Cascade: contextual sarcasm detection in online discussion forums. arXiv:1805.06413 (2018)
18. Ieracitano, C., Adeel, A., Gogate, M., Dashtipour, K., Morabito, F.C., Larijani, H., Raza, A., Hussain, A.: Statistical analysis driven optimized deep learning system for intrusion detection. In: International Conference on Brain Inspired Cognitive Systems, pp. 759–769. Springer (2018)
19. Ieracitano, C., Adeel, A., Morabito, F.C., Hussain, A.: A novel statistical analysis and autoencoder driven intelligent intrusion detection approach. Neurocomputing (2019)
20. Ieracitano, C., Mammone, N., Bramanti, A., Hussain, A., Morabito, F.C.: A convolutional neural network approach for classification of dementia stages based on 2D-spectral representation of EEG recordings. Neurocomputing **323**, 96–107 (2019)
21. Ieracitano, C., Mammone, N., Hussain, A., Morabito, F.C.: A novel multi-modal machine learning based approach for automatic classification of EEG recordings in dementia. Neural Netw. (2019)
22. Kirilenko, A.P., Stepchenkova, S.O., Kim, H., Li, X.: Automated sentiment analysis in tourism: comparison of approaches. J. Travel Res. **57**(8), 1012–1025 (2018)
23. Li, B., Dimitriadis, D., Stolcke, A.: Acoustic and lexical sentiment analysis for customer service calls. In: ICASSP 2019-2019 IEEE International Conference on Acoustics, Speech and Signal Processing (ICASSP), pp. 5876–5880. IEEE (2019)
24. Minaee, S., Azimi, E., Abdolrashidi, A.: Deep-sentiment: sentiment analysis using ensemble of CNN and Bi-LSTM models. arXiv:1904.04206 (2019)
25. Nakayama, M., Wan, Y.: Is culture of origin associated with more expressions? An analysis of yelp reviews on japanese restaurants. Tour. Manag. **66**, 329–338 (2018)
26. Onishi, A., Natsume, K.: Overlapped partitioning for ensemble classifiers of P300-based brain-computer interfaces. PloS One **9**(4), e93045 (2014)
27. Öztürk, N., Ayvaz, S.: Sentiment analysis on Twitter: a text mining approach to the syrian refugee crisis. Telemat. Inform. **35**(1), 136–147 (2018)
28. Peng, H., Ma, Y., Li, Y., Cambria, E.: Learning multi-grained aspect target sequence for Chinese sentiment analysis. Knowl. Based Syst. **148**, 167–176 (2018)
29. Poria, S., Hussain, A., Cambria, E.: Concept extraction from natural text for concept level text analysis. In: Multimodal Sentiment Analysis, pp. 79–84. Springer (2018)
30. Poria, S., Hussain, A., Cambria, E.: Sentic patterns: sentiment data flow analysis by means of dynamic linguistic patterns. In: Multimodal Sentiment Analysis, pp. 117–151. Springer (2018)
31. Rogers, A., Romanov, A., Rumshisky, A., Volkova, S., Gronas, M., Gribov, A.: RuSentiment: an enriched sentiment analysis dataset for social media in Russian. In: Proceedings of the 27th International Conference on Computational Linguistics, pp. 755–763 (2018)
32. Shuai, Q., Huang, Y., Jin, L., Pang, L.: Sentiment analysis on Chinese hotel reviews with Doc2Vec and classifiers. In: 2018 IEEE 3rd Advanced Information Technology, Electronic and Automation Control Conference (IAEAC), pp. 1171–1174. IEEE (2018)
33. Sohangir, S., Wang, D., Pomeranets, A., Khoshgoftaar, T.M.: Big data: deep learning for financial sentiment analysis. J. Big Data **5**(1), 3 (2018)

Insects Image Classification Through Deep Convolutional Neural Networks

Francesco Visalli, Teresa Bonacci, and N. Alberto Borghese

Abstract We present and discuss results of the application of a deep convolutional network model developed for the automatic recognition of images of insects. The network was trained using transfer learning on an architecture called MobileNet, specifically developed for mobile applications. To fine tune the model, a grid-search on hyperparameters space was carried out reaching a final accuracy of 98.39% on 11 classes. Fine-tuned models were validated using 10-fold cross validation and the best model was integrated into an Android application for practical use. We propose solving the "open set" problem through feed-back collected with the application itself. This work also led to the creation of a well-structured image dataset of some important species/genera of insects.

Keywords Visual recognition · Insect classification · Deep learning · Convolutional neural networks

1 Introduction

Insects are the largest class of the animal kingdom in our planet counting over a million of species. Within this vast taxonomic group of animals we find many vectors of various pathogens responsible for diseases and zoonosis; some of these insects infest foodstuffs and stored products, others cause the loss of entire crops and alter the quality of cultivated products and their derivatives. In order to limit their impact,

F. Visalli · N. A. Borghese (✉)
Department of Computer Science, Università degli Studi di Milano, Milan, Italy
e-mail: alberto.borghese@unimi.it

F. Visalli
e-mail: francesco.visalli@studenti.unimi.it

T. Bonacci
DiBEST Department, Università della Calabria, Rende, Italy
e-mail: teresa.bonacci@unical.it

A. Esposito et al. (eds.), *Progresses in Artificial Intelligence and Neural Systems*, Smart Innovation, Systems and Technologies 184,
https://doi.org/10.1007/978-981-15-5093-5_21

a timely intervention is often essential. For this reason, tools that would allow the immediate classification of specific species of insects are needed.

This problem can be recasted as an image classification task, that is an instance of supervised learning problems. In classification tasks, we train a model $f : X \rightarrow Y$ on a set of labeled data, called training set, composed of pairs (x, y) where $y \in Y$ is the correct label for the instance $x \in X$. A supervised learning algorithm modifies progressively $f(.)$ to minimize an error, associated to misclassification and specific for the task so that the correct label $y \in Y$ can be assigned to any input instance of $x \in X$.

Convolutional Neural Networks (CNNs) are kinds of multi-layer neural network architectures specialized in finding patterns within images. They have become the state of the art for image classification since 2012, when AlexNet [12] won the ImageNet Large Scale Visual Recognition Competition[1] (ILSVRC) beating others no neural network methods and achieving an error rate of 16.4% on top-5 [16].

Since then, the enormous strand of learning algorithms working on networks endowed with many layers (named deep networks) has begun for the solution of visual recognition tasks. These models obtain remarkable results thanks also to the increase in GPUs computing power that allows the execution of complex learning algorithms in reasonable amount of time.

We propose here a particular CNN fine-tuned on a MobileNet [8] architecture, that was itself pre-trained on ImageNet. MobileNet is an architecture designed specifically for mobile applications. This architecture was chosen for its simplicity, because it can run on smart phones and for future experiments.

Because there are no specific databases of insects, we built our own dataset, that counts 13,588 images (belonging to 11 classes) and we leveraged this dataset to train MobileNet. Different hyper-parameters were used. The models obtained were validated through 10-fold cross validation and the best model was integrated into an Android application.

Thanks to this application we have been able to propose an operational transversal solution to an open problem in visual recognition: the "open set" recognition [19]. This occurs when we have to recognize if an image belongs to one of the specified classes or it belongs to a different external class. This information can be collected by the users themselves who take a picture of insects from their smart-phone, see the current classification proposal of the system with the associated degree of confidence, and, in case the insect has not been correctly classified, he/she can report this to the system that can refine, in batches, the classification output.

Finally, in order to investigate the nature of the features that the network learnt, we tested our model on a small dataset of insects that do not belong to the classes identified, including various insects similar to those present in the dataset on which the network was trained.

[1] http://www.image-net.org/challenges/LSVRC/.

2 Related Work

CNNs trained on large datasets such as ImageNet, that include a large number of images of insects, have already been proved to be capable of classifying them. However, to the best of our knowledge, there are few existing works [7, 14], specialized in the automatic classification of images of these invertebrates. Moreover, there are not satisfying works dealing with the "open set" problem, that for insects that belong to millions of species, is a real issue.

In addition, there was the need to create an accurate and well-structured dataset targeted to classification of insects of medical and agronomic interest, taking into account that at certain levels of detail even the taxonomist expert needs a thorough morphological study for the identification of species.

3 Methodology

3.1 Transfer Learning

Training a neural network from scratch is a difficult task which requires a huge amount of resources like computational power, a lot of data and time; nevertheless it frequently leads to overfitting. These kinds of neural networks are also rich in hyperparameters that need to be tuned. To solve these issues we trained our network using transfer learning [4, 15, 20]. This is a machine learning method that allows the knowledge transfer between domains; in neural networks knowledge is embedded in the network weights. The idea behind transfer learning on CNNs is that features learned in the first layers are often the same regardless of the domain. The more similar are domains, the more are the shared features, the less is the training needed.

We exploited two aspects of transfer learning: first we trained, on the top of a MobileNet [8], a Softmax classifier. Such classifier allows interpreting its output as a degree of similarity of the image with one of the identified 11 classes. Then, we fine-tuned MobileNet specializing it on the features of insects of our interest.

3.2 MobileNet

We chose the MobileNet architecture because it was designed for mobile development. In particular, it can do the forward pass on client side. Furthermore, its simplicity makes it a good candidate for future experiments and developments.

MobileNet is a 28 layers CNN that uses depth-wise separable convolutional layers instead of classical convolutional layers. Depth-wise separable convolution operation splits the convolution operation in two phases: first, the input is filtered in a depth-wise convolutional layer, then the output of the first phase is combined in a separate point-wise convolutional layer. In classical convolution these two steps are merged into one. This split reduces the number of operations required for the convolution without degrading the performances. This is particularly suitable to RGB images where convolution can be computed separately on the three different channels. For more details on depth-wise separable convolution we refer to the original paper [8].

The first layer of the net is a classic convolutional layer, then there are 26 layers alterning depth-wise convolution and point-wise convolution. Between a convolution operation and another one, there is a batch normalization operation followed by a ReLU6 activation function The ReLU 6 is a variant of the classical ReLU which makes the network more robust than regular ReLU when using low-precision computation [11].

The last layer of the network is a fully connected layer preceded by a global average pooling operation. The final classification is made by a Softmax classifier.

4 Training

4.1 Dataset

There are a lot of challenges that any algorithm of image classification has to face off including: image occlusion, illumination conditions, image deformation, viewpoint variation, scale variation, intra-class variation, background clutter and so on. In supervised learning we try to tackle these issues through a data-driven approach. To correctly address such problems, the dataset on which the algorithm learns should be as various as possible and cover all possible variants.

Our dataset is composed of 11 classes. Each class counts a mean of 1000 images (except one that was not present in the laboratory when the dataset was created) for a total amount of 13,588 images. We collected all images for insects of our interest from Google Images through a script, and then we cleaned the results by hand. We integrated these images with photos took in the laboratory (Fig. 1). The classes included into the dataset are the following: *Brachinus* sp., *Chrysolina* sp., *Chrysomela* sp., *Cucujus* sp., *Graphosoma* sp., *Cucujus* sp., *Leptinotarsa decemlineata*, *Nezara viridula*, *Pyrochroa* sp., *Rhynchophorus ferrugineus*, *Vespa crabro*, *Vespula* sp. The classes were chosen because they are dangerous to agriculture, dangerous to other animal and plant species or simply because they are common.

The dataset was splitted, in a random but reproducible way,[2] into training set and test set using 90% and 10% of images for each class, respectively.

[2]https://cs230.stanford.edu/blog/split/.

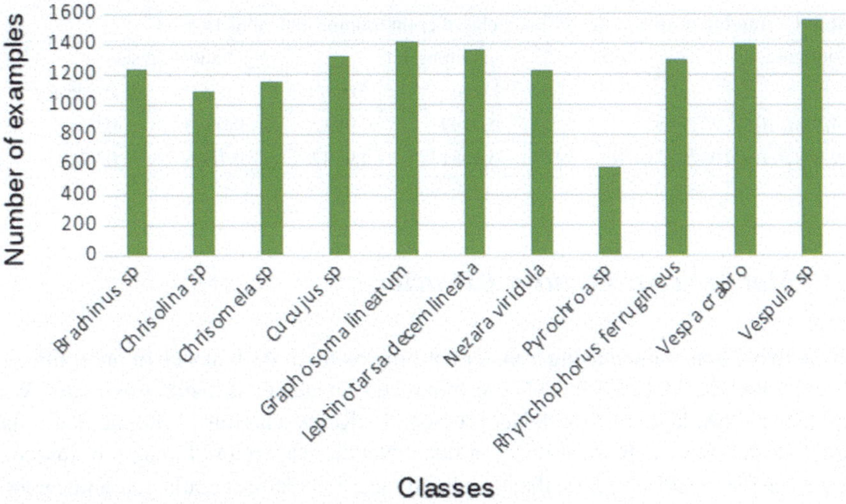

Fig. 1 Number of examples within the classes

4.2 Training Settings

In order to train our network we leveraged Keras[3] with TensorFlow[4] as back-end. Keras is an open-source, high level library, written in Python, for deep learning. It provides several models. We focused on CNN models pre-trained on ImageNet. In particular, we leveraged the MobileNet architecture described above.

In order to achieve the best accuracy, we set the multiplier width and resolution to 1. We used the cross entropy function as loss function that is the best choice using Softmax classifiers.

The dataset was constructed in such a way that the number of examples within the classes is balanced (Fig. 1). Moreover, given the nature of the problem, we used the accuracy as metric to evaluate the results.

Every training run ends with early stopping, 10% of training set images for each class was used to build the validation set. We monitored the loss on the validation set and stopped the training process after 4 epochs that the loss did not decrease anymore.

Models were trained and validated on a NVIDIA GeForce GTX 950M.

[3]https://keras.io/.

[4]https://www.tensorflow.org/.

Table 1 Training results of the Softmax classifier on training and validation set

Configuration	Epochs	Training set		Validation set	
		Loss	Accuracy	Loss	Accuracy
$(Adam, 10^{-4}, 2^3)$	5	0.0599	0.9842	0.0908	0.9737
$(RMSProp, 10^{-4}, 2^4)$	10	0.0081	0.9982	0.1066	0.9742

4.3 MobileNet as Features Extractor

The network was trained using transfer learning method. As first step of our training, we exploited the MobileNet CNN pre-trained on ImageNet as features extractor. We replaced the last layer of MobileNet provided by Keras with our 11 outputs Softmax classifier. As previously mentioned, in ImageNet there are a lot of images of insects, so we can reasonably suppose that the already learned features could guarantee good results already at this step.

We froze all the weights of the network and did a grid-search for tuning hyper-parameters in the following space:

$$\text{learning rate: } \{log_{10}b = c\}, c \in \{-6, \ldots, -1\}$$
$$\text{batch size: } \{2^a\}, a \in \{3, \ldots, 6\}$$
$$\text{optimizers: } \{Adam, RMSProp\}$$

Because the learning rate is a multiplier, it is usually searched in logarithmic space, while the batch size is commonly set as power of 2 for efficient computation. We adopted two optimizers here: Adam algorithm [10] is one of the best choice to optimize the gradient descent in CNNs and RMSprop [3] is the optimizer used in the original MobileNet paper. The hyper-parameters of the optimizers are the default ones suggested by Keras.

Let's introduce the tuple $(optimizer, learning\,rate, batchsize)$ to define a configuration of hyperparameters. Table 1 shows results about the two best models calculated on the training and validation set. The best hyper-parameters configurations are: $(Adam, 10^{-4}, 2^3)$ and $(RMSProp, 10^{-4}, 2^4)$. Training results are interesting already in this phase. The best results are obtained with batch sizes in $\{2^3, 2^4\}$ and with learning rate in $\{10^{-5}, 10^{-4}, 10^{-3}\}$. Learning rate of 10^{-1} was too high, whereas a learning rate of 10^{-6} was too slow in convergence. The grid-search took about 3 days.

4.4 MobileNet: Fine Tuning

In this phase we fine-tuned the best models obtained in the previous step leaving the weights of the network free to vary. In such way, the network can learn features for

Table 2 Results of the fine tuning on training and validation set

Configuration	Epochs	Training set		Validation set	
		Loss	Accuracy	Loss	Accuracy
$(Adam, 10^{-5}, 2^3)$	13	0.0041	0.9995	0.0448	0.9845
$(RMSProp, 10^{-4}, 2^3)$	7	5.2790e−04	0.9997	0.0460	0.9860

the classification of the insects specific of this application. Since we have a fairly large dataset we fine-tuned all the layers of MobileNet.

Results of the previous phase suggested us to use a small batch size and a small learning rate. We searched hyper-parameters through grid-search in the following space:

$$\text{learning rate: } \{log_{10}b = c\}, c \in \{-6, \ldots, -3\}$$
$$\text{batch size: } \{2^a\}, a \in \{3, 4, 5\}$$
$$\text{optimizers: } \{Adam, RMSProp\}$$

The best configuration are $(Adam, 10^{-5}, 2^3)$ and $(RMSProp, 10^{-4}, 2^3)$ both of them obtained from the fine tuning of $(Adam, 10^{-4}, 2^3)$. Results about the two best models calculated on the training and validation set are presented in Table 2. The grid-search on both models took about three and a half days.

5 Validation

We validated the two models obtained from fine tuning phase through 10-fold cross validation. 10 blocks of images were randomly extracted from the initial training set. The blocks are composed of 1/10 of the total images for each class. These blocks were used in turn as a validation set using each time the 9/10 of the remaining images as training set.

The cross validation error of $(Adam, 10^{-5}, 2^3)$, obtained by averaging the single validation loss for each run, is 0.0093, whereas the cross validation error of $(RMSProp, 10^{-4}, 2^3)$ is 0.0177. Therefore, the best model is obtained by fine tuning $(Adam, 10^{-4}, 2^3)$ (i.e. the model obtained by training only the Softmax classifier from the first phase) with hyper-parameters $(Adam, 10^{-5}, 2^3)$. The cross-validation on both models took about one day.

6 Results and Discussion

The model proposed here is based on stacking convolutional layers one on top of the other, adding eventually a pooling layer in between. Although convolutional neural networks have been popularized inside the deep-learning domain, they were proposed in the early nineties by the group of Slotine [18] in the domain of radial basis function networks and further developed inside a hierarchical framework with real-time learning by the group of Borghese [2, 5, 6]. Similar concepts are also well-known in the mathematical domain where functional approximation through function bases is largely adopted.

The accuracy calculated on the test set with the best model is 98.39%. This should not come as a surprise given the large number of parameters implemented by such networks. The particular nature of the task and the fact of having built a custom dataset does not allow us to compare results with any existing work.

We know that as the number of classes increases, the accuracy of the network may deteriorate. However, there are many solutions that we could adopt. We could change the network architecture: MobileNet belongs to the first generation of "mobile CNNs" as well as ShuffleNet [22]. We could leverage more sophisticated networks like MobileNet V2 [17] or ShuffleNet V2 [13]. We could use the "Squeeze-and-Excitation" (SE) blocks in our MobileNet, introduced in SENet [9] that is the winner of the task of image classification of ILSVRC 2017.

In all cases, the claim would be that the numerosity of the dataset should be increased to improve the recognition rate. This is a mantra of all deep-learning algorithms and it resembles the mantra of classical Artificial Intelligence, for which any artificial intelligence would approach human intelligence provided that as complex enough local function is implemented inside it. We remark that some peculiar characteristics of human brain and reasoning are missing in these pictures.

6.1 Application

The classification network was integrated into an Android application (Fig. 2) to field test the model and for practical use. It allows loading a picture from the phone's memory or take a picture on the spot. Once the image is chosen, in order to get a better classification, the application allows selecting the portion that contains the insect.

Results of the classification are presented displaying the thumbnail prototypical image of each class and the probability value associated. This would tackle also the problem of "open set" recognition: what would happen if we try to classify an insect that is not part of the dataset? The model, due to the nature of the classification task, would give a result based on the most similar matching between the image and the features that the network learnt.

Fig. 2 Screenshot of the application, example of classification

There are two possible cases: the insect is not into the dataset or the image is misclassified. In both cases, we will provide a feedback mechanism with which the user can send a report containing the image and the top-5 / top-10 results. In this way if the insect is not into the dataset, we could decide to insert it. Otherwise, if the algorithm misclassifies the image, we will have the opportunity to investigate why the image was not correctly classified and to integrate the dataset with different images for a future training. Therefore, we can say that the thumbnail provides a self-evaluation mechanism for the user and allows increasing the size of the database with annotated images easily.

6.2 *Test on the Features*

Neural networks, particularly CNNs, are a sort of black box. While the first layer features are human readable, those of deeper levels are hard to understand. Some efforts have been done to try to understand how CNNs work internally, that is catalogued under "open deep-neural network" research stream [21]. However, no results are provided up to now on internal codes used by such networks. This shortcoming is shared with classical neural networks for which the output of hidden layers was almost never explored with a few remarkable exception (e.g. [23]). We remark here that, given the large number of parameters in the network, the same results can be

Fig. 3 On the left a training class (*Cucujus* sp.), on the right an external class (*Pediacus depressus*) mostly classified as *Cucujus* sp. because of its shape

obtained with different outputs of the hidden layers [1] and it is not clear yet if there is any hidden output that is biologically plausible, contains a certain code and it is common across different people brains, or hidden output can vary largely from individual to individual.

We classified 19 external classes of images that were similar to those into the training set. Results were predictable: the network looks for shape/color matchings between taining and external classes (Fig. 3). The external classes are the following: *Adalia bipunctata*, *Aelia acuminate*, *Anchomenus dorsalis*, *Carpocoris pudicus*, *Corizus hyoscyami*, *Curculionidae*, *Dolichovespula* sp., *Eurygaster maura*, *Leistus* (*Pogonophorus*), *Lema daturaphila*, *Lilioceris* sp., *Nebria* (*Eunebria*) sp., *Pediacus depressus*, *Polistes* sp., *Pyrochroidae*, *Sulcopolistes* sp., *Tenthredo notha*, *Tenthredo scrophulariae*, *Vespa orientalis* (cf. Fig. 3). Further investigation are needed in order to derive some insights on the features that the network learnt.

7 Conclusions

We built a custom dataset of images of insects on which we fine-tuned a mobile CNN: MobileNet. The best models from fine tuning phase were validated through 10-fold cross validation and the one with less cross validation error was tested on the test set obtaining 98.39% of accuracy. Finally, the network was integrated into an Android application. We have tried to give a transversal contribution to the open problem of "open set" recognition. We tried to classify a dataset composed of 19 external classes similar to those present into the training set in order to investigate the black box infrastructure of CNNs.

References

1. Borghese, N.A., Arbib, M.A.: Generation of temporal sequences using local dynamic programming. Neural Netw. **8**(1), 39–54 (1995). https://doi.org/10.1016/0893-6080(94)00053-O
2. Borghese, N.A., Ferrari, S.: Hierarchical RBF networks and local parameters estimate. Neurocomputing **19**, 259–283 (1998)
3. Dauphin, Y.N., de Vries, H., Chung, J., Bengio, Y.: RMSProp and equilibrated adaptive learning rates for non-convex optimization. CoRR arXiv:1502.04390 (2015)
4. Donahue, J., Jia, Y., Vinyals, O., Hoffman, J., Zhang, N., Tzeng, E., Darrell, T.: DeCAF: a deep convolutional activation feature for generic visual recognition. In: Proceedings of the 31th International Conference on Machine Learning, ICML 2014, Beijing, China, JMLR Workshop and Conference Proceedings, vol. 32, pp. 647–655. JMLR.org, 21–26 June 2014. http://proceedings.mlr.press/v32/donahue14.html
5. Ferrari, S., Bellocchio, F., Piuri, V., Borghese, N.A.: A hierarchical RBF online learning algorithm for real-time 3-D scanner. IEEE Trans. Neural Netw. **21**(2), 275–285 (2010). https://doi.org/10.1109/TNN.2009.2036438
6. Ferrari, S., Maggioni, M., Borghese, A.: Multiscale approximation with hierarchical radial basis functions networks. IEEE Trans. Neural Netw. (a publication of the IEEE Neural Networks Council) **15**, 178–188 (2004). https://doi.org/10.1109/TNN.2003.811355
7. Glick, J., Miller, K.: Insect classification with heirarchical deep convolutional neural networks convolutional neural networks for visual recognition (CS231N) (2016)
8. Howard, A.G., Zhu, M., Chen, B., Kalenichenko, D., Wang, W., Weyand, T., Andreetto, M., Adam, H.: MobileNets: efficient convolutional neural networks for mobile vision applications. CoRR arXiv:1704.04861 (2017)
9. Hu, J., Shen, L., Sun, G.: Squeeze-and-excitation networks. In: 2018 IEEE Conference on Computer Vision and Pattern Recognition, CVPR 2018, Salt Lake City, UT, USA, pp. 7132–7141. IEEE Computer Society, 18–22 June 2018. https://doi.org/10.1109/CVPR.2018.00745. http://openaccess.thecvf.com/content_cvpr_2018/html/Hu_Squeeze-and-Excitation_Networks_CVPR_2018_paper.html
10. Kingma, D.P., Ba, J.: Adam: a method for stochastic optimization. In: Bengio, Y., LeCun, Y. (eds.) 3rd International Conference on Learning Representations, ICLR 2015, San Diego, CA, USA, Conference Track Proceedings, 7–9 May 2015. arXiv:1412.6980
11. Krizhevsky, A.: Convolutional deep belief networks on CIFAR-10 (2010)
12. Krizhevsky, A., Sutskever, I., Hinton, G.E.: Imagenet classification with deep convolutional neural networks. In: Bartlett, P.L., Pereira, F.C.N., Burges, C.J.C., Bottou, L., Weinberger, K.Q. (eds.) Advances in Neural Information Processing Systems 25: 26th Annual Conference on Neural Information Processing Systems 2012. Proceedings of a meeting held 3–6 Dec 2012, Lake Tahoe, Nevada, United States, pp. 1106–1114 (2012). http://papers.nips.cc/paper/4824-imagenet-classification-with-deep-convolutional-neural-networks
13. Ma, N., Zhang, X., Zheng, H., Sun, J.: ShuffleNet V2: practical guidelines for efficient CNN architecture design. In: Ferrari, V., Hebert, M., Sminchisescu, C., Weiss, Y. (eds.) Computer Vision—ECCV 2018—15th European Conference, Munich, Germany, Proceedings, Part XIV. Lecture Notes in Computer Science, vol. 11218, pp. 122–138. Springer, 8–14 Sept 2018. https://doi.org/10.1007/978-3-030-01264-9_8
14. Martineau, M., Conte, D., Raveaux, R., Arnault, I., Munier, D., Venturini, G.: A survey on image-based insect classification. Pattern Recognit. **65**, 273–284 (2017). https://doi.org/10.1016/j.patcog.2016.12.020
15. Razavian, A.S., Azizpour, H., Sullivan, J., Carlsson, S.: CNN features off-the-shelf: an astounding baseline for recognition. In: IEEE Conference on Computer Vision and Pattern Recognition, CVPR Workshops 2014, Columbus, OH, USA, pp. 512–519. IEEE Computer Society, 23–28 June 2014. https://doi.org/10.1109/CVPRW.2014.131

16. Russakovsky, O., Deng, J., Su, H., Krause, J., Satheesh, S., Ma, S., Huang, Z., Karpathy, A., Khosla, A., Bernstein, M.S., Berg, A.C., Li, F.: ImageNet large scale visual recognition challenge. Int. J. Comput. Vis. **115**(3), 211–252 (2015). https://doi.org/10.1007/s11263-015-0816-y
17. Sandler, M., Howard, A.G., Zhu, M., Zhmoginov, A., Chen, L.: MobileNetV2: inverted residuals and linear bottlenecks. In: 2018 IEEE Conference on Computer Vision and Pattern Recognition, CVPR 2018, Salt Lake City, UT, USA, pp. 4510–4520. IEEE Computer Society, 18–22 June 2018. https://doi.org/10.1109/CVPR.2018.00474. http://openaccess.thecvf.com/content_cvpr_2018/html/Sandler_MobileNetV2_Inverted_Residuals_CVPR_2018_paper.html
18. Sanner, R.M., Slotine, J.E.: Gaussian networks for direct adaptive control. IEEE Trans. Neural Netw. **3**(6), 837–863 (1992). https://doi.org/10.1109/72.165588
19. Scheirer, W.J., de Rezende Rocha, A., Sapkota, A., Boult, T.E.: Toward open set recognition. IEEE Trans. Pattern Anal. Mach. Intell. **35**(7), 1757–1772 (2013). https://doi.org/10.1109/TPAMI.2012.256
20. Yosinski, J., Clune, J., Bengio, Y., Lipson, H.: How transferable are features in deep neural networks? In: Ghahramani, Z., Welling, M., Cortes, C., Lawrence, N.D., Weinberger, K.Q. (eds.) Advances in Neural Information Processing Systems 27: Annual Conference on Neural Information Processing Systems 2014, Montreal, Quebec, Canada, pp. 3320–3328, 8–13 Dec 2014. http://papers.nips.cc/paper/5347-how-transferable-are-features-in-deep-neural-networks
21. Yosinski, J., Clune, J., Nguyen, A.M., Fuchs, T.J., Lipson, H.: Understanding neural networks through deep visualization. CoRR arXiv:1506.06579 (2015)
22. Zhang, X., Zhou, X., Lin, M., Sun, J.: ShuffleNet: an extremely efficient convolutional neural network for mobile devices. In: 2018 IEEE Conference on Computer Vision and Pattern Recognition, CVPR 2018, Salt Lake City, UT, USA, pp. 6848–6856. IEEE Computer Society, 18–22 June 2018. https://doi.org/10.1109/CVPR.2018.00716. http://openaccess.thecvf.com/content_cvpr_2018/html/Zhang_ShuffleNet_An_Extremely_CVPR_2018_paper.html
23. Zipser, D., Andersen, R.A.: A back-propagation programmed network that simulates response properties of a subset of posterior parietal neurons. Nature **331**, 679–684 (1988)

Neural Networks and Pattern Recognition in Medicine

A Nonlinear Autoencoder for Kinematic Synergy Extraction from Movement Data Acquired with HTC Vive Trackers

Irio De Feudis, Domenico Buongiorno, Giacomo Donato Cascarano, Antonio Brunetti, Donato Micele, and Vitoantonio Bevilacqua

Abstract How the human central nervous system (CNS) copes with the several degrees of freedom (DoF) of the muscle-skeletal system for the generation of complex movements has not been fully understood yet. Many studies in literature have stated that likely the CNS does not independently control DoF but combines few building blocks that consider the synergistic actuation of each DoF. Such building blocks are called synergies. Synergies have been defined both at muscle level, i.e. muscle synergies, and kinematic level, i.e. kinematic synergies. Kinematic synergies consider the synergistic movement of several human articulations during the performance of a complex task, e.g. a reaching-grasping task. The principal component analyses (PCA) is the most used approach in literature for the kinematic synergy extraction. However, the PCA only considers linear correlations among DoFs which can be considered as the most-simple model of inter-joint coupling. In this work, we have extracted synergies from kinematics data (five upper limb angles) acquired during 12 different reaching movements with a tracking system based on the HTC Vive Trackers. After the extraction of the upper-limb joint angles with the OpenSim software, the kinematic synergies have been extracted using nonlinear under-complete autoencoders. Different models of nonlinear autoencoders were investigated and evaluated with R^2 index and normalized reconstruction error. The results showed that 4 synergies were enough for describing the 0.973 ± 0.005 (R^2 index of log sigmoid model) and 0.979 ± 0.004 (R^2 index of tan sigmoid model) of the movement variance for the entire experiment with respectively a Normalized Reconstruction Error (E_{RMS}) of 0.03 ± 0.005 and 0.034 ± 0.004. Comparing the non-linear autoencoders (AE) with the standard linear PCA it emerged that the AE performance are comparable with the PCA results. However, more experiments are needed to perform a deep comparison on a dataset including more joint angles.

I. De Feudis · D. Buongiorno · G. D. Cascarano · A. Brunetti · D. Micele · V. Bevilacqua (✉)
Department of Electrical and Information Engineering, Polytechnic University of Bari, Bari, Italy
e-mail: vitoantonio.bevilacqua@poliba.it

I. De Feudis · D. Buongiorno · G. D. Cascarano · A. Brunetti · V. Bevilacqua
Apulian Bioengineering s.r.l., Via delle Violette n°14, Modugno, BA, Italy

A. Esposito et al. (eds.), *Progresses in Artificial Intelligence and Neural Systems*, Smart Innovation, Systems and Technologies 184,
https://doi.org/10.1007/978-981-15-5093-5_22

1 Introduction

Object reaching and grasping are some of the most frequent tasks among the activities of daily living. Such action, that is perceived as an easy task, actually requires a series of sensorimotor transformations which map the target image impressed on the retina onto a specific muscle activation pattern resulting in the movement of the hand towards the target. The human upper limb is characterized by 30 degrees of freedom (DoFs), including the fingers DoFs, for exploring the surrounding environment and interact with it. Such degrees of freedom are actuated by more than fifty skeletal muscles. It turns out that the human brain has to cope with a large number of actuation patterns to control our upper limb. Then, the human central nervous system (CNS) has to solve two different redundant problems at kinematic level and at muscle level [1].

Many studies regarding the motor control of humans and some animals (e.g. frog and monkey) have introduced the concept of motor primitives [1]. This concept concerns the idea that several DoFs of the motor system are not independently controlled but instead are synergistically controlled by using a defined set of spatiotemporal patterns called synergies. Synergies have been defined at two main stages: muscles synergies if the patterns act at the muscle level, and kinematic synergies if the synergy patterns act the joints level [2].

Several mathematical methods have been proposed for synergies extraction: Principal component analysis (PCA), singular value decomposition (SVD), non-negative matrix factorization (NMF) and artificial neural network (ANN). However, NMF is commonly used to extract muscle synergies [3], whereas PCA is frequently used to derive kinematic synergies [4]. Concerning the kinematic synergies, the advantage of PCA is that it captures covariant thus intuitively coupling DoFs. However, PCA only considers linear correlations among DoFs which can be considered as the most-simple model of inter-joint coupling.

In this work, we investigated several both linear and nonlinear autoencoder topologies used to extract kinematic synergies of the upper limb during reaching and grasping tasks. In particular, we are interested at investigating whether the reduction ability of the autoencoders outperforms the PCA in terms of the amount of data variability described by a reduced set of components/patterns/synergies. We asked three healthy subjects to perform reaching/grasping tasks within an immersive virtual environment. The task consisted in reaching several books placed on two shelves at different height, grasping the book by closing the hand, and moving the book on a desktop. We also have designed and developed a tracking system based on the HTC Vive trackers [5] that has been used both to track the subject's hand during the tasks and to reconstruct the movement of the skeletal system of the upper limb. To achieve a more natural interaction between the subject and the virtual game, the Leap Motion device has been used to detect the opening and closing action of the hand in order to trigger the grasping of the book.

The work is structured as follows: Sect. 2 introduces the materials and methods of the experiment; Sect. 3 presents the experimental results and Sect. 4 discusses the results and concludes the paper.

2 Materials and Methods

2.1 Participants

Three healthy male subjects between 28 and 30 years old were involved in this study. They were all right-handed with normal or corrected-to-normal vision and with no known motor deficit. Participants were informed about experiment of the study and gave their consent.

2.2 Experimental Setup

Each subject was asked to perform reaching/grasping task in a 3D virtual environment while wearing the HTC Vive Headset and a low-cost tracking system. The virtual reality scenario was developed using the Unity 3D framework and the virtual reality HTC Vive platform. The HTC Vive trackers have been used to track the upper limb movements and accordingly control a virtual hand in the VR. The Leap Motion device has been used to detect the opening and the closing of the user's hand, thus allowing the interaction with the virtual objects placed in the virtual scenario.

2.3 Virtual Reality Scenario

The virtual reality scenario implements and simulates the interaction between the human upper limb and virtual books. In the scenario the user is sitting close to a table in front of two bookshelves. He is asked to retrieve some books that randomly appear at six possible positions (see Fig. 1a) with two different orientations (see Fig. 1b). The different height of the shelves, the positions and orientations of the books modulate the complexity and variability of the action. The experimental session is composed of 12 independent trials (six positions by two orientations) that are randomly repeated three times. Each trial considers the interaction with a book that must be moved from the bookshelf to the top of a desktop. Each trial is composed of a sequence of sub actions as follows:

1. the user moves his hand at the starting position ("START" button as in Fig. 1a), then, a book located at a random position and orientation on the bookcase appears;

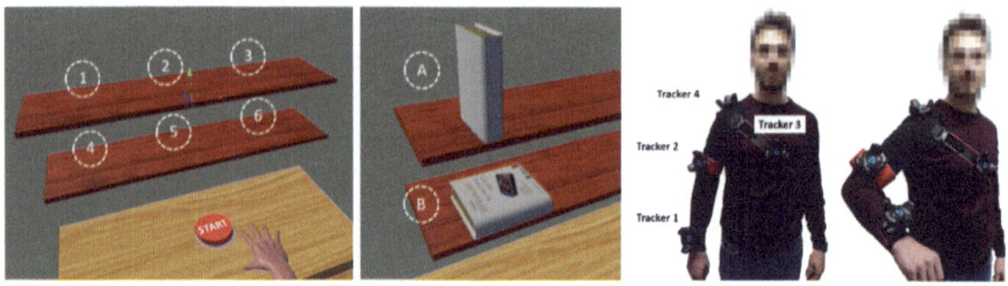

(a) (b) (c)

Fig. 1 **a** The Virtual reality scenario. Numbers from 1 to 6 indicate the possible book position on the shelves. **b** The virtual reality scenario. A to B indicate the two possible book orientation. **c** A participant of the study wearing the accessories of the tracking system

2. then, the user is asked to reach and grasp the book by closing his hand (the book will be "grabbed" only if the hand closes on the book);
3. once the book is grabbed, a light spot is activated in correspondence of the return position (the same for starting);
4. the user must place the book at the right location and open the hand.
5. finally, the user goes back in the rest position to trigger both the end of the current trial and the start of a new one.

2.4 *The Tracking System Based on the HTC Vive Trackers*

The low-cost tracking system was developed using the HTC Vive platform but in particular by means of HTC Vive Trackers. Such system has been used to track the position and the orientation of four Vive trackers placed at specific positions of the user's upper limb, thus allowing the interaction with the VR and reconstruction of the complete movement in terms of joint angles.

The low-cost tracking system has been designed for the acquisition of the angles of the shoulder, elbow and forearm joints, i.e. the three shoulder rotation angles defined as in the work of Holzbaur et al. [6], the elbow flexion/extension angles and the prono-supination angle. The system is based on the virtual reality HTC Vive platform and, in particular on, the Vive Tracker device (Fig. 1c) that is a battery-powered tracking device that allows the acquisition of the full pose (position and orientation) of a rigid body on which it is fixed.

The tracking system employs four HTC Vive trackers, elastic belts with plastic supports for the trackers and a metallic stick with a known length (Fig. 1c). The four markers are positioned at the wrist, the arm, the sternum and above the acromion. The stable positioning of the trackers is ensured by means of elastic bands and appropriate 3D printed plastic supports.

A calibration procedure is needed to individuate the relative position of 8 landmark points of the skeletal system respect to the markers: ulna styloid process, radio styloid process, medial epicondyle, lateral epicondyle, xiphoid process, sternal extremities of the two clavicles and acromion. After positioning the trackers 1, 2 and 3, the tracker 4 is fixed to the stick (Fig. 1c) to identify the relative positions of the landmark points respect to a specific marker. In particular, the ulna styloid process and the radio styloid process are referred respect to the tracker 1; the medial epicondyle and the lateral epicondyle are referred respect to the tracker 2; the xiphoid process and the sternal extremities of the two clavicles are referred to the tracker 3. The acquisition of the relative positions is performed by placing the extremity of the stick (that has a known position respect to the tracker 4) above the landmark points. Finally, the tracker 4 is positioned with a plastic support above the acromion at a defined distance.

After the calibration phase, the subjects of the study, wearing the markers accessories (Fig. 1c) and HTC Vive headset, were immerged into the virtual scenario where they execute the task explained in the previous paragraph.

2.5 Joint Angles Extraction

During the entire experiment session, a custom-made software was used to record the pose of all trackers at 90 Hz rate. Then, it was possible to reconstruct the trajectory of each land-mark point given its relative position to the specific tracker. Finally, the articulation angles were extracted by running an inverse kinematic procedure on a scaled version of the upper limb model developed by Holzbaur et al. [4] using the OpenSim software [7].

For each participant of the study, it was built a dataset of joint angles extracted during the execution of the trials over the experiment. More in details having:

- 5 joints angles ($\varphi_1, \varphi_2, \varphi_3$ three shoulder rotation angles, φ_4 elbow angle and φ_5 prono-supination angle);
- $t = [20000–25000]$ number of samples recorded during the action at 90 Hz;
- $m = 12$ reaching-grasping-pulling actions.

The result was a matrix D as in (1).

$$D = \begin{bmatrix} \varphi_1^1(1) \dots \varphi_5^1(1) \dots \varphi_1^1(t_{max}) \dots \varphi_5^1(t_{max}) \\ \varphi_1^2(1) \dots \varphi_5^2(1) \dots \varphi_1^2(t_{max}) \dots \varphi_5^2(t_{max}) \\ \vdots \quad \vdots \quad \vdots \quad \ddots \quad \vdots \quad \vdots \quad \vdots \\ \varphi_1^m(1) \dots \varphi_5^m(1) \dots \varphi_1^m(t_{max}) \dots \varphi_5^m(t_{max}) \end{bmatrix} \tag{1}$$

Furthermore, trigger system implemented in the virtual scenario labelled automatically the samples basing on action phases (reaching–pulling).

2.6 Nonlinear Autoencoder for Kinematic Synergy Extraction

Autoencoder is an artificial neural network designed with the purpose of coding the input variables into a latent space from which reconstructing them as accurately as possible. This unsupervised learning technique is mostly used for dimensionality reduction. An autoencoder (AE) has a hidden layer that generates a coded representation h of the input x. AE is composed of two main parts: an encoder that codifies the input into the code ($h = e(x)$) and a decoder that reconstruct it, ($r = d(h)$). There are different kinds of AE that differ for internal structure of the network and training modalities [8]: undercomplete AE, regularized AE, sparse AE and denoising AE.

An undercomplete autoencoder is an AE able to extract the most representative features contained in the input data. Such property is ensured by setting the dimension of the code h to a value smaller than the size of the input x. Such network's bottleneck should force the AE to learn some sort of structure that exists in the input data, e.g. correlation among input signals.

In this study, we investigated different models of nonlinear undercomplete autoencoders to extract the spatial kinematic synergies of the human upper limb from joint angles while executing a reaching task in a 3D space. After that, we compared the result with the most used technique in dimensionality reduction, Principal Component Analysis.

The general structure of the autoencoder is shown in Fig. 2. All models, we designed, had one hidden layer and N neurons, $N \in [1, \ldots, 4]$ that produce the synergy activations named s_i with $\in [1, \ldots, 4]$ depending on the number of neurons. Neurons' activation functions ($f = g$) were chose between *log sigmoid* and *tan sigmoid*.

We evaluated the performance of all possible autoencoders generated from these specifics. In the design of the autoencoder linear neurons were avoided because a linear autoencoder would have provided only a linear transformation that would have

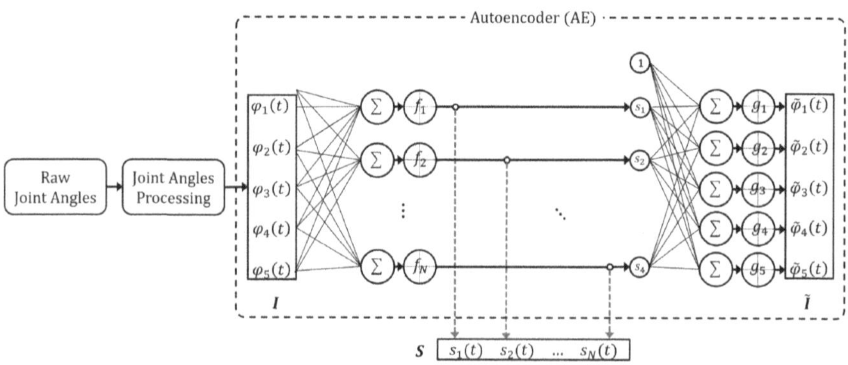

Fig. 2 Undercomplete autoencoder topology

been equivalent to Principle Components Analysis. The output layer of every model had the same dimension as the input layer as a typical AE.

A pre-processing phase composed of the following 4 steps was needed to prepare the dataset of each subject: (1) outliers' removal to remove some bad samples; (2) low-pass filtering (0.01 Hz with Kaiser window) to remove the noise on the data, (3) segmentation of the samples that encoded reaching phases, (4) normalization in joint angles' range. We used the pre-processed datasets to train, validate and test all autoencoders.

The AE networks have been implemented in MATLAB and trained using a gradient descent with momentum and adaptive learning rate algorithm and considering 1000 training epochs. Given a train set, the AE training was repeated 20 times with different initial weights. Then we chose the best one among the 20 AEs featuring the minimum correlation index among the synergy activations. Such index was computed as the sum of the elements of the absolute upper triangular matrix extracted by the correlation matrix of $s_1(t), \ldots, s_n(t)$.

The multivariate R^2 index [9] and Normalized Root Mean Square Error (E_{RMS}) were computed in order to evaluate the synergy extraction performance of every model of autoencoder. The multivariate R^2 index is the fraction of total variation accounted by the synergy reconstruction and then is an indicator with E_{RMS} of the goodness of reconstruction ability of the autoencoder.

2.7 *Autoencoder Versus PCA*

Since the autoencoders are typically used for dimensionality reduction, the designed autoencoders models were compared with the most used technique in literature for kinematic synergies extraction, Principal Components Analysis (PCA) [10].

PCA is a linear transformation that projects a set of multivariate data on a new orthonormal coordinate system with unity vectors aligned with the directions of largest variance. These vectors are so called Principal Components (PCs). Although the number of PCs is equal to the number of variables often only the first few components are needed to explain most of the variance of the data set [11].

It was applied PCA on dataset of each subject; cumulative explained variance and Normalized E_{RMS} by PCs were computed in order to compare the PCA with Autoencoders.

3 Results

We evaluated and compared encoding ability of nonlinear autoencoder models in terms of the multivariate R^2 index computed between the input joint angles dataset of each subject (see Fig. 3) and the reconstructed one and Normalized Reconstruction Error (E_{RMS}) (see Fig. 4).

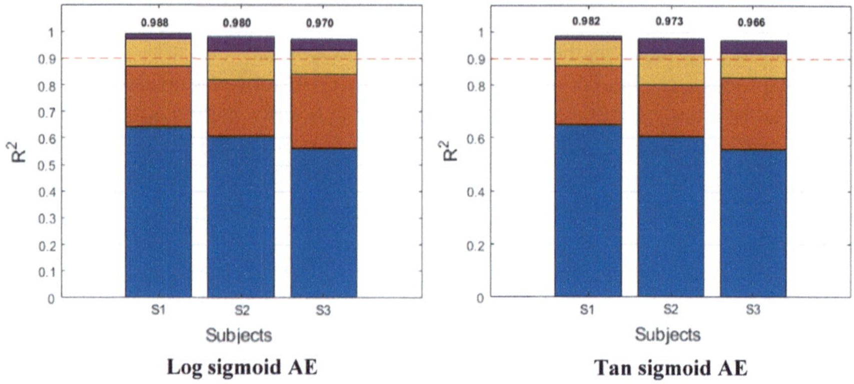

Fig. 3 R^2 index of log sigmoid and tan sigmoid AE models calculated for each subject

Fig. 4 Normalized reconstruction error of the two AE models calculated for each subject

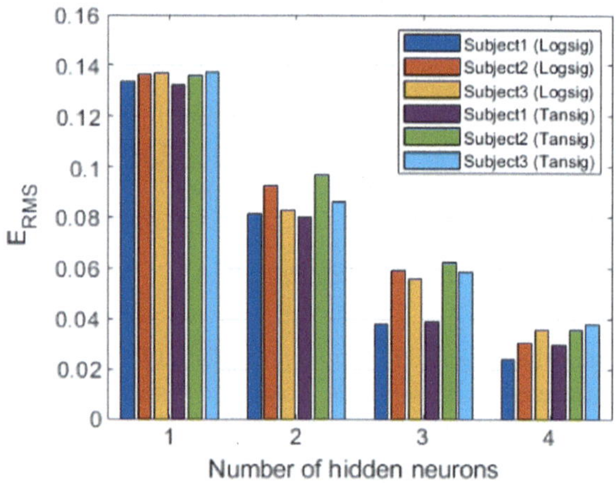

Table 1 reports that 3 synergies were enough for describing the 0.942 ± 0.013 (R^2 index of log sigmoid model) and 0.936 ± 0.015 (R^2 index of tan sigmoid model)

Table 1 R^2 of autoencoder models compared with normalized explained variance of PCA. Reconstruction error (normalized E_{RMS}) of autoencoders models and PCA comparison

# Syn	R^2/normalized explained variance			Normalized reconstruction error (E_{RMS})		
	Autoencoder (*Log sigmoid*)	Autoencoder (*Tan sigmoid*)	PCA	Autoencoder (*Log sigmoid*)	Autoencoder (*Tan sigmoid*)	PCA
1	0.601 ± 0.026	0.603 ± 0.029	0.604 ± 0.037	0.136 ± 0.002	0.135 ± 0.003	0.135 ± 0.001
2	0.841 ± 0.014	0.833 ± 0.019	0.841 ± 0.018	0.086 ± 0.006	0.088 ± 0.008	0.086 ± 0.005
3	0.942 ± 0.013	0.936 ± 0.015	0.945 ± 0.021	0.05 ± 0.011	0.053 ± 0.012	0.05 ± 0.009
4	0.979 ± 0.005	0.974 ± 0.004	0.986 ± 0.009	0.03 ± 0.006	0.034 ± 0.004	0.025 ± 0.001

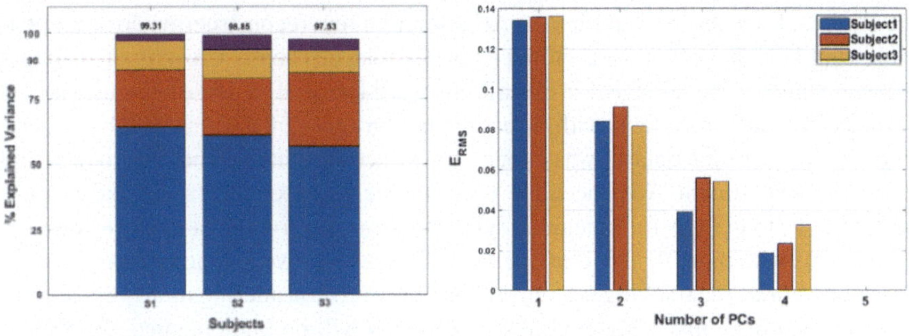

Fig. 5 Cumulative explained variance by PCs and E_{RMS} of PCA for each subject

of the movement variance for the entire experiment with respectively a Normalized Reconstruction Error (E_{RMS}) of 0.05 ± 0.011 and 0.053 ± 0.012.

It was applied the PCA to the same dataset to extract kinematic synergies with a linear transformation and the Explained Variance % and the Normalized Reconstruction Error (E_{RMS}) of each principal component were calculated.

The results of PCA are reported in Fig. 5 with cumulative explained Variance percentage of PCs across the 3 subjects and Normalized Reconstruction Error (E_{RMS}). They show that 3 components are needed to represent task data; an index threshold of 90% variance is enough.

Table 1 reports the performance comparison between autoencoder models and PCA in term of mean of the Normalized explained variance (NEV) and mean of the Normalized Reconstruction Error on the three subjects. In order to have a fair comparison, the NEV percentage was normalized in range 0–1 then it is a matter of fact that number of synergies are represented by hidden neurons in autoencoders and principle components in PCA. The result shows how nonlinear autoencoder models are accurate as PCA in term of reconstruction of the input never overcoming it.

4 Discussion and Conclusion

The same complex reaching movement can be executed with several arm trajectories due to the high redundancy of the arm skeletal system. This redundancy can be simplified into kinematic synergies. We extracted synergies from kinematics data of 12 different reaching movements by means of nonlinear autoencoder obtaining accurate performances. Different models of AE were investigated and evaluated with two metrics, R^2 index and Normalized E_{RMS}. The results showed that 3 synergies were enough for describing the 0.942 ± 0.013 (R^2 index of *log sigmoid* model) and 0.936 ± 0.015 (R^2 index of *tan sigmoid* model) of the movement variance for the entire experiment with respectively a Normalized Reconstruction Error (E_{RMS}) of 0.05 ± 0.011 and 0.053 ± 0.012. The AE reconstruction ability was compared with the PCA and results showed that the AE commit a low reconstruction error comparable with PCA but never overcoming its performance. The PCA results showed that 3 components were enough for describing the 94.48% ± 2.1% of the movement variance for the entire task with a reconstruction error of 0.05 ± 0.009. A linear transformation of the dataset was enough to extract accurately synergies to describe the entire experiment. We speculate that nonlinear autoencoders can have better performance than PCA for kinematic synergy extraction from data of more complex movements with more than 5 joint angles tracked. More investigations are needed. In the future works, the tracking system will be improved considering more joint angles, virtual scenarios—involving complex movements will be implemented, number of test subjects will be increased, other AE models will be investigated, and they will be used also for movement prediction. Furthermore, future studies will investigate the use of synergies as performance index of the motor activity in neurodegenerative disorders [12–15].

Acknowledgments This work has been supported by the Italian project RoboVir (BRIC INAIL-2016).

References

1. Bernstein, N.: The Co-ordination and Regulation of Movements (1966)
2. Buongiorno, D., Barone, F., Berger, D.J., Cesqui, B., Bevilacqua, V., D'Avella, A., Frisoli, A.: Evaluation of a pose-shared synergy-based isometric model for hand force estimation: towards myocontrol. In: Biosystems and Biorobotics (2017). https://doi.org/10.1007/978-3-319-46669-9_154
3. Tresch, M.C., Cheung, V.C.K., d'Avella, A.: Matrix factorization algorithms for the identification of muscle synergies: evaluation on simulated and experimental data sets. J. Neurophysiol. **95**, 2199–2212 (2006). https://doi.org/10.1152/jn.00222.2005
4. Bockemühl, T., Troje, N.F., Dürr, V.: Inter-joint coupling and joint angle synergies of human catching movements. Hum. Mov. Sci. **29**, 73–93 (2010). https://doi.org/10.1016/j.humov.2009.03.003
5. VIVE^{TM} | VIVE Tracker: https://www.vive.com/us/vive-tracker/
6. Holzbaur, K.R.S., Murray, W.M., Delp, S.L.: A model of the upper extremity for simulating musculoskeletal surgery and analyzing neuromuscular control. Ann. Biomed. Eng. **33**, 829–840 (2005). https://doi.org/10.1007/s10439-005-3320-7
7. Delp, S.L., Anderson, F.C., Arnold, A.S., Loan, P., Habib, A., John, C.T., Guendelman, E., Thelen, D.G.: OpenSim: open-source software to create and analyze dynamic simulations of movement. IEEE Trans. Biomed. Eng. **54**, 1940–1950 (2007). https://doi.org/10.1109/TBME.2007.901024
8. Goodfellow, I., Bengio, Y., Courville, A.: Deep learning (2016)
9. d'Avella, A., Portone, A., Fernandez, L., Lacquaniti, F.: Control of fast-reaching movements by muscle synergy combinations. J. Neurosci. **26**, 7791–7810 (2006). https://doi.org/10.1523/JNEUROSCI.0830-06.2006
10. Burns, M.K., Patel, V., Florescu, I., Pochiraju, K. V., Vinjamuri, R.: Low-dimensional synergistic representation of bilateral reaching movements. Front. Bioeng. Biotechnol. **5** (2017). https://doi.org/10.3389/fbioe.2017.00002
11. Cooper, R.I., Manly, B.F.J.: Multivariate statistical methods: a primer. J. R. Stat. Soc. Ser. A **150**, 401 (2006). https://doi.org/10.2307/2982053

12. Bevilacqua, V., D'Ambruoso, D., Mandolino, G., Suma, M.: A new tool to support diagnosis of neurological disorders by means of facial expressions. In: MeMeA 2011—2011 IEEE International Symposium on Medical Measurements and Applications, Proceedings. pp. 544–549. IEEE (2011). https://doi.org/10.1109/MeMeA.2011.5966766
13. Bortone, I., Trotta, G.F., Brunetti, A., Cascarano, G.D., Loconsole, C., Agnello, N., Argentiero, A., Nicolardi, G., Frisoli, A., Bevilacqua, V.: A novel approach in combination of 3D gait analysis data for aiding clinical decision-making in patients with Parkinson's disease. In: Lecture Notes in Computer Science (including subseries Lecture Notes in Artificial Intelligence and Lecture Notes in Bioinformatics), pp. 504–514 (2017). https://doi.org/10.1007/978-3-319-63312-1_44
14. Triggiani, A.I., Bevilacqua, V., Brunetti, A., Lizio, R., Tattoli, G., Cassano, F., Soricelli, A., Ferri, R., Nobili, F., Gesualdo, L., Barulli, M.R., Tortelli, R., Cardinali, V., Giannini, A., Spagnolo, P., Armenise, S., Stocchi, F., Buenza, G., Scianatico, G., Logroscino, G., Lacidogna, G., Orzi, F., Buttinelli, C., Giubilei, F., Del Percio, C., Frisoni, G.B., Babiloni, C.: Classification of healthy subjects and Alzheimer's disease patients with dementia from cortical sources of resting state EEG rhythms: a study using artificial neural networks. Front. Neurosci. **10** (2017). https://doi.org/10.3389/fnins.2016.00604
15. Buongiorno, D., Bortone, I., Cascarano, G.D., Trotta, G.F., Brunetti, A., Bevilacqua, V.: A low-cost vision system based on the analysis of motor features for recognition and severity rating of Parkinson's Disease. BMC Med. Inform. Decis. Mak. **19**, 243 (2019). https://doi.org/10.1186/s12911-019-0987-5

Neural Feature Extraction for the Analysis of Parkinsonian Patient Handwriting

Vincenzo Randazzo, Giansalvo Cirrincione, Annunziata Paviglianiti, Eros Pasero, and Francesco Carlo Morabito

Abstract Parkinson's is a disease of the central nervous system characterized by neuronal necrosis. Patients at the time of diagnosis have already lost up to 70% of the neurons. It is essential to define early detection techniques to promptly intervene with an appropriate therapy. Handwriting analysis has been proven as a reliable method for Parkinson's disease diagnose and monitoring. This paper presents an analysis of a Parkinson's disease handwriting dataset in which neural networks are used as a tool for analyzing the problem space. The goal is to check the validity of the selected features. For estimating the data intrinsic dimensionality, a preliminary analysis based on PCA is performed. Then, a comparative analysis about the classification performances of a multilayer perceptron (MLP) has been conducted in order to determine the discriminative capabilities of the input features. Finally, fifteen temporal features, capable of a more meaningful discrimination, have been extracted and the classification performances of the MLP trained on these new datasets have been compared with the previous ones for selecting the best features.

V. Randazzo (✉) · A. Paviglianiti · E. Pasero
DET, Politecnico di Torino, Turin, Italy
e-mail: vincenzo.randazzo@polito.it

A. Paviglianiti
e-mail: annunziata.paviglianiti@polito.it

E. Pasero
e-mail: eros.pasero@polito.it

G. Cirrincione
Lab. LTI, University of Picardie Jules Verne, Amiens, France
e-mail: giansalvo.cirrincione@usp.ac.fj

SEP, University of South Pacific, Suva, Fiji Islands

F. C. Morabito
DICEAM, Mediterranea University of Reggio Calabria, Reggio Calabria, Italy
e-mail: morabito@unirc.it

A. Esposito et al. (eds.), *Progresses in Artificial Intelligence and Neural Systems*, Smart Innovation, Systems and Technologies 184,
https://doi.org/10.1007/978-981-15-5093-5_23

1 Introduction

Neurodegenerative diseases (NDD) [1] are a group of diseases of the central nervous system characterized by neuronal necrosis, which leads to an inevitable and irreversible damage of brain functions. The causes of the onset are still unclear [2]. For sure, several factors, such as genetic or environment, contribute to one another in giving rise to the pathology [3]. NDD follow a progressive course that is phenotypically highlighted when the anatomical brain damage is in an advanced stage: on average, the patient at the time of diagnosis has already lost up to 70% of the neurons, thus reducing the possibility of therapeutic intervention effectively [4]. It is essential to define reliable early detection techniques to promptly intervene with appropriate therapy that can be more effective the more the neuronal destruction mechanism is in the early stages. The disabling forms arising from NDD, such as Alzheimer's, Parkinson's, Huntington's chorea and Amyotrophic Lateral Sclerosis (ALS), are characterized by the slow and progressive loss of one or more functions of the nervous system. Parkinson's disease (PD) [5, 6] is a degenerative disease of the central nervous system that affects muscle control, and therefore can influence movement, speech and posture. It is often characterized by muscle stiffness, tremor, slowing of physical movement, and in extreme cases, loss of physical movement. From a pathological point of view, it does not exist a reliable method for an objective and quantitative diagnosis of Parkinson's disease.

Human beings' skills are strongly related to their state of health; indeed, cognitive functions are closely linked to aging processes. Particularly, calligraphy and speech are motor control tasks performed by our brain, therefore the degradation of these abilities implies a neurological deterioration. Handwriting signals are useful for diagnostic and disease monitoring applications. Several tests [7], e.g. house drawing, can be performed in order to check the status of an NDD disease. One of the most effective studies for Parkinson's disease diagnosis concerns the analysis of patient calligraphy [8]. Indeed, it is usually characterized by the development of micrographia, which is a reduction in the size of the writing, and other deficits regarding geometry, kinematics, pressure patterns and air movement [9, 10].

Feature extraction and feature selection techniques have been used to process handwriting signals. A popular approach for PD detection from handwriting consists in extracting kinematic features, which can be either a single value or a sequence of values extracted through time [11]. On one side, feature transformation strategies, such as principal component analysis (PCA) [12] and independent component analysis [13], involve a transformation of the original inputs and produce a set of new variables. On the other hand, feature selection approaches reduce the dimensionality of the input data, removing the irrelevant features and retaining the original interpretations of inputs. A comparative analysis of these techniques and their application to handwriting of people affected from Parkinson's disease is presented in [14]. Bakar et al. [15] proposes an experimental analysis of ANOVA [16], which is a technique used to determine whether differences in two or more datasets are statistically significant. Taleb et al. [17] suggests another feature selection approach based on Support

Vector Machine (SVM), with Radial Basis Function (RBF) as kernel [18], which is used as a classifier to predict class labels, in particular to discriminate the task samples into two classes (PD and healthy).

In classification applications, attributes selected from initial dataset are given as input to the classification algorithms. According to [19] attributes that can better distinguish between classes (high-level attributes) are more important than the others in terms of performance. In the ReliefF algorithm [20], attributes are selected according to their suitability with target function; the principle is similar to the basic rules of k-NN algorithm. Liogienė and Tamulevičius [21] proposes a simple and fast feature selection algorithm, Sequential forward selection (SFS), based on a greedy search algorithm. It extracts the subset of features by maximizing the efficiency of the feature subset.

2 Experimental Setup

The dataset has been built collecting data from 36 Parkinsonian subjects (18m and 18f, aged between 33 and 83 years old) and 10 healthy subjects (6m and 4f, aged between 49 and 67 years old) recruited at the Matarò Hospital in Barcelona. Every sick person was observed before and after the daily drug (L-dopa COMT catecolo–metal transferasi) administration. Unfortunately, we do not have access to patient clinical information such as Parkinson's disease rating scale part III, levodopa equivalent daily dose, etc. All the patients were right-handed. 22 of these had attended primary school (21 PD/1 Healthy), 17 secondary school (9 PD/8 Healthy), 6 University (5 PD/1 Healthy) and one had not attended any academic studies. Participants were individually tested in a laboratory free of auditory and visual disturbances.

At the beginning of the experiment, the study was explained to the participants and then they underwent a task concerning the writing of the sentence "*La casa de Barcelona es preciosa*" (in Spanish, the native language of the participants). Handwriting collection and analysis has been performed using a digitizing tablet with an ink pen. This approach has an advantage over the classic method based on handwriting and posterior scanning: the machine, actually, can record the pen pressure on the tablet and acquire the information even "in the air", that is, where there is no contact between the pen and the surface. The data acquisition was made by means of a tablet, specifically an Intuos Wacom digitizer, which acquired 100 samples per second (total number of samples amounts to around 244 K). The acquired features are the same of [22]: X and Y pen positions (the spatial coordinates), Altitude (the angle between the pen and the tablet surface along the vertical), Azimuth (the horizontal angle between the pen and the tablet surface) and the pen pressure on the tablet surface.

3 The Proposed Approach

This paper presents an analysis of a Parkinson's disease handwriting dataset in which neural networks are used to describe the problem. The goal is not the classification in itself, but the validity of the corresponding selected features. Indeed, it is assumed that the best description of the phenomenon should correspond to the best possible classification. In this sense, it can be argued that neural networks are here used as a tool of exploratory data analysis.

This study requires a preliminary analysis (here based on a linear one), in order to have a first insight on the database and, particularly, on its intrinsic dimensionality.

4 Linear Analysis of the Dataset

The manifold of the proposed dataset has been deeply analyzed using the Principal Component Analysis (PCA) in order to understand its intrinsic dimensionality and select the best feature subset. The former has been studied using Pareto diagrams [23], the latter with biplots [24]. The whole dataset (*H-Pre-Post*), i.e. healthy subjects together with sick patients before and after the drug treatment, has been projected using PCA. The corresponding Pareto diagram is shown in Fig. 1. It displays the explained data variance by the principal component (PC). The bars represent the associated singular values. Figure 1 shows the importance of the first four components. They explain 88.47% and suggest the intrinsic dimensionality of the manifold is around five.

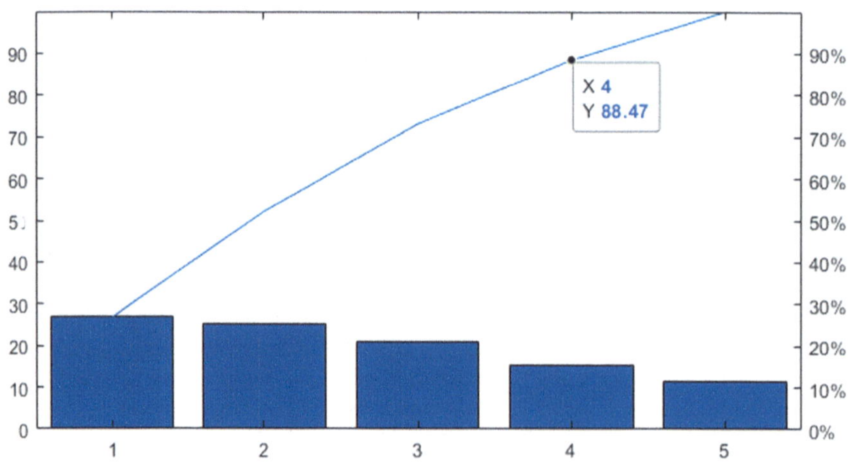

Fig. 1 Pareto diagram on whole dataset, *H-Pre-Post*

4.1 Biplots

Additional information from the linear analysis of data can be retrieved from a biplot. It is a graphic representation which allows to display, at the same time, both samples and variables of a data matrix. By means of PCA, it is possible to show both the data projected into the principal component space together with the input variable directions. Figure 2 shows the biplot computed on the whole dataset after being projected with PCA. Although data appear to be clustered along the third principal component, it is not clear which are the features that discriminate and explain the three clusters of data (healthy, pre-treatment, post-treatment).

In order to determine which is this subset of features, three new datasets have been created:

1. *H-Pre*: Healthy and pre-treatment subjects.
2. *H-Post*: Healthy and post-treatment subjects.
3. *Pre-Post*: Pre-treatment and post-treatment subjects.

The former, *H-Pre*, has been analyzed in Fig. 3 (left). It can be noticed that the first two input variables (blue directions 1 and 2 in the figure) are nearly parallel to the first two axis, PC1 and PC2, while the rest is explained by the last principal component, PC3. This behavior can be clearly understood by looking at Fig. 3 (right), which is a zoom near the origin. Here, it is evident that the first two components of the PCA projection represent the first two input variables (X and Y pen positions); in fact, it is possible to directly read the original subject handwriting "*La casa de Barcelona es preciosa*". Although the direction of maximum variance, i.e. PC1, obviously follows

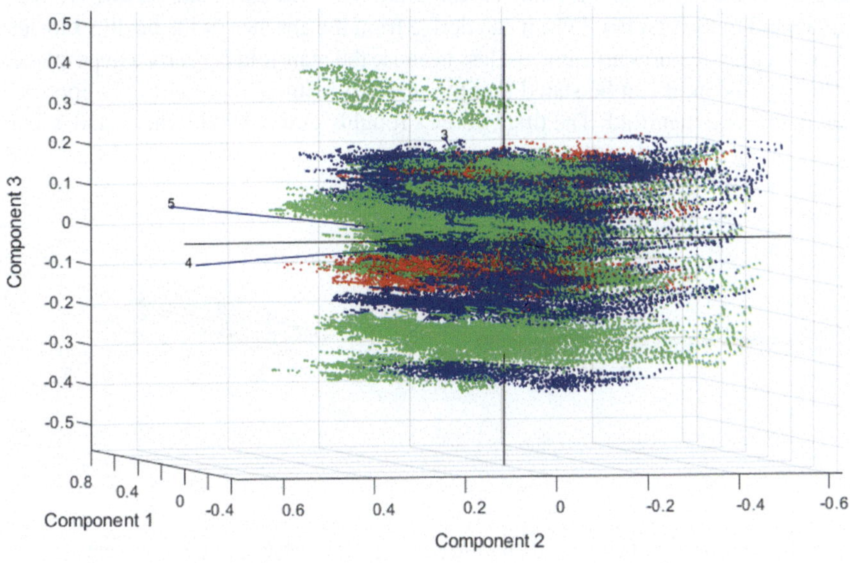

Fig. 2 Biplot on *H-Pre-Post*: healthy (red), pre-treatment (green), post-treatment (blue)

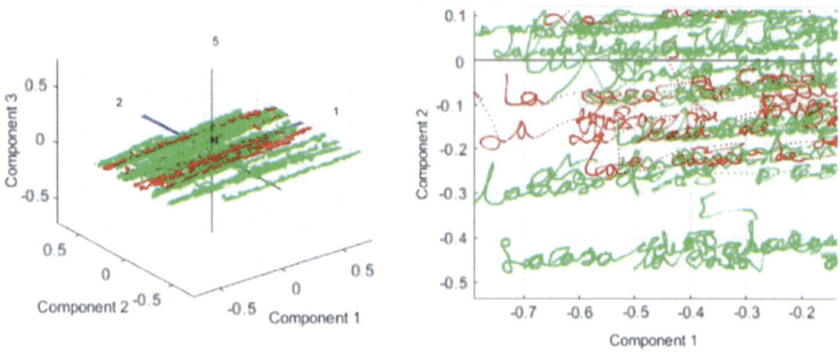

Fig. 3 Biplot on *H-Pre*: healthy (red), pre-treatment (green): whole (left), zoom (right)

the X component (writing from left to right), the most significative feature is the Y pen position; indeed, as shown in Fig. 3, this direction clearly discriminates between the healthy and the pre-treatment clusters.

Figure 4 (left) shows the biplot for the second subset *H-Post*. The first two input features behave as in the previous case, while, in this case, the remaining three are, also, meaningful for distinguishing between the clusters. Indeed, Fig. 4 (right), which is the Z-view of the same biplot, shows that the clusters are linearly separated along PC3.

The biplot for the last subset, *Pre-Post*, is shown in Fig. 5. As in the previous cases, the first feature (X pen position), is able to fully explain the clusters while along the second one (Y pen position) is possible to discriminate between the clusters. The main difference with the previous cases is that their directions are slightly rotated with regard to the first two PCs; it can derive from the absence of the healthy cluster. The remaining features are quite useless because the manifold is nearly a hyperplane.

In conclusion, it can be stated that the selected features represent only approximately the data manifold. The first two PCs roughly coincide with the X and Y pen

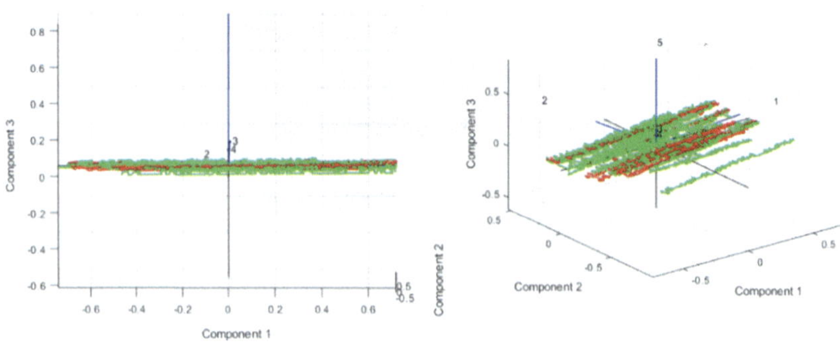

Fig. 4 Biplot on H-*Post*: healthy (red), post-treatment (green): whole (left), Z-view (right)

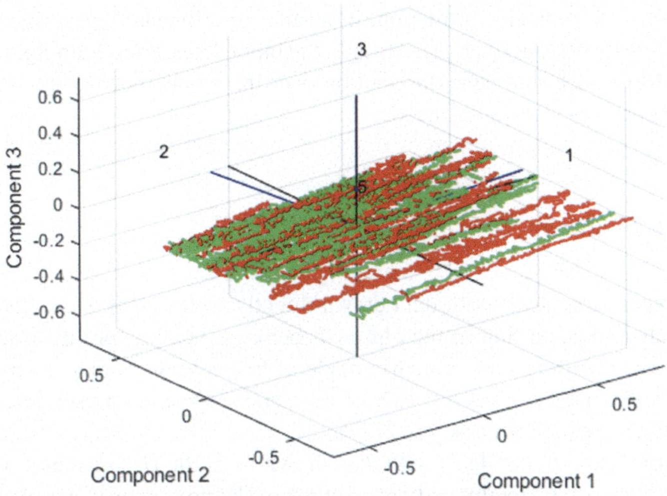

Fig. 5 Biplot on *Pre-Post*: pre-treatment (red), post-treatment (green)

positions, which is obvious because most variance in writing is in these two directions. Hence, the most meaningful information should stem from the other three components, which, as seen in Figs. 2 and 4, do not discriminate well enough. It can be argued that Y pen position ability to discriminate among clusters is related to vertical micrographia and with the activation of interphalangeal and metacarpophalangeal joints. This idea is worth to be further deepened and it will be explored in a future work.

5 Neural Classification

A comparative analysis about the classification performances of a multilayer perceptron (MLP) has been conducted in order to determine the discriminative capabilities of the input features. The MLP has been chosen because it is well-suited for pattern recognition [23]. At this purpose, it has a single hidden layer, composed of twenty neurons, and output units equipped with the soft-max activation function [23]. Because of the use of the cross-entropy error function, they yield the probability of membership for the following classes: healthy, pre-treatment, post-treatment. The input layer is mapped one-to-one to the input features; hence, it is always composed of five neurons. The MLP has been trained, by using the Scaled Conjugated Gradient technique [23], both on the whole dataset (three-neurons output layer) and on three subsets (two-neurons output layer) defined in the previous section; then, for each of these training sets, fifteen statistical features, based on the temporal behavior, have been extracted and fed to other MLPs to check their classification performances. Due to the absence of clinical information, in all the experiments, labels (healthy, pre-treatment, post-treatment) were used to split the input dataset into training, validation and test subsets such that their distribution over the labels (healthy, pre-treatment, post-treatment) was always balanced.

5.1 Raw Features

The first experiment deals with data drawn directly from *H-Pre-Post*. Each record has been labelled according to the cluster it belongs: healthy, pre-treatment, post-treatment. The resulting set is a matrix made of five columns and as many rows as the number of samples (~244 K). 70% of this set, i.e. the training set, has been fed to the MLP. The overall accuracy is 77.9%.

The second experiment deals with data drawn from the *H-Pre* subset. Only two labels have been used: healthy and pre-treatment. The input matrix has about 134 K samples; as before, 70% is used for training and the rest is divided in equal parts between test and validation sets. An overall accuracy of 95.9% is reached. This classification is very accurate, which is obvious because healthy and sick patients have a significantly different motor control and, so, handwriting.

The experimental setup for the MLP trained on the *H-Post* dataset (~129 K examples) is the same as before: two output classes (healthy and post) and 15% of input data used, respectively, for testing and validating. Compared to the previous experiment, the overall test performance decrease to 95.0%. However, this is not a negative result; indeed, it suggests that, after drug treatment, some patients have recovered enough to be confused with the healthy ones.

The last experiment regards the MLP trained on the *Pre-Post* subset. The dataset is made of around 224 K records. Two class labels have been chosen: pre-treatment and post-treatment. The classification is worsened with regard to the previous methods (83.2%). Obviously, this is the most difficult pair of classes to be discriminated. All the patients are sick; as a consequence, their handwritings have similar characteristics. Unfortunately, Parkinson's disease treatments are not very effective yet, so, even after drug administration, improvements are quite limited especially when the pathology is, already, in an advanced stage. Another possible way of interpreting it, is that, maybe, patients are in early stages of PD, therefore the effect of levodopa is not so significant.

Resuming, it can be observed that the healthy state is the easiest to classify, because it is based on very peculiar values of the features. It can be used as a basis for determining if the post-treatment state tends to an improvement for the patient, in the sense that data post drug administration yield values of the features closer to the healthy state ones.

5.2 Temporal Features

The data manifold analysis in Sect. 4 and the previous Sect. 5.1 have proven that the initial set of features was not able to distinguish properly the three clusters of subjects. Therefore, a new set of features, capable of a more meaningful discrimination have been proposed. The idea is to exploit their temporal content; fifteen temporal features have been extracted from each record of the four previous datasets (*H-Pre-Post, H-Pre, H-Post, Pre-Post*). The selected features are the following: mean, max value, root mean square (RMS), square root mean (SRM), standard deviation, variance, shape factor (with RMS), shape factor (with SRM), crest factor, latitude factor, impulse factor, skewness, kurtosis, normalized 5th central moment, normalized 6th central moment. Then, the comparative analysis about the classification performances of the multilayer perceptron has been repeated for each of the four new datasets: *H-Pre-PostT, H-PreT, H-PostT and Pre-PostT.*

For all the following experiments the chosen MLP has a single hidden layer, composed of forty neurons and an input layer of fifteen units. The rest of the setup is the same as the previous section.

The first experiment deals with data drawn directly from *H-Pre-PostT*. Each record has been labelled according to the cluster it belongs: healthy, pre-treatment, post-treatment. The resulting set is a matrix made of five columns and as many rows as the number of samples (~244 K). 70% of this set, i.e. the training set, has been fed to the MLP. The overall accuracy is 99.3%, that is an 27% increase.

The second experiment deals with data drawn from the *H-PreT* subset. As before, only two labels have been used: healthy and pre-treatment. The input matrix has the same size of the raw case (*H-Pre*); again, 70% of data are used for training and the rest is divided in equal parts between test and validation sets. Despite this classification is more accurate (99.2%) than its corresponding raw case, the overall accuracy is not significantly improved (3%). The considerations done for *H-Pre* also hold for this experiment.

In the third experiment, the MLP has been trained using the *H-PostT* dataset (~129 K examples). The experimental setup is the same as before: two output classes (healthy and post) and 15% of input data used, respectively, for testing and validating. The overall test reaches its maximum (100%) with an increase of 5.3%. It is worth of notice that, in this case, the network does not confuse patients who have recovered with the healthy ones. It may suggest that even if the handwritings are closer to normality, the temporal features are now able to discriminate from the healthy case.

The validity of the proposed approach is proved by the last experiment, which regards the MLP trained on the *Pre-PostT* subset. Two class labels have been chosen: pre-treatment and post-treatment. A dataset hard to cluster (83.2% of accuracy) like *Pre-Post*, is now perfectly learnt (100% of accuracy) by the classifier, with an increase of performance of more than the 20%.

Some resuming considerations (see Table 1) can be added. Considering that the input layer requires fewer units in case of raw features and the neural network is fully connected, the use of temporal features requires more *epochs* for training.

Table 1 MLP performance and classification

	# Epochs	Final error	% Training	% Test
H-Pre-Post	990	0.18	77.8	77.9
H-Pre-PostT	1000	0.01	99.3	99.3
H-Pre	629	0.57	96.0	95.9
H-PreT	831	0.013	99.3	99.2
H-Post	497	0.07	94.8	95
H-PostT	1000	0.0008	100	100
Pre-Post	972	0.175	83.5	83.2
Pre-PostT	1000	0.0004	100	100

However, the final training error is several orders of magnitude smaller than in the raw case. This observation is enforced by the classification rates and proves that the temporal model represents better the database (the cross-entropy error yields the correlation between data and model). Hence, the temporal features describe better the phenomenon. This approach justifies the medical consideration of the importance of the temporal behavior in the handwriting.

6 Conclusions

Parkinson's disease is hard to diagnose timely. Indeed, when symptoms are evident, 70% of neurons are already compromised. Techniques for early detection are essential to intervene with appropriate therapies. Handwriting analysis has proved to be a reliable tool for Parkinson's disease diagnose. Starting from a Parkinson's disease database collected at Matarò Hospital in Barcelona, multiple features sets have been extracted and compared in order to select the best feature subset. The PCA-based analysis has shown that the dataset lays on a five-dimensional manifold and the raw features have been studied. Then, a comparative analysis based on an MLP has proved temporal features to be both a reliable model of the Parkinson's disease dataset and more effective in discriminating the different sub-clusters, with an upper bound performance of 100% and a final training error of 0.0004.

Future works will deal with a more accurate analysis of the post-treatment cluster, in order to assess the response of the patient. Also, a non-linear study can be performed for determining the shape of the cluster manifolds in a more accurate way.

Acknowledgments A special thanks to Prof. Marcos Faundez-Zanuy of the Escola Universitària Politècnica de Mataró and Prof. Anna Esposito of Università degli Studi della Campania for providing the original dataset.

References

1. Heemels, M.: Neurodegenerative diseases. Nature **539**(7628) (2016)
2. Gitler, A., Dhillon, P., Shorter, J.: Neurodegenerative disease: models, mechanisms, and a new hope. Dis. Models Mech. 499–502 (2017)
3. Cambier, J., Masson, M., Cambier, H.: Neurologia. Elsevier
4. Rizzi, B., Mori, P., Scaglioni, A., Mazzucchi, A., Rossi, M.: La malattia di Parkinson Guida per pazienti e familiari. Fondazione Don Gnocchi (2013)
5. Parkinson, J.: An essay on the shaking palsy. J. Neuropsychiatr. Clin. Neurosci.
6. Pahwa, R., Lyons, K., Koller, W.: Handbook of Parkinson's Disease: Neurological Disease and Therapy. Marcel Dekker Inc., New York (2003)
7. Larner, A.: Addenbrooke's cognitive examination-revised (ACE-R) in day today clinical practice. Age Age. **36**(6) (2007)
8. Lepelley, M., Thullier, F., Bolmont, B., Lestienne, F.: Age-related differences in sensorimotor representation of space in drawing by hand. Clin. Neurophysiol. (2010)
9. Pinto, S., Velay, J.: Handwriting as a marker for PD progression: a shift in paradigm. Neurodegener. Disease Manag. **5**(5) (2015)
10. Rosenblum, S., Samuel, M., Zlotnik, S., Erikh, I., Schlesinger, I.: Handwriting as an objective tool for Parkinson's disease diagnosis. J. Neurol. **260**(9) (2013)
11. Drotár, P., Mekyska, J., Rektorová, I., Masarová, L., Smékal, Z., Faundez-Zanuy, M.: A new modality for quantitative evaluation of Parkinson's disease: in-air movement. In: 13th IEEE International Conference on BioInformatics and BioEngineering, Chania (2013)
12. Hotelling, H.: Analysis of a complex of statistical variables into principal components. J. Educ. Psychol. **24** (1933)
13. Lu, W., Rajapakse, J.C.: Approach and applications of constrained ICA. IEEE Trans. Neural Netw. **16**(1) (2005)
14. Drotár, P., Mekyska, J., Smékal, Z., Rektorová, I., Masarová, L., Faundez-Zanuy, M.: Contribution of different handwriting modalities to differential diagnosis of Parkinson's disease. In: 2015 IEEE International Symposium on Medical Measurements and Applications (MeMeA) Proceedings, Turin (2015)
15. Bakar, Z.A., Ispawi, D.I., Ibrahim, N.F., Tahir, N.M.: Classification of Parkinson's disease based on multilayer perceptrons (MLPs) neural network and ANOVA as a feature extraction. In: 2012 IEEE 8th International Colloquium on Signal Processing and Its Applications, Melaka
16. Zoubek, L.: Introduction to Educational Data Mining Using MATLAB
17. Taleb, C., Khachab, M., Mokbel, C., Likforman-Sulem, L.: Feature selection for an improved Parkinson's disease identification based on handwriting. In: 2017 1st International Workshop on Arabic Script Analysis and Recognition (ASAR), Nancy (2017)
18. Vapnik, V.: Statistical Learning Theory. Willey, London (1998)
19. Peker, M., Arslan, A., Şen, V., Çelebi, F.V., But, A.: A novel hybrid method for determining the depth of anesthesia level: combining ReliefF feature selection and random forest algorithm (ReliefF + RF). In: 2015 International Symposium on Innovations in Intelligent SysTems and Applications (INISTA), Madrid (2015)
20. Kononenko, I.: Estimating attributes: analysis and extensions of RELIEF. In: European Conference on Machine Learning, Catania (1994)
21. Liogienė, T., Tamulevičius, G.: SFS feature selection technique for multistage emotion recognition. In: 3rd IEEE Workshop on Advances in Information, Electronic and Electrical Engineering (AIEEE), Riga (2015)
22. Faundez-Zanuy, M., Hussain, A., Mekyska, J., Sesa-Nogueras, E., Monte-Moreno, E., Esposito, A., Chetouani, M., Garre-Olmo, J., Abel, A., Smekal, Z., Lopez-de-Ipiña, K.: Biometric applications related to human beings: there is life beyond security. Cogn. Comput. **5**(1) (2013)
23. Bishop, C.M.: Neural Networks for Pattern Recognition. Oxford University Press (1995)
24. Gower, J., Lubbe, S., Le Roux, N.: Understanding Biplots. Wiley (2010)

Discovering Hierarchical Neural Archetype Sets

Gabriele Ciravegna, Pietro Barbiero, Giansalvo Cirrincione, Giovanni Squillero, and Alberto Tonda

Abstract In the field of machine learning, coresets are defined as subsets of the training set that can be used to obtain a good approximation of the behavior that a given algorithm would have on the whole training set. Advantages of using coresets instead of the training set include improving training speed and allowing for a better human understanding of the dataset. Not surprisingly, coreset discovery is an active research line, with several notable contributions in literature. Nevertheless, restricting the search for representative samples to the available data points might impair the final result. In this work, neural networks are used to create sets of virtual data points, named *archetypes*, with the objective to represent the information contained in a training set, in the same way a coreset does. Starting from a given training set, a hierarchical clustering neural network is trained and the weight vectors of the leaves are used as archetypes on which the classifiers are trained. Experimental results on several benchmarks show that the proposed approach is competitive with traditional coreset discovery techniques, delivering results with higher accuracy, and showing a greater ability to generalize to unseen test data.

G. Ciravegna
University of Siena, Siena, Italy
e-mail: gabriele.ciravegna@unifi.it

P. Barbiero (✉) · G. Squillero
Politecnico di Torino, Torino, Italy
e-mail: pietro.barbiero@polito.it

G. Squillero
e-mail: giovanni.squillero@polito.it

G. Cirrincione
University of the South Pacific, Suva, Fiji
e-mail: nimzoexin59@gmail.com

A. Tonda
Université Paris-Saclay, Saclay, France
e-mail: alberto.tonda@inra.fr

UMR 782 INRA, Thiverval-Grignon, France

A. Esposito et al. (eds.), *Progresses in Artificial Intelligence and Neural Systems*, Smart Innovation, Systems and Technologies 184,
https://doi.org/10.1007/978-981-15-5093-5_24

1 Introduction

The concept of *coreset*, stemming from computational geometry, has been redefined in the field of machine learning (ML) as the subset of input samples of minimal size, for which a ML algorithm can obtain a good approximation of the behavior the algorithm itself would normally have on the whole set of input samples. Differently speaking, a coreset can be seen as a fundamental subset of a target training set that is sufficient for a given algorithm to deliver good results, or even the same results it would have if trained on the whole training set [1]. While this definition might appear generic, it must be noted that it encompasses significantly different tasks, ranging from classification, to regression, to clustering, whose performance is measured in entirely different ways. Practical applications of coresets include: obtaining a better understanding of the data, drastically reducing the number of data points a human expert has to analyze; and considerably speeding up training time of ML algorithms.

Coreset discovery is an active research line, and specialized ML literature reports a substantial number of approaches: Greedy Iterative Geodesic Ascent (GIGA) [2], Frank-Wolfe (FW) [3], Forward Stagewise (FSW) [4], Least-angle regression (LAR) [5, 6], Matching Pursuit (MP) [7], and Orthogonal Matching Pursuit (OMP) [8]. Often such algorithms require the user to specify the number N of desired points in the coreset; or assume, for simplicity, that a coreset is independent from the task and/or the algorithm selected for that task. As ML algorithms employ different techniques to accomplish the same goals, however, it is reasonable to assume that they would need coresets of different size and shape to operate at the best of their possibilities. Furthermore, restricting the search of coresets to the data points actually available in the given dataset might impair the final result: as coresets can be seen as a *summary* of the information contained in a dataset, it is conceivable that a set of virtual points might better represent the information in an even more concise way. In the following, we call these virtual data points *archetypes*.

Starting from these two considerations, a neural network approach to archetype set discovery for classification tasks is proposed. Starting from a training set, a GH-ARCH (Growing Hierarchical Archetypes) algorithm is used to find the archetype sets best representing the geometry of the dataset. GH-ARCH is a variant of the recently published GH-EXIN neural network [9] which performs hierarchical clustering by building a divisive hierarchical tree in an incremental and self-organized way. Indeed, in this work it will be shown how cluster centroids are the best virtual points to represent the information contained in a given dataset. Moreover, the hierarchical architecture of the network allows to select the best trade-off between the number of points used and the accuracy of the ML algorithm, by simply stopping the network when a desired resolution level is reached.

Experimental results on classification benchmarks show that the proposed approach is able to best several state-of-the-art coreset discovery algorithms in literature, obtaining results that also allow the classifier to generalize better on an unseen test set of the same benchmark.

2 Background

In computational geometry, coresets are defined as a small set of points that approximates the shape of a larger point set. The concept of coreset in ML is extended to intend a subset of the (training) input samples, such that a good approximation to the original input can be obtained by solving the optimization problem directly on the coreset, rather than on the whole original set of input samples [1].

Finding coresets for ML problems is an active line of research, with applications ranging from speeding up training of algorithms on large datasets [10] to gaining a better understanding of the algorithm's behavior. Unsurprisingly, a considerable number of approaches to coreset discovery can be found in the specialized literature. In the following, a few of the main algorithms in the field, that will be used as a reference during the experiments, are briefly summarized: Greedy Iterative Geodesic Ascent (GIGA) [2], Frank-Wolfe (FW) [3], Forward Stagewise (FSW) [4], Least-angle regression (LAR) [5, 6], Matching Pursuit (MP) [7] and Orthogonal Matching Pursuit (OMP) [8].

BLR is based on the idea that finding the optimal coreset is too expensive. In order to overcome this issue, the authors use a k-clustering algorithm to obtain a compact representation of the data set. In particular, they claim that samples that are bunched together could be represented by a smaller set of points, while samples that are far from other data have a larger effect on inferences. Therefore, the BLR coreset is composed of few samples coming from tight clusters plus the outliers.

The original FW algorithm applies in the context of maximizing a concave function within a feasible polytope by means of a local linear approximation. Section 4 refers to the Bayesian implementation of the FW algorithm designed for core set discovery. This technique, described in [11], aims to find a linear combination of approximated likelihoods (which depends on the core set samples) that is similar to the full likelihood as much as possible.

GIGA is a greedy algorithm that further improves FW. In [2], the authors show that computing the residual error between the full and the approximated likelihoods by using a geodesic alignment guarantees a lower upper bound to the error at the same computational cost.

FSW [4], LAR [5, 6], MP [7] and OMP [8] were all originally devised as greedy algorithms for dimensionality reduction. The simplest is FSW which projects high-dimensional data in a lower dimensional space by selecting one at a time the feature whose inclusion in the model gives the most statistically significant improvement. MP instead includes features having the highest inner product with a target signal, while its improved version OMP at each step carries an orthogonal projection out. Similarly, LAR increases the weight of each feature in the direction equiangular to each one's correlations with the target signal. All these procedures could be applied to the transpose problem of feature selection, that is approximation of core sets. Often these algorithms start from the assumption that the coreset for a given dataset will be independent from the ML pipeline used, but this premise might not always be correct. For example, the optimization problem underlying a classification task might vary considerably depending on the ML algorithm used.

3 GH-ARCH

The proposed approach exploits neuron weight vectors of a clustering network as a set of virtual points, *archetypes*. These can better and more concisely represent the entire dataset rather than a single set of real data points. Neurons, in fact, are placed in such a way to provide the best topological representation of data and each neuron summarizes the information provided by its Voronoi set, the compact set of points assigned to it.

The use of a hierarchical divisive neural network makes it possible to provide at the same time multiple archetypes set at different level of resolution—each one corresponding to a level of the network. In fact, basic clustering algorithms are capable of producing only a single representation of the data. Furthermore, the proposed neural network, GH-ARCH, automatically suggests the best representation to be used by a given classifier. After finding an entire level of neurons (archetype set), GH-ARCH trains the classifier on the weight vectors of the leaf neurons: in case the accuracy overcomes $A_{\min}$, the minimum accuracy archetype sets needs to satisfy (user-defined parameter), the algorithm returns the current list of archetypes sets; otherwise, the current set of archetypes is inserted in the list and a further layer is added to the network.

As previously introduced, GH-ARCH is a modified version of the GH-EXIN algorithm, devised ad-hoc for archetype set discovery. The major difference between the two algorithms consists in the final goal of the network and on the indices to be minimized: while GH-EXIN is a network which focuses on finding biclusters and minimize a biclustering quantization index, GH-ARCH attempts to minimize the heterogeneity and maximize the purity of the clusters, in order to group points which are both close to each other and belonging to the same class. The way in which data is divided at deeper and deeper levels, on the other hand, follows the same algorithm and it is briefly explained in the following. The overall proposed approach is summarized in Fig. 1.

3.1 Neural Clustering

Neural clustering algorithms build a hierarchical (divisive) tree whose vertices correspond to its neurons, as shown in Fig. 2. The architecture of the network is data-driven, the number of neurons and the number of layers are automatically determined according to the data. Each neuron is associated to a weight vector whose dimensionality corresponds to the input space. For each father neuron, a neural network is trained on its set of data (Voronoi set). The related cardinality is shown in Fig. 2, next to each vertex. The children nodes are the neurons of the associated neural network, and define a partition of the father Voronoi set. For each leaf, the procedure is repeated. The initial structure of the neural network is always a pair of neurons (seed), linked by an edge, with age set to zero. Multiple node creation and pruning determines the

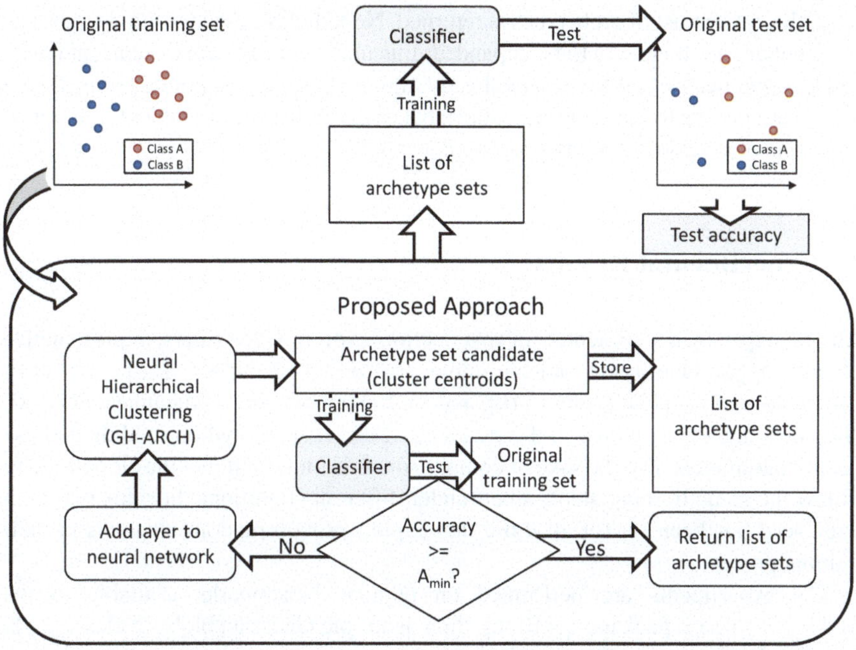

Fig. 1 Scheme of the proposed approach. $A_{\min}$ is a user-defined parameter

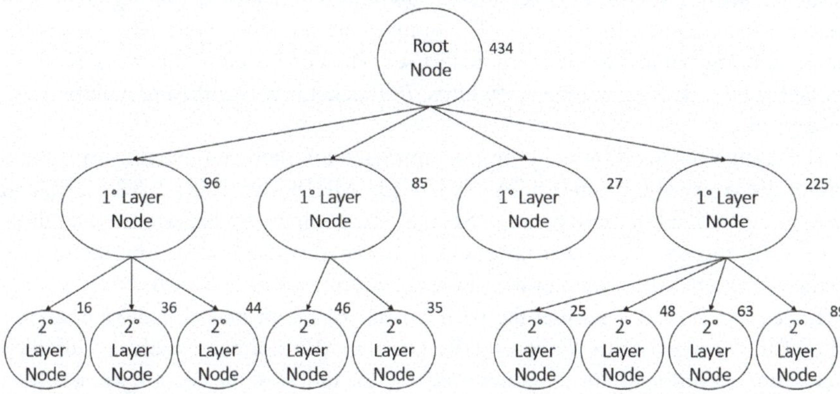

Fig. 2 GH-ARCH tree

correct number of neurons of each network. For further details regarding GH-EXIN neuron creation and deletion we remind to [12, 13] were it has been fully explained and compared with other SOA algorithms.

After repeating this procedure for all the leaves of a single layer, recalling Fig. 1, a given classifier is trained on the weight vectors of the leaves found by GH-ARCH so far. As previously stated, in case the accuracy obtained on a test set is higher than

$A_{\min}$, the current list of archetypes is returned. Nonetheless, the algorithm stops also in case there are no leaves to be expanded: this may occur in case the current purity and heterogeneity of all leaves are already high. Lastly, in case points grouped by a leaf do not belong to the same class, the label of the archetype of that leaf is assigned by means of a majority voting procedure.

4 Experimental Results

All the experiments presented in this section exploit 4 classifiers, representative of both hyperplane-based and ensemble, tree-based classifiers: `Bagging` [14], `RandomForest` [15], `Ridge` [16], and `SVC` (Support Vector Machines) [17]. All classifiers are implemented in the `scikit-learn`[1] [18] Python module and use default parameters. For the sake of comparison, it is important that the classifier will follow the same training steps, albeit under different conditions. For this reason, a fixed seed has been set for all those that exploit pseudo-random elements in their training process.

The experiments are performed on popular benchmarks available in the `scikit-learn` package: (i) Blobs, three isotropic Gaussian blobs (3 classes, 400 samples, 2 features); (ii) Circles, a large circle containing a smaller one (2 classes, 400 samples, 2 features); (iii) Moons, two interleaving half circles (2 classes, 400 samples, 2 features); (iv) Iris [19] (3 classes, 150 samples, 4 features); (v) Digits [20] (10 classes, 1797 samples, 64 features). The samples are randomly split between a 66%-sample training set and a 33%-sample test set. The code used in this work is freely available in the BitBucket repository https://bitbucket.org/evomlteam/evolutionary-archetypes.

The results obtained by the proposed approach are then compared against the 6 coreset discovery algorithms GIGA [2], FW [3], MP [8], OMP [8], LAR [5, 6], and FSW [4], described in more detail in Sect. 2. The comparison is performed on three metrics: (i) coreset size (lower is better); (ii) classification accuracy on the test set (higher is better); (iii) running time of the algorithm (lower is better).

Tables 1, 2, 3, and 4 present the performance of the proposed approach against state-of-the-art algorithms for coreset discovery in ML literature. Tables are divided in columns according to the classifier used. On the first row, the accuracy reached on whole original dataset is also reported.

With regard to test accuracy, GH-ARCH not only outperforms the other techniques by far, but it is sometimes able to increase the performance obtained training the same classifier with all the training samples available. This means that the decision boundaries generated by the classifier using the neural archetypes may generalize even better than those generated using the whole training set, as shown in Fig. 3. These results suggest that the performance of ML classifiers is not just a function of the *size* of the training set (as Big Data and Deep Learning often claim) but also a function

[1] scikit-learn: Machine Learning in Python, http://scikit-learn.org/stable/.

Table 1 Blobs dataset. Coreset size, accuracy on test set and running time (seconds) of the considered classifiers and coreset algorithms

		RandomForest			Bagging			SVC			Ridge		
Algorithm	Core type	Size	Accuracy	Avg. time	Size	Accuracy	Avg. time	Size	Accuracy	Avg. time	Size	Accuracy	Avg. time
All samples		265	**0.9185**		265	0.8963		265	**0.9407**		265	0.8963	
GH-ARCH	4 level	29	**0.9185**	0.9	29	**0.9259**	0.9	27	0.9259	1.0	30	0.8889	0.8
	3 level	27	**0.9185**		27	0.9185		25	0.9185		26	**0.9185**	
GIGA		3	0.6296	0.01	3	0.8889	0.01	3	0.8889	0.01	3	0.8519	0.01
FW		4	0.5185	3	4	0.6593	3	4	0.5852	3	4	0.8815	3
MP		5	0.4047	4	5	0.6296	4	5	0.3333	4	5	0.5778	4
FS		5	0.7481	4	5	0.7481	4	5	0.3333	4	5	0.6296	4
OP		4	0.4074	0.01	4	0.6148	0.01	4	0.3333	0.01	4	0.5852	0.01
LAR		3	0.8074	0.01	3	0.8074	0.01	3	0.5185	0.01	3	0.5185	0.01

Table 2 Circles dataset. Coreset size, accuracy on test set and running time (seconds) of the considered classifiers and coreset algorithms

		RandomForest			Bagging			SVC			Ridge		
Algorithm	Core type	Size	Accuracy	Avg. time	Size	Accuracy	Avg. time	Size	Accuracy	Avg. time	Size	Accuracy	Avg. time
All samples		266	**0.9552**		266	**0.9478**		266	**0.9851**		266	0.5000	
GH-ARCH	4 level	23	0.9254	0.8	23	0.9403	0.8	26	**0.9851**	1.0	26	0.5000	0.8
	3 level	15	0.6418		15	0.5970		24	0.9776		25	0.5000	
GIGA		2	0.5970	0.01	2	0.5746	0.01	2	0.6343	0.01	2	0.6364	0.01
FW		5	0.5597	3	5	0.5000	3	5	0.5000	3	5	0.5000	3
MP		3	0.5000	4	3	0.5224	4	3	0.5000	4	3	**0.6567**	4
FS		4	0.6567	4	4	0.6194	4	4	0.6119	4	4	0.6269	4
OP		3	0.5000	0.01	3	0.4851	0.01	3	0.6418	0.01	3	0.5448	0.01
LAR		2	0.5522	0.01	2	0.6194	0.01	2	0.5970	0.01	2	0.5970	0.01

Table 3 Moons dataset. Coreset size, accuracy on test set and running time (seconds) of the considered classifiers and coreset algorithms

		RandomForest			Bagging			SVC			Ridge		
Algorithm	Core type	Size	Accuracy	Avg. time	Size	Accuracy	Avg. time	Size	Accuracy	Avg. time	Size	Accuracy	Avg. time
All samples		266	**0.9328**		266	**0.9254**		266	**0.9179**		266	**0.8134**	
GH-ARCH	3 level	28	0.8209	0.8	26	0.7687	1.0	23	0.8358	0.7	26	0.7687	0.9
	2 level	17	0.7313		17	0.8358		16	0.8433		17	0.7985	
GIGA		2	0.4254	0.01	2	0.2463	0.01	2	0.4701	0.01	2	0.4701	0.01
FW		6	0.6493	3	6	0.6493	3	6	0.5299	3	6	0.6866	3
MP		3	0.5149	4	3	0.5821	4	3	0.5896	4	3	0.6642	4
FS		2	0.5149	4	2	0.2313	4	2	0.6119	4	2	0.6119	4
OP		2	0.5149	0.01	2	0.2463	0.01	2	0.6493	0.01	2	0.6493	0.01
LAR		3	0.5149	24	3	0.2388	24	3	0.5224	24	3	0.5896	24

Table 4 Iris dataset. Coreset size, accuracy on test set and running time (seconds) of the considered classifiers and coreset algorithms

		RandomForest			Bagging			SVC			Ridge		
Algorithm	Core type	Size	Accuracy	Avg. time	Size	Accuracy	Avg. time	Size	Accuracy	Avg. time	Size	Accuracy	Avg. time
All samples		99	**0.9608**		99	**0.9608**		99	0.9412		99	**0.8824**	
GH-ARCH	3 level	23	0.9412	0.2	23	0.9412	0.2	23	0.9020	0.1	23	0.8431	0.1
	2 level	14	0.9412		14	0.9216		14	0.8824		14	0.8431	
GIGA		7	0.9216	0.01	7	0.6667	0.01	7	**0.9804**	0.01	7	0.8431	0.01
FW		15	0.8824	3	15	0.8627	3	15	0.9412	3	15	0.8235	3
MP		14	0.9412	4	14	0.8627	4	14	0.9216	4	14	0.7255	4
FS		7	0.6667	4	7	0.7059	4	7	0.6471	4	7	0.6275	4
OP		5	0.7059	0.01	5	0.5294	0.01	5	0.7843	0.01	5	0.8235	0.01
LAR		4	0.5294	22	4	0.6863	22	4	0.6471	22	4	0.7059	22

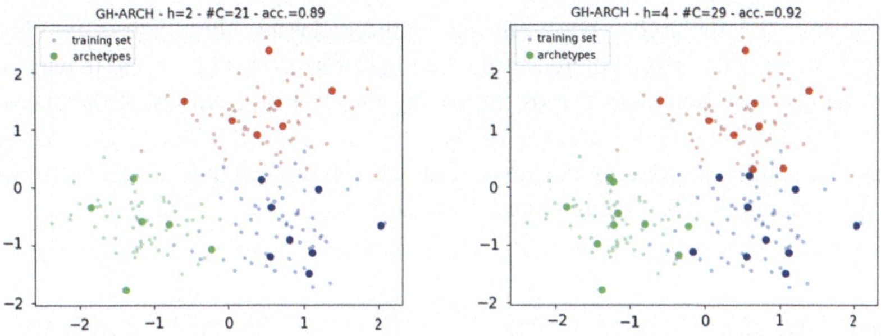

(a) Archetypes extracted by GH-ARCH at the 2nd level of the network

(b) Archetypes extracted by GH-ARCH at the 4th level of the network

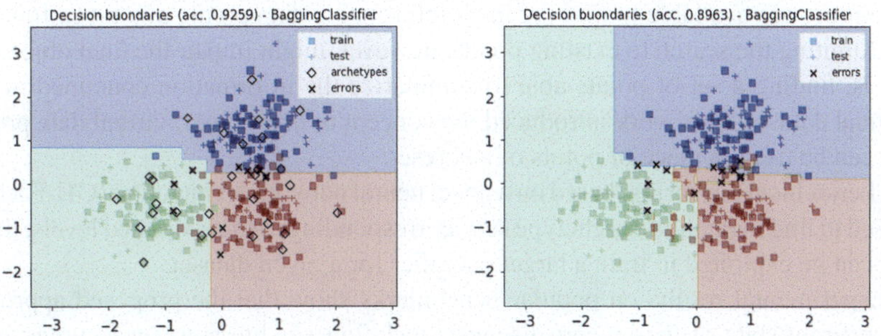

(c) Decision boundaries using the archetypes.

(d) Decision boundaries using the whole training set.

Fig. 3 GH-ARCH on the Blobs dataset using the Bagging classifier

of the *mutual position* of the training samples in the feature space. By exploiting the original training set and by relaxing the constraint of sample positions, GH-ARCH generates a new, smaller data set suited for each classifier in order to provide the best generalization ability.

Overall, the number of points used by GH-ARCH is generally higher compared to the other state-of-the-art algorithms. Still, the number of data points used is generally an order of magnitude smaller than the original training set. The time taken by all algorithms is also comparable, with GH-ARCH resulting among the fastest algorithms on all datasets but Digits. At last, sometimes the final archetype set may not be the best set found by the network. As shown in Table 3, the archetype set found at the 3rd level performs worse than the one found at the 2nd level for most classifier. For this reason, GH-ARCH returns the entire list of possible archetypes set produced by each layer, together with its accuracy, so that a human expert may choose the preferred configuration.

In Fig. 3, the process of creation of the archetypes can be observed at different levels of the network and how these points may also outperform the whole training set. GH-ARCH, after the second level of training (Fig. 3a already distributes correctly the neurons—i.e. the archetypes—among the three classes. Nonetheless, two layers below (Fig. 3b), GH-ARCH increases the number of neurons ($21 \rightarrow 29$) in particular along the decision boundaries, thus improving the overall accuracy of the classifier ($90 \rightarrow 92$). Of notable importance is also the fact that GH-ARCH is capable of recognizing outliers, reducing the noise of the dataset present in Fig. 3d which causes a bad classification when using the whole training set.

5 Conclusions

Coreset discovery is a research line of utmost practical importance, and several techniques are available to find the most informative data points in a given training set. Limiting the search to existing points, however, might impair the final objective, that is, finding a set of points able to summarize the information contained in the original dataset. This work introduced the concept of *archetypes*, virtual data points that can be used in place of points of a coreset.

Hierarchical clustering, based on a novel neural network architecture (GH-EXIN), is used to find meaningful archetype sets, corresponding to the estimated levels; these sets can be exploited to train a target classifier for a given dataset.

Experimental results on popular benchmarks show that the proposed approach outperforms state-of-the-art core discovery techniques in literature on accuracy, generality, and computational time.

Future work will extend the current results to regression problems, and explore new methodologies to improve the archetype sets even further.

References

1. Bachem, O., Lucic, M., Krause, A.: Practical coreset constructions for machine learning (2017). arXiv:1703.06476
2. Campbell, T., Broderick, T.: Bayesian coreset construction via greedy iterative geodesic ascent. In: International Conference on Machine Learning (ICML) (2018). https://arxiv.org/pdf/1802.01737.pdf
3. Clarkson, K.L.: Coresets, sparse greedy approximation, and the Frank-Wolfe algorithm. In: ACM Transactions on Algorithms (2010). http://citeseerx.ist.psu.edu/viewdoc/download?doi=10.1.1.145.9299&rep=rep1&type=pdf
4. Efroymson, M.A.: Multiple regression analysis. In: Mathematical Methods for Digital Computers (1960)
5. Efron, B., Hastie, T., Johnstone, I., Tibshirani, R.: Least angle regression. Ann. Statis. **32**(2), 407–451 (2004). https://arxiv.org/pdf/math/0406456.pdf
6. Boutsidis, C., Drineas, P., Magdon-Ismail, M.: Near-optimal coresets for least-squares regression. Technical Report (2013). https://arxiv.org/pdf/1202.3505.pdf

7. Mallat, S., Zhang, Z.: Matching pursuits with time-frequency dictionaries. IEEE Trans. Signal Process. **42**(12), 3397–3415 (1993)
8. Pati, Y., Rezaiifar, R., Krishnaprasad, P.: Orthogonal matching pursuit: recursive function approximation with applications to wavelet decomposition. In: Proceedings of 27th Asilomar Conference on Signals, Systems and Computers, pp. 40–44 (1993). http://ieeexplore.ieee.org/document/342465/
9. Barbiero, P., Ciravegna, G., Piccolo, E., Cirrincione, G., Cirrincione, M., Bertotti, A.: Neural biclustering in gene expression analysis. In: 2017 International Conference on Computational Science and Computational Intelligence (CSCI), pp. 1238–1243, Dec. 2017
10. Tsang, I.W., Kwok, J.T., Cheung, P.-M.: Core vector machines: fast SVM training on very large data sets. J. Mach. Learn. Res. **6**(Apr), 363–392 (2005)
11. Campbell, T., Broderick, T.: Automated Scalable Bayesian Inference via Hilbert Coresets (2017). http://arxiv.org/abs/1710.05053
12. Cirrincione, G., Ciravegna, G., Barbiero, P., Randazzo, V., Pasero, E.: The GH-EXIN neural network for hierarchical clustering. Neural Netw. **121**, 57–73 (2020). http://www.sciencedirect.com/science/article/pii/S0893608019302060
13. Ciravegna, G., Cirrincione, G., Marcolin, F., Barbiero, P., Dagnes, N., Piccolo, E.: Assessing discriminating capability of geometrical descriptors for 3D face recognition by using the GH-EXIN neural network, pp. 223–233. Springer, Singapore (2020). https://doi.org/10.1007/978-981-13-8950-4_21
14. Breiman, L.: Pasting small votes for classification in large databases and on-line. Mach. Learn. **36**(1–2), 85–103 (1999)
15. Breiman, L.: Random forests. Mach. Learn. **45**(1), 5–32 (2001)
16. Tikhonov, A.N.: On the stability of inverse problems. Dokl. Akad. Nauk SSSR **39**(5), 195–198 (1943)
17. Hearst, M.A., Dumais, S.T., Osman, E., Platt, J., Scholkopf, B.: Support vector machines. IEEE Intell. Syst. Appl. **13**(4), 18–28 (1998)
18. Pedregosa, F., Varoquaux, G., Gramfort, A., Michel, V., Thirion, B., Grisel, O., Blondel, M., Prettenhofer, P., Weiss, R., Dubourg, V., Vanderplas, J., Passos, A., Cournapeau, D., Brucher, M., Perrot, M., Duchesnay, E.: Scikit-learn: machine learning in Python. J. Mach. Learn. Res. **12**, 2825–2830 (2011)
19. Fisher, R.A.: The use of multiple measurements in taxonomic problems. Ann. Eugen. **7**(2), 179–188 (1936)
20. Dheeru, D., Karra Taniskidou, E.: UCI Machine Learning Repository (2017). http://archive.ics.uci.edu/ml

1-D Convolutional Neural Network for ECG Arrhythmia Classification

Jacopo Ferretti, Vincenzo Randazzo, Giansalvo Cirrincione, and Eros Pasero

Abstract Automated electrocardiogram analysis and classification is nowadays a fundamental tool for monitoring patient heart activity and, consequently, his state of health. Indeed, the main interest is detecting the arise of cardiac pathologies such as arrhythmia. This paper presents a novel approach for automatic arrhythmia classification based on a 1D convolutional neural network. The input is given by the combination of several databases from Physionet and is composed of two leads, LEAD1 and LEAD2. Data are not preprocessed, and no feature extraction has been performed, except for the medical evaluation in order to label it. Several 1D network configurations are tested and compared in order to determine the best one w.r.t. heart-beat classification. The test accuracy of the proposed neural approach is very high (up to 95%). However, the goal of this work is also the interpretation not only of the results, but also of the behavior of the neural network, by means of confusion matrix analysis w.r.t. the different arrhythmia classes.

J. Ferretti
Dipartimento di Scienze Chirurgiche, Universitá degli Studi di Torino, Turin, Italy
e-mail: jacopo.ferretti@unito.it

J. Ferretti · V. Randazzo (✉) · E. Pasero
DET, Politecnico di Torino, Turin, Italy
e-mail: vincenzo.randazzo@polito.it

E. Pasero
e-mail: eros.pasero@polito.it

G. Cirrincione
Lab. LTI, University of Picardie Jules Verne, Amiens, France
e-mail: giansalvo.cirrincione@usp.ac.fj

SEP, University of South Pacific, Suva, Fiji Islands

A. Esposito et al. (eds.), *Progresses in Artificial Intelligence and Neural Systems*,
Smart Innovation, Systems and Technologies 184,
https://doi.org/10.1007/978-981-15-5093-5_25

1 Introduction

Electrocardiogram (ECG) is the electrical signal produced by heart contraction, which is recorded by physicians to monitor the heart state of health and, consequently, the person it belongs. The standard procedure uses an electrocardiograph with ten electrodes placed on specific points of a human body, which acquire up to twelve different signals, called *LEADS*. As explained in [1], an healthy ECG, shown in Fig. 1, presents six fiducial points (P, Q, R, S, T, U) which are correlated to the four principal stages of activity of a cardiac cycle: isovolumic relaxation, inflow, isovolumic contraction, ejection. This path should repeat itself constantly over the time; otherwise, a person suffers from arrhythmia. Cardiac arrhythmia is one of the most common disease people are killed by, therefore normal and abnormal ECG signal automatic classification is raising more and more interest in the scientific community.

Several approaches have been already proposed in literature. The most famous algorithm for automatic QRS-complex detection within an ECG signal is described in [2], while [3] uses Support Vector Machines at the same purpose. In [4] and [5] fuzzy and artificial neural networks are used, respectively, for ECG analysis. Cardiac arrhythmia is studied using hidden Markov models in [6]. Wavelet transformation and artificial neural network for arrhythmia detection is presented in [7, 8] and [9] show two possible approaches for atrial fibrillation recognition.

Recently, a novel class of techniques based on Convolutional Neural Networks (CNN) is gathering the attention of the scientific community thanks to its capability in automatically learning the intrinsic patterns from the data; indeed, this approach can avoid the need of manual feature engineering and it can also infer hidden intrinsic patterns more effectively. Inspired by the human mind visual cortex, CNN consists of multiple layers, each of which owns a small subset of neurons to process portions of the input data. These subsets are tiled to introduce region overlap, and the process is repeated layer by layer to achieve a high level abstraction of the original dataset as shown in Fig. 2. An application to arrhythmia detection can be found in [10]. Advanced machine learning techniques like CNNs have already been extensively used in biomedical field with various application—such as in the classification of EEG recordings in dementia [11, 12]—with very promising results. A particular, quite interesting, class of convolutional neural network is the 1D-CNN, which takes as input data a single stream (i.e. signal), e.g. ECG, and slides a kernel along it in search of particular patterns, as shown in Fig. 3. Applications to heart disease classification and biometric identification are presented in [13] and [14], respectively. In this paper different 1D-CNNs are applied to the MIT-BIH database [15] in order to test which configuration yields the best performances in ECG classification of arrhythmia. However, the goal of this work is also the interpretation not only of the results, but also of the behavior of the neural network, by means of confusion matrix analysis w.r.t. the different arrhythmia classes. Moreover, in this study all of the available classes of arrhythmia are used for the classification whereas in other similar studies—such as [16, 17]—only the most common ones are present.

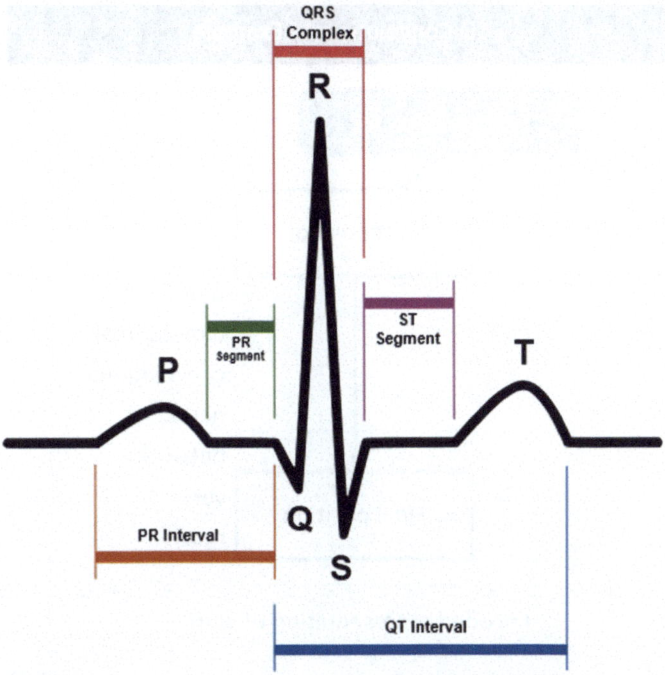

Fig. 1 Example of an healthy ECG

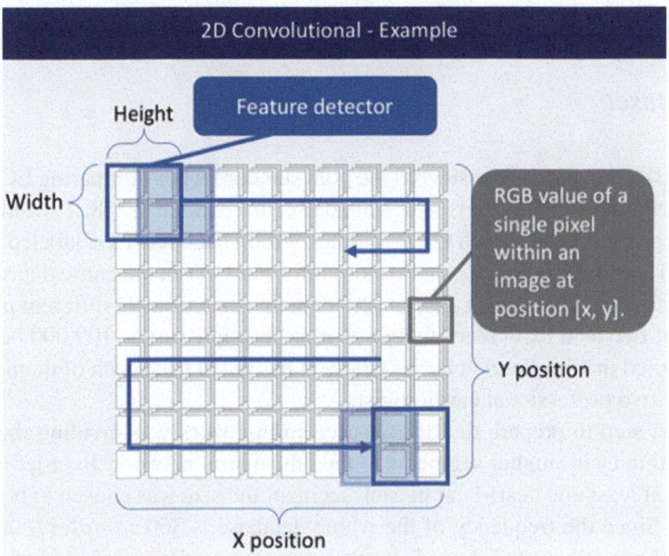

Fig. 2 2D convolutional neural network

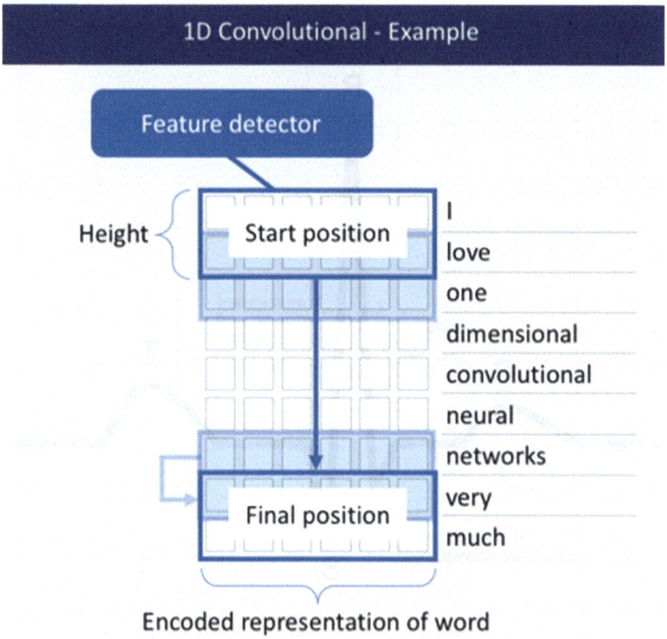

Fig. 3 1D convolutional neural network

2 Methodology

2.1 Dataset

The MIT-BIH database is considered the gold standard when comparing ECG classification techniques. Indeed, it is widespread used in research [1, 18, 19] and it covers a wide range of diseases. Each QRS complex within each record is labeled; hence, a supervised learning approach is quite straightforward. Also, the entire dataset is very well documented. The chosen dataset [20] contains data from 48 different patients in the form of two lead ECG recording of 30 min. Its approximate 109,000 heart-beats are distributed in 16 different classes (showed in Table 1) and each of them has been labeled by two professional cardiologists.

The first step to prepare the dataset used in this work was dividing the over 31 $million$ samples in smaller segments to feed the neural network. In order to be sure to include at least one heart-beat in each segment the size was chosen to be between 1 and 2 s. Since the frequency of the whole database is 360 $samples/s$ a segment size of 500 $samples$ was selected. Furthermore an overlapping factor of 10% was chosen in order to increase the final number of segments (data augmentation).

Table 1 Heart-beat labels and their meaning

Label	Meaning	Label	Meaning
/	Paced beat	R	Right bundle branch block beat
A	Atrial premature beat	S	Supraventricular premature beat
E	Ventricular escape beat	V	Premature ventricular contraction
F	Fusion of ventricular and normal beat	!	Ventricular flutter wave
J	Nodal premature beat	a	Aberrated atrial premature beat
L	Left bundle branch block beat	e	Atrial escape beat
N	Normal beat	f	Fusion of paced and normal beat
Q	Unclassifiable beat	j	Nodal escape beat

Each segment was then normalized in a range of $[-1, +1]$, and the appropriate label was assigned to the segment. Finally, the dataset was randomly divided in training dataset and validation dataset with a ratio of 90%/10%, respectively.

2.2 *1D-CNN*

A Convolutional Neural Network (CNN) is a class of neural network where a filter—commonly called kernel—is passed (convoluted) along data in order to learn particular patterns. These pattern extracted grow in complexity along with the depth of the network. Namely, deeper networks extracts more elaborate features.

In Fig. 4 there is an example of the most commonly used 2D-CNN where a kernel of width and height $3x3$ is passed across an image, or a generic numerical matrix, producing an output filtered image. The convolution starts by superimposing the kernel with part of the image; then, the corresponding elements are multiplied and summed with each other and the results is the new element of the output matrix. Finally, the kernel is moved and the process is repeated for all the elements of the input matrix. It is important to note that because of how the convolution works, the output matrix is smaller than the input one depending on the size of the kernel.

In 1D-CNN the procedure is analogous. The only difference is that filters and signals are mono-dimensional, and thus the kernel can only slide in one direction.

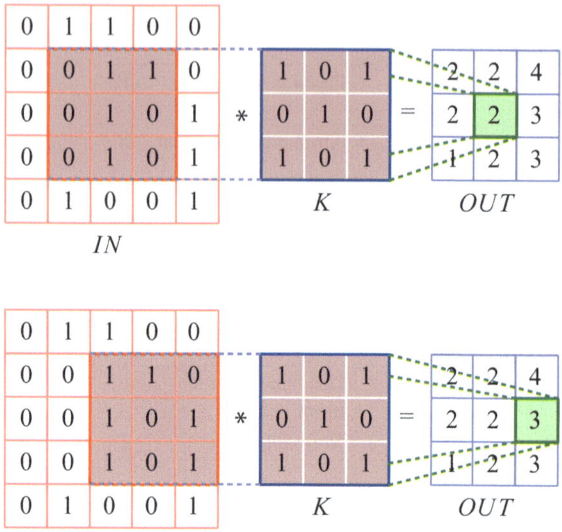

Fig. 4 Example of two passages of a 2D CNN

2.3 *Google Colab*

Albeit this work didn't needed to elaborate vasts amount of data, the training of a deep CNN can require a tremendous amount of time if performed on a low end machine. For this reason, all the experiments were performed on Google Colab, where it was available a virtual server with a resourceful GPU (Nvidia Tesla k80), which greatly helped in speeding up the training process.

3 Experiments

To assess the classification quality of a 1D-CNN on the MIT-BIH dataset, several configurations of the network have been tested and compared. The number of layers, the size and the number of the filters, the dropout rate, together with the activation function have been varied to determine the best architecture. Among the different possibilities, only the 4 most representative examples, w.r.t. the classification performances, are reported together with their topology. Table 2 resumes the results of the selected experiments.

To begin, a simple configuration, say Net1, with around 65 K parameters has been tested. It was made of a first convolutional layer of 16 filters with a kernel size of 32, followed by a max pooling layer and a softmax classifier. Net1 has reported a training and testing accuracy equal to 92% and 91%, respectively.

The second experiment deals with a more complex network (257 K parameters), called Net2. It consisted of a first convolutional layer of 64 filters with a kernel

Table 2 Accuracy values for the four most representative network architectures

	Training accuracy (%)	Test accuracy (%)	Total parameters
Net 1	92	91	65,056
Net 2	96	94	257,104
Net 3	96	94	533,072
Net 4	98	95	1,266,768

size of 8, followed by a max pooling layer and a softmax classifier. Overall training and testing accuracy were equal to 96% and 94%, respectively, thus improving the previous classification performances.

Third experiment was conducted using a deeper architecture (Net3) made of three convolutional layers with growing number of filters—64, 128, 256—and decreasing kernel size—32, 16, 8—each followed by a pooling layer. The convolutional layers feed a 128 neurons fully-connected layer which finally flows into the softmax classifier. Despite the number of parameters doubled (533 K) w.r.t. the previous experiment, the performances remained roughly the same.

In order to improve the classification, in the last experiment a much more complex configuration (1200 K parameters) was implemented, Net4. Figure 5 shows the architecture detail: it is a 5-layer CNN with 2 fully connected layers and 1 fully connected softmax classifier. The latter experiment yields the best results w.r.t. classification performances; indeed, it reached 98% and 95%, training and testing accuracy, respectively.

3.1 Results Analysis

Net4 is the best architecture resulting from the experiments. Despite several attempts to increase the classification rate both in training and test sets, the network did not improved its performances. Therefore, the confusion matrix of the latter experiment has been analysed in order to deepen the response of Net4 to the different classes of the input dataset.

Analyzing the confusion matrix in Fig. 6 there are a few observation to be made. First, class F, which is the fusion of ventricular and normal beat, is sometimes mistaken with class N (Normal beat) or class V (Premature ventricular contraction); F is, by definition, the fusion of the other two classes, therefore if the analyzed window is not perfectly aligned with the whole series of heart-beat, these classes are virtually unrecognizable. A possible solution could be the window expanding; unfortunately, this approach is not feasible because it would prevent the recognition of the other classes.

It can also be observed that class e (Atrial escape beat) is spread across multiple classes, but not class e itself. The main reason for this behaviour is that class e is

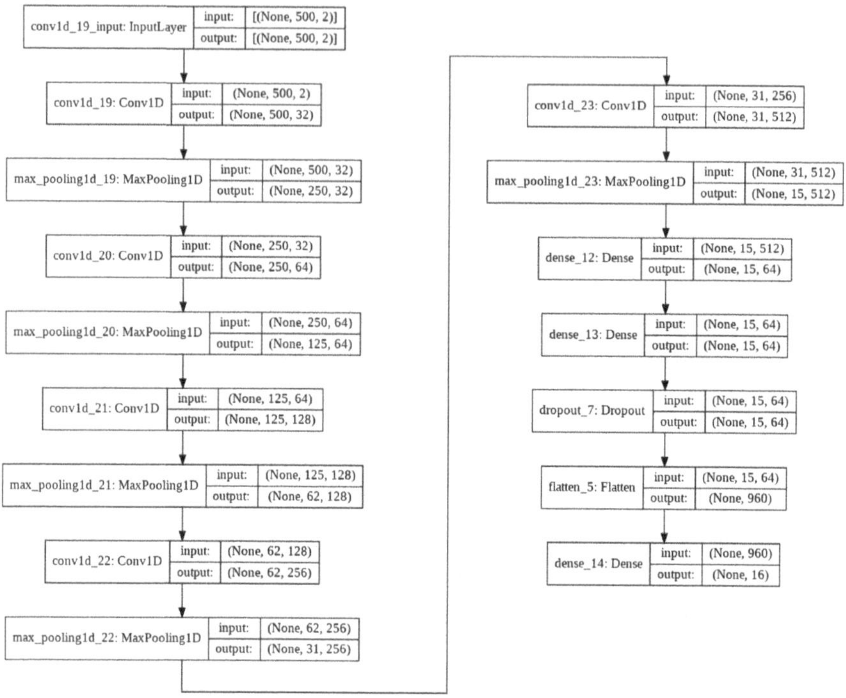

Fig. 5 Net4 architecture

the least represented class in the whole dataset, then Net4 could not be able to train properly on its recognition. However, since class *e* is very similar to class *A* (Atrial premature beat), the net was able to partially classify it as that.

Class *S* (Supraventricular premature beat) is completely misinterpreted as class *V* (Premature ventricular contraction) and requires further investigation since they are two very different patterns.

The last remark is about the class *Q* (Unclassifiable beat). This class is a special case because, by definition, it does not have a specific pattern. In fact, it represent an heart-beat that the cardiologists discarded or were not able to classify due to noise, uncertainties, alteration, etc. It is interesting to note how the network classified most (49%) of those heart-beats as class *N* (Normal heartbeat). Albeit the net clearly classified this class as a wrong heart-beat, we cannot exclude, in advance, the fact that it was responding to some specific pattern of the correctly trained classes. Therefore a further investigation is required for each of class *Q* heart-beats, to assert if the classification was right.

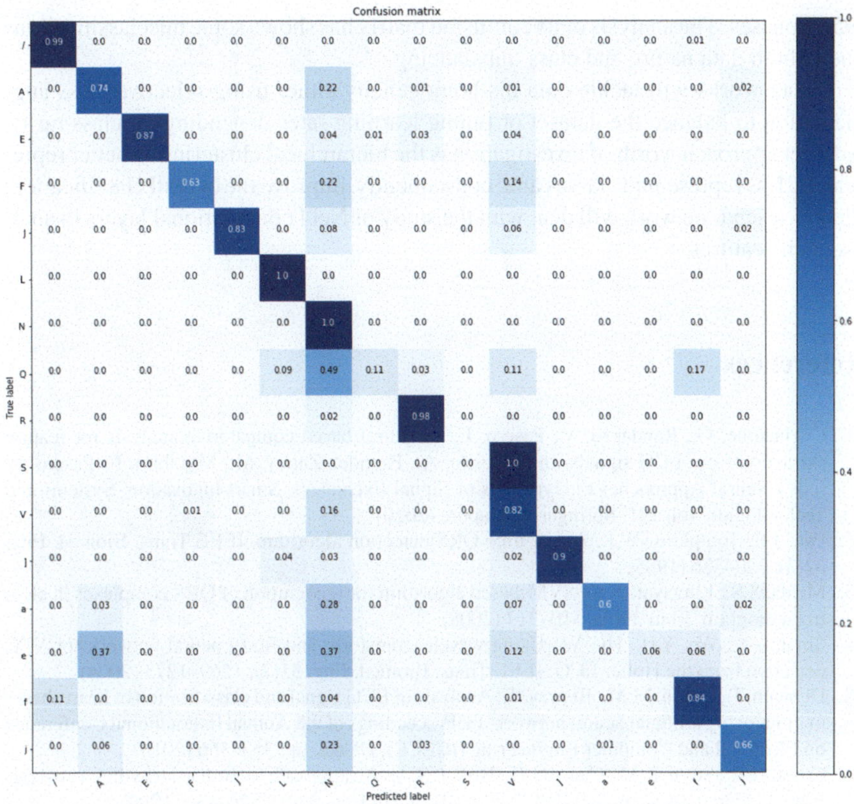

Fig. 6 Confusion matrix obtained with Net4

Finally, the most influential flaw of the network was the class unbalance in the dataset. Almost 40% of the whole dataset examples were of class *N*, while other classes were only represented by a very small amount of examples (e.g. class *e* only counted less than 2% of the dataset). Of course, this dis-homogeneity of class representations had a decisive impact on the results of the training.

4 Conclusions

Automated ECG classification represents a promising technique to improve physicians diagnostic performances on cardiac diseases. Several techniques have been already proposed in literature at this purpose. This paper presents a novel approach based on 1D-CNN. Different network architectures have been tested and compared; among these, Net4 has reached the highest accuracy both in training (98%) and test

(95%) phases. The analysis of its confusion matrix has shown some misclassifications due to both data nature and class unbalancing.

Future works will tackle data dis-homogeneity either using selective class augmentation to balance the dataset or tuning learning rates depending on class rarity. Another approach worth of investigation is the hierarchical clustering to better represent the less represented classes and, consequently, improve the overall classification. Finally, a separate work will deal with the study of Net4 convolutional layers to analyse their features.

References

1. Cirrincione, G., Randazzo, V., Pasero, E.: A neural based comparative analysis for feature extraction from ECG signals. In: Esposito, A., Faundez-Zanuy, M., Morabito, F., Pasero, E. (eds.) Neural Approaches to Dynamics of Signal Exchanges. Smart Innovation, Systems and Technologies, vol. 151. Springer, Singapore (2020)
2. Pan, J.P., Tompkins, W.J.: A real-time QRS detection algorithm. IEEE Trans. Biomed. Eng. **32**(3), 320–326 (1985)
3. Mehta, S.S., Lingayat, N.S.: SVM-based algorithm for recognition of QRS complexes in electrocardiogram. Irbm **29**(5), 310–317 (2008)
4. Shyu, L.Y., Wu, Y.H., Hu, W.: Using wavelet transform and fuzzy neural network for VPC detection from the Holter ECG. IEEE Trans. Biomed. Eng. **51**(7), 1269–1273 (2004)
5. Debnath, T., Hasan, M.M., Biswas, T.: Analysis of ECG signal and classification of heart abnormalities using artificial neural network. In: Proceedings of 9th Annual International Conference on Electrical and Computer Engineering (ICECE), Dhaka, pp. 353–356 (2016)
6. Coast, D.A., Stern, R.M., Cano, G.G., Briller, S.A.: An approach to cardiac arrhythmia analysis using hidden markov models. IEEE Trans. Biomed. Eng. **37**(9), 826–836 (1990)
7. Ranaware, P.N., Deshpande, R.A.: Detection of Arrhythmia based on discrete wavelet transform using artificial neural network and support vector machine. In: Proceedings of 11th Annual International Conference on Communication and Signal Processing (ICCSP), Beijing, pp. 1767–1770 (2016)
8. Artis, S.G., Mark, R.G., Moody, G.B.: Detection of atrial fibrillation using artificial neural networks. In: Computers in Cardiology 1991, Proceedings, pp. 173–176. IEEE (1991)
9. Clifford, G.D., Liu, C.Y., Moody, B., Lehman, L., Silva, I., Li, Q., Johnson, A.E.W., Mark, R.G.: Af classification from a short single lead ECG recording: the physionet computing. In: Cardiology Challenge (2017)
10. Hannun, A.Y., Rajpurkar, P., Haghpanahi, M., Tison, G.H., Bourn, C., Turakhia, M.P., Ng, A.Y.: Cardiologist-level arrhythmia detection and classification in ambulatory electrocardiograms using a deep neural network. In: Nature Medicine, vol. 25, pp. 65–69 (2019)
11. Ieracitano, C., Mammone, N., Bramanti, A., Hussain, A., Morabito, F.C.: A convolutional neural network approach for classification of dementia stages based on 2D-spectral representation of EEG recordings. Neurocomputing **323**, 96–107 (2019). ISSN 0925-2312
12. Ieracitano, C., Mammone, N., Hussain, A., Morabito, F.C.: A novel multi-modal machine learning based approach for automatic classification of EEG recordings in dementia. Neural Netw. **123**, 176–190 (2020). ISSN 0893-6080
13. Karhe, R.R., Badhe, B.: Arrhythmia detection using one dimensional convolutional neural network. Int. Res. J. Eng. Technol. (IRJET) **05**(08) (2018)
14. Karhe, R.R., Badhe, B.: Heart disease classification using one dimensional convolutional neural network. Int. J. Innov. Res. Electr. Electron. Instrum. Control. Eng. **06**(06) (2018)
15. MIT-BIH Arrhythmia Database: https://www.physionet.org/physiobank/database/mitdb/. Last accessed 19 April 2019

16. Li, D., Zhang, J., Zhang, Q., Wei, X.: Classification of ECG signals based on 1D convolution neural network. In: 2017 IEEE 19th International Conference on e-Health Networking, Applications and Services (Healthcom), Dalian, pp. 1–6 (2017). https://doi.org/10.1109/HealthCom.2017.8210784.
17. Kiranyaz, S., Ince, T., Gabbouj, M.: Real-time patient-specific ECG classification by 1D convolutional neural networks. IEEE Trans. Bio-Med. Eng. **63** (2015). https://doi.org/10.1109/TBME.2015.2468589
18. Moody, G.B., Mark, R.G.: The impact of the MIT-BIH Arrhythmia Database. IEEE Eng. Med. Biol. **20**(3), 45–50 (2001)
19. Goldberger, A.L., Amaral, L.A.N., Glass, L., Hausdorff, J.M., Ivanov, P.C., Mark, R.G., Mietus, J.E., Moody, G.B., Peng, C.K., Stanley, H.E.: PhysioBank, PhysioToolkit, and PhysioNet: components of a new research resource for complex physiologic signals. Circulation **101**(23), 215–220 (2000)
20. MIT-BIH Arrhythmia Database Directory: https://www.physionet.org/physiobank/database/html/mitdbdir/mitdbdir.htm

Understanding Abstraction in Deep CNN: An Application on Facial Emotion Recognition

Francesca Nonis, Pietro Barbiero, Giansalvo Cirrincione, Elena Carlotta Olivetti, Federica Marcolin, and Enrico Vezzetti

Abstract Facial Emotion Recognition (FER) is the automatic processing of human emotions by means of facial expression analysis [1]. The most common approach exploits 3D Face Descriptors (3D-FD) [2], which derive from depth maps [3] by using mathematical operators. In recent years, Convolutional Neural Networks (CNNs) have been successfully employed in a wide range of tasks including large-scale image classification systems and to overcome the hurdles in facial expression classification. Based on previous studies, the purpose of the present work is to analyze and compare the abstraction level of 3D face descriptors with abstraction in deep CNNs. Experimental results suggest that 3D face descriptors have an abstraction level comparable with the features extracted in the fourth layer of CNN, the layer of the network having the highest correlations with emotions.

1 Introduction

1.1 Facial Emotion Recognition and Deep Learning

Facial Emotion Recognition (FER) is an active line of research in the human-computer interaction domain, due to its potential in many real-time applications, such as surveillance, security and communication. Different architectures of deep neural networks have been proposed, such as Convolutional Neural Networks (CNNs), which have been applied in several research fields, including health care [4] and cybersecurity [5]. Most of the existing algorithms exploit 2D features extracted

F. Nonis (✉) · P. Barbiero · E. C. Olivetti · F. Marcolin · E. Vezzetti
Politecnico di Torino, Torino, Italy
e-mail: francesca.nonis@polito.it

G. Cirrincione
University of the South Pacific, Suva, Fiji
e-mail: nimzoexin59@gmail.com

A. Esposito et al. (eds.), *Progresses in Artificial Intelligence and Neural Systems*, Smart Innovation, Systems and Technologies 184,
https://doi.org/10.1007/978-981-15-5093-5_26

from images to predict emotions. Albeit computational expensive, 3D feature-based approaches have produced more robust and accurate models thanks to their information supplement [6]. In recent years, CNNs have been successfully employed in large-scale image classification systems and to overcome the hurdles in facial expression classification. The first studies on 3D FER have appeared only in the last decade, thanks to the publication of the first public databases suitable for this objective [7]. In this research field, state of the art is currently represented by a few interesting neural-based approaches. In [8], the authors presented a novel deep fusion CNN for subject-independent multimodal 2D+3D FER. A 3D facial expression recognition algorithm using CNNs and landmark features/masks, exploiting 3D geometrical facial models only, has been proposed in [9]. Finally, a deep CNN model merging RGB and depth map latent representation has been designed in [10] for facial expression learning.

1.2 Understanding Abstraction in Deep CNN

In mathematics, abstraction refers to the process of extracting the underlying structure, properties or patterns from observations, removing case specific information, and building high-level concepts that can be profitably applied in unseen but equivalent environments [11, 12]. Similarly to animals and human beings, deep neural networks process raw signals by building abstract representations that can be used to generalize to new data. The deeper the layer, the higher the abstraction level. Such abstract representations are encoded in the numeric values of the network weights. The cross-correlation operation [13, 14] used in convolutional layers of deep CNNs does not change the data type provided in input. Therefore, when deep CNNs are applied to images, the abstract features extracted by the neural network can be visualized and manually analyzed by domain experts. Based on results and developments in previous studies, the purpose of the present work is to analyze and compare the abstraction level of 3D face descriptors with abstraction in deep CNNs.

2 Data

The data set used in this work was obtained from the Bosphorus 3D facial database [15]. The database contained both 3D facial expression shapes and 2D facial textures up to 54 scans in various poses, expressions and occlusion conditions. Such samples were obtained from 105 different subjects with different racial ancestries and gender (for a total of 4666 face scans). In the following, only two sets of expressions have been considered from the original database. The expressions of the first set were based on Action Units (AU) of the Facial Action Coding System (FACS) [16].

The second set, instead, was composed of facial expressions corresponding to the 6 basic emotions (happiness, surprise, fear, sadness, anger and disgust) plus the neutral expression. The resulting data set was composed of 453 images. Among them, 299 were faces having a neutral expression while the others were almost evenly split into the 6 universal emotions.

3 Methods

The CNN utilized in the following experiments was AlexNet [17]. The input layer of the network requires RGB images having size 227×227. For such purpose, grayscale depth maps (see Fig. 2) extracted from the Bosphorus database have been cropped, and converted to RGB images by replicating the grayscale channel.

3.1 3D Face Descriptors

One of the most common techniques for the analysis of human emotions using facial expressions exploits 3D Face Descriptors (3D-FD). 3D-FDs can be generated from depth maps by means of mathematical operators. In this study, the first principal curvature (k1), the shape index (S), the mean curvature (H), the curvedness (C) and a second fundamental form coefficients (f) have been used [2, 18]. In Fig. 1 three geometrical 3D-FDs for happiness, sadness and surprise emotions are shown (Fig. 2).

3.2 Transfer Learning

When the sample size is small, training a deep CNN from scratch may be time consuming as well as resulting in poor performances or overfitting. Better and satisfying performances can be easily obtained through transfer learning approaches [19]. Given the small amount of samples in the Bosphorus data set, the 3D facial emotion recognition has been performed by using a pretrained AlexNet model. The CNN was fine-tuned using the Bosphorus data set and trained for classifying images into the 7 universal emotions.

3.3 Correlation Analysis

The maximum activations of the fine-tuned network were calculated. The corresponding filters were manually visualized and analyzed by domain experts to understand the abstraction process of the network. As expected [20], first layers tend to detect

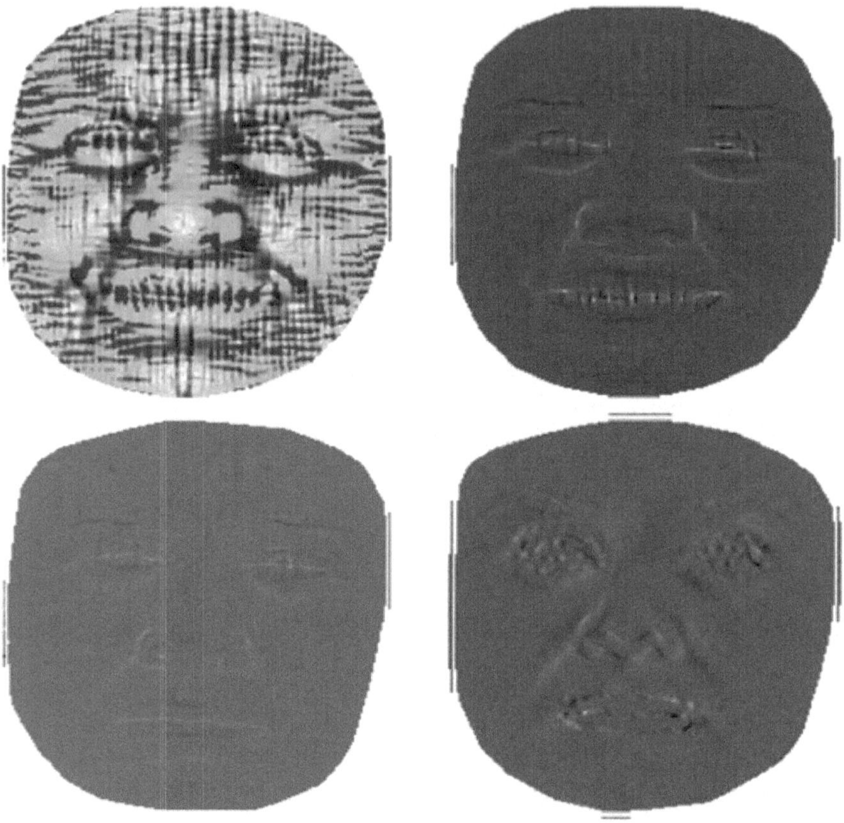

Fig. 1 Geometrical 3D face descriptors: S **(top left)**, k1 **(top right)**, H **(bottom left)**, and f **(bottom right)**

simple patterns like edges, while channels in deeper layers tend to focus on more complex and abstract features like nose and mouth. In order to assess the abstraction level of 3D face descriptors the Pearson correlation coefficient ρ has been used [21]. Pearson's ρ has been computed between each 3D face descriptor and CNN filter activations using three different images representing happy, sad and surprise emotions.

3.4 Symbolic Regression

In order to extend the correlation analysis to more complex models, *symbolic regression* [22, 23] has been exploited. Symbolic regression is a multi-objective regression analysis for exploring the space of mathematical expressions to find optimal models

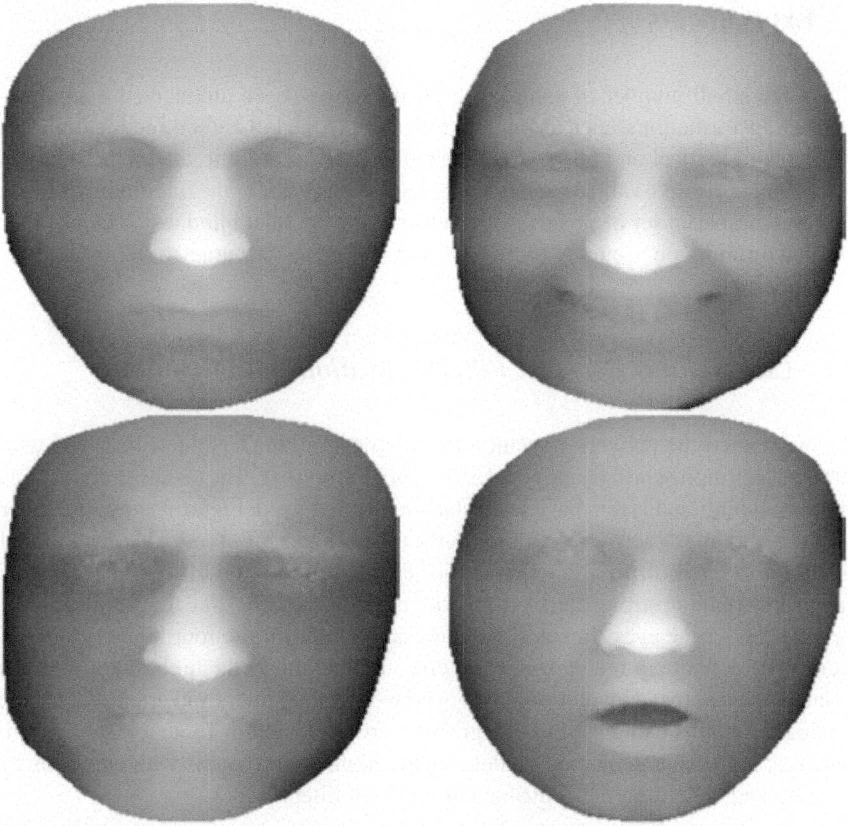

Fig. 2 Grayscale depth maps: neutral expression **(top left)**, happiness **(top right)**, sadness **(bottom left)**, and surprise **(bottom right)** emotions

in terms of accuracy and simplicity. In this experimental setting, symbolic regression was used to assess the abstraction level of 3D face descriptors. Symbolic regression has the advantage of returning human-readable models, that can later be interpreted and explained. For this task, the commercial evolutionary-based software Eureqa Formulize[1] was employed. The software has been used to find mathematical expressions involving CNN filter activations (obtained from images representing happy, sad and surprise emotions) which were highly correlated with 3D face descriptors.

[1]Eureqa Formulize is developed by Nutonian, Inc. https://www.nutonian.com/products/eureqa/.

4 Experiments

Given the small number of samples, the data set has been augmented using geometric transformations, such as random reflection in the left-right direction, uniform scaling, and vertical and horizontal translation [24]. In order to assess the network performance, a cross-validation procedure has been applied to the fine-tuning process [25] and a random set of images were selected for the final blind test. The network reached a validation accuracy of 82.67% and a blind test accuracy of 82.09%.

4.1 Correlations Between Filter Activations and Emotions

Having trained the network for emotion classification, the CNN has been fed with three images representing happy, sad and surprise emotions. The resulting activations of the convolutional layers have been statistically analyzed. Figure 3 shows the filter with the maximum activation in the fourth (*conv4*) and fifth (*conv5*) convolutional layers for sad and surprise emotions. The selected filter of the fifth layer highlights image areas having strong correlations with emotion patterns, like the mouth and the wrinkles under the eyes. Besides, the most active filter of the fourth layer does not seem to detect human-recognizable patterns. Table 1 shows the highest correlations found in the last two convolutional layers between single filters and emotion images. Analogous results using symbolic regression are presented in Table 2. As expected, symbolic regression generated models having higher correlations with emotions by merging and weighting the contribute of different filters.

4.2 Correlations Between Filter Activations and 3D Face Descriptors

A similar correlation analysis has been performed between filter activations and 3D face descriptors. Pearson's ρ increased from the input to the output layer of the CNN culminating in the fourth convolutional layer (see Fig. 4). On the contrary, in the fifth layer correlations between filter activations and descriptors dropped Fig. 4. This result suggests that 3D face descriptors correspond to an abstraction level comparable with the fourth layer of the network.

4.3 Abstraction Level of 3D Face Descriptors

The fourth layer of the network was the one having the highest correlations with emotions. Besides, the above experiments show how 3D face descriptors correspond

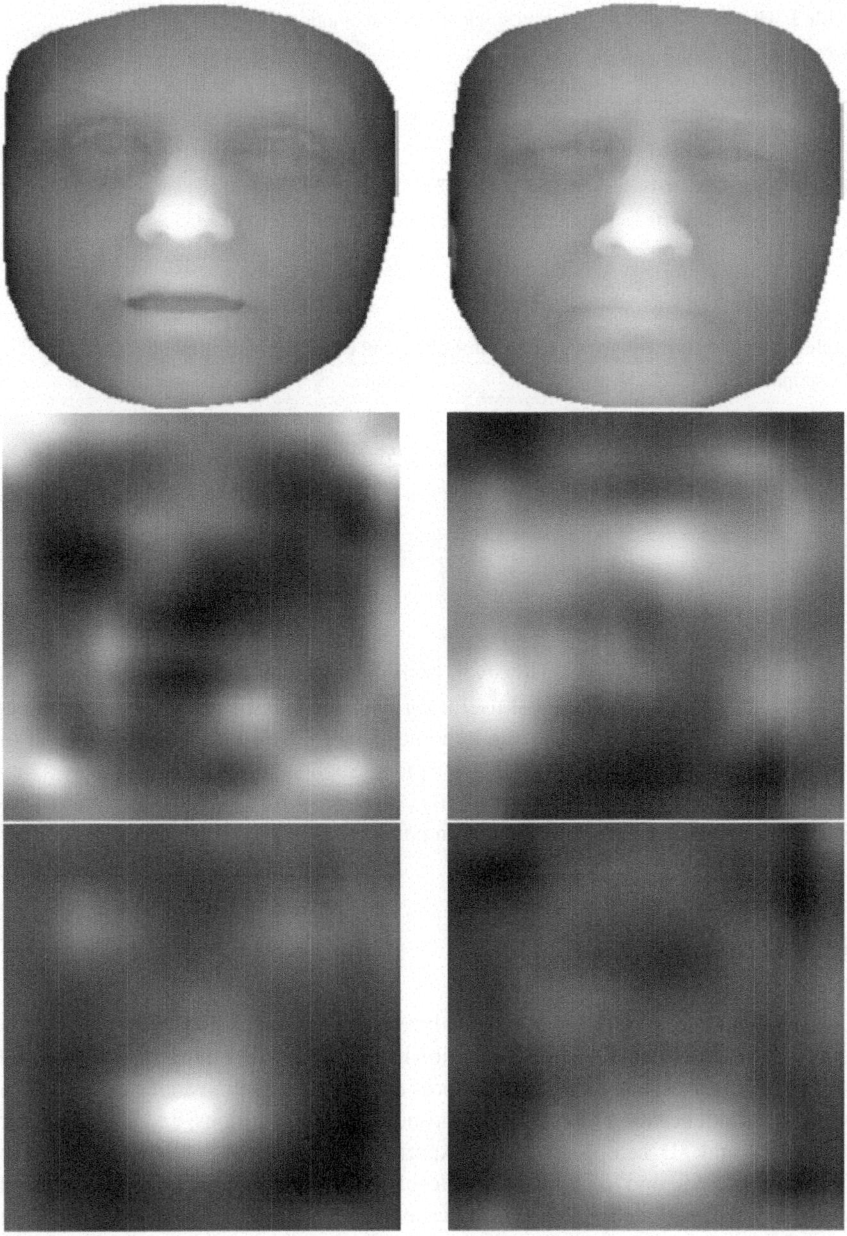

Fig. 3 Channel with the largest activation: *conv4* **(middle)** and *conv5* **(bottom)**, compared to the original image **(top)**

Table 1 Highest correlations between single filters and emotions

Descriptor	Conv4			Conv5		
	Happy	Sadness	Surprise	Happy	Sadness	Surprise
C	0.7482	0.7223	−0.7008	−0.5322	0.5205	0.4722
f	0.7430	0.7313	−0.7007	0.5507	0.5301	0.4754
H	**0.7464**	**0.7355**	**−0.7027**	**0.5773**	0.5311	**0.4929**
k1	0.7450	0.7338	−0.7030	0.5724	**0.5333**	0.4906
S	−0.5229	−0.5283	−0.4863	0.4538	0.3642	0.3896

Table 2 Highest correlations between Conv4 filters and emotions using symbolic regression

Descriptor	Happy	Sadness	Surprise
C	0.8327	0.8149	−0.7958
f	0.8222	0.8201	−0.8077
H	0.8406	0.8265	**−0.8184**
k1	**0.8498**	**0.8329**	−0.8078
S	−0.6578	−0.6329	−0.6307

to a similar abstraction level. However, both for the CNN and from a human point of view the fifth layer is the most useful for emotion classification (compare filters in Fig. 3). These results support the hypothesis that CNNs have a superior level of abstraction with respect to 3D face descriptors. Such superior level may play a key role in transforming features that are highly correlated with emotions (as conv4 filters and descriptors) into useful classification patterns.

5 Conclusions

The main purpose of this work was to analyze the differences between the abstraction level of 3D face descriptors with abstraction in deep CNNs. For this aim, a pre-trained deep CNN was fine-tuned on the Bosphorus data set. Correlation analyzes have been performed both between filter activations and universal emotions, and between filter activations and 3D face descriptors. Experimental results suggested that 3D face descriptors correspond to an abstraction level comparable with the features extracted in the fourth layer of the CNN. However, both for the network and from a human point of view the most useful features for emotion recognition correspond to the fifth layer activations. Such features may play a key role in transforming features that are highly correlated with emotions into useful classification patterns. Future steps consist of continuing and deepening the activation and correlation analyses to better understand abstraction in deep CNN.

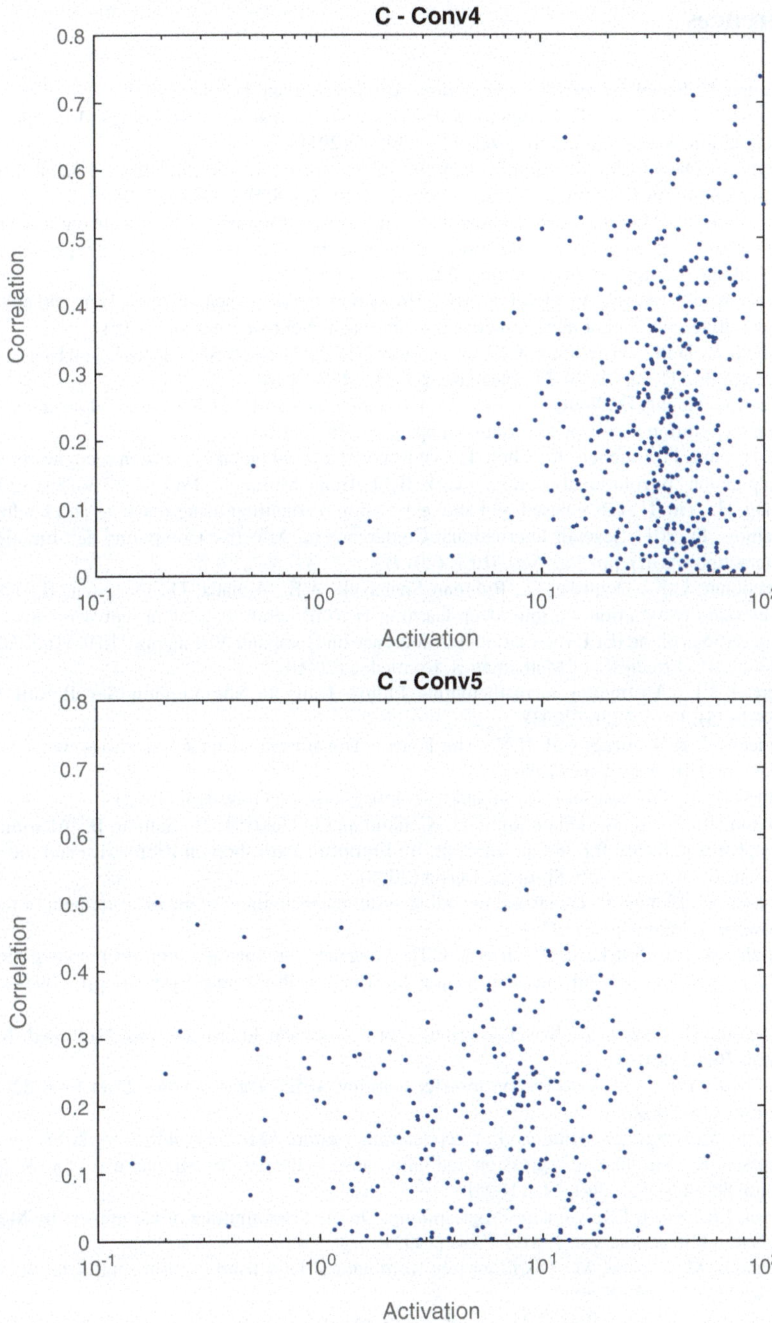

Fig. 4 Relationship between filter activations and correlation values in *conv4* **(top)** and *conv5* **(bottom)**. Correlation analysis has been performed between filter activations and descriptors

References

1. Ekman, P.: Facial expression and emotion. Am. Psychol. **48**(4), 384 (1993)
2. Vezzetti, E., Marcolin, F.: Geometrical descriptors for human face morphological analysis and recognition. Robot. Auton. Syst. **60**, 928–939, 06 (2012)
3. Nonis, F., Dagnes, N., Marcolin, F., Vezzetti, E.: 3d approaches and challenges in facial expression recognition algorithmsa literature review. Appl. Sci. **9**(18), 3904 (2019)
4. Ieracitano, C., Mammone, N., Bramanti, A., Hussain, A., Morabito, F.C.: A convolutional neural network approach for classification of dementia stages based on 2d-spectral representation of EEG recordings. Neurocomputing **323**, 96–107 (2019)
5. Ieracitano, C., Adeel, A., Morabito, F.C., Hussain, A.: A novel statistical analysis and autoencoder driven intelligent intrusion detection approach. Neurocomputing (2019)
6. Huynh, P., Tran T.-D., Kim, Y.-G.: Convolutional Neural Network Models for Facial Expression Recognition Using 3D-BUFE Database, pp. 441–450 (2016)
7. Chen, Z., Huang, D., Wang, Y., Chen, L.: Fast and light manifold CNN based 3d facial expression recognition across pose variations, pp. 229–238 (2018)
8. Li, H., Sun, J., Zongben, X., Chen, L.: Multimodal 2d+ 3d facial expression recognition with deep fusion convolutional neural network. IEEE Trans. Multimed. **19**(12), 2816–2831 (2017)
9. Yang, H., Yin, L.: CNN based 3d facial expression recognition using masking and landmark features. In: 2017 Seventh International Conference on Affective Computing and Intelligent Interaction (ACII), pp. 556–560. IEEE (2017)
10. Oyedotun, O.K., Demisse, G., Rahman Shabayek, A.E., Aouada, D., Ottersten, B.: Facial expression recognition via joint deep learning of RGB-depth map latent representations. In: Proceedings of the IEEE International Conference on Computer Vision, pp. 3161–3168 (2017)
11. Russell, B.: Principles of Mathematics. Routledge (2009)
12. Ferrari, P.L.: Abstraction in mathematics. Philos. Trans. R. Soc. London. Ser. B Biol. Sci. **358**(1435), 1225–1230 (2003)
13. Bracewell, R.N., Bracewell, R.N.: The Fourier Transform and Its Applications, vol. 31999. McGraw-Hill, New York (1986)
14. Papoulis, A.: The Fourier Integral and Its Applications. McGraw-Hill (1962)
15. Savran, A., Alyüz, N., Dibeklioğlu, H., Çeliktutan, O., Gökberk, B., Sankur, B., Akarun, L.: Bosphorus database for 3d face analysis. In: European Workshop on Biometrics and Identity Management, pp. 47–56. Springer, Berlin (2008)
16. Friesen, E., Ekman, P.: Facial action coding system: a technique for the measurement of facial movement. Palo Alto **3** (1978)
17. Krizhevsky, A., Sutskever, I., Hinton, G.E.: Imagenet classification with deep convolutional neural networks. In: Advances in Neural Information Processing Systems, pp. 1097–1105 (2012)
18. Marcolin, F., Vezzetti, E.: Novel descriptors for geometrical 3d face analysis. Multimed. Tools Appl. **76**, 07 (2016)
19. Pan, S.J., Yang, Q.: A survey on transfer learning. IEEE Trans. Knowl. Data Eng. **22**(10), 1345–1359 (2009)
20. LeCun, Y., Bengio, Y., Hinton, G.: Deep learning. Nature **521**(7553), 436–444 (2015)
21. Pearson, K.: Vii. note on regression and inheritance in the case of two parents. Proc. R. Soc. Lond. **58**(347–352), 240–242 (1895)
22. Koza, J.R., Koza, J.R.: Genetic Programming: On the Programming of Computers by Means of Natural Selection, vol. 1. MIT Press (1992)
23. Schmidt, M., Lipson, M.: Distilling free-form natural laws from experimental data. Science **324**(5923), 81–85 (2009)
24. Shorten. C., Khoshgoftaar, T.M.: A survey on image data augmentation for deep learning. J. Big Data **6**(1), 60 (2019)
25. Stone, M.: Cross-validatory choice and assessment of statistical predictions. J. R. Stat. Soc. Ser. B (Methodological) **36**(2), 111–133 (1974)

Computational and Methodological Intelligence in Economics and Finance

Exploration and Exploitation in Optimizing a Basic Financial Trading System: A Comparison Between FA and PSO Algorithms

Claudio Pizzi, Irene Bitto, and Marco Corazza

Abstract When coping with complex global optimization problems, often it is not possible to obtain either analytical or exact solutions. Therefore, one is forced to resort to approximate numerical optimizers. With this aim, several metaheuristics have been proposed in the literature and the primary approaches can be traced back to biology and physics. On one hand, there exist bio-inspired metaheuristics that imitate the Darwinian evolution of species (like, for instance, Genetic Algorithms) or the behaviour of group of social organisms (like, for instance, Ant Colony Optimization). On the other hand, there exist physics-inspired metaheuristics that mimic physical laws (like, for instance, gravitation and electromagnetism). In this work, we take into account the Fireworks Algorithm and the Particle Swarm Optimization in order to compare their exploration and exploitation capabilities. In particular, the investigation is performed considering as complex global optimization problem the estimation of the parameters of the technical analysis indicator Bollinger Bands in order to build effective financial trading systems, similarly to what proposed in [3].

1 Introduction

In finance, as well as in many other science fields, global optimization methods play an important role. For instance, portfolio selection and effective parametrization of financial trading systems (FTSs) require the solution of non simple global optimization problems. In last years, the interest of scholars in solving such non

C. Pizzi (✉) · I. Bitto · M. Corazza
Department of Economics, Ca' Foscari University of Venice,
Cannaregio 873, 30121 Venice, Italy
e-mail: pizzic@unive.it

I. Bitto
e-mail: irenebitto@live.it

M. Corazza
e-mail: corazza@unive.it

A. Esposito et al. (eds.), *Progresses in Artificial Intelligence and Neural Systems*,
Smart Innovation, Systems and Technologies 184,
https://doi.org/10.1007/978-981-15-5093-5_27

elementary problems is particularly increased and, as consequence, many solution approaches which do not need the smoothness of the optimization problem functions—as required by the classical optimization methods—have been proposed. Some of these solution approaches, namely the metaheuristics which refer to evolutionary computational principles, have recently grown their relevance both in the academic world (see, for instance, [6]) and in the industrial and the financial ones.

In this paper, we consider two optimization metaheuristics both based on the so-called Swarm Intelligence[1] (SI), the Fireworks Algorithm (FA) and the Particle Swarm Optimization (PSO), with the aim to compare their capabilities to explore the solution space and, when a promising solution space area is identified, to compare their abilities to exploit that area. In particular, we apply these metaheuristics to optimally estimate the three parameters of a simple FTS based on a single technical analysis indicator, the Bollinger Bands (BBs). Note that, although the related global optimization problem may appear simple given its low dimensionality, it is, anyway, complex to solve due to the presence of an highly nonlinear objective function and of integer variables, as it will be clear in Sect. 3.

The remainder of this paper is organized as follows. In the next section, we briefly introduce the basics FA and PSO. In Sect. 3, we first present the FTS and then we specify the related global optimization problem. In Sect. 4, we apply the two metaheuristics for optimally estimating the parameters of the simple aforementioned FTS and we present and compare their performances. Finally, in Sect. 5 we conclude with some remarks.

2 A Recall on FA and PSO

In this section, we provide a brief qualitative presentation of FA and PSO.

As common features of these two metaheuristics, we recall that both are characterized by a population of agents and that each agent constitutes a possible solution of the investigated global optimization problem. Therefore, an agent represents a point of the solution space. This population evolves over time[2] according to the specific rules of the used metaheuristic, in order to possibly improve the "quality" of the agents themselves. The "quality" of each agent, that is of each potential solution, is evaluated by the so-called fitness function: lower/higher the fitness score in case of minimization/maximization, better the agent-solution. Finally, the evolution ends when a pre-established stop criterion is satisfied.

More specifically, regarding FA, it is a metaheuristic introduced in [7] that searches for the global optimum by mimicking firework explosions. In each iteration, FA first selects a set of positions, that is points of the solution space, where the fireworks have

[1] Briefly, it is a kind of intelligence whose problem-solving abilities take advantage of the interactions of simple information-processing units.

[2] The passage of time is represented by the iterations of the algorithm that implements the considered metaheuristic.

to explode. Then, after the explosion of each firework, FA appropriately determines a set of positions of the solution space where to locate the sparks around the various fireworks. Finally, for all the fireworks and the sparks so determined, the fitness score is evaluated. The positions with the best fitness scores are selected as positions of the fireworks of the next iteration. Note that the original FA suffers some drawbacks, so in our investigation we use the Enhanced Fireworks Algorithm, introduced in [8], that overcomes such limitations.

As far as PSO concerns, it is a metaheuristic introduced in [4] that searches for the global optimum by imitating the way in which swarms of birds or flocks of fishes move in the space. Birds or fishes, generically called particles, represent points of the solution space which "move" in this space accordingly to an appropriate velocities. In each iteration, for all the particles the fitness score is evaluated. Each particle has a memory in which its best personal position ever reached, in terms of fitness score, and the best global position ever reached by the swarm, again in terms of fitness score, are stored. If the fitness score of the current position is better than that of the best personal position, the former becomes the new best personal position. Furthermore, if such a fitness score is even better than that of the best global position, the former becomes even the new best global position. Finally, PSO evolves the swarm updating the position of each particle through the application of a velocity which is function of the current position of the particle itself, of its best personal position and of the best global position.

Note that an important difference between FA and PSO consists in the number of positions which are evaluated in each iteration: it varies for FA while it is constant for PSO.

3 The FTS and the Global Optimization Problem

In order to investigate the performances of the considered metaheuristics, we consider a simple FTS based on a single technical analysis indicator, the BBs (see, for instance, [5]).

This indicator takes into account a suitable moving average of the price of the investigated financial asset, the so-called central line, and two suitable symmetric bands with respect to this central line. Both the bands are characterized by the same amplitude, which provides a "measure" of the volatility of the asset price. In its standard version, BBs depends on two parameters: n, the (integer) number of terms to use for calculating the moving average, and m, the (real) number indicating the amplitude of the bands in terms of moving standard deviation. Formally, one has

$$\begin{cases} CL(t) = MA(t, n) \\ BBup(t) = CL(t) + m \cdot MSD(t, n) \\ BBdn(t) = CL(t) - m \cdot MSD(t, n) \end{cases},$$

where $CL(t)$ indicates the central line at time t, $MA(t, n)$ is the moving average calculates at time t using the last n observations, $BBup(t)$ denotes the upper band at time t, $MSD(t, n)$ is the moving standard deviation calculated at time t using the last n observations, and $BBdn(t)$ is the lower band at time t. As evident, this indicator implicitly assumes that positive financial asset price variations follow the same distribution of the negative ones.

But, the study of financial asset prices time series very often shows a negative autocorrelation in financial returns as, for instance, outlined in [1]. Therefore, it is generally accepted the asymmetric distribution of financial asset price variations. For these reasons, in this paper we generalize the standard BBs allowing that the distance of $BBup$ and $BBdn$ from the central line may be different, that is allowing that $|BBup(t) - CL(t)| \neq |BBdn(t) - CL(t)|$. In particular, this generalization is implemented by modifying as follows the standard BBs: $BBup(t) = CL(t) + m.up \cdot MSD(t, n)$ and $BBdn(t) = CL(t) - m.dn \cdot MSD(t, n)$, where $m.up$ and $m.dn$ indicate the upper amplitude and the lower one, respectively.

The correct estimation of n, $m.up$ and $m.dn$ is crucial for our FTS, as the buy or sell signals are generated by the crossing of the bands by the asset price. In particular, the trading rule produced by the BBs is

$$signal(t) = \begin{cases} -1 \text{ if } P(t) > BBup(t) \bigwedge P(t-1) < BBup(t-1) \\ +1 \text{ if } P(t) < BBdn(t) \bigwedge P(t-1) > BBdn(t-1) \\ signal(t-1) \text{ otherwise} \end{cases}, \quad (1)$$

where -1 indicates the sell signal, $+1$ denotes the buy signal, and $P(t)$ is the financial asset price at time t. Note that in each instant time $t = 1, \ldots, T$ of the trading period, the gain $G(t)$ is given by the previous gain multiplied by the performance, in terms of return, obtained by the FTS from $t - 1$ to t.

As far as the FTS is concerned, its purpose consists in maximizing the cumulative trading gain at the end of the trading period, that is $G(T)$, where

$$G(t) = G(t-1)\left[1 + signal(t-1) \cdot r(t)\right], \text{ with } t = 0, \ldots, T \text{ and } G(0) = G,$$

in which $r(t)$ is the log-return of the financial asset at time t, and G denotes the starting invested capital.

Therefore, the global optimization problem to cope with is

$$\max_{n, m.up, m.dm} G(T)$$

$$\text{s.t.} \begin{cases} n \in \{1, 2, \ldots, 99, 100\} \cdot \\ m.up \in [1, 10] \\ m.dn \in [1, 10] \end{cases} \quad (2)$$

Note that this constrained global optimization problem is strongly nonlinear due to the objective function and is NP-hard due to the constraint of integrality on n. Therefore, it is an optimization problem complex to solve.

4 Applications and Results

In this section, we apply the two presented metaheuristics to approximately solve the constrained global optimization problem (2) and we present the obtained results. Note that FA and PSO were originally designed for solving unconstrained global optimization problems. So, in order to cope with our constrained one, we reformulate the later as proposed in [2], using a standard exact penalty scheme to transform it into an equivalent unconstrained one.

Regarding the time series of the (closing) prices, we consider five important stocks which are components of the S&P 500 index, that cover different economic sectors: Bank of America Corporation (BAC), Boeing Company (BC), Booking Holdings Inc. (BKNG), Microsoft Corporation (MC), and NIKE Inc. (NKE).

Then, as the trading concerns, we take into account three different periods: the first, from January 02, 2002 to December 29, 2006 (1259 observations); the second, from January 03, 2006 to December 30, 2011 (1511 observations); the third, from January 03, 2011 to December 30, 2015 (1257 observations). The first half of each of these periods is used to carry out the training of the investigated FTS, while the second half is used to validate the same FTS. Note that using prices time series of five different stocks and three different trading periods allow us to consider a relative wide range of situations. In fact, as a matter of fact, the three periods show different behaviours of the financial market, which include: first, a "varied period" characterized by an initial downtrend phase followed by a reprise and by a subsequent lateral phase (2002–2006); then, a strong crisis phase preceded and followed by uptrend phases (2006–2011); finally, a strong uptrend phase (2011–2015). Consequently, the three validation (or out-of-sample) periods are quite different among them as well: the first is mainly characterized by the 2007–08 financial crisis; the second shows a strong uptrend phase; the third presents an uptrend phase followed by a lateral one. Figures 1, 2 and 3 depict the behaviours of the prices of the considered stocks both in the training period (black line) and in the validation one (red line). It is evident that the evolutions of the stock prices do not necessarily follow that of the financial market as a whole. Finally, note that we consider neither transaction costs nor other frictional aspects and that short-selling is not practiced.

Finally, as for the settings of the two metaheuristics regard, for FA we have considered 5 fireworks, 50 sparks per firework, and 200 iterations, and for PSO we have considered 10 particles and 200 iterations. The other parameters of both the metaheuristics have been set following the indication of the prominent literature. The application of each of the two metaheuristics has been replicated 100 times.

Now, we proceed to the analysis of the results, which are summarized in Table 1. The results obtained by the two FTSs optimized by FA and by PSO, respectively,

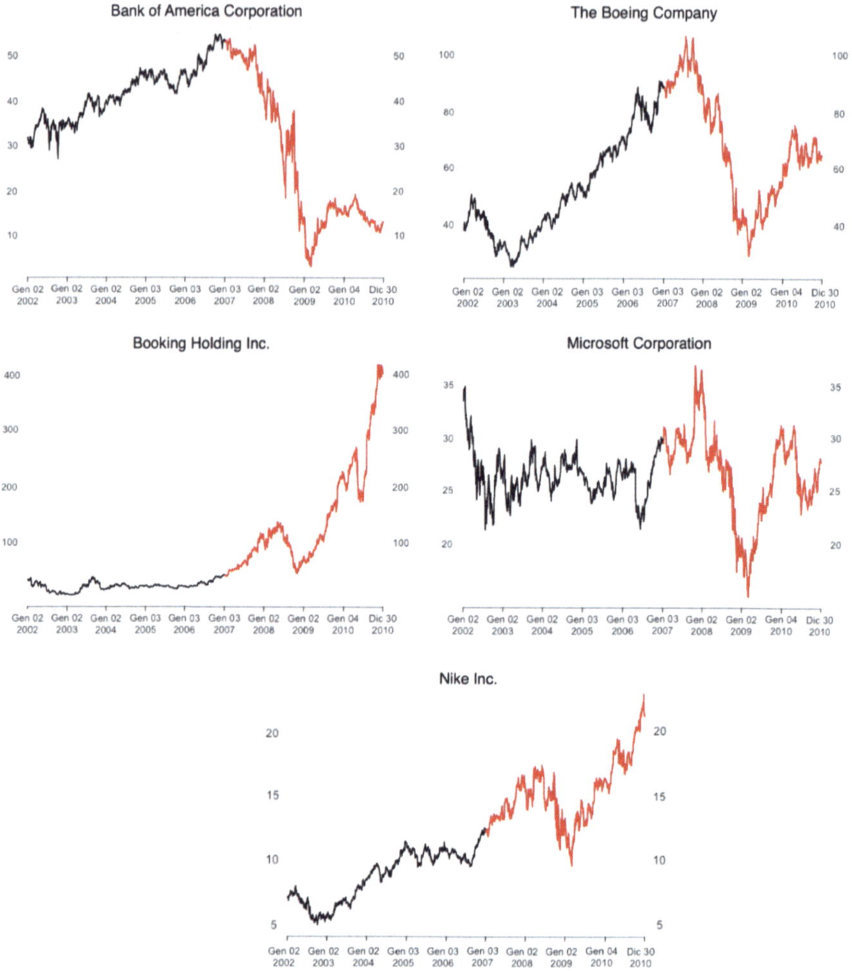

Fig. 1 Closing prices time series of the five stocks from January 02, 2002 to December 30, 2010 (black line in the training period and red line in the validation period)

are compared with the results obtained by an FTS whose parameters are those most commonly used by the professional practice, that is $n = 20, m.up = 2$ and $m.dn = 2$. With regard to the fitness score, the FTSs set by FA and by PSO show always greater valuations than that of the FTS set following the standard parametrization (Column 4, Table 1). Note that in some cases FA and PSO reach similar fitness scores even the respective optimized parameters are different, especially in the last trading period (Columns 4 and 5–7, Table 1). This may be due to the flattening of the objective function near the point of constrained maximum. Note also that, in each of the two optimized FTSs, $m.up$ and $m.dn$ are generally different. This tends to confirm the conjecture we previously advanced following which the bands of the technical

Fig. 2 Closing prices time series of the five stocks from January 02, 2006 to December 30, 2015 (black line in the training period and red line in the validation period)

analysis indicator BBs should be asymmetric. In particular, in the 80.00% of the considered cases, one has $m.up > m.dn$, which indicates a noticeable prevalence of right asymmetry in the stock price variations. Finally, it is noteworthy that a meaningful operational difference between FA and PSO consists in the number of points of the solution space which are evaluated through the fitness function. In fact, FA optimizes the fitness function evaluating, on average, more than the double of points of the solution space evaluated by PSO (Column 3, Table 1). This evidence is graphically exemplified in Fig. 4 for the Microsoft Corporation stock during the first training period. In this figure, the density of the number of evaluated points of the solution space restricted to $m.up$ and $m.dn$ is estimated (warmest the color, greater

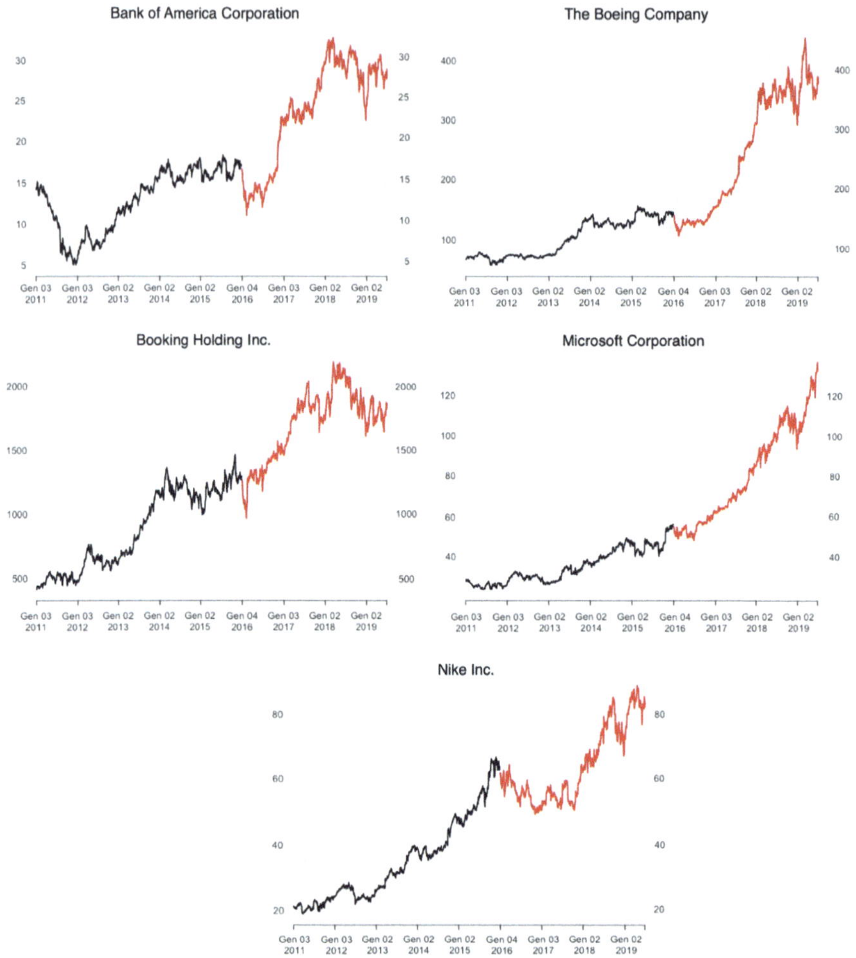

Fig. 3 Closing prices time series of the five stocks from January 02, 2011 to June 30, 2019 (black line in the training period and red line in the validation period)

the number of evaluations). This peculiarity of FA with respect to PSO allows it to better perform the exploitation. However, this greater exploitation of FA does not seem to have particular positive impacts on the performance of the related FTS.

As far the financial performances of FA and of PSO in the validation periods are concerned, the results are presented in terms of average annualized rate of returns. In particular, in each case the FTSs are evaluated using a one-year out-of-sample period (Column 8, Table 1) and a four-years out-of-sample period (Column 10, Table 1). In Columns 9 and 11 of the same table, the number of close operations, that is *bought-and-sold*, is reported for the one-year out-of-sample period and the four-years out-of-sample one, respectively. Generally, the FTSs whose parameters are set by the

Table 1 Results obtained by the two FTSs optimized by FA and by PSO, respectively, and by the FTS set following the standard indications (Stand.) from the professional practice

Stock	Approach	N	Fitness score	Parameters			Out-of-sample rate of return			
				n	m.up	m.dn	One year		Four years	
Trading periods: 2002–2006										
BAC	FA	1532	1.761	5	2.068	2.03	−0.054	(14)	0.382	(43)
	PSO	910	1.816	5	2.073	2.070	−0.054	(14)	0.382	(43)
	Stand.		0.509	20	2.000	2.000	−0.120	(2)	−0.690	(10)
BC	FA	1052	2.532	96	3.791	2.830	0.000	(0)	0.008	(1)
	PSO	600	2.532	96	5.000	2.824	0.000	(0)	0.008	(1)
	Stand.		0.352	20	2.000	2.000	−0.025	(3)	−0.117	(12)
BKNG	FA	715	7.356	5	2.155	1.566	2.108	(11)	5.958	(40)
	PSO	990	8.955	5	2.152	1.305	2.223	(12)	7.496	(45)
	Stand.		0.082	20	2.000	2.000	0.325	(3)	0.207	(10)
MC	FA	1169	1.242	83	2.546	2.288	0.246	(1)	0.279	(3)
	PSO	380	1.242	84	2.546	2.267	0.246	(1)	0.279	(3)
	Stand.		0.133	20	2.000	2.000	0.010	(2)	−0.402	(9)
NKE	FA	1210	2.669	32	3.874	2.979	0.000	(0)	1.031	(3)
	PSO	720	2.669	32	3.885	2.944	0.000	(0)	1.031	(3)
	Stand.		0.314	20	2.000	2.000	0.064	(3)	0.951	(15)
Trading periods: 2006–2011										
BAC	FA	1605	5.485	6	2.405	2.730	0.044	(3)	0.709	(14)
	PSO	440	4.627	6	1.953	2.592	0.013	(3)	0.248	(19)
	Stand.		−0.830	20	2.000	2.000	0.383	(5)	1.291	(17)
BC	FA	805	1.498	98	4.740	3.037	0.107	(1)	1.169	(1)
	PSO	800	1.941	5	1.041	1.432	0.151	(18)	0.507	(73)
	Stand.		0.182	20	2.000	2.000	0.109	(3)	0.564	(22)
BKNG	FA	1060	48.479	6	3.170	2.516	−0.118	(2)	0.870	(4)
	PSO	450	37.889	6	3.158	2.258	−0.091	(2)	1.196	(5)
	Stand.		0.579	20	2.000	2.000	0.129	(4)	0.664	(12)
MC	FA	738	1.382	28	3.018	3.401	0.000	(0)	0.446	(4)
	PSO	350	1.382	28	3.035	3.460	0.000	(0)	0.446	(4)
	Stand.		−0.415	20	2.000	2.000	−0.122	(3)	0.075	(11)
NKE	FA	740	3.972	27	2.551	1.669	0.089	(3)	0.854	(11)
	PSO	360	3.972	27	2.559	1.676	0.089	(3)	0.854	(11)
	Stand.		1.705	20	2.000	2.000	0.022	(5)	0.369	(11)
Trading periods: 2011–2015										
BAC	FA	659	1.880	9	3.508	3.179	0.963	(2)	1.335	(3)
	PSO	190	1.814	10	4.743	3.225	0.999	(6)	1.396	(18)
	Stand.		0.358	20	2.000	2.000	0.424	(3)	0.851	(10)
BC	FA	764	1.563	54	3.802	3.068	0.000	(0)	0.000	(0)
	PSO	250	2.830	25	3.464	3.207	0.000	(0)	0.000	(0)
	Stand.		0.807	20	2.000	2.000	0.227	(2)	0.799	(10)

(continued)

Table 1 (continued)

Stock	Approach	N	Fitness score	Parameters			Out-of-sample rate of return			
				n	$m.up$	$m.dn$	One year		Four years	
BKNG	FA	941	3.541	36	1.247	1.182	0.596	(12)	1.541	(37)
	PSO	360	4.343	5	2.960	2.542	0.644	(16)	1.370	(42)
	Stand.		0.148	20	2.000	2.000	0.700	(4)	1.469	(12)
MC	FA	568	2.485	6	1.043	1.038	0.435	(4)	2.491	(20)
	PSO	280	2.638	5	1.911	1.000	0.492	(4)	2.036	(17)
	Stand.		0.148	20	2.000	2.000	0.146	(4)	0.687	(11)
NKE	FA	991	2.647	31	3.301	1.961	0.088	(2)	0.589	(6)
	PSO	660	2.587	28	3.258	1.913	0.056	(2)	0.490	(6)
	Stand.		0.556	20	2.000	2.000	0.042	(4)	0.516	(11)

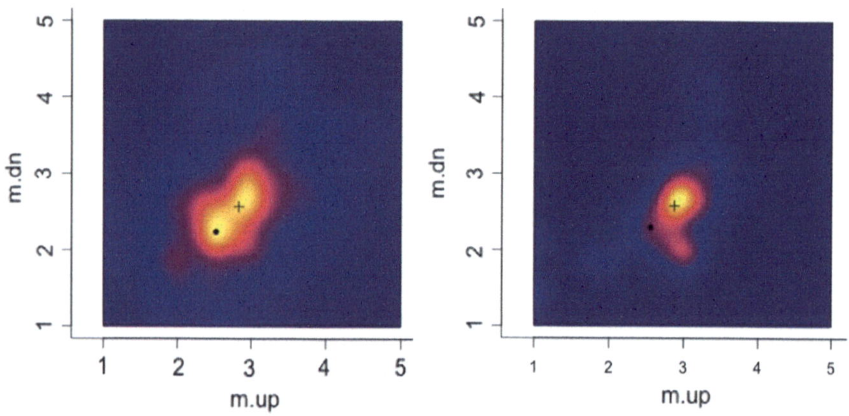

Fig. 4 Coverage of the solution space restricted to $m.up$ and $m.dn$ by FA (graph on the left) and by PSO (graph on the right) for the Microsoft Corporation stock. Training period: 2002–2006. The symbol "•" denotes the best global optimum, that is the best of the various best global solutions reached over the 100 replications, and the symbol "+" indicates the mean best global optimum, that is the mean of the various best global solutions reached over the 100 replications

two metaheuristics perform better than that set with standard parameters proposed by the professional practice. In particular: in the one-year validation period, FA-based and PSO-based FTSs perform better than the standard FTS in the 60.00% and in 66.67% of the cases, respectively; in the four-years validation period, FA-based and PSO-based FTSs perform better than the standard FTS in the 86.67% and in 73.33% of the cases, respectively. Finally, note that in five cases over thirty, the FA-based and PSO-based FTSs do not generate *buy* or *sell* signal in the respective out of sample periods ("(0)" in Columns 9 and 11, Table 1), while the standard FTS performs trades, obtaining negative or close-to-zero rates of return. In our opinion, this shows the good exploration and exploitation capabilities of the formers.

5 Some Final Remarks

In this paper, we have considered two metaheuristics for optimization, FA and PSO, in order to estimate the optimal parameters of the technical analysis indicator BBS for building effective FTSs. Generally, the FA-based and the PSO-based FTSs perform better than the standard one. These results highlight the good exploration and exploitation abilities of FA and PSO. In particular, a greater attitude to the exploitation of FA with respect to that of PSO is underlined. Our future research goals consists in checking FA, PSO and other metaheuristics for optimization when applied to FTSs more articulated than that considered here.

References

1. Cont, R.: Empirical properties of asset returns: stylized facts and statistical issues. Quant. Financ. **1**, 223–236 (2001)
2. Corazza, M., Fasano, G., Gusso, R.: Particle swarm optimization with no-smooth penalty reformulation, for a complex portfolio selection problem. Appl. Math. Comput. **224**, 611–624 (2013)
3. Corazza, M., Parpinel, F., Pizzi, C.: Can PSO improve TA-based trading systems?. In: Esposito, A., Faundez-Zanuy, M., Morabito, F.C., Pasero, E. (eds.) Neural Advances in Processing Nonlinear Dynamic Signals. Smart Innovation, Systems and Technologies, vol. 102, pp. 277–288. Springer (2019)
4. Kennedy, J., Eberhart, R.: Particle swarm optimization. In: Proceedings of ICNN'95—International Conference on Neural Networks, IV, 1942–1948 (1995)
5. Murphy, J.J.: Technical Analysis of the Financial Markets: A Comprehensive Guide to Trading Methods and Applications: Study Guide. New York Institute of Finance (1999)
6. Simon, D.: Evolutionary Optimization Algorithms. Wiley (2013)
7. Tan, Y., Zhu, Y.: Fireworks algorithm for optimization. In: Tan, Y., Shi, Y., Tan, K.C. (eds.) Advances in Swarm Intelligence—First International Conference, ICSI 2010—Proceedings, Part I. Lecture Notes in Computer Science, vol. 6145, pp. 355–364 (2010)
8. Zheng, S., Janecek, A., Tan, Y.: Enhanced fireworks algorithm. In: 2013 IEEE Congress on Evolutionary Computation, 2069–2077 (2013)

SDOWA: A New OWA Operator for Decision Making

Marta Cardin and Silvio Giove

Abstract Ordered Weighted Aggregation operators (OWA) are widely analyzed and applied to real world problems, given their appealing characteristic to reflect human reasoning, but are enable in the basic definition to include *importance* weights for the criteria. To obviate, some extensions were introduced, but we show how none of them can satisfy completely a set of required properties. Thus we introduce a new proposal, the Standard Deviation OWA (SDOWA) which conversely satisfy all the listed properties and seems to be more convincing then other ones.

1 Introduction

Several classes of Aggregation Operators, suitable functions designed for aggregating information, are proposed in the multi criteria literature, see [1]. Among them, we recall the Ordered Weighted Aggregation operators were introduced by Yager [15] and widely applied in Decision Making problems, for their intuitive meaning, and easiness of implementation. Depending on the choice of suitable parameters, many classes of different operators can be obtained [16], we limit to quote S-OWA operators [17], and many others [8, 10]. OWA operators consists in a *positional* set of weights, that is, a non negative weight is assigned to the *ordered* (and normalized) values of the criteria, see below for a more detailed explanation. Using this framework, and suitably tuning the OWA weights, many aggregation (and non linear) operators can be obtained, the MIN operator, the MAX operator, the MEDIAN, the simple arithmetic averaging, k-th order statistic, and others [15–17]. Anywise, in the basic OWA formulation no *importance* weight can be included, i.e. weights that are attached to each criterion and can represent the relative importance of it. To obviate, some extensions of OWA were introduced. Starting from the early papers

M. Cardin · S. Giove (✉)
Department of Economics, University of Venice, Venice, Italy
e-mail: sgiove@unive.it

A. Esposito et al. (eds.), *Progresses in Artificial Intelligence and Neural Systems*, Smart Innovation, Systems and Technologies 184,
https://doi.org/10.1007/978-981-15-5093-5_28

of [19, 21], Torra introduced the WOWA operator [13], while other extensions were subsequently formulated [10, 22].

After having identified a set of rational and desired properties, we firstly verify that the existing extensions of OWA do not satisfy all of them, and thus an original modification is proposed, namely the *standard deviation*—OWA, SDOWA for brevity, which on the other side, completely satisfies the listed properties.

2 Aggregation Operators

An Aggregation Operators (AO for brevity) is a function $F(\mathbf{x}) : [0, 1]^n \rightarrow [0, 1]$ which combines different *normalized* arguments[1] (criteria), the vector $\mathbf{x}$, into a single numerical datum. Some recent contributions include both methodological aspects and practical suggestions especially, see among other [1, 2]. AO include include as particular cases the Simple Arithmetic Averaging (SAW), the Weighted Averaging (WA), the Geometrical Averaging (GA), but also the harmonic mean, the median the mode, the k-order statistics, the Generalized Mean (GM). Even if SAW and WA are the most common cases, due to their simplicity, as remarked by many contributions, its requires the assumption of strong hypothesis, the *Preferential Independence Axiom*, see [7]. Roughly speaking, this axiom, as a consequence of the linearity in the aggregation function implies no interactions among the criteria, thus a complete *compensability*, i.e. a low value of a criterion can be compensated by an high value of another. This property is not desirable in many cases, and is not shared by GA (a total non compensative approach). For this reason and even to encounter the necessity to be as most general as possible in the Decision Maker's preference elicitation, many other methods where proposed, as Ordered Weighted Aggregation (OWA), Non Additive Measures (NAM) in conjunction with the Choquet or the Sugeno integral [4, 9]. The two latter approaches are very general, but a price is required, given that a NAM is a set-valued function which requires 2^n parameters, an unrealistic huge number as soon as n the number of the criteria overpasses $5, 56$. Conversely, OWA, as NAM, can represent a wide set of aggregation algorithm, moving from MIN up to MAX passing through SAW, but is unable to include relative importance of the criteria, as WA does. To this purpose, various attempts were done to include in OWA the possibility to manage even criteria importance, some of them will be briefly resumed in what follows. Most of them, as WOWA or OWAWA, requires a set of positional weights, as for OWA, and an other set of weights, the importance weights, thus doubling the number of parameters with respect to WA or OWA. Thus, for instance, if $n = 5$, NAM requires $2^5 = 32$ parameters, WA and OWA 5 parameters, and OWAWA requires 10 parameters.[2] Such extensions of OWA which include

[1] Values in between zero and one. In many cases, the normalization of the original data is obtained normally by a values function, see [7].

[2] Both OWA and WOWA are particular cases of NAM, see [12], 1citeTorra-3.

bot positional and cardinal information, can be a good trade-off between easiness of elicitation parameters and the requirement to be as general as possible.

In this contribution we firstly list a set of rational and desirable properties to be fulfilled by an Aggregation Operators, and then we analyze the existing extensions of OWA, showing that the none of them satisfy all the required properties. Subsequently, we introduce a novel OWA extension designed in a way to satisfy all the properties.

Partially following the notation used in [8] let be:

(a) $\mathbf{x} = (x_1, x_2, \ldots, x_n)$ vector of (normalized) criteria, $x_i \in [0, 1]$
(b) $w_1, w_2, \ldots, w_n$ vector of positive weights, $\sum_{i=1}^{n} w_i = 1$
(c) $p_1, p_2, \ldots, p_n$ vector of positive weights, $\sum_{i=1}^{n} p_i = 1$
(d) $x_\sigma = (x_{\sigma(1)}, x_{\sigma(2)}, \ldots, x_{\sigma(n)}$ if σ is a permutation of $\{1, 2, \ldots, n\}$.

3 The WA and the OWA Operators

Definition 1 WA operator. Given the non negative set of weights $p_1, p_2, \ldots, p_n$, the WA operator (Weighed Averaging) is defined as:

$$WA^p(x_1, x_2, \ldots, x_n) = \sum_{i=1}^{n} p_i x_i \tag{1}$$

WA is *compensative*, a low value of a criterion can be compensated by an other one. This is due to the satisfaction of the Preferential Independence Axiom [2], but as a consequence, it cannot represents synergic or conflicting interactions among the criteria. Among them, let us recall the Non Additive Measures (*Fuzzy Measures*) and the *Choquet* integral, see [2, 9], a very general approach but requiring an huge number of parameters to be elicited. Conversely, OWA aggregation operator, [15–17], see also [20] for a complete review, overpasses this limit, often undesired in many real world applications, see also [18, 19]. As for WA, any OWA requires n parameters to be elicited.[3]

Definition 2 OWA operator. An OWA operator is defined as follows [15, 16]:

$$OWA^w(x_1, x_2, \ldots, x_n) = \sum_{i=1}^{n} w_i x_{\sigma(i)} \tag{2}$$

being σ a permutation of the index set $\{1, 2, \ldots, n\}$ such that $x_{\sigma(1)} \geq x_{\sigma(2)} \geq \cdots \geq x_{\sigma(n)}$, thus it is linear w.r.t. the *ordered* values of the criteria.

We observe that despite the WA, the OWA weights are *positional*, in the sense that....

[3]Many other methods exist, as Non Additive Measures (NAM) and the Choquet integral, [2, 7], but usually requires many parameters, and their elicitation is resource and time consuming.

Due to the criteria permutation, OWA is a not linear operator, and with different values of **w** different operators can be obtained. For instance if $w_1 = 1, w_i = 0, \forall i = 2, \ldots, n, OWA^w(x_1, x_2, \ldots, x_n) = MAX_i(x_1, x_2, \ldots, x_n)$, if $w_n = 1, w_i = 0, \forall i = 1, \ldots, n-1$, $OWA^w(x_1, x_2, \ldots, x_n) = MIN_i(x_1, x_2, \ldots, x_n)$. Let us remark that the former and the latter are the two extreme cases, on the other side, the intermediate case is given by the SAW, obtained if $w_i = \frac{1}{n}, \forall i = 1, \ldots, n$: $OWA^w(x_1, x_2, \ldots, x_n) = \frac{1}{n} \sum_i^n x_i$.

OWA can also implement the median, the k-th order statistic, the Hurwicz operator (linear combination of average and MIN) and other ones. An OWA operator can be *andness*-type (*orness*-type) if the aggregated value is more or less close to the minimum (maximum) of its arguments. The following *orness* index is defined as:

$$Orness(OWA^w) = \sum_{i=1}^{n} w_i \frac{n-1}{n-1} \tag{3}$$

We obtain *Orness* = 1 in the *optimistic* case, corresponding to the MAX case, the logic quantifier *At least one*; while *Orness* = 0, in the *pessimistic* case) corresponding to the MIN case, the quantifier *All*.

Some procedures were proposed to elicit the OWA weights, using constrained optimizing techniques, as the ones obtained from minimum entropy approach [6], or directly from sampled data [1, 5].

An other approach to obtain the OWA parameters makes use of the so called *Regular Increasing Monotonic Quantifier*, RIM for brevity, a non decreasing function $Q(z) : [0, n] \rightarrow [0, 1]$, $Q(0) = 0$, $Q(1) = 1$, see [17].[4] The usefulness of a RIM is particularly appreciated by WOWA, one of the most commonly used extension of OWA, see below. Given a RIM, the OWA weights are obtained as:

$$w_i = Q\left(\frac{i}{n}\right) - Q\left(\frac{i-1}{n}\right) \tag{4}$$

Sometimes it can be useful to define a family of RIM, depending on one or more parameters. One of the most popular is the following one:

$$Q(z) = z^{\alpha} \tag{5}$$

where $\alpha \geq 0$ [2, 16, 19]. Finally, let us remark that the parameter α suggests the tendency to *optimism/pessimism* of the Decision Maker; an *optimistic* tendency appears if $\alpha \rightarrow +\infty$, that is, an *orness-type* behaviour, while the opposite if $\alpha \simeq 0$, if the Decision Maker is characterized by a *pessimistic* tendency, that is, an *andness* behavior.[5] An *andness* measures how much prone to pessimism a Decision Maker is, it

[4]A linguistic *term-set* as "it exists", "most", "for all" etc. can be represented by a RIM, see [19].

[5]*Andness*, or full *non compensation*, means as usual in multi criteria, the property of an aggregation operator to be close to the MIN of its arguments; the contrary for *orness*-tendency, or full *compensation* [1, 2].

equals one iff the Decision Maker is totally *andness*-prone, that is, he shows a *conjunctive* behavior, considering the MIN as aggregation of the criteria, thus looking only to the worst case. The other extreme case is when $andness = 0$, a complete disjunctive behaviour, and the Decision Maker considers only the best value of the criteria, the MAX of them.[6] In the case of the quantifier $Q(z) = z^{\alpha}$, the parameter α completely characterizes the *andness*-tendency of the Decision Maker, and different values of α permit to represent a wide range of a Decision Maker preference structure, from totally *non compensative* to *complete compensative*, and passing through the neutral case, when $andness = orness = 0.5$.

4 Discussion

In what follows, we list two sets of characterizing properties for an operator $F(\mathbf{x})$. The firs set, Basic Properties, includes the properties that are normally considered as advisable for an aggregation operators, while in what follows, we introduce other properties which characterizes in particular any extensions of OWA with importance weights.

Definition 3 Basic Properties

(1) **B1**. Compensativeness: $min(\mathbf{x}) \leq F(\mathbf{x}) \leq max(\mathbf{x})$
(2) **B2**. Monotonicity: $\mathbf{x} \geq \mathbf{y} \Rightarrow F(\mathbf{x}) \geq F(\mathbf{y})$
(3) **B3**. Idempotency: $F(x, x, \ldots, x) = x$.

The *border* conditions $F(0, 0, \ldots, 0) = 0$, $F(1, 1, \ldots, 1) = 1$ are immediate consequences of idempotency. It is straightforward to check that SAW and WA satisfy Properties **B1**, **B2**, **B3**.[7]

5 OWA Extensions

In this section, we consider some extensions of the OWA operator, designed to include importance weights. In particular, we briefly introduce the HWA, the WOWA and the OWAWA operators. Let $\mathbf{w}$ and $\mathbf{p}$ be two sets of weights, $\mathbf{N} = \{1, 2, \ldots, n\}$ and $\mathbf{x}$ the vector of normalized criteria. Extensions of WOWA usually consider a function depending on two sets of weights, the first p regards the *cardinal* characteristic (importance weights), while the other, w refers to the *ordinal* characteristics (OWA weights). Apart the properties **B1**,**B2**, **B3** above, it is advisable that any function

[6]Clearly, the *orness* index is defined by $1 - andness$.

[7]Other properties are introduces by some Authors, as *homogeneity* and *symmetry*, but they are with minor significance for our purpose, see [8].

F_p^w which mixes together *cardinal* and *ordinal* weights, satisfies the following other added properties[8].

Definition 4 Added Properties

(1) **A1**. Internal boundness: $MIN(OWA^w(\mathbf{x}), WA_p(\mathbf{x})) \leq F_p^w(\mathbf{x}) \leq MAX(OWA^w(\mathbf{x}, WA_p(\mathbf{x}))$
(2) **A2**. Coherence: if $OWA^w(\mathbf{x}) = WA_p(\mathbf{x}) = K$ then $F_p^w(\mathbf{x}) = K$
(3) **A3**. Collapsing: if $w_1 = \cdots = w_n$ then $F_p^w(\mathbf{x}) = OWA_p(\mathbf{x})$, if $p_1 = \cdots = p_n$ then $F_p^w(\mathbf{x}) = WA^w(\mathbf{x})$.

5.1 The HWA Operator

If σ a permutation of **N** such that $p_{\sigma(1)}x_{\sigma(1)} \geq \cdots \geq p_{\sigma(n)}x_{\sigma(n)}$, the HWA operator, introduced by Xu [22], is defined as:6

Definition 5

$$HWA_p^w(x_1, x_2, \ldots, x_n) = \sum_{i=1}^{n} w_i(np_{\sigma(i)}x_{\sigma(i)}) \tag{6}$$

This operator consists of an OWA with weights **w** applied to the product of the criteria and the *importance* weights **p** (the term n is just a normalizing factor) this we can also write:

$$HWA_p^w(x_1, x_2, \ldots, x_n) = OWA(w_1(np_{\sigma(1)}x_{\sigma(1)}, \ldots, w_n(np_{\sigma(n)}x_{\sigma(n)}) \tag{7}$$

Llamazares considers the two following examples [8]. Given **p** = (0.5, 0.2, 0.2, 0.1), **w** = (0, 0.5, 0.5, 0):

$$HWA_p^w(10, 10, 10, 10) = 8 \tag{8}$$

being $np_ix_i = (20, 8, 8, 4)$, while:

$$HWA_p^w(10, 5, 6, 8) = 2.4 + 2 = 4.4 \tag{9}$$

being $np_ix_i = (20, 8, 8, 4) = (20, 4, 4.8, 3.2)$.

Thus the Author concludes that HWA_p^w is not idempotent (first example: $8 < 10$) neither compensative (second example: $4.4 < MIN(10, 5, 6, 8)$.

[8]In [8] these properties are implicitly used in the discussion of WOWA operators, see point 1. and 2. at page 386 of the quoted reference.

5.2 The WOWA Operator

The WOWA operator, introduced by Torra [13, 14] is defined as:

Definition 6 WOWA definition

$$WOWA_p^w = (x_1, x_2, \ldots, x_n) = \sum_{i=1}^{n} w_i x_i \tag{10}$$

being $w_i = Q(\sum_{j=1}^{i} p_{\sigma(j)}) - Q(\sum_{j=1}^{i-1} p_{\sigma(j)})$, and $Q(z)$ a Regular Increasing Monotonic Quantifier.[9]

In the case of the RIM $Q(z) = z^\alpha$, to elicit w_i and α, a numerical approach based on a grid based algorithm was proposed by Cardin and Giove [3], and subsequently applied to an environmental risk problem.

5.3 The OWAWA Operator

The OWAWA operator, introduced by Merigò [11], given two sets of weights $p_{\sigma(i)}$ and w_i makes a linear combination of an WA with weights w_i and a OWA with (*positional*) weights $p_{\sigma(i)}$.

Definition 7

$$OWAWA_p^w(\mathbf{x}) = \alpha WA^p(\mathbf{x}) + (1-\alpha) OWA^w(\mathbf{x}) \tag{11}$$

being α a parameter in [0, 1] assigned by the Decision Maker. Clearly, an α value close to one means a more tendency to WA, the converse when α is close to zero. After some manipulation, the OWAWA operator can be also written as [8, 11]:

$$SDOWA_p^w(\mathbf{x}) = \sum_{i=1}^{n} v_i x_{\sigma(i)} \tag{12}$$

being $v_i = \alpha p_{\sigma(i)} + (1-\alpha) w_i$. This operator represents an other attempt to compact together positional and cardinal attitudes. Moreover, it is very intuitive and easy to implement; differently from WOWA, it requires only two sets of weights, w_i and $p_{\sigma(i)}$.[10]

---

[9] We observe in parenthesis that both OWA and WOWA can be represented by suitable Non Additive Measures [13, 14].

[10] Two sets of weights can be sufficient for a WOWA too, but an interpolation algorithm is required, see [12].

5.4 The IWA Operator

Both WOWA, HWA and OWAWA suffer of some drawbacks which will be later described. Trying to obviate, Llamazares [8] proposed the aggregation function $L_p^w(x_1, \ldots, x_n)$ which maintain the relationship among the initial *importance* weights, that is $\frac{\rho(i)}{\rho(j)} = \frac{p(i)}{p(i)}$:

$$IWA_p^w(\mathbf{x}) = \sum_{i=1}^{n} \rho_i x_{\sigma(i)} \tag{13}$$

being $\sigma(i)$ a permutation such that $x_{\sigma(1)} \geq \cdots \geq x_{\sigma(n)}$. To maintain the relationship above among the weights, the parameters $\rho(1), \ldots, \rho(n)$ has to be computed as follows:

$$\rho(i) = \frac{w_i p_{\sigma(i)}}{\sum_{j=1}^{n} w_j p_{\sigma(j)}} \tag{14}$$

Thus:

$$IWA_p^w(\mathbf{x}) = \sum_{i=1}^{n} \frac{w_i p_{\sigma(i)} x_{\sigma(i)}}{\sum_{j=1}^{n} w_j p_{\sigma(j)}} \tag{15}$$

As below discussed into more detail, even if this operator bypasses some shortcomings, nevertheless, apart other minor drawbacks, it fails to satisfy monotonicity and other properties, see [8] Example 5 page 390 and following, see even [10].

6 Characterization

In this Paragraph, let us analyze the above operators in the light of the defined Properties. All the above introduced operators are continuous. WA_p and OWA^w satisfies **B.1**, **B.2**, **B.3**. For what it concerns the OWA extensions, the following results can be proved [8]:

2.1 HWA_p^w do not satisfy **B.1**, **B.3**
2.2 $WOWA$ do not satisfy **A.1**, **A.2**
2.3 $OWAWA_p^w$ do not satisfy **A.3**
2.4 IWA_p^w do not satisfy **B.2**.

Thus no one of the above defined OWA extensions can be considered completely satisfactory. Nevertheless OWAWA seems to be the most compelling, but, apart the dissatisfaction of **A.3**, the assignment of the parameter α, which gives more or less importance to the *ordinal* component respect to the *cardinal* one, can be unclear. To obviate, in the next Paragraph we propose an improvement of the OWAWA operator, such that all the introduces Properties will comply, and, at the same time, the parameter α will depend only on the weight vector **p** and **w**.

7 A New Proposal; The SDOWA Operator

The characterization of OWA extensions introduced above, showed that none of them satisfy all the set of desired properties. HWA and IWA violate monotonicity (HWA fails even idempotency). Again WOWA operator, probably the most commonly used, [2, 14], requires the definition of a Regular Increasing Monotone Quantifier $Q(z)$ which cannot be simply obtained by the OWA weights **w** but through a suitable interpolating procedure [12].[11] To this purpose, in this Paragraph we introduce a new operator which, on the other side, satisfies all of them. We start observing that the OWAWA operator introduced by Merigò [11] seems to be the most promising, also given its natural appealing, being a simple linear convex combination of OWA and WA. Nevertheless, as previously remarked, it reveals two weakness points, namely:

(a) it do not satisfy **A.3**, the *Collapsing* Property,
(b) there is no an easy way to assign the value of α, the weight of the linear combination. To overpass such limitations, we modify the OWAWA operator in the following way. Namely, the two points a) and b) are in some sense interdependent, given that it is straightforward to observe that the value of α needs to depend on the vectors **p** and **w**, if the *Collapsing* Property has to be satisfied. To justify, it is sufficient to consider the fact that if **p** (or **w**) has equal components, α has to be null (equal to one), given the linear formulation of the OWAWA operator. Thus necessarily α has to be a function of **p** and **w**: $\alpha = G(\mathbf{p}, \mathbf{w})$. This result solves even the point b); no effort is now required to the Decision Maker, to assess the value of α, usually discouraging, given the intrinsic difficult to answer; the value of α is itself obtained *directly* and *implicitly* by the weights. Namely, if a Decision Maker assignees all equal values to **p**, he is indifferent among them, and this means that he gives no importance to the *cardinal* structure. The converse is true if he assignees all equal values to **w**; he is indifferent w.r.t. the *ordinal* structure. If **A.3** has to be satisfied, the function G needs to verify: $G(\mathbf{p}, \mathbf{w}) = 0$ when the vector **p** has equal components, and $G(\mathbf{p}, \mathbf{w}) = 1$, on the other side, when the vector **w** has equal components. More in general, the value of $G(\mathbf{p}, \mathbf{w})$ has to be more close to zero when **p** has a greatest *dispersion* than the vector **w**.

A natural choice for $G(\mathbf{p}, \mathbf{w})$ is to express it as the standard deviation of **p**, normalized by the sum of the two standard deviations of **p** and **w**. Thus, being $sd(\mathbf{y})$ the standard deviation of the vector **y**, and $G(\mathbf{p},\mathbf{w}) = \frac{sd(\mathbf{p})}{sd(\mathbf{p})+sd(\mathbf{w})}$, we can formulate the following.

[11] A part other drawback, WOWA sometimes returns counterintuitive results [8].

Definition 8

$$SDOWA_p^w(\mathbf{x}) = G(\mathbf{p},\mathbf{w})WA^p(\mathbf{x}) + (1 - G(\mathbf{p},\mathbf{w}))OWA^w(\mathbf{x}) \quad (16)$$

It is immediate to verify that this operator overpasses all the limitations of the other ones previously defined. Namely, it satisfies **B.1**, **B.2**, **B.3** because once assigned **p** and **w**, the parameter α is a constant thus SDOWA becomes a linear combination of WA and WOWA, as OWAWA is. Again, it satisfies **A.1**, **A.2**, **A.3**; **A.1** and **A.2** because for a fixed value of α, it collapses to an OWAWA. The Property **A.3**, the only that OWAWA does not verify, is satisfied by SDOWA by construction. Thus we can conclude:

Proposition 1 *The operator $SDOWA_p^w$ satisfies both **B.1**, **B.2**, **B.3** and **A.1**, **A.2**, **A.3**.*

To enhance understanding of our proposal, let us compare some examples proposed in the literature with SDOWA method. To comment the lackness of **A.3** for OWAWA, Llamazares proposed the following example [8]: $\mathbf{w} = (0, 0.5, 0.5, 0)$, $\mathbf{p} = (0.25, 0.25, 0.25, 0.25$. Choosing[12] $\alpha = 0.5$, if $\mathbf{x} = (10, 2, 2, 2,)$ we have:

$$\begin{aligned} OWAWA_{\mathbf{p}}^{\mathbf{w}}(10, 2, 2, 2) =& 0.5 \cdot WA_{\mathbf{p}}(10, 2, 2, 2) \\ &+ 0.5 \cdot OWA_{\mathbf{w}}(10, 2, 2, 2) = 2 + 1 = 3 \end{aligned} \quad (17)$$

thus $OWAWA_{\mathbf{p}}^{\mathbf{w}}(10, 2, 2, 2) \neq OWA^{\mathbf{w}}(10, 2, 2, 2)$ and *Collapsing* (**A.3**) is not satisfied.

On the other side, it is $sd(\mathbf{p}) = 0$, $sd(\mathbf{w}) = 0.083$ thus $G(\mathbf{p},\mathbf{w}) = 0$, $1 - G(\mathbf{p},\mathbf{w}) = 1$ and $SDOWA_{\mathbf{p}}^{\mathbf{w}}(10, 2, 2, 2) = OWA^{\mathbf{w}}(10, 2, 2, 2)$.

8 Conclusions and Future Work

We proposed a new extension of OWA, a multi criteria aggregation operator widely used in many real world applications. We firstly analyzed the existing extensions proposed in the literature, characterizing each of them on the basis of a set of desired Properties. After having showed that none of them is completely satisfying, we introduced a new extension, the SDOWA operator which on the other side verifies all the requested properties. The SDOWA operator is a linear combination of a WA and an OWA, where the coefficient of the combination depends on the standard deviation of the two vectors of weights characterizing the WA (cardinal structure) and the OWA (ordinal structure). As a next step, we intend to extend our result to some not linear combination of WA and OWA, and in parallel to characterize our new operator in terms of Non Additive Measure.

[12]With $\alpha = 0.5$ the Author claims that the two sets of weights receive the same *importance*, but as just commented above, it is unclear the real meaning of such a choice.

References

1. Beliakov, A., Pradera, A., Calvo, T.: Aggregation Functions: A Guide for Practicioners. Springer, Heidelberg (2007)
2. Calvo, T., Mayor, G., Mesiar, R.: Aggregation Operators: New Trends and Applications. Springer, Heidelberg (2002)
3. Cardin, M., Giove, S.: A grid-based optimisation algorithm for parameter elicitation in WOWA operators: an application to risk assessment. Recent Advances in Neural Networks Models and Applications. Smart Innovations and Systems Technology, vol. 37, pp. 208–215. Springer International Publishing, Cham (2015)
4. Couceiro, M., Marichal, J.-L.: Characterizations of discrete Sugeno integrals as lattice polynomial functions. In: Proceedings of the 30th Linz Seminar on Fuzzy Set Theory (LINZ2009), pp. 17–20 (2009)
5. Filev, D., Yager, R.R.: On the issue of obtaining OWA operator weights. Fuzzy Sets Syst. **94**, 157–169 (1998)
6. Fuller, R.M.: On obtaining OWA operator weights: a sort survey of recent developments. In: IEEE International Conference on Computational Cybernetics (2007)
7. Klement, E.P., Mesiar, R., Pap, E.: Triangular Norms. Kluwer Academic Publishers (2000)
8. Llamazares, B.: On generalization of weighet means and OWA operators. In: Proceedings of EUSLAT-LFA 2011 (2011)
9. Marichal, J.: An axiomatic approach of the discrete Choquet integral as a tool to aggregate interacting criteria. IEEE Trans. Fuzzy Syst. **8**(6), 800–807 (2000)
10. Merigò, J.M.: On the use of the OWA operator in the weighted average and its application in decision making. In: Proceedings of World Congress on Engineering, pp. 82–87 (2009)
11. Merigò, J.M.: A unified model between the weighted average and the induced OWA operator. Expert Syst. Appl. **38**, 11560–11572 (2011)
12. Torra, V., Lv, Z.: On the WOWA operator and its interpolation function. Int. J. Intell. Syst. **24**, 1039–1056 (2009)
13. Torra, V.: The weighted OWA operator. Int. J. Intell. Syst. **12**, 153–166 (1997)
14. Torra, V.: On some relationships between the WOWA operator and the Choquet integral. In: Proceedings of the Seventh Conference on Information Processing and Management of Uncertainty in Knowledge-Based Systems (IPMU'98), pp. 818–824, Paris, France (1998)
15. Yager, R.R.: On ordered weighted averaging aggregation operators in multicriteria decision making. IEEE Trans. Syst. Man Cybern. **18**, 183–190 (1988)
16. Yager, R.R.: Applications and extensions of OWA aggregation. Int. J. Man Mach. Stud. **37**, 103–132 (1992)
17. Yager, R.R.: Families of OWA operators. Fuzzy Sets Syst. **59**, 125–148 (1993)
18. Yager, R.R., Kacprzyk, J. (eds.): The Ordered Weighted Averaging Operators, Theory and Applications. Kluwer Academic Publisher (1997)
19. Yager, R.R.: Including importances in OWA aggregation using fuzzy systems modelling. IEEE Trans. Fuzzy Syst. **6**, 286–294 (1998)
20. Yager, R.R., Kacprzyk, J., Beliakov, J. (eds.): Recent Developments in the Ordered Weighted Avereging Operators: Theory and Practice. Springer, Berlin (2011)
21. Yager, R.R.: On the inclusion of importance in OWA aggregations. In: Yager, R.R., Kacprzyk, J.(eds.) The Ordered Weighted Avereging Operators: Theory and Applications, Kluwer Academic Publishers (1997); Yusoff, B., Merigò, J.M., Ceballos, D.: OWA-based aggregation operations in Multi-Expert MCDM model. Econ. Comput. Econ. Cybern. Stud. Res. **2**, 51 211–230 (2017)
22. Xu, Z., Da, Q.L.: An overview of operators for aggregating information. Int. J. Intell. Syst. **18**, 953–969 (2003)

A Fuzzy Approach to Long-Term Care Benefit Eligibility; an Italian Case-Study

Ludovico Carrino and Silvio Giove

Abstract We propose a fuzzy approach to quantify a cash-benefit for older people in need of *Long-Term* Care, e.g., affected by limitations in daily-living activities. Many approaches exist at national or regional level in Europe, and most legislation determine eligibility to public care-programs using rule-based approaches which aggregate basic health-outcomes into main pillars and then into eligibility categories. Population ageing and improvements in longevity make access to care a crucial problem for Western economies. In this paper we focus on Italy, where public-care eligibility is decentralized at regional level and often based on check-lists, and in particular on the Toscana region. We investigate the extent to which the existing legislation violates basic properties of monotonicity and continuity, thus potentially increasing inequity in care access. We then propose the introduction of a fuzzy approach to the eligibility determination, which allows for smoother results and reduced inequality.

1 Introduction

While both longevity and health conditions have largely improved in the last century in many developed countries, disease-free life-expectancy indicators have increased at a much lower pace and a significant degree of health inequality is emerging among different socioeconomic groups [1]. The rate of older people in need of Long-Term Care has risen due to a higher prevalence of conditions and to a higher number of disorders limiting the autonomy of individuals [2–4]. In order to postpone the onset

L. Carrino (✉)
Department of Global Health and Social Medicine, King's College London, London Aldwych 40,, UK
e-mail: ludovico.carrino@kcl.ac.uk

L. Carrino · S. Giove
Department of Economics, Ca' Foscari University, Venice, Cannaregio 873, Italy
e-mail: sgiove@unive.it

A. Esposito et al. (eds.), *Progresses in Artificial Intelligence and Neural Systems*, Smart Innovation, Systems and Technologies 184,
https://doi.org/10.1007/978-981-15-5093-5_29

of severe disability and reduce social exclusion in older age, policy makers have focused on implementing programs of formal care [3].[1] Such programs vary greatly across OECD countries in terms of the services offered (in cash, in-kind; domiciliary-care vs institutional-care), and aim to provide accessible, equal, and adequate care coverage [5].

Previous studies have highlighted that a crucial determinant of (in)equity in care-access and coverage is played by eligibility rules, which are policy tools defining the target population in 'need-of-care': they represent a compulsory gateway to receive home-care benefits, either in-kind or in-cash [6–8]. However, although recent evidence highlighted the extent to which eligibility rules impact the coverage of care systems and potentially care expenditure [7, 9], the literature has overlooked the role of eligibility algorithms in determining horizontal equity in care-access (e.g., people with similar needs should receive similar amounts of care support) and vertical equity in care-access (e.g., people with different needs should receive different amounts of care support).

This paper investigates how a rule-based approach which determines eligibility for cash-for-care schemes may inadvertently impact equity in care-access. Moreover, we are among the first to simulate the implementation of a Fuzzy Inference System (FIS) as an eligibility Decision System, and discuss the benefits of this approach. We select Italy's Toscana region as a case study, due to its comprehensive need-assessment design (it accounts for several dimensions of loss-of-autonomy such as cognitive, functional and mental health) and its eligibility algorithm which is particularly suitable for a FIS application. We show that the existing legislation introduces sharp discontinuities in the relationship between the cash-allowance and the individual health status, which can in turn result in a failure of both horizontal and vertical equity. Indeed, the existing eligibility rules imply that a marginal change in health conditions may result in large changes in the allowance. Moreover, as the evaluation classifies individual in five broad need-of-care categories, a decline in health condition may not result in an increase in care-allowance.

We then show that implementing a FIS decision system allows for an increased granularity and smoothness in the eligibility determination, and reduce the undesired properties of the current legislation.

Our contribution is relevant under several perspectives. First, this is the among the first analysis that explicitly investigates how eligibility algorithms affect equity in care coverage. This is particularly important given the ongoing policy debate on the trade-off between public budget sustainability and adequate care-provision [10]. Second, we introduce a novel strategy adopting a more flexible fuzzy system

---

[1] *Formal-care* includes all care services that are provided in the context of formal regulations, such as through contracted services, mostly by trained care workers, that can be paid out of pocket or through reimbursement by public (or, less often, by private) institutions. What characterizes formal care-provision is its acknowledgment by the Social or Health departments at the proper governmental level. *Informal-care* is, conversely, a term that refers to the unpaid assistance provided by partners, adult children and other relatives, friends or neighbors who hold a significant personal relationship with the care recipient.

in the field of Long-Term Care eligibility determination. Third, although our case-study is necessarily restricted to a specific European region (Italy's Toscana), we pave the road for a larger investigation on how the equality embedded by European care-legislations might be enhanced by the adoption of a FIS approach.

2 Care Eligibility: The Italian Case Study

Long Term Care (LTC) is defined as a range of services required by persons who cannot cope with basic Activities of Daily Living (ADL) and instrumental Activities of Daily Living (iADL) due to a reduced physical and/or cognitive capacity [8].

In European countries, eligibility for LTC is largely determined based on the evaluation of functional (ADL and iADL), cognitive and mental health limitations. Legislations define an eligibility algorithm which summarizes single health outcomes into an index of need of care. Such algorithm is often highly nonlinear, and its characteristics vary greatly across countries (for a review, see [7]). The Italian public LTC is based on in-kind or in-cash programs which are mostly region-based and not harmonized, in terms of both the services provided and the eligibility rules [7].[2] In 2006, the Italian government established a National Fund to be allocated to regions in order to provide in-cash or in-kind LTC support (FNNA, *Fondo Nazionale Non Autosufficienza*). Moreover, several Regions chose to complement the FNNA with a similar Regional Fund for LTC (FRNA, Fondo Regionale per la Non Autosufficienza).

2.1 A Case Study: Italy's Toscana

Toscana's main regional Long-Term Care programme PAC (*Progetto per l'assistenza continua alla persona non autosufficiente*, Long-term care for non-autonomous individuals) was introduced in 2010 with the regional law D.G.R. n.370 (March 22, 2010). The PAC is financed by the FRNA fund (regional law 66/2008) and encompasses both benefits in-cash (aimed at sharing the costs of hiring a private professional care-giver) and in-kind (nursing-care by public medical professionals) for adults aged 65 or higher. The programme is means-tested, since the household income is taken into account when defining the amount-of-care to be supplied/reimbursed or the cash-benefit to be allocated [7]. The PAC is managed at the district level, where a Multi-disciplinary Evaluation Unit (*Unità di Valutazione Multidisciplinare*, UVM), composed of a doctor, a nurse and a social assistant, is responsible for the assessment-of-need of the elderly applicants and for the definition of a Personalized Plan of Assistance, which regulates the care-services to be supplied.

[2] A nation-wide cash benefit, the *Indennità di Accompagnamento* (IA), is available to individuals classified as *invalid*. Yet, there is no nationwide guideline as to how to assess and evaluate such outcome.

Table 1 Definition of functional dependency, Toscana's PAC

Dependency in BADL	Description	BADL scale
Light	Full dependency in 2 BADL *or* light/heavy dependency in 3 BADL	8–14
Moderate	Full dependency in 3 BADL *or* light/heavy dependency in 4 + BADL	15–20
Severe	Heavy dependency in roughly all BADL	21–24

In Toscana, need of care is assessed through a multi-dimensional approach in three main domains: Functional limitations, Cognition, Behavior/depression disorders. Within each domain, the loss-of-autonomy is categorized as either Light, Medium or Severe. Depending on the combination of the scores obtained in the domains, an individual is assigned in one of five eligibility classes, which correspond to a specific cash-allowance. This makes it an ideal case study for a Fuzzy Inference System application, as we will discuss later. Similar eligibility algorithms are implemented, to various extents, in several European countries (see, for example, the existing legislation in France, for the APA programme [7]). However, the Toscana system is particularly suitable for a case-study, as the eligibility algorithm is more transparent and relatively simpler with respect to the French one.

2.1.1 Functional Impairment Assessment

Functional autonomy is evaluated through the Basic Activities of Daily Living scale (BADL), a Katz-adapted list of activities-of-daily-living included in the Minimum Data Set for Home Care (MDS-HC) assessment method [11]. The BADL has seven items, each evaluated on a five-step scale, from 0 (independence) to 4 (full assistance required) according to the need of care required in the last seven days. The BADL score ranges from 0 to 24. The degree of functional limitation is determined as follows (Table 1).

2.1.2 Cognitive Impairment Assessment

Cognitive impairment is measured through the application of Eric Pfeiffer's Short Portable Mental Status Questionnaire [12]. The questionnaire includes questions on, for example, time orientation (current day of the week, current date in full), space orientation (name of the current location, phone number), age and birthdate, knowledge of the current and former Pope or President of the Republic, own mother's maiden name, numeric questions. The answers are then recorded and an overall score is attributed, depending on the number of mistakes, so that individuals are classified as "non-impaired or lightly impaired", "moderately impaired" and "severely impaired", as follows [13] (Table 2).

Table 2 Definition of cognitive dependency, Toscana's PAC

Cognitive dependency	Short portable mental score
Light	0–4
Moderate	5–7
Severe	8–10

Table 3 Definition of behavioral/depressive issues, Toscana's PAC

Behavioral/depression risk	Short portable mental score
Light	0–3
Moderate	4–7
Severe	8–12

2.1.3 Behavioral/Depression Disorders

Depression- and Behavior-assessment follow the guidelines from MDS-HC. Depression (mood) assessment consists in a list of questions about whether the patient exhibits: (i) a feeling of sadness depression or death-wishes; (ii) persistent anger with self or others; (iii) expressions of what appears to be unrealistic fears; (iv) repetitive health complaints (obsessive concerns); (v) repetitive anxious complaints; (vi) sad, pained, worried facial expressions; (vii) recurrent crying, tearfulness; (viii) withdrawal from activities of interest; (ix) reduced social interaction; Instances when client exhibited behavioral symptoms. Behavior-assessment deals with the occurrence of: (i) wandering; (ii) verbally abusive behavioral symptoms; (iii) physically abusive behavioral symptoms; (iv) other behavioral symptoms; (v) resisting care/taking medications/injections/ADL assistance/eating/changes in position. The assessment results in a score between 0 (low behavioral/depression risk) and 12 (high risk). Individuals are then categorized as "lightly disturbed", "moderately disturbed", "severely disturbed", as described in Table 3.

2.1.4 Eligibility Rules

By combining the functional, cognitive and the behavioral/depression scores, individuals are categorized in 5 ISO-groups, representing five homogeneous levels of need-of-care (see [13]). Group 5 corresponds to the most severe profiles, while group 1 gathers individuals who have at most a light deficit in the three domains.

Table 4 explains in details the eligibility rules, i.e. how the ISO-groups are defined (see [13, 14]).

Table 4 ISO-eligibility groups, Toscana

ISO-GROUP	Functional deficit								
	Light			Moderate			Severe		
	Behav. deficit			Behav. deficit			Behav. deficit		
Cognitive deficit	L	M	S	L	M	S	L	M	S
L	1	2	3	2	3	4	4	4	5
M	2	2	3	3	3	4	4	4	5
S	3	3	4	3	4	5	4	5	5

The eligibility rules are as follows:

- Age should be at least 65 years
- Yearly household income should be lower than € 25000[3]
- ISO-GROUP should be 3 or higher[4]

For those eligible, the amount of the in-kind or the in-cash allowance ranges between a minimum and a maximum depending on individuals' income (ISEE). As we are interested in a representative individual, we will consider the average benefit amount, as follows

- ISO-GROUP 3: € 140 [€80 – €200]
- ISO-GROUP 4: €240 [€170 – €310]
- ISO-GROUP 5: €355 [€260 – €450]

For instance, an average-earning individual (satisfying age and income constraints above defined) with moderate functional deficit, medium cognitive limitations, and low behavioral/depression issues respectively, would be classified in the ISO-GROUP 3, with a monthly allowance of €140.

3 The Proposed Modified Fuzzy Approach

As many other LTC systems in Europe, the Toscana system allocates individuals in five eligibility classes (ISO-GROUP). Subdivision in classes is a popular strategy in welfare-benefit systems, as it can be useful for practical purposes. However, it suffers from some undesired drawbacks. Namely, the *crisp* border between contiguous classes implies sharp discontinuities ("jumps") in the output: a small marginal change in one basic health-indicator can shift an individual to the next ISO-GROUP, with a significant variation in the cash-benefit (e.g., switching from ISO3 to ISO4 increases

[3]See, e.g., the regulation of the Casentino district, at http://www.uc.casentino.toscana.it/regolamenti/disposizioni-attuative-anno-2013.pdf.

[4]The UVM can, in principle, decide to allow some benefit for individuals in groups 1 and 2 (Regional law D.G.R. n.370, Attachment A).

the monthly benefit from €140 to €240 for an average-earning individual). Such a sharp discontinuity in the benefit allocation has no clear economic justification, and may be perceived as a driver of inequity in care-access. Moreover, it can incentivize strategic and, in extreme cases, illegal behaviors.

On the other hand, as this method pools together many individuals in the same ISO-GROUP (assigning them the same benefit), it neglects the fact that, even within the same group, some profiles may be characterized by more severe limitations than others.

For such reasons, we claim that ISO-GROUP clustering does not allow for an adequate degree of granularity and smoothness, to guarantee (i) *strong monotonicity* of benefit-eligibility to health; and (ii) *pseudo-continuity* of benefit-eligibility to health.

Let us clarify the previous points with an example of three hypothetical individuals:

- individual A, with a score of 15 in the physical scale (medium), 0 in the cognitive scale (low), and 4 in the behavioral/depression scale (medium);
- individual B, with a score of 20 in the physical scale (medium), 7 in the cognitive (medium) scale and 7 in the behavioral/depression scale (medium);
- individual C with a score of 20 in the physical scale (medium), 7 in the cognitive scale (medium), and 8 in the behavioral/depression scale (severe).

Individuals A and B would both be classified in the ISO-GROUP 3, and would thus get the same monetary amount, Nevertheless, individual A has a lower need-of-care, as she has no cognitive impairment, and she lies at the lower bound of the "medium dependency" category for both Functioning and Behavior/depression dimensions. Conversely, individual B fares much worse than A, lying at the upper bound of the "medium dependency" category in all the dimensions. Albeit A and B are characterized by different degrees of loss-of-autonomy, the eligibility rule is insensitive to such a worsening in health conditions. Thus, the legislation does not satisfy the (strong) *monotonicity* assumption, and risks to inadvertently contribute to care-access inequality.

Consider now individual C, who has the same clinical profile as B, but has a worse behavior/depression score by just one point. This marginal increase makes C eligible for ISO-GROUP-4 benefits, which means an average monthly allowance of €250. As a marginal increment in one dimension causes a large change in the monetary outcome, the eligibility rules violate the *pseudo-continuity* property.

It is important to note that such issues would arise, to different extents, for most LTC programs in Europe, as most of them allocate people in classes based on the score they fare in several health dimensions [7].

We argue that a Fuzzy-Logic Inference System (FIS) can enhance both the *granularity* and the *smoothness* of the eligibility rules, basing on the existing ISO-GROUP clustering (see [15, 16] for further details). This way, a personalized benefit can be assigned ad hoc to each eligible person.

Pseudo-continuity is linked to granularity, while monotonicity to smoothness. Through a FIS, monotonicity can be obtained by using a Sugeno-type with L-R type and unimodal fuzzy numbers [17, 18], as *triangular* fuzzy numbers, with MIN t-norm, defined on the universe set of each of the 3 ISO-GROUP class. Pseudo-continuity can be obtained by differentiating together the output of each rule, this increasing its granularity. That is, instead of the (discrete and natural) score between 0 and 5, each cell of the rule block will be directly assigned the economic benefit.[5] By assigning a specific monetary amount to each cell, we would realize the highest granularity.

3.1 Structure of the Proposed FIS

We hereby describe an example of a FIS tailored for this type of problem, whose parameters are based on the Toscana legislation. In order to enhance granularity and smoothness in the eligibility rules, which correspond to monotonicity and pseudo-continuity, we make use of a *zero-order* Sugeno model (aka as TSK, Takagi-Sugeno-Kang model) with MIN t-norm and trapezoidal/triangular membership [19].

This is realized using trapezoidal fuzzy numbers (rather than triangular). Moreover, in order to avoid a complete departure from the actual legislation, we do not force the maximum granularity, thus some cells of the rule block contain the same level of allowance (for instance, in the first Table, the amount €140 appears in the second row, third column, but also in the third row second column). For each of the three input variables, Functional Deficit (Func), Cognitive Deficit (Cogn) and Behavioral/depression Deficit (Behav) we used three membership functions (trapezoidal fuzzy numbers), corresponding to the linguistic terms-sets Low, Medium and Severe, which are the actual terms used in the legislation (Table 4), and represented in Fig. 1. Again, to increase the granularity, we modified the values in the Table 4, substituting the class label (natural numbers 1 up to 5) with the direct value of the benefit, inferred for each class from the average values in Table 3, suitably modified to differentiate the elements within classes. The results are reported in Table 6. By way of example, Fig. 2 reports the rule surface corresponding to the second and the third health variables (Cognitive and Behavioral scores).

4 A Numerical Example

The proposed zero-order Sugeno FIS was tested with some simulated cases. To this purpose, we evaluated the system with the 3 hypothetical profiles described above, and a fourth one (case D) characterized by a worse health status. The profiles are

[5]The exact monetary value of the benefit in each cell needs to be determined by the Public Authority. This phase will would require participatory decision methods (focus group, brainstorming, questionnaires). In this paper, the values allocated to each cell are purely indicative.

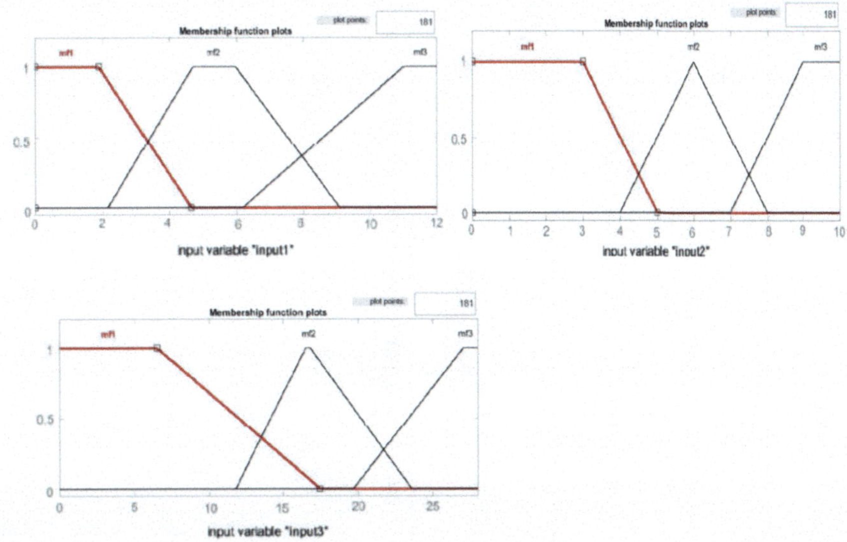

Fig. 1 Membership functions for the three input variables Behav, Cogn, Func. (Input1, Input2, Input3 respectively)

Table 5 Monetary amounts assigned to four hypothetical profiles

	Actual legislation	FIS-system
Profile A	€140 (ISO-3)	€79.3
Profile B	€140 (ISO-3)	€ 206.38
Profile C	€240 (ISO-4)	€237.6
Profile D	€355 (ISO-5)	€281.11

characterized by the following scores in the three main variables capturing loss-of-autonomy in the Behavior/Depression, Cognition, and Functioning domains:

1. Case A (4, 0, 15)
2. Case B (7, 7, 20)
3. Case C (8, 7, 20)
4. Case D (8, 7, 22)

The input activation of the Sugeno FIS and the corresponding output for each rule is reported in Fig. 3 for case A. Similar results, available on request, are obtained for cases B, C, D.

The FIS then provides the monetary amounts that the four clinical profiles would be eligible to, keeping income constant (at the average level).

Results in Table 5 show that, unlike the original Toscana's legislation, a FIS can implement a set of eligibility rules which allocate care-allowances to different clinical profiles, by satisfying pseudo-continuity and monotonicity. In the original set-rules,

Table 6 Output of the Sugeno FIS (net average benefit for each class)

ISO-GROUP	Functional deficit								
	Light			Moderate			Severe		
	Behav. deficit			Behav. deficit			Behav. deficit		
Cognitive deficit	L	M	S	L	M	S	L	M	S
L	0	0	100	0	140	240	200	240	300
M	0	0	140	140	180	280	240	280	355
S	100	140	280	140	280	355	280	355	400

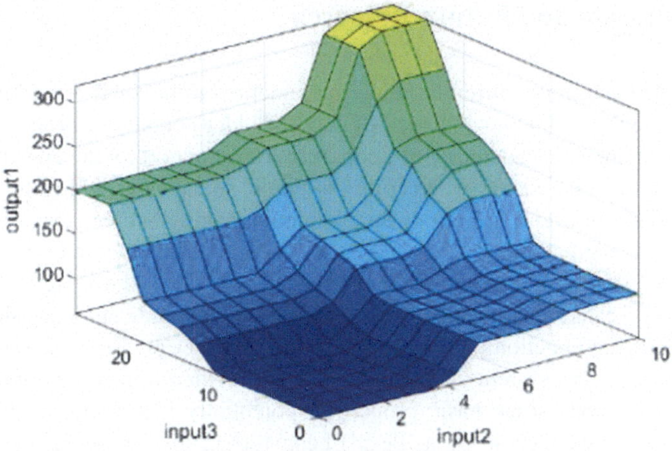

Fig. 2 Output surface for the second and the third variables

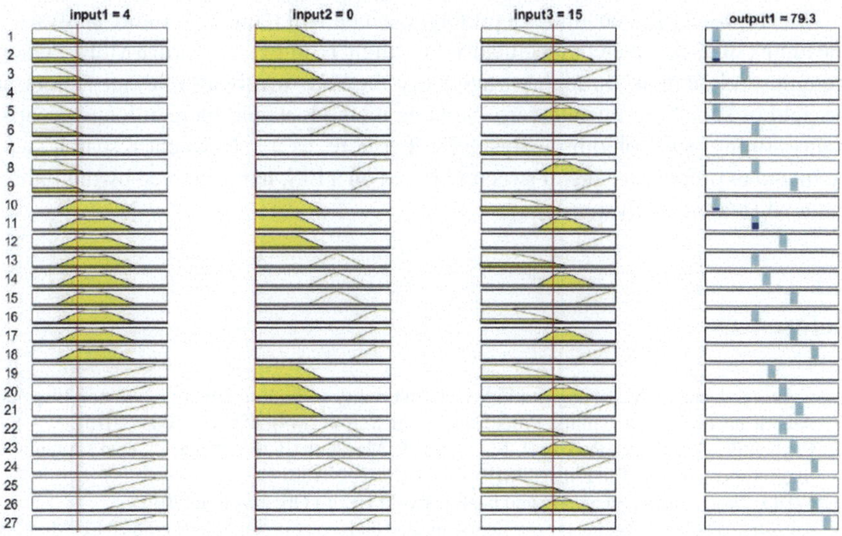

Fig. 3 Activation rules and output for case A

individual A would be allocated the same allowance as individual B, although being characterized by a healthier profile. Under the FIS rules, individual B would get a consistently higher allowance than individual A.

Similarly, individual C, who is just marginally different than individual B, is allocated a largely different allowance in the original rules. Conversely, the FIS rules assign her only an increment of around €30 in the cash-benefit.

5 Conclusion and Future Research

Various approaches are currently being implemented at national or regional level, to ameliorate the wellbeing and need of care for older people in Europe. Concerns related to, on one hand, adequacy of Long-Term Care support for dependent individuals and, on the other hand, sustainability of public social-care and health programs are particularly relevant, in light of the population ageing and enhanced longevity. In Italy, most regions have stablished cash-for-care schemes based on rule-based approaches. Among the most encompassing eligibility algorithms, we focus on the case of the Toscana region, which aggregates basic health indicators into three main pillars measuring Functioning, Cognitive and Behavior/Depression outcomes. After having analyzed the system currently in use, we verified how the eligibility rules are likely to violate some basic properties, potentially increasing inequality and incentivizing strategic (and even illegal) behaviors. Thus, to increase granularity and smoothness in the Decision System, we introduce a Fuzzy Inference System (FIS) to compensate, at least partially, the undesired characteristics of the currently implemented rules.

The proposed FIS constitutes a prototype which will require, in future analysis, a fine tuning of its parameter. Specifically, this might require to perform a Multi-Person preference elicitation, through participatory methods involving relevant Actors in the field of social- and health care. Suitable methods would include focus groups, brainstorming, and conjoint analysis. Furthermore, as a subsequent research step, we intend to propose a general structure based on a FIS, to be adopted by the Italian National Healthcare System.

References

1. Case, A., Deaton, A.: Rising morbidity and mortality in midlife among white non-hispanic Americans in the 21st century. Proce. Nat. Acad. Sci. **112**(49), 15078–15083 (2015)
2. Rechel, B., Grundy, E., Robine, J.-M., Cylus, J., Mackenbach, J.P., Knai, C., et al.: Ageing in the European union. The Lancet **381**(9874), 1312–1322 (2013)
3. WHO.: World report on ageing and health: World Health Organization (2015)
4. Eurostat.: The 2015 Ageing report: Economic and budgetary projections for the 28 EU Member States (2013–2060) (2015)
5. Gori, C., Fernandez, J.-L.: Long-term Care Reforms in OECD Countries. Policy Press (2015)
6. Muir, T.: Measuring social protection for long-term care (2017)
7. Brugiavini, A., Carrino, L., Orso, C.E., Pasini, G.: Vulnerability and Long-term Care in Europe: an Economic perspective. Palgrave MacMillan, London (2017)
8. Colombo, F., Llena-Nozal, A., Mercier, J., Tjadens, F.: OECD Health Policy Studies Help Wanted? Providing and Paying for Long-Term Care: Providing and Paying for Long-Term Care: OECD Publishing (2011)
9. Carrino, L., Orso, C.E., Pasini, G.: Demand of long-term care and benefit eligibility across European countries. Health Economics (2018)
10. OECD.: Preventing Ageing Unequally: OECD Publishing (2017)

11. Morris, J., Fries, B., Steel, K., Ikegami, N., Bernabei, R., Carpenter, G., et al.: Comprehensive clinical assessment in community setting: applicability of the MDS-HC. J. Am. Geriatr. Soc. **45**(8), 1017–1024 (1997)
12. Pfeiffer, E.: A short portable mental status questionnaire for the assessment of organic brain deficit in elderly patients. J. Am. Geriatr. Soc. **23**(10), 433–441 (1975)
13. Profili, F., Razzanelli, M., Soli, M., Marini, M.: Il bisogno socio-sanitario degli anziani in Toscana: i risultati dello studio epidemiologico di popolazione BiSS. Documenti dell'Agenzia Regionale di Sanità della Toscana. Disponibile presso (2009) www.ars.toscana.it/c/document_library/get_file
14. Visca, M., Profili, F., Federico, B., Damiani, G., Francesconi, P., Fortuna, P. et al.: La Ricerca AGENAS. La presa in carico degli anziani non autosufficienti. I Quaderni Di Monitor **30**(a), 145–183 (2012)
15. Kukolj, D.: Design of adaptive Takagi–Sugeno–Kang fuzzy models. Appl. Soft Comput. **2**(2), 89–103 (2002)
16. Takagi, T., Sugeno, M.: Fuzzy identification of systems and its applications to modeling and control. In: Readings in Fuzzy Sets for Intelligent Systems (pp. 387–403). Elsevier (1993)
17. Chen, S.-J., Hwang, C.-L.: Fuzzy multiple attribute decision making methods. In: Fuzzy multiple attribute decision making (pp. 289–486). Springer (1992)
18. Beliakov, G., Pradera, A., Calvo, T.: Aggregation Functions: A Guide for Practitioners. Springer (2007)
19. Klement, E.P., Mesiar, R., Pap, E.: Triangular Norms. 2000. Kluwer Academic Publishers, Dordrecht (2000)

On Fuzzy Confirmation Measures of Fuzzy Association Rules

Emilio Celotto, Andrea Ellero, and Paola Ferretti

Abstract Many researchers from different sciences focused their attention on quantifying the degree to which an antecedent in a rule supports a conclusion. This long-standing problem results to be particularly interesting in the case of fuzzy association rules between a fuzzy antecedent and a fuzzy consequence: in fact, rules become much more flexible in describing information hidden in the data and new interestingness measures can be defined in order to assess their relevance. This implies, in particular, a new definition of support and of confidence of the association rule. In this framework, we focus on fuzzy confirmation measures defined in terms of confidence. In this way, it is possible to propose new fuzzy confirmation measures in a setting that allows their comparison with reference to some potential properties.

1 Introduction

A possible way to mine hidden patterns contained in a dataset is to detect inductive rules of the kind $E \Rightarrow H$, that relate the values of a set of attributes, the *premise* E, with the values of another set of attributes, the *conclusion* H. A strong relationship, for example when a large number of records in a database that possess a (set of)

Authors are listed in alphabetical order. All authors contributed equally to this work, they discussed the results and implications and commented on the manuscript at all stages.

E. Celotto · A. Ellero (✉)
Department of Management, Ca' Foscari University of Venice, Venice, Italy
e-mail: ellero@unive.it

E. Celotto
e-mail: emi.web@tin.it

P. Ferretti (✉)
Department of Economics, Ca' Foscari University of Venice, Venice, Italy
e-mail: ferretti@unive.it

A. Esposito et al. (eds.), *Progresses in Artificial Intelligence and Neural Systems*, Smart Innovation, Systems and Technologies 184,
https://doi.org/10.1007/978-981-15-5093-5_30

attribute(s) E possess also a (set of) attribute(s) H, highlights a regularity in the dataset, i.e. the association turns out to be relevant since exceptions to it are *rare* (see [1]).

One way to assess the quality of inductive rules makes use of Bayesian Confirmation Measures (BCMs) that evaluate the effect of an evidence E on the probability of conclusion H, basically using their probabilities $Pr(H)$, $Pr(E)$ and the conditional probability $Pr(H|E)$. The literature has deeply explored confirmation measures when the involved sets are *crisp* (see e.g. [4, 5, 7]) fixing their analytical properties and relationships (see, e.g., [6, 11]).

When dealing with a real database, better results in modelling relationships among data could be obtained in a fuzzy sets scenario, i.e. considering fuzzy association rules $A \Rightarrow B$, with A and B a couple of fuzzy sets. Fuzzy association rules allow a record to possess an attribute with a certain *degree* and are much more flexible in describing information contained in the data. In this case the measurement of the quality of the emerging associations needs an update of the concept of BCM, namely Fuzzy Confirmation Measures (FCMs) for fuzzy association rules, defined by Glass [8].

In this paper we will recall in Sect. 2 the definitions of Bayesian and of Fuzzy Confirmation Measures. In Sect. 3 we discuss possible ways to relate existing FCMs and to suggest new measures. Concluding remarks are presented in Sect. 4.

2 Bayesian and Fuzzy Confirmation Measures

Inductive rules $E \rightarrow H$ represent relations in which the knowledge of E supports conclusion H, relations that may be supported by data with different strength. To evaluate the intensity with which data support a rule, the so called *interestingness* of the rule, a possible choice is represented by Bayesian Confirmation Measures (BCMs) which are defined by means of prior probability $Pr(H)$, posterior probability $Pr(H|E)$ and the probability of antecedent E, $Pr(E)$. In fact the knowledge of E may change the knowledge about H, since E may confirm H when $Pr(H|E) > Pr(H)$, or disconfirm H when $Pr(H|E) < Pr(H)$.

The natural definition of a Bayesian measure of confirmation becomes then (see, e.g., [4, 7]).

Definition 1 A function c of antecedent E and conclusion H, is a *Bayesian Confirmation Measure* (BCM) when

$$
\begin{array}{ll}
c(E,H) > 0 \text{ if } Pr(H|E) > Pr(H) & \textit{(confirmation case)} \\
c(E,H) = 0 \text{ if } Pr(H|E) = Pr(H) & \textit{(neutrality case)} \\
c(E,H) < 0 \text{ if } Pr(H|E) < Pr(H) & \textit{(disconfirmation case)}
\end{array}
$$

As an example of BCM consider, e.g.,

$$d(E,H) = Pr(H|E) - Pr(H) \qquad K(E,H) = \frac{Pr(E|H) - Pr(E|\neg H)}{Pr(E|H) + Pr(E|\neg H)} \tag{1}$$

defined by Carnap in [3] and by Kemeny and Oppenheim in [12], respectively.

Fuzzy Confirmation Measures (FCMs), instead, are necessary when fuzzy association rules $A \Rightarrow B$ are considered, which are rules where a fuzzy antecedent A has a fuzzy consequence B. Let us be more precise. Given a set of records $X = \{x_1, x_2, \ldots, x_n\}$ we assume that they possess a set $\mathcal{A}$ of attributes, being each attribute A_i $(i = 1, \ldots, m)$ in $\mathcal{A}$ a fuzzy set defined by a membership function $A_i : X \to [0, 1]$; $A_i(x_j)$ measures the degree to which the i-th attribute A_i applies to record x_j. The standard negator $n(A_i) : X \to [0, 1]$ defined by $n(A_i(x_j)) = 1 - A_i(x_j)$ identifies, instead, the fuzzy complement set $n(A_i)$ of the fuzzy set A_i.

The strength of a fuzzy association rule $A \Rightarrow B$ between antecedent A and consequence B, can be evaluated by measures like *support* (*supp*) and *confidence* (*conf*):

$$supp(A \Rightarrow B) = \frac{\sum_{x \in X} A(x) \otimes B(x)}{|X|} \qquad conf(A \Rightarrow B) = \frac{\sum_{x \in X} A(x) \otimes B(x)}{\sum_{x \in X} A(x)}$$

where $\otimes$ denotes a t-norm used to define the intersection of fuzzy sets (see [8]).[1]

The definition of confidence for the fuzzy association rule $A \Rightarrow B$ is used by Glass [8] to propose a definition of Fuzzy Confirmation Measure. It is based on the observation that, in the crisp case, the support can be considered as an estimate of the probability of the intersection of A and B while confidence can be considered as the fuzzy counterpart of $Pr(B|A)$ (see [8]). Moreover Glass adopts the confidence of the default rule $T \Rightarrow A$ as the fuzzy counterpart of the estimate of $Pr(A)$.[2]

The idea of considering confidence as the fuzzy counterpart of probability requires some caution: in fact, focusing on the notion of fuzzy (in)dependence Glass in [8] observed that the product based fuzzy confidence measure is a suitable candidate in extending the notion of positive/negative dependence and independence of a BCM in an analogous concept in the fuzzy framework. In fact, with the aim of setting the definition of a confirmation measure of fuzzy association rules, first of all Glass [8] focused on new notions of fuzzy (in)dependence: in doing so the definition of probabilistic (in)dependence, which is generally assumed in the context of BCMs, is restated for the fuzzy framework thus obtaining two new definitions: the definition of weak dependence and that of strong dependence. Recall, in fact, that in Definition 1 of Bayesian confirmation measure the sign condition on $c(E, H)$ is related to the sign of $Pr(H|E) - Pr(H)$ which admits different formulations that are equivalent from the logical point of view but represent different perspectives on confirmation

[1]Throughout the paper, the formulas are assumed to be well defined, i.e. we assume as granted that denominators do not vanish.

[2]Here T is the totally true constant, that is $T(x) = 1 \; \forall x \in X$.

(as discussed by Greco et al. in [11]). More precisely, the alternative formulations are

a. $Pr(H|E) > Pr(H)$ (*Bayesian confirmation*)
b. $Pr(H|E) > Pr(H|\neg E)$ (*strong Bayesian confirmation*)
c. $Pr(E|H) > Pr(E)$ (*likelihoodist confirmation*)
d. $Pr(E|H) > Pr(E|\neg H)$ (*strong likelihoodist confirmation*).

In the fuzzy framework the corresponding inequalities are equivalent under specific conditions. To deal with these conditions let us first recall the definition of weak dependence.

Definition 2 A fuzzy confidence measure satisfies weak dependence if

(i) $conf(A \Rightarrow B) > conf(T \Rightarrow B)$ if $conf(B \Rightarrow A) > conf(T \Rightarrow A)$
(ii) $conf(A \Rightarrow B) = conf(T \Rightarrow B)$ if $conf(B \Rightarrow A) = conf(T \Rightarrow A)$
(iii) $conf(A \Rightarrow B) < conf(T \Rightarrow B)$ if $conf(B \Rightarrow A) < conf(T \Rightarrow A)$

provided the default rules $T \Rightarrow A$ and $T \Rightarrow B$ have no null fuzzy confidence measure.

All t-norm based fuzzy confidence measures fulfil weak dependence, given that $conf(T \Rightarrow B) = \sum_{x \in X} B(x)/|X|$. Moreover, the fuzzy context involves the necessity of a further dependence definition, as follows.

Definition 3 A fuzzy confidence measure satisfies strong dependence if it satisfies weak dependence and

(i) $conf(A \Rightarrow B) > conf(T \Rightarrow B)$ if $conf(A \Rightarrow B) > conf(n(A) \Rightarrow B)$
(ii) $conf(A \Rightarrow B) = conf(T \Rightarrow B)$ if $conf(A \Rightarrow B) = conf(n(A) \Rightarrow B)$
(iii) $conf(A \Rightarrow B) < conf(T \Rightarrow B)$ if $conf(A \Rightarrow B) < conf(n(A) \Rightarrow B)$

provided the default rules $T \Rightarrow A$ and $T \Rightarrow B$ have fuzzy confidence measures neither 0 nor 1.

Glass proved that the product based fuzzy confidence measure

$$conf(A \Rightarrow B) = \frac{\sum_{x \in X} A(x) \cdot B(x)}{\sum_{x \in X} A(x)} \tag{2}$$

satisfies strong dependence.

In this way, it is possible to prove that a product t-norm based confidence ensures the equivalence among the four fuzzy perspectives:

a. $conf(A \Rightarrow B) > conf(T \Rightarrow B)$
b. $conf(A \Rightarrow B) > conf(n(A) \Rightarrow B)$
c. $conf(B \Rightarrow A) > conf(T \Rightarrow A)$
d. $conf(B \Rightarrow A) > conf(n(B) \Rightarrow A)$

provided $conf(T \Rightarrow A)$ and $conf(T \Rightarrow B)$ are neither 0 nor 1.

The definition of fuzzy confirmation measure proposed by Glass in [8] is based on the product t-norm and is the following.

Definition 4 A fuzzy confirmation measure of the degree to which a fuzzy set A confirms a fuzzy set B is a real-valued function c_f such that

(i) $c_f(A, B) > 0$ if $conf(A \Rightarrow B) > conf(n(A) \Rightarrow B)$ (*confirmation case*)
(ii) $c_f(A, B) = 0$ if $conf(A \Rightarrow B) = conf(n(A) \Rightarrow B)$ (*neutrality case*)
(iii) $c_f(A, B) < 0$ if $conf(A \Rightarrow B) < conf(n(A) \Rightarrow B)$ (*disconfirmation case*)

where $conf$ is the product based fuzzy confidence measure (2).

In this way, we can focus on fuzzy confirmation measures which are defined by adapting the corresponding Bayesian confirmation measures to the fuzzy environment. As benchmark examples we consider four fuzzy confirmation measures, whose BCM counterparts have been used in *jMAF* (see [2, 13]), a well-established software for Rough Sets based Decision Support Systems. The following measures are in fact easily linked to known confirmation measures[3]

$$G(A, B) = \log\left[\frac{conf(B \Rightarrow A)}{conf(n(B) \Rightarrow A)}\right] \tag{3}$$

which is connected to Good proposal [9],

$$K(A, B) = \frac{conf(B \Rightarrow A) - conf(n(B) \Rightarrow A)}{conf(B \Rightarrow A) + conf(n(B) \Rightarrow A)} \tag{4}$$

which corresponds to the BCM defined by Kemeny and Oppenheim [12], and

$$Z(A, B) = \begin{cases} Z_1(A, B) = \dfrac{conf(A \Rightarrow B) - conf(T \Rightarrow B)}{1 - conf(T \Rightarrow B)} & \text{in case of confirmation} \\ Z_2(A, B) = \dfrac{conf(A \Rightarrow B) - conf(T \Rightarrow B)}{conf(T \Rightarrow B)} & \text{otherwise.} \end{cases} \tag{5}$$

The last measure was defined in the BCM framework by Rescher [14] and further analysed in [4] and [11]. Lastly, it is possible to suggest the definition of the function

$$GSS(A, B) = \begin{cases} GSS_1(A, B) & \text{in case of confirmation} \\ GSS_2(A, B) & \text{otherwise} \end{cases} \tag{6}$$

where

$$GSS_1(A, B) = \frac{conf(T \Rightarrow B) - conf(n(A) \Rightarrow B)}{conf(T \Rightarrow B)}$$

and

$$GSS_2(A, B) = \frac{conf(T \Rightarrow B) - conf(n(A) \Rightarrow B)}{1 - conf(T \Rightarrow B)}$$

whose corresponding BCM was defined by Greco, Słowiński and Szczęch in [11].

[3] In the following, we use the same notation for both BCM and the corresponding FCM.

Glass in [8] proved that G, K and Z are FCMs, analogously it is possible to prove that the suggested function $GSS(A, B)$ is a fuzzy confirmation measures.

Theorem 1 *The measure $GSS(A, B)$ is a fuzzy confirmation measure.*

Proof We give the proof for the confirmation case. By definition, $GSS(A, B) > 0$ if and only if

$$conf(T \Rightarrow B) > conf(n(A) \Rightarrow B).$$

The assumption of a product based fuzzy confidence measure ensures the satisfaction of strong dependence, namely it is

$$conf(T \Rightarrow B) > conf(n(A) \Rightarrow B)$$

if and only if

$$conf(n(n((A)) \Rightarrow B) > conf(n(A) \Rightarrow B)$$

that is

$$conf(A \Rightarrow B) > conf(n(A) \Rightarrow B).$$

Analogous observations can be used to prove the sign conditions for the case of neutrality and of disconfirmation.

□

Note that with the help of some algebraic manipulation G and K can be expressed in terms of $conf(T \Rightarrow B)$ and $conf(A \Rightarrow B)$ only:

$$G(A, B) = \log \frac{conf(A \Rightarrow B)[1 - conf(T \Rightarrow B)]}{conf(T \Rightarrow B)[1 - conf(A \Rightarrow B)]}$$

$$K(A, B) = \frac{conf(A \Rightarrow B) - conf(T \Rightarrow B)}{conf(A \Rightarrow B) - 2conf(A \Rightarrow B)conf(T \Rightarrow B) + conf(T \Rightarrow B)}$$

while $GSS(A, B)$ can be written in terms of $conf(A \Rightarrow B)$, $conf(T \Rightarrow B)$ and $conf(T \Rightarrow A)$:

$$GSS_1(A, B) = \frac{conf(T \Rightarrow A)}{1 - conf(T \Rightarrow A)} \frac{conf(A \Rightarrow B) - conf(T \Rightarrow B)}{conf(T \Rightarrow B)}$$

$$GSS_2(A, B) = \frac{conf(T \Rightarrow A)}{1 - conf(T \Rightarrow A)} \frac{conf(A \Rightarrow B) - conf(T \Rightarrow B)}{1 - conf(T \Rightarrow B)}.$$

3 Relationships Between Fuzzy Confirmation Measures

The discovering of relationships among fuzzy confirmation measures can help in better understanding which of them are more suitable for specific purposes or can even be useful to suggest the definition of new measures, as it was already observed for classical confirmation measures (see, e.g., [7]). For example, K (see Eq. (4)) can be seen as a weighted harmonic mean of the expressions of Z (Eq. (5)) in case of confirmation (the expression indicated by Z_1) and disconfirmation (expression Z_2).

At this aim, let us extend the definition of function Z_1 also to the case of disconfirmation, and that of Z_2 to the case of confirmation: observe that this way Z_1 and Z_2 with their definition extended to both confirmation and disconfirmation cases are two FCMs since they satisfy Definition 4. We can readily write K as a weighted harmonic mean of those functions, with weights $w = conf(A \Rightarrow B)$ and $1 - w$, respectively [4]:

$$K(A, B) = \left(\frac{conf(A \Rightarrow B)}{Z_1(A, B)} + \frac{1 - conf(A \Rightarrow B)}{Z_2(A, B)}\right)^{-1} . \tag{7}$$

The way in which the expressions of K and Z can be linked seems to allow insights into the meaning of those measures. In fact, K can be interpreted as a synthesis of the two expressions for Z putting higher weight to the confirmation formula when the confidence of the rule $A \Rightarrow B$ is high, and to the disconfirmation formula when confidence is low. We can observe that K, as an harmonic mean, will be more stable with respect to extreme values of Z_1 or Z_2 (when $conf(A \Rightarrow B)$ is either rather close to 1 or to 0).

By simple substitution we also obtain another way to express K:

$$K(A, B) = \frac{conf(A \Rightarrow B) - conf(T \Rightarrow B)}{conf(A \Rightarrow B) \cdot (1 - conf(T \Rightarrow B)) + (1 - conf(A \Rightarrow B)) \cdot conf(T \Rightarrow B)}$$

where a factor depending on the confidence of $T \Rightarrow A$ is now required.

Likewise, given any pair of fuzzy confirmation measures fcm_1 and fcm_2, their weighted harmonic mean

$$\left(\frac{w}{fcm_1(A, B)} + \frac{1 - w}{fcm_2(A, B)}\right)^{-1} \tag{8}$$

with $w \in [0, 1]$, satisfies the sign requirements of Definition (4), i.e., it is a new fuzzy confirmation measure.

To give a further example we can consider the fuzzy confirmation measure GSS (see Eq. (6)) which is defined by two different expressions in the cases of confirmation

---

[4] As a matter of fact, when the evidence A disconfirms conclusion B, i.e. $conf(A \Rightarrow B) < conf(T \Rightarrow B)$, both Z_1 and Z_2 assume a negative value: strictly speaking their harmonic mean is not defined, but the proposed link among measures holds, with the same meaning. In the neutrality case we have the boundary values $K = Z = 0$ and their link cannot be defined by a harmonic mean.

and disconfirmation: by extending the domains of functions GSS_1 and GSS_2 to both the cases of confirmation and disconfirmation, we obtain a new FCM setting

$$GSS_{AB}(A,B)=\left(\frac{conf(A\Rightarrow B)}{GSS_1(A,B)}+\frac{1-conf(A\Rightarrow B)}{GSS_2(A,B)}\right)^{-1}. \tag{9}$$

where the weight assigned to the first expression is $w=conf(A\Rightarrow B)$. Twiddling with the two FCMs we can consider instead $w=1-conf(A\Rightarrow B)=conf(A\Rightarrow n(B))$ and obtain a new FCM

$$GSS_{AnB}(A,B)=\left(\frac{1-conf(A\Rightarrow B))}{GSS_1(A,B)}+\frac{conf(A\Rightarrow B)}{GSS_2(A,B)}\right)^{-1} \tag{10}$$

whose definition is, again, not far from measure K. In fact K can be written as

$$K(A,B)=\frac{1-conf(T\Rightarrow A)}{conf(T\Rightarrow A)}\left(\frac{1-conf(A\Rightarrow B))}{GSS_1(A,B)}+\frac{conf(A\Rightarrow B)}{GSS_2(A,B)}\right)^{-1} \tag{11}$$

i.e., K can be expressed also in terms of GSS_1 and GSS_2, but the role played by confidence is now upset with respect to Eq. (7): a higher weight is assigned to the confirmation formula when the confidence of the rule is low and to the disconfirmation formula when confidence is high.

Once recognised that K can be expressed as a weighted harmonic mean of Z_1 and Z_2, or even of GSS_1 and GSS_2, it seems interesting to explore other possible relationships among FCMs. For example, the FCM inspired by the simplest confirmation measure (see Carnap [3]), i.e., the difference fuzzy confirmation measure $d(H,E)=conf(A\Rightarrow B)-conf(T\Rightarrow B)$, can be expressed as the harmonic mean of Z_1 and Z_2. In Table 1 we collect some FCMs that can be obtained as a weighted harmonic mean of Z_1 and Z_2: they are obtained by setting the weight w equal to $conf(A\Rightarrow B)$, $1-conf(A\Rightarrow B)$, $conf(T\Rightarrow B)$, $1-conf(T\Rightarrow B)$, $conf(T\Rightarrow A)$, $1-conf(T\Rightarrow A)$, respectively. While d and K correspond to well-known BCMs, the others are examples of new FCMs.

If we define

$$\alpha=conf(T\Rightarrow A),\qquad \beta=conf(T\Rightarrow B),\qquad \gamma=conf(A\Rightarrow B)$$

all the above defined weighted harmonic means can be expressed in easier to read formulas in terms of α, β, γ only. For example, the FCM $K(A,B)$ proposed in (4) admits the following equivalent definition

$$K(\alpha,\beta,\gamma)=\frac{\gamma-\beta}{\gamma+\beta-2\gamma\beta}. \tag{12}$$

Table 1 Fuzzy Confirmation Measures as harmonic means of Z_1 and Z_2

FCM	Weight w	FCM formula
d	1/2	$2[conf(A \Rightarrow B) - conf(T \Rightarrow B)]$
K	$conf(A \Rightarrow B)$	$\frac{conf(A \Rightarrow B) - conf(T \Rightarrow B)}{conf(A \Rightarrow B) - 2 \cdot conf(A \Rightarrow B) \cdot conf(T \Rightarrow B) + conf(T \Rightarrow B)}$
Z_{AnB}	$1 - conf(A \Rightarrow B)$	$\frac{conf(A \Rightarrow B) - conf(T \Rightarrow B)}{2 \cdot conf(A \Rightarrow B) \cdot conf(T \Rightarrow B) - conf(A \Rightarrow B) - conf(T \Rightarrow B) + 1}$
Z_B	$conf(T \Rightarrow B)$	$\frac{1}{2} \frac{conf(A \Rightarrow B) - conf(T \Rightarrow B)}{conf(T \Rightarrow B) \cdot (1 - conf(T \Rightarrow B))}$
Z_{nB}	$1 - conf(T \Rightarrow B)$	$\frac{conf(A \Rightarrow B) - conf(T \Rightarrow B)}{2 \cdot (conf(T \Rightarrow B))^2 - 2 \cdot conf(T \Rightarrow B) + 1}$
Z_A	$conf(T \Rightarrow A)$	$\frac{conf(A \Rightarrow B) - conf(T \Rightarrow B)}{-2 \cdot conf(T \Rightarrow B) \cdot conf(T \Rightarrow A) + conf(T \Rightarrow B) + conf(T \Rightarrow A)}$
Z_{nA}	$1 - conf(T \Rightarrow A)$	$\frac{conf(A \Rightarrow B) - conf(T \Rightarrow B)}{2 \cdot conf(T \Rightarrow B) \cdot conf(T \Rightarrow A) - conf(T \Rightarrow B) - conf(T \Rightarrow A) + 1}$

Table 2 Fuzzy Confirmation Measures as harmonic means of GSS_1 and GSS_2

FCM	Weight w	FCM formula
$GSS_{1/2}(\alpha, \beta, \gamma)$	1/2	$2 \frac{\alpha}{1-\alpha} (\gamma - \beta)$
$GSS_{AB}(\alpha, \beta, \gamma)$	γ	$\frac{\alpha}{1-\alpha} \left[\frac{\gamma-\beta}{1-\gamma-\beta+2\gamma\beta}\right]$
$GSS_{AnB}(\alpha, \beta, \gamma)$	$1 - \gamma$	$\frac{\alpha}{1-\alpha} \left[\frac{\gamma-\beta}{\gamma+\beta-2\gamma\beta}\right]$
$GSS_B(\alpha, \beta, \gamma)$	β	$\frac{\alpha}{1-\alpha} \left[\frac{\gamma-\beta}{1+2\beta-2\beta^2}\right]$
$GSS_{nB}(\alpha, \beta, \gamma)$	$1 - \beta$	$\frac{\alpha}{1-\alpha} \left[\frac{\gamma-\beta}{2\beta(1-\beta)}\right]$
$GSS_A(\alpha, \beta, \gamma)$	α	$\frac{\alpha}{1-\alpha} \left[\frac{\gamma-\beta}{1-\alpha-\beta+2\alpha\beta}\right]$
$GSS_{nA}(\alpha, \beta, \gamma)$	$1 - \alpha$	$\frac{\alpha}{1-\alpha} \left[\frac{\gamma-\beta}{\alpha+\beta-2\alpha\beta}\right]$

Of course, the goal in writing the new FCMs as functions of α, β, γ is to simplify the notation. For example, in the case of weighted harmonic means of GSS_1 and GSS_2 the weight γ defines the FCM

$$GSS_{AB}(\alpha, \beta, \gamma) = \frac{\alpha}{1-\alpha} \left[\frac{\gamma - \beta}{1 - \gamma - \beta + 2\gamma\beta}\right] \tag{13}$$

that can be written as

$$GSS_{AB}(A, B) = \frac{conf(T \Rightarrow A)}{1 - conf(T \Rightarrow A)} \cdot \left[\frac{conf(A \Rightarrow B) - conf(T \Rightarrow B)}{1 - conf(A \Rightarrow B) - conf(T \Rightarrow B) + 2conf(A \Rightarrow B)conf(T \Rightarrow B)}\right]. \tag{14}$$

Table 2 contains the weighted harmonic means of GSS_1 and GSS_2 obtained using the same set of weights w used in Table 1.

4 Conclusions

Fuzzy Confirmation Measures are the Fuzzy counterpart of Bayesian Confirmation Measures and the large number of Bayesian Confirmation Measures available in the literature accordingly suggests possible definitions of corresponding FCMs. We explored a possible way to find ties among different FCMs by rewriting some of them as weighted harmonic means of other FCMs. Working properly on weights, the approach simplifies the definition of new FCMs which are more apt to be chosen in relation to specific requirements.

Besides the harmonic mean, also different (weighted) means could be considered and, again, choosing the weight so as to calibrate the FCM in order to possess specific properties. More in general one could also exploit the large variety of aggregation functions in order to obtain measures which satisfy desired properties (see, e.g., [10]), which are research directions that could be explored. Testing the definitions of new FCMs within a dataset, would be a way to illustrate how to properly choose the weights depending on their required features.

References

1. Agrawal, R., Imieliński, T., Swami, A.: Mining association rules between sets of items in large databases. In: Buneman P., Jajodia S. (eds.) ACM SIGMOD International Conference on Management of Data. SIGMOD 1993, vol. 22(2), pp. 207–216. Rec. ACM, New York, NY, USA (1993)
2. Błaszczyński, J., Greco, S., Matarazzo, B., Słowiński, R.: jMAF—dominance-based rough set data analysis framework. In: Skowron, A., Suraj, Z. (eds.) Rough Sets and Intelligent Systems—Professor Zdzisław Pawlak in Memoriam. ISRL, vol. 42, pp. 185–209. Springer, Heidelberg (2006)
3. Carnap, R.: Logical Foundations of Probability, 2nd edn. University of Chicago Press, Chicago (1962)
4. Crupi, V., Festa, R., Buttasi, C.: Towards a grammar of Bayesian confirmation. In: Suárez, M., Dorato, M., Rédei, M. (eds.) Epistemology and Methodology of Science, pp. 73–93. Springer, Dordrecht (2010)
5. Fitelson, B.: The plurality of Bayesian measures of confirmation and the problem of measure sensitivity. Philos. Sci. **66**, 362–378 (1999)
6. Fitelson, B.: Likelihoodism, Bayesianism, and relational confirmation. Synthese **156**, 473–489 (2007)
7. Geng, L., Hamilton, H.J.: Interestingness measures for data mining: a survey. ACM Comput. Surv. **38**(3), 1–32 (2006)
8. Glass, D.H.: Fuzzy confirmation measures. Fuzzy Sets Syst. **159**(4), 475–490 (2008)
9. Good, I.J.: Probability and the Weighing of Antecedent. Hafners, New York (1950)
10. Grabisch, M., Marichal, J.-L., Mesiar, R., Pap, E.: Aggregation Functions. Cambridge University Press, Cambridge (2009)

11. Greco, S., Słowiński, R., Szczęch, I.: Properties of rule interestingness measures and alternative approaches to normalization of measures. Inform. Sci. **216**, 1–16 (2012)
12. Kemeny, J., Oppenheim, P.: Degrees of factual support. Philos. Sci. **19**, 307–324 (1952)
13. Laboratory of Intelligent Decision Support Systems of the Poznan University of Technology. http://idss.cs.put.poznan.pl/site/software.html
14. Rescher, N.: A theory of antecedent. Philos. Sci. **25**, 83–94 (1958)

Q-Learning-Based Financial Trading: Some Results and Comparisons

Marco Corazza

Abstract In this paper, we consider different financial trading systems (FTSs) based on a Reinforcement Learning (RL) methodology known as Q-Learning (QL). QL is a machine learning method which real-time optimizes its behavior in relation to the responses it gets from the environment as a consequence of its acting. In the paper, first we introduce the essential aspects of RL and QL which are of interest for our purposes, then we present some original and differently configurated FTSs based on QL, finally we apply such FTSs to eight time series of daily closing stock returns from the Italian stock market.

1 Introduction

In this paper we investigate the effectiveness of simple automated financial trading systems (FTSs) based on a machine learning technique known as Q-Learning (QL). QL belongs to the family of the so-called Reinforcement Learning (RL) methodologies. Briefly, these methodologies concern an agent, in our case a FTS, dynamically interacting with an environment, in our case a financial market. During this interaction, the agent perceives the state of the environment and takes an action, in our case to trade a given asset. In its turn, the environment, on the basis of this action, provides a negative or a positive local reward, in our case an appropriately measured investor's loss or gain. This process aims at the detection of a policy, in our case a trading strategy, that permits the maximization over time of an appropriate function of overall reward.

The remainder of the paper is organized as follows. In the next section, we give a brief review of the literature on RL-based FTSs and we describe the elements of novelties we take into account. In Sect. 3, we synthetically present the essential aspects of RL and of QL which are of interest to our purposes. In Sect. 4, we present

M. Corazza (✉)
Department of Economics, Ca' Foscari University of Venice, Cannaregio 873, 30121 Venice, Italy
e-mail: corazza@unive.it

A. Esposito et al. (eds.), *Progresses in Artificial Intelligence and Neural Systems*, Smart Innovation, Systems and Technologies 184,
https://doi.org/10.1007/978-981-15-5093-5_31

our QL-based FTSs and we provide the results of their applications to eight real time series of stock daily closing returns from the Italian stock market, all belonging to the FTSE MIB basket. In Sect. 5, we give some final remarks.

2 Review of the Literature and Our Elements of Novelties

Among the contributions presented in literature, we recall the following ones.

In [5] the Authors compare various FTSs based on different neurocomputing methods, among which a hybrid one constituted by QL combined with a supervised Artificial Neural Network. Also in [12] the Authors consider a hybrid method constituted by the Adaptive Network Fuzzy Inference System supplemented by the RL paradigm. In [8] the Authors proposes a RL-based asset allocation strategy able to utilize the temporal information coming both from a given stock and from the fund over that stock. In [10] two stock market timing predictors are presented: an actor-only RL and and actor-critic RL. Also in [2] an actor-critic RL-based FTS is proposed, but in a fuzzy version. In [9] the Authors use different RL-based high-frequency trading systems in order to optimally manage the individual orders and the trade execution strategies. Finally, in [6] the Authors propose a FTS based on an Artificial Neural Network structured in two modules: the first one permits the detection of the financial market conditions through a Deep Learning approach, the second one employs the knowledge so detected to make trading decisions through RL.

With respect to this literature, in this paper we do the following:

- In the vast majority of the papers, the classical Sharpe ratio is used as local reward function. Given that this financial ratio is a generally accepted benchmark, here we utilize again it, calculated over the last $L \in \mathbb{N}$ trading days. But, we also use other two local reward functions: the average logarithmic return calculated over the last L trading days; the ratio between the sum calculated over the last L trading days of the logarithmic returns and the sum calculated over the last L trading days of the absolute value of the logarithmic returns. By doing so, we aim to verify whether the widespread choice of the Sharpe ratio as local reward function is confirmed or less. Note that to the best of our knowledge, the newly proposed local reward function have never been used in this context;
- As state variables describing the financial market, we consider the last closing returns and the current state. Generally, this is not the case for several FTSs to which more or less refined state variables are provided. We make this choice in order to check the performance capability of the QL-based FTSs also starting from basic information. Notice that the closing returns are continuous over $\mathbb{R}$.

Finally, note that in order to test the operational capability of the considered FTSs we explicitly take into account the transaction costs.

3 Some Basics on RL and QL

In this section, we synthetically recall the main aspects of RL and of QL (see for technical deepenings [1, 3] and [7]).

3.1 RL

An RL-based agent is characterized by four elements: a policy; a local reward function; a value function; a description of the environment. In a few details:

- A policy specifies the agent's behavior at each time instant. It is a mapping from the set of perceived environmental states to the set of the actions to be taken when the agent finds itself in those states;
- A local reward function represents the agent's goal at each time instant. It translates each states-action pair into a number which indicates to the agent the desirability of that same state-action pair;
- The agent's overall reward is the summation of all the local rewards—possibly discounted—the agent can expect to accumulate over the future starting from the state in which it is. The expectation of such a summation defines the value function. Note that the local reward function determines the immediate desirability of a state-action pair, while the value function indicates the long term desirability of a state-action pair considering the probable future state-action pairs and their future rewards;
- A description of the environment has the purpose to represent the behavior of the environment itself at each time instant. This description directs the agent in the next state and to the next local reward on the basis of the current state-action pair.

Formalizing what is stated above, at each discrete time instant $t = 0, 1, \ldots, N$, the agent and the environment interact between them. In the time instant t, the former receives a representation of the latter as a vector state $s_t \in \mathcal{S}$, where $\mathcal{S}$ is the set of available vector states. On the basis of s_t, a vector action $a_t \in \mathcal{A}(s_t)$ is selected by the agent, where $\mathcal{A}(s_t)$ is the set of available actions in the state s_t. In the next time instant $t + 1$, the agent will be in a new state s_{t+1} and will receive a local reward $r_{t+1} \in \mathbb{R}$ as response to the action a_t.

As the agent's goal is the maximization of the value function, at each time instant t the agent selects the action which maximizes the expected value of the summation of all the local rewards, suitably discounted, she can obtain in the future from the state in which she is. This overall reward, R_t, is defined as:

$$R_t = \sum_{k=0}^{+\infty} \gamma^k r_{t+k},$$

where $\gamma \in [0, 1]$ is the discount rate.

The agent's learning target consists in detecting a policy able to maximize the expected value of R_t through the interaction with the environment on the basis of hers experience. To do this, in each time instant t, the agent, starting from the state s_t, has to calculate the following action-value functional equation for each possible action:

$$Q^{\pi}(s, a) = \mathbb{E}_{\pi}(R_t | s_t = s, a_t = a),$$

where π indicates the policy, s is the state and a is the action.

A fundamental property of this functional equation is the fulfillment of the following recursive relationship, known as Bellman equation, which ensures the existence of a unique solution:

$$Q^{\pi}(s, a) = \mathbb{E}_{\pi}\left(r_{t+1} + \gamma Q^{\pi}(s_{t+1}, a_{t+1}) | s_t = s, a_t = a\right).$$

The agent, in order to learn a policy able to maximize the expected value of R_t, has to select in each time instant t the action that satisfies the following relationship, called greedy policy:

$$Q^{*}(s, a) = \max_{\pi}(Q^{\pi}(s, a)).$$

It is possible to compute the action-value functional equation in each time instant t through an iterative method which is called policy evaluation. It starts with an arbitrary initialization $Q_0^{\pi}(s, a)$ of $Q^{\pi}(s, a)$. Then, in each successive iteration, the action-value functional equation is approximated using the Bellman equation through an update rule:

$$Q_{k+1}^{\pi}(s, a) = \mathbb{E}_{\pi}\left(r_{t+1} + \gamma Q_k^{\pi}(s_{t+1}, a_{t+1}) | s_t = s, a_t = a\right).$$

Finally, it may be useful to verify if there exists another policy that, if followed, is able to perform better than the currently followed one. This process is known as policy improvement.[1]

3.2 QL

QL learns the value of the optimal state-action pair, $Q^{*}(s, a)$, transitioning from a state-action pair to another one. This learning method is an online control approaches. It is an online method because it performs the updates of the action-value functional equation at the end of each step without waiting for the terminal one. It is a control method as it performs actions to reach its purpose, that is the detection of the optimal

[1] In Sect. 4, we specify the policy improvement we consider in the applications.

state-action functional equation estimate $Q^*(s, a)$. Furthermore, QL is also an off-policy method, that is the agent learns the value of the state-action pair independently from the taken action because the updates of such value are done regardless the current action, but with respect to the action which maximizes the value of the next state-action pair. Therefore, the specific QL update rule is:

$$Q_{k+1}(s_t, a_t) = Q_k(s_t, a_t) + \alpha \left[r_{k+1} + \gamma \max_a (Q_k(s_{t+1}, a_{t+1})) - Q_k(s_t, a_t) \right], \quad (1)$$

where $\alpha \in (0, 1]$ is the so-called learning rate which determines the degree of update.

4 The FTSs and Their Applications

In this section, we design and implement differently configured FTSs through the use of QL. We start by specifying the action variable, the local reward functions, the state variable vector, the approximator of the value function, and the policy improvement we use in the applications.

Regarding the possible actions of the FTSs, we utilize the three following ones:

$$a_t = \begin{cases} -1 \text{ (sell or stay-short-in-the-market signal)} \\ 0 \text{ (stay-out-from-the-market signal)} \\ 1 \text{ (buy or stay-long-in-the-market signal)} \end{cases},$$

in which the "stay-out-from-the-market" implies the closing of whatever previously open position (if any). Note that in most of the prominent literature only the "sell or ..." and the "buy or ..." actions are considered.

As concerns the local reward functions, detecting a successfull one is likely more an art than a science. So, in this paper we take into account the three following local reward functions in order to check whether different ways to assess the reward can meaningfully affect the performances of the FTSs:

- The well known Sharpe ratio calculated at time instant t, SR_t, very often used in several RL-based FTSs, that is:

$$SR_t = \frac{\mathbb{E}_L (g_{t-1-L}, \dots, g_{t-1})}{\sqrt{\mathbb{V}ar_L (g_{t-1-L}, \dots, g_{t-1})}} \in \mathbb{R},$$

  where $\mathbb{E}_L(\cdot)$ and $\mathbb{V}ar_L(\cdot)$ indicate respectively the sample mean and the sample variance calculated over the last L stock market days, and $g_t = a_{t-1}e_t - \delta\,|a_t - a_{t-1}|$ indicates the net-of-transaction cost logarithmic rate of return obtained by the same FTS at t as a consequence of the action taken by the same FTS in the previous time instant, in which $\delta > 0$ indicates the transaction costs in terms of percentage.[2]

- The average logarithmic rate of return calculated at time instant t, $ALRR_t$, that is:

$$ALRR_t = \mathbb{E}_L(g_{t-1-L}, \ldots, g_{t-1}) \in \mathbb{R}.$$

Note that, to the best of our knowledge, $ALRR_t$ is little used in QL-based FTSs;

- The ratio calculated at time instant t between the sum of the last L logarithmic rates of return and the sum of the absolute values of the same logarithmic rates of return, $OVER_t$, that is

$$OVER_t = \frac{\sum_{i=0}^{L-1} g_{t-i}}{\sum_{i=0}^{L-1} |g_{t-i}|} \in [-100\%,\ 100\%].$$

Qualitatively, given the maximum performance achievble during the considered time period, $OVER_t$ can be interpreted as the percentage of the performance achieved by the FTS during the same time period. Note that, again to the best of our knowledge, in this paper we propose the first use of $OVER_t$ in QL-based FTSs.

Finally, as we would like local reward functions reacting quickly enough to the actions of the FTSs, we calculate SR_t, $ALRR_t$ and $OVER_t$ by using only the last $L = \{5, 10\}$ stock market days. Furthermore, we set δ equal to the realistic percentage 0.15% per transaction.

Concerning the state variables, recalling that we are interested in checking the performance capability of our FTSs starting from basic information, we simply use the last N logarithmic rates of return of the asset to trade and the current state. So, the state variable vector at the time t is:

$$s_t = (e_{t-N+1},\ e_{t-N+2},\ \ldots,\ e_t, a_{t-1}),$$

where $e_\tau = \ln(p_\tau / p_{\tau-1})$, in which p_τ indicates the price of the asset at time instant τ. In particular, as we would like to develop FTSs reacting quickly enough also to new information, in our application we consider only the last $N = \{2, 4\}$ stock market days.

For what concerns the approximator of the quantity $Q_k(s_{t+1}, a_{t+1})$, both linear and non-linear ones are proposed in the literature (see for instance [3]).[3] In this paper, following the same philosophy of simplicity yet used for the choice of the state variable vector, we consider the following linear approximator:

---

[2]For simplicity's sake, in the following of the paper we use only the term "net" for the expression "net-of-transaction cost".

[3]Note that the need to specify such an approximator is due to the fact that some of the state variables, namely the logarithmic rates of return, are continuous.

$$Q_k(s_{t+1}, a_{t+1}; \theta_k) \approx \theta_{k,0} + \sum_{n=1}^{N} \theta_{k,n}\phi(e_{t-1+n}) + \theta_{k,N+1}\phi(a_t),$$

where $\theta_k = (\theta_{k,0}, \theta_{k,1}, \ldots, \theta_{k,N}, \theta_{k,N+1})$ is the vector of the parameters at the k-th iteration of the update rule,[4] and $\phi(\cdot)$ is the logistic function which plays the role of squashing function.[5]

Finally, as regards the policy improvement, as stated in Sect. 3.1, it is an exploration approach which verifies whether there exists another policy, with respect to the currently followed one, that is able to perform better than the latter. The policy improvement we consider in this paper is:

$$a_t = \begin{cases} \arg\max_{a_t} Q(s_t, a_t; \theta_k) & \text{with probability } 1-\epsilon \\ u & \text{with probability } \epsilon \end{cases},$$

where $\epsilon \in \{5\%, 25\%\}$, and $u \sim \mathcal{U}\{-1, 0, 1\}$. Note that the values of ϵ are respectively close to those suggested by the literature (5%) and not-so-used (25%).

5 Applications and Results

In this section, we provide the results of the applications of the above-specified FTSs to eight time series of daily closing stock returns, all components of the Italian FTSE MIB index.

Before starting, we still have to face a question related to the stochasticity of QL. As clear from the previous section, the performance of the QL-based FTSs are affected both by the random inizialitazion of the parameters θ_k and by the use of a random policy improvement. In order to manage such a stochasticity, each FTS is runned 250 times so that 250 actions a_t are produced in each time instant t. Of course, in each of these time instants only one operational action is needed. Therefore, we have to suitably aggregate the produced 250 actions. We perform it following two different approaches:

- Analogously to what done in [11], the operational action is generated through the following decisional rule:

$$a_{OA,t} = \begin{cases} -1 \text{ if } \overline{a}_t \in [-1, -0.\overline{3}) \\ 0 \text{ if } \overline{a}_t \in [-0.\overline{3}, 0.\overline{3}] \\ 1 \text{ if } \overline{a}_t \in (0.\overline{3}, 1] \end{cases}, \qquad (2)$$

[4]When $k = 0$, the parameters are randomly initialized following a $\mathcal{U}(-1, 1)^{N+2}$.

[5]Note that, in order to determine the optimal parameters, we perform a mean square error minimization through a gradient descent-based method.

where $a_{OA,t}$ indicates the operational action in the time instant t, and $\overline{a}_t$ is the sample mean at the time instant t of the 250 actions produced in the same time instant;

- As the produced 250 actions at each time instant can be considered independent and identically distributed random variables, from basic statistics one obtaines $\overline{a}_t \xrightarrow{d} \mathcal{N}(m_t, s_t^2)$, where $\mathcal{N}(\cdot, \cdot)$ is the Gaussian probability density function. Therefore, it is possible to perform the following bilateral t-test:

$$\begin{cases} H_0 : m_t = 0 \\ H_1 : m_t \neq 0 \end{cases}.$$

On the basis of this test, the following decisional rule for the generation of the operational action is proposed:

$$a_{OA,t} = \begin{cases} 0 & \text{if } H_0 \text{ is accepted} \\ \text{sign}(m_t) & \text{if } H_0 \text{ is rejected} \end{cases}. \tag{3}$$

The confidence level used in the applications is the usual 95%. Note that, to the best of our knowledge, this decisional rule has never been used before.

Summarizing, we consider three different local reward functions (SR_t, $ALRR_t$ and $OVER_t$), two different values of L (5 and 10), two different values of N (2 and 4), two different values of ϵ (5% and 25%) and two different aggregation rules of the produced actions ((2) and (3)). Taking into account all the possible combinations of local reward functions, parameters and aggregation rules, we get 48 differently configured FTSs. Finally, note that in the update rule 1, also parameters α and γ have to be set. For both, we consider values widely used in the literature, namely $\alpha = 5\%$ and $\gamma = 95\%$.

With regard to the time series of daily closing stock returns, we consider A2A S.p.A. (A2A), Assicurazioni Generali S.p.A. (AG), Brembo S.p.A. (B), Mediaset S.p.A. (MS), Saipem S.p.A. (S), Telecom Italia S.p.A. (TI), UniCredit S.p.A. (UC), and Unipol Gruppo S.p.A. (UG), from January 2, 2007 to January 31, 2019, just over twelve stock market years. These stocks well represent important sectors of the Italian economy. Note that we have voluntarily made coincide the starting date of the time series with the starting date of the 2007 crisis in order to check the robustness of our FTSs.

In order to evaluate the performances of each of such FTSs, we use the following indicators:

- "$\overline{g}$": the annualized net average logarithmic rate of return obtained by the FTS;
- "$\uparrow$": the percentage of times in which the net equity line performed by the FTS, C_t, with $t = N + 1, \ldots, T$, is greater than or equal to the invested starting capital C_0, where $C_t = C_{t-1}(1 + g_t)$. "$\uparrow$" can be interpreted as a measure of starting capital preservation;

Table 1 Percentages associated to the assessment measure "$\overline{g} \geq 0\%$"

		A2A (%)	AG (%)	B (%)	MS (%)	S (%) (%)	TI (%)	UC (%)	UG (%)	All (%)
SR	S	77.34	0.00	96.09	21.09	17.19	26.56	92.97	100.00	53.91
	F	22.66	100.00	3.91	78.91	82.81	73.44	7.03	0.00	46.09
ALLR	S	42.19	1.56	97.66	36.72	69.53	9.38	82.81	100.00	54.98
	F	57.81	98.44	2.34	63.28	30.47	90.63	17.19	0.00	45.02
OVER	S	0.00	100.00	0.00	100.00	100.00	100.00	100.00	100.00	75.00
	F	100.00	0.00	100.00	0.00	0.00	0.00	0.00	0.00	25.00
$L = 5$	S	39.06	33.85	64.58	53.65	57.81	47.92	93.23	100.00	61.26
	F	60.94	66.15	35.42	46.35	42.19	52.08	6.77	0.00	38.74
$L = 10$	S	40.63	33.85	64.58	51.56	66.67	42.71	90.63	100.00	61.33
	F	59.38	66.15	35.42	48.44	33.33	57.29	9.38	0.00	38.67
$N = 2$	S	36.98	33.85	63.54	49.48	59.90	44.27	94.27	100.00	60.29
	F	63.02	66.15	36.46	50.52	40.10	55.73	5.73	0.00	39.71
$N = 4$	S	42.71	33.85	65.63	55.73	64.58	46.35	89.58	100.00	62.30
	F	57.29	66.15	34.38	44.27	35.42	53.65	10.42	0.00	37.70
$\epsilon = 5\%$	S	50.52	34.38	66.67	51.56	54.17	53.65	88.54	100.00	62.43
	F	49.48	65.63	33.33	48.44	45.83	46.35	11.46	0.00	37.57
$\epsilon = 25\%$	S	29.17	33.33	62.50	53.65	70.31	36.98	95.31	100.00	60.16
	F	70.83	66.67	37.50	46.35	29.69	63.02	4.69	0.00	39.84
(3)	S	42.19	33.33	64.58	54.17	61.46	42.19	93.75	100.00	61.46
	F	57.81	66.67	35.42	45.83	38.54	57.81	6.25	0.00	38.54
(4)	S	37.50	34.38	64.58	51.04	63.02	48.44	90.10	100.00	61.13
	F	62.50	65.63	35.42	48.96	36.98	51.56	9.90	0.00	38.87
Global	S	39.84	33.85	64.58	52.60	62.24	45.31	91.93	100.00	61.30
	F	60.16	66.15	35.42	47.40	37.76	54.69	8.07	0.00	38.70

- "$\nearrow$": the percentages of time in which $C(t + N + 1 + 21) \geq C(t + N + 1)$, with $t = 0, 21, \ldots, \left(\left\lfloor \frac{T-N-1}{21} \right\rfloor - 1\right) 21, \left\lfloor \frac{T-N-1}{21} \right\rfloor 21$. "$\nearrow$" can be interpreted as a measure of non-decreaseness of C_t detected with monthly frequency.[6]

Given the huge amount of results, we synthesize them in Tables from 1, 2 and 3. In particular, with reference to all the investigated FTSs:

- In Table 1, we provide the percentages of success (rows labelled "S") and of failure (rows labelled "F") of the assessment measure "$\overline{g} \geq 0\%$". These percentages are calculated conditionally with respect to the local reward functions, to the values of N, L and ϵ, and to the aggregation rules of the produced actions (see the corresponding rows). Finally, these percentages are calculated unconditionally and overall (rows labelled "Global").
- In Table 2, we present the percentages of success and of failure of the assessment measure "$\uparrow \geq 50\%$". These percentages are calculated as in Table 1.
- In Table 3, we give the percentages of success and of failure of the assessment measure "$\nearrow \geq 50\%$". These percentages are calculated as in Tables 1 and 2.

[6] In this context, "annualized" and "monthly" have to be meant as referring to the stock market year and to the stock market month, respectively.

Table 2 Percentages associated to the assessment measure "↑≥ 50%"

		A2A (%)	AG (%)	B (%)	MS (%)	S (%)	TI (%)	UC (%)	UG (%)	All (%)
SR	S	85.16	0.00	92.97	97.66	33.59	26.56	95.31	100.00	66.41
	F	14.84	100.00	7.03	2.34	66.41	73.44	4.69	0.00	33.59
ALLR	S	80.47	5.47	89.06	98.44	88.28	11.72	86.72	100.00	70.02
	F	19.53	94.53	10.94	1.56	11.72	88.28	13.28	0.00	29.98
OVER	S	100.00	100.00	0.00	100.00	3.13	100.00	100.00	100.00	75.39
	F	0.00	0.00	100.00	0.00	96.88	0.00	0.00	0.00	24.61
$L = 5$	S	90.10	34.38	60.94	99.48	30.73	46.35	94.27	100.00	69.53
	F	9.90	65.63	39.06	0.52	69.27	53.65	5.73	0.00	30.47
$L = 10$	S	86.98	35.94	60.42	97.92	52.60	45.83	93.75	100.00	71.68
	F	13.02	64.06	39.58	2.08	47.40	54.17	6.25	0.00	28.32
$N = 2$	S	85.42	34.38	58.85	99.48	41.67	45.31	96.88	100.00	70.25
	F	14.58	65.63	41.15	0.52	58.33	54.69	3.13	0.00	29.75
$N = 4$	S	91.67	35.94	62.50	97.92	41.67	46.88	91.15	100.00	70.96
	F	8.33	64.06	37.50	2.08	58.33	53.13	8.85	0.00	29.04
$\epsilon = 5\%$	S	91.67	36.98	63.02	100.00	32.81	55.73	90.10	100.00	71.29
	F	8.33	63.02	36.98	0.00	67.19	44.27	9.90	0.00	28.71
$\epsilon = 25\%$	S	85.42	33.33	58.33	97.40	50.52	36.46	97.92	100.00	69.92
	F	14.58	66.67	41.67	2.60	49.48	63.54	2.08	0.00	30.08
(3)	S	87.50	34.90	60.42	98.44	41.67	43.75	95.31	100.00	70.25
	F	12.50	65.10	39.58	1.56	58.33	56.25	4.69	0.00	29.75
(4)	S	89.58	35.42	60.94	98.96	41.67	48.44	92.71	100.00	70.96
	F	10.42	64.58	39.06	1.04	58.33	51.56	7.29	0.00	29.04
Global	S	88.54	35.16	60.68	98.70	41.67	46.09	94.01	100.00	70.61
	F	11.46	64.84	39.32	1.30	58.33	53.91	5.99	0.00	29.39

The main evidences which are detectable from Tables 1 to 3 are the following ones:

- Generally, the performances of our RL-based FTSs are enough satisfactory. In fact, with reference to the assessment measure "$\overline{g} \geq 0\%$", the percentages of configurations for which the annualized net average logarithmic rate of return is not negative is equal to 61.30% (see Table 1). The stocks that contribute most to this result[7] are B, S, UC and UG (see again Table 1). Then, with reference to the assessment measure "↑≥ 50%", the percentages of cases for which the net cumulative profit line C_t, with $t = N + 1, \ldots, T$, is greater than the invested starting capital C_0 is equal to 70.61% (see Table 2). In order to critically evaluate this results, we highlight that whenever $a_t = 0$, with $t = 1, \ldots, T - 1$, we get that $C_{t+1} = C_t$, and consequently "↑≥ 50%" increases. The stocks that contribute most to this result are A2A, B, MS, UC and UG (see again Table 2). Finally, with reference to the assessment measure "↗≥ 50%", the percentages of cases for which $C(t + N + 1 + 21) \geq C(t + N + 1)$, with $t = 0,\ 21,\ \ldots,\ \left(\left\lfloor \frac{T-N-1}{21} \right\rfloor - 1\right) 21,\ \left\lfloor \frac{T-N-1}{21} \right\rfloor 21$, is equal to 46.00% (see Table 3). This shows that

[7]From here on in, by the expression ≪[...] *stocks that contribute most to this result* [...]≫, or equivalent, we mean stocks whose percentages of succes are greater than or equal to 60%.

Table 3 Percentages associated to the assessment measure "$\nearrow\geq$ 50%"

		A2A (%)	AG (%)	B (%)	MS (%)	S (%)	TI (%)	UC (%)	UG (%)	All (%)
SR	S	29.69	7.81	96.09	2.34	35.16	40.63	11.72	85.94	38.67
	F	70.31	92.19	3.91	97.66	64.84	59.38	88.28	14.06	61.33
ALLR	S	33.59	18.75	84.38	31.25	53.13	30.47	39.84	78.13	46.19
	F	66.41	81.25	15.63	68.75	46.88	69.53	60.16	21.88	53.81
OVER	S	0.00	50.00	0.00	78.13	21.09	100.00	67.19	72.66	48.63
	F	100.00	50.00	100.00	21.88	78.91	0.00	32.81	27.34	51.37
$L = 5$	S	22.40	22.92	58.85	34.90	30.21	59.38	43.75	79.17	43.95
	F	77.60	77.08	41.15	65.10	69.79	40.63	56.25	20.83	56.05
$L = 10$	S	19.79	28.13	61.46	39.58	42.71	54.69	35.42	78.65	45.05
	F	80.21	71.88	38.54	60.42	57.29	45.31	64.58	21.35	54.95
$N = 2$	S	22.92	37.50	64.06	44.27	35.94	52.60	46.88	68.23	46.55
	F	77.08	62.50	35.94	55.73	64.06	47.40	53.13	31.77	53.45
$N = 4$	S	19.27	13.54	56.25	30.21	36.98	61.46	32.29	89.58	42.45
	F	80.73	86.46	43.75	69.79	63.02	38.54	67.71	10.42	57.55
$\epsilon = 5\%$	S	40.10	34.38	66.67	43.23	54.69	74.48	48.96	86.46	56.12
	F	59.90	65.63	33.33	56.77	45.31	25.52	51.04	13.54	43.88
$\epsilon = 25\%$	S	2.08	16.67	53.65	31.25	18.23	39.58	30.21	71.35	32.88
	F	97.92	83.33	46.35	68.75	81.77	60.42	69.79	28.65	67.12
(3)	S	63.54	65.63	79.17	73.44	75.00	83.85	73.96	89.58	75.52
	F	36.46	34.38	20.83	26.56	25.00	16.15	26.04	10.42	24.48
(4)	S	15.63	19.27	61.98	28.65	27.08	48.44	32.29	79.17	39.06
	F	84.38	80.73	38.02	71.35	72.92	51.56	67.71	20.83	60.94
Global	S	23.27	27.51	61.38	38.86	38.17	58.10	41.18	79.55	46.00
	F	76.73	72.49	38.62	61.14	61.83	41.90	58.82	20.45	54.00

our FTSs are endowed with no particular capabilities to generate non-decreasing net cumulative profit lines, despite the compelling values taken by the other two assessment measures;

- Also from the standpoint of the local reward functions and of the aggregation rules of the produced actions, the performances of our RL-based FTSs are enough satisfactory. In fact, with reference to the three assessment measures, their (conditional) percentages of success are very often greater than 50% (see Tables 1, 2 and 3). In particular, with regard to the local reward functions, OVER appears preferable with respect to the other two ones. We recall that, to the best of our knowledge, in this paper we propose the first application of OVER to QL-based FTSs. Furthermore, it is noteworthy to highlight that generally OVER shows behaviors meaningfully different from those shown by the other local reward functions. Finally, with regard to the aggregation rule of the produced actions, the fact that the aggregation rule (3) has percentages of success greater than those of the aggregation rule (2) only in one of the three assessment measures, makes the aggregation rule (3) slightly preferable to use;
- Concerning the setting of L, N and ϵ to take into account, the (conditional) performances of the QL-based FTSs coming from the use of different values of these

parameters are more or less equivalent. It indicates that, generally, the investment opportunities which are present in the investigated of financial markets are quickly exploited by our FTSs. Furthermore, our FTSs also show to be able to exploit such opportunities starting from basic state variables like the (logarithmic) rate of returns and the current action.

6 Some Final Remarks

In this paper, first we have designed and developed some original FTSs based on the machine learning technique QL; then we have applied these FTSs to eight time series of daily closing stock returns from the Italian stock market; finally we have presented the generally satisfactory results coming from such applications. Many questions remain to be explored. Among the main ones:

- The choice of the (logarithmic) rate of returns and of the current action as state variables has been deliberately simple. Once checked the capability of our FTSs to satisfactorily perform also starting from basic information, now we are starting to work for specifying more refined state variables;
- SR, ALRR and OVER have performed quite satisfactory as local reward functions. Nevertheless, all three suffer some financial limitation which make them incapable to appropriately measure the performance-risk profiles of advanced FTSs (like ours) when applied to the complexity of real financial markets. Therefore, currently we are starting to consider new and not standard local reward functions;
- Finally, in order to deepen the assessment about the capabilities of our FTSs, we have to apply them to more and more stocks coming from different financial markets.

References

1. Barto, A.G., Sutton, R.S.: Reinforcement Learning: An Introduction, 2nd edn. The MIT Press (2018)
2. Bekiros, S.D.: Heterogeneous trading strategies with adaptive fuzzy actor-critic reinforcement learning: a behavioral approach. J. Econ. Dyn. Control. **34**(6), 1153–1170 (2010)
3. Bertsekas, D.P., Tsitsiklis, J.N.: Neuro-Dynamic Programming. Athena Scientific (1996)
4. Brent, R.P.: Algorithms for Minimization Without Derivatives. Prentice-Hall (1973)
5. Casqueiro, P.X., Rodrigues, A.J.L.: Neuro-dynamic trading methods. Eur. J. Oper. Res. **175**(3), 1400–1412 (2006)
6. Deng, Y., Bao, F., Kong, Y., Ren, Z., Dai, Q.: Deep Direct Reinforcement Learning for financial signal representation and trading. IEEE Trans. Neural Netw. Learn. Syst. **28**(3), 653–664 (2017)
7. Gosavi, A.: Simulation-Based Optimization. Parametric Optimization Techniques and Reinforcement Learning. Springer, (2015)
8. Jangmin, O., Lee, J., Lee, J.W., Zhang, B.-T.: Adaptive stock trading with dynamic asset allocation using reinforcemnt learning. Inform. Sci. **176**(15), 2121–2147 (2006)

9. Kearns, M., Nevmyvaka, Y.: Machine learning for market microstructure and high frequency trading. In: Easley, D., López de Prado, M., O'Hara, M. (eds.) High-Frequency Trading—New Realities for Traders, Markets and Regulators, pp. 91–124. Risk Books (2013)
10. Li, H., Dagli, C.H., Enke, D.: Short-term stock market timing prediction under reinforcement learning schemes. In: Proceedings of the 2007 IEEE Symposium on Approximate Dynamic Programming and Reinforcement Learning, pp. 233–240 (2007)
11. Moody, J., Saffel, M.: Learning to trade via direct reinforcement. IEEE Trans. Neural Netw. **12**(4), 875–889 (2001)
12. Tan, Z., Quek, C., Cheng, P.Y.K: Stock trading with cycles: a financial application of ANFIS and reinforcement learning. Expert. Syst. Appl. **38**(5), 4741–4755 (2011)

Financial Literacy and Generation Y: Relationships Between Instruction Level and Financial Choices

Iuliana Bitca, Andrea Ellero, and Paola Ferretti

Abstract The heuristics used by investors in their process of decision-making tend to find short cuts and simplified roads to complicated answers, often unintentionally forgetting to be rational in contrast with market efficiency assumptions. We conducted a survey on about 250 young people (18–27 years old) concerning their financial literacy and economic choices, given an education level which is predominantly very high (73% enrolled in a bachelor degree, 80% took part to at least some basic finance or economics courses). More precisely, the survey was designed to study the influence of financial-economic literacy on the flaws occurring in financial decisions of young people (the so called generation Y): biases, overconfidence, framing. The results of the survey give an insight into the behaviour of a new and educated generation in typical economic decision frameworks, which could be a useful tool for stakeholders. In fact, being aware of the psychological component of the financial decision is a key factor to better understand and manage risk.

Authors are listed in alphabetical order. All authors contributed equally to this work, they discussed the results and implications and commented on the manuscript at all stages.

I. Bitca
Transaction Advisory Services, Ernst & Young GmbH, Frankfurt, Germany
e-mail: iuliana.bitca@gmail.com

A. Ellero (✉)
Department of Management, Ca' Foscari University of Venice, Venice, Italy
e-mail: ellero@unive.it

P. Ferretti (✉)
Department of Economics, Ca' Foscari University of Venice, Venice, Italy
e-mail: ferretti@unive.it

A. Esposito et al. (eds.), *Progresses in Artificial Intelligence and Neural Systems*, Smart Innovation, Systems and Technologies 184,
https://doi.org/10.1007/978-981-15-5093-5_32

1 Introduction

Human behaviour is complicated and irrational sometimes. The assumption that all investors act in a rational way and maximise their utility becomes partially untrue considering the psychological part of the decision making process. Behavioural finance incorporates psychological knowledge into classical finance; the behaviour of *Homo Economicus* is thus revisited from a perspective that is more usual in social sciences.

The heuristics of the modern investor need to be studied carefully. A heuristic is a simple and general rule we employ to solve a specific category of problems, including situations involving a high degree of risk-taking and uncertainty [8]. In the process of decision-making, investors tend to find short cuts and simplified roads toward the more complicated answer, often ignoring some overt facts in front of them. This occurs in cases when there is too much information and not enough time to process it. This investor behaviour is in direct contrast with the theory of market efficiency, which implies that all investors are rational. Another common heuristic is being overconfident with past experiences and taking decisions based on past cases, ignoring new information that makes the situation more complex, which is risky for investors because it generalizes one's belief to similar products, which are different in many ways. The question is not to blame investors for poorly managing the risk of the securities, but to study to which extent these biases can influence one's decision, and if the influence is significant, how to prevent it from happening. Since risk is present in almost all financial decisions, the study of biases and psychology of risk are crucial. Heuristics can indeed be a problem in modern finance, influencing in different ways investors around the world, sometimes independently of the years of experience they gained. After a brief recall of the notions that are more characteristic in behavioural finance (Sect. 2), we present (Sect. 3) the survey conducted on Generation Y on their financial literacy and behaviour in risky investment decisions. We recall then (Sect. 4) the main ideas of Dominance-based Rough Set Approach (DRSA), a multicriteria methodology. DRSA is then used (Sect. 5) to detect ties between financial literacy and investment decisions making of the respondents.

2 Behavioural Finance Studies

As a discipline having individual choices as one of its fundamental principal, finance surprisingly pays little attention to the individual. In the financial industry, a human decision can influence one's portfolio, but taking into consideration the sum of all these decisions, you get the performance of an entire market [10]. Humans prove to behave in a different way, which was the objective of the Prospect Theory: to study if the rational assumptions of Expected Utility Theory hold. In fact, Prospect Theory actually proves that people tend to value the losses differently than the same amount (in absolute numbers) of gains when they evaluate risky situations. Especially, people tend to accept a zero loss (which is zero gain too) over a certain amount of gain that contains a small probability of loss. This asymmetry of weights was subjected to several financial and psychological experiments that proved the statement above [6]. Individuals consider a new investment as separated from their existing wealth, which means that the cumulative utility assumption of Expected Utility Theory does not hold. Mental accounting is a relatively new concept in consumer behaviour, which contends that individuals separate the current and future assets in different accounts, offering each different utility levels [11]. One of the fundamental differences between these theories, besides the rationality or irrationality assumption is that prospect theory is a theory of average behaviour [1].

Overconfidence is one of the most studied biases in behavioural finance. It's the most common and obvious so far. Usually to measure overconfidence, researchers use the miscalibration test, consisting in a true/false question followed by "How sure are you that your answer is correct?" For example, in a study conducted to understand if Decision Support System can help lowering the bias of investors [3], people were asked several basic finance questions. One of these questions was "If interest rates rise, the price of a bond will rise." This is false. Next question was "How sure are you that your answer is correct?". The study showed that 43% of participants were overconfident, meaning they thought they had more correct answers than it was in reality. This test can give an idea of how overconfident investors are and how this is going to influence their future decisions and risk perception.

A person's perception of risk is easily influenced by the way a question is posed, positive outcomes and negative ones are viewed differently by investors; people tend to answer positively in a frame that includes a thorough description of an appealing success [13]. The formatting bias is interconnected with the Prospect Theory: the correct answer to a fair question should not depend on how the question was phrased [12]. Historically, the average return on equity is much higher than the average return on federal bonds, or default-free debt. Ambiguity aversion consists in detesting unknown risk over the known risk [9]. People don't like when the probability of getting a return is even partially unknown; investor are risk averse, but equity returns minus bond returns, has been about 6% on average for the past century. The level of uncertainty is what makes an investment worth its return. There is not a unique way to calculate the risk of a portfolio, and behavioural finance considers that risk can be considered as subjective. Cavezzali et al. [4] suggest that risk taking is closely related to the level of financial literacy of investors, proving that people who possess some basic knowledge of finance are positively correlated with the risk taking behaviour while tending to take simplified decisions on diversification. The survey behind the paper was based on a questionnaire sent to 208 U.S. individuals, and once it settled that complex strategies of diversifications are not going to be used, it moves to the next question to be discussed: how financial literacy influences the risk taking behaviour. The study finds that there is a positive correlation between financial literacy and risk taking, for example people that are more skilled take higher risks, using all the opportunities the market is offering [4].

3 Description of the Survey

The survey is mainly inspired by Cavezzali et al. [4] and Bhandari et al. [3]. The questionnaire is reported in the Appendix and can be roughly divided into 3 parts. The first part aims at determining whether respondents have some familiarity with general economic and financial concepts, if they are overconfident about their expertise in this respect, and their tendency to take risks. For what concerns the latter aspect, two types of questions investigate if respondents are risk averse or not. The first one (Q5) is asking to advise a friend on forming a portfolio out of three types of securities, ordered from less risky to most risky. The choices consist of bonds, equity and flexible funds, and the respondent should assign a certain percentage to each choice forming a full 100% portfolio. The second question (Q6) invites the respondent to create a portfolio with the same available choices but for him. The differences enlighten whether people perceive differently the risk when they have to give an advice and when they consider an investment for themselves. Additionally, the order of the two questions were changed so that a part of the participants got the questionnaire where they first had to give an advice to a friend while for the second part of the group the first question about allocating the investment for themselves preceded the advise to their friend.

The second part of the survey is concentrated on a kind of framing effect; two versions of the questionnaire present two different graphs, originated from the same S&P500 data, the first version displays the graph of last month's returns while the second one shows last year's data. Participants are asked to say if they would advise a friend to invest in a portfolio whose return is displayed in a graph. The results should show how people are biased by the way information is presented to them; the graphs show the same investment opportunity differing only in the length of the considered time interval, but people choose to invest when they see a longer, apparently stable, sequence of prices and tend to refuse to invest when they see just a month of prices which appear to be less stable. We end this second part of the questionnaire asking participants to describe their level of risk acceptance (above average, on average, totally risk averse). The answer to this question allows a comparison with the portfolio choice described above while its position in the questionnaire does not allow the respondent to easily connect them, allowing for potentially conflicting answers.

The last part of the survey gathers some basic demographic information in order to explain more effectively the background of the participants, their gender, age, nationality and current education level, referring in particular to courses in finance or economics.

The questionnaire was administered through Google Forms to people contacted via a social network (Facebook) and it was accessible for a period of 10 days in June 2017. Data were collected anonymously and on a voluntary basis. The answers were not editable by the survey participants. More than 300 questionnaires were collected but their number was narrowed down to 271, removing partially empty surveys. Although participation in the survey was not restricted to a predefined age range, 93% of the target group was young people aged between 18 and 27, i.e. young adults belonging to the so-called Generation Y. Among participants, 66% are male and 34% are female; the education level is in average very high having 73% currently enrolled in a bachelor degree, 13% in a master degree and 1% with doctoral degree, while 14% has a high school education. The large majority of respondents (80%) took also part in a course on finance or economics, so we can state that a large part of the survey was conducted on people who are quite familiar with concepts like risk taking or financial portfolio creation (see Table 1).

Concerning the asset allocation choices of the participants, the respondents are requested to suggest an investment choice first to a (fictitious) friend and then to take a similar decision for themselves. The results interestingly show that people tend to be less cautious when they are not involved in risk taking: up to 29% flexible risky funds are chosen for the portfolio of a friend but only 26% tapping into their own wallet (see Table 2). This (slight) difference appears to have the opposite sign when the order of the two questions is reversed (see Table 3).

The choices appear therefore to be (slightly) biased by a framing effect. Moreover, when facing the second question, some of them rethink the answer given and become more prudent, indifferently if the portfolio they are forming is for friends or for themselves. In any case, the target population of the survey generally tends to allocate

Table 1 Demographics of participants

Gender		Age		Education		Econ/Fin courses	
Male	66%	18–27	93%	Ph.D.	1%	Yes	80%
Female	34%	28–36	6%	Master	13%	No	20%
		>36	1%	Bachelor	73%		
				High school	14%		

Table 2 Asset allocation decisions, a choice among Bonds, Equities and Flexible funds; Questionnaire 1

Questionnaire 1	Choose for a friend		Choose for yourself	
	Bonds	36%	Bonds	38%
	Equity	35%	Equity	36%
	Flexible funds	29%	Flexible funds	26%

Table 3 Asset allocation decisions, a choice among Bonds, Equities and Flexible funds; Questionnaire 2

Questionnaire 2	Choose for a friend		Choose for yourself	
	Bonds	35%	Bonds	38%
	Equity	35%	Equity	36%
	Flexible funds	30%	Flexible funds	26%

almost equally between the three options, with a slight preference for more secure assets as bonds and equities.

Overconfidence, makes investors assume that some gains obtained investing are due to their knowledge and not due to other factors, in other words it's a feeling of (illusory) superiority. Most of the studies conducted on this bias focused on people who were already working in the financial context and the bias was more evident in experienced workers rather than in beginners.

This survey asked, to Generation Y people, questions similar to those used in [3], (Q1, Q2, Q3). The complete results are reported in Table 4. The most important result obtained is the fact that overconfidence is almost absent (questions Q1.1, Q2.1, Q3.1). In 2 out of 3 questions concerning overconfidence, the percentage of people being sure of their answer is smaller than the percentage of people answering correctly. This reveals that even though people knew the correct answer, they were not confident enough to confirm it another time. Besides, only 49% of all the respondents think their knowledge of finance is above average, this figure being found much lower in other studies on overconfidence [2]. A direct suggestion of this result can be that individuals acquire overconfidence with experience and they do not possess it when they are young. Perhaps, it may be that if investors would be more informed, they would (try to) be less biased.

Another question studying the framing effect was question Q7. The respondent to the two versions of the survey, were required to answer to exactly the same question, namely if they would advise a friend to invest in a portfolio which has a trend displayed on a graph, though the graph was different for the two versions, reporting the plot of S&P500 prices during last year and, respectively, during last month. The results, reported in Table 5, confirm that also Generation Y students are in fact biased

Table 4 Financial literacy (percentages corresponding to correct answers to the first three questions are underlined) and overconfidence

Answer	Q1 (%)	Q2 (%)	Q3 (%)	Q4 (%)
True/Yes	39	<u>73</u>	35	49
False/No	<u>48</u>	15	<u>30</u>	51
Don't know	13	12	35	
Overconfidence: at least 80% sure to be correct	41	45	51	

Table 5 Framing effect

Do you suggest to invest?	With last <u>year</u> S&P data (113 respondents) (%)	With last <u>month</u> S&P data (158 respondents) (%)
Yes	70	37
No	11	52
Don't know	19	11

by the framing effect. Facing the yearly graph 70% of the respondents chose to advise their friend to invest, but when shown the monthly graph, which in fact is just a small part of the previous graph, individuals were more cautious and 52% advised not to invest in that portfolio. Performing a t-test, the null hypothesis that *there is no difference in the answers* is rejected at a 1% significance level for both answers *yes* and *no*. It is therefore possible to conclude that the framing effect can push the investor in excessive risk taking depending on the formulation of the question.

4 Ties Between Financial Literacy and Investment Decisions: Dominance-Based Rough Set Approach

By means of the Dominance-based Rough Set Approach (DRSA) [5], a multicriteria methodology, we will try to link financial literacy to (possibly biased) decisions in practical investment problems. DRSA allows detecting "decision rules" (*if* conditions *then* decisions) [7] which represent a method to detect patterns in a data set.

A decision rule can be seen as a sequence of n condition attributes $c_1, \ldots, c_n$, and m decision attributes $d_1, \ldots, d_m$, that can be written as $c_1, \ldots, c_n \Rightarrow d_1, \ldots, d_m$ or in short $\mathcal{C} \rightarrow \mathcal{D}$.

The strength $\sigma(\mathcal{C}, \mathcal{D})$ of the decision rule $\mathcal{C} \rightarrow \mathcal{D}$ is represented by the ratio between the support of the decision rule and the cardinality of the whole set $\mathcal{U}$ of questionnaires, that is

$$\sigma(\mathcal{C}, \mathcal{D}) = \frac{|\mathcal{C} \cap \mathcal{D}|}{|\mathcal{U}|}.$$

With reference to the same decision rule $\mathcal{C} \rightarrow \mathcal{D}$, the certainty factor $cer(\mathcal{C}, \mathcal{D})$ and the coverage factor $cov(\mathcal{C}, \mathcal{D})$ are defined as

$$cer(\mathcal{C}, \mathcal{D}) = \frac{|\mathcal{C} \cap \mathcal{D}|}{|\mathcal{C}|} \qquad cov(\mathcal{C}, \mathcal{D}) = \frac{|\mathcal{C} \cap \mathcal{D}|}{|\mathcal{D}|}.$$

In our analysis, we focus on two particular decision attributes, that is on the percentage of bond investment in the portfolio suggested by the respondent to a friend and on the percentage of bond investment in the portfolio chosen by the respondent for herself.

In Tables 6 and 7 we report the most significant (with respect to Support, Certainty and Coverage measures) rules obtained by applying DRSA to the survey database.

For example, Rules 2 and 4 suggest that if a respondent is risk neutral (i.e., the answer to Q8 is *Take average financial risks expecting to earn average returns*) and her education is of a high level, then she will invest at least 10% in bonds but if she has to suggest a strategy to a friend, this component of the portfolio will be higher (at least 20%). Rules 3 and 6 instead tell us that the respondents with no finance nor economics skills invest certainly less than 70% in bonds when suggesting a strategy to a friend, but less, i.e., no more than 60% if the choice concerns themselves.

Table 6 Behavioural rules: decision attribute = *% of bonds suggested to your FRIEND*

RULE	IF	THEN *% of bonds suggested to your FRIEND*	Support	Certainty	Coverage
1	Answer to Q1 correct **and** Education ≥ Master	≥30%	15	1	0.183
2	Risk neutral **and** Education ≥ Master	≥20%	14	1	0.09
3	Answer to Q1 correct **and** Fin/Econ studies: NO	≤70%	15	1	0.07

Table 7 Behavioural rules: decision attribute = *% of bonds YOU choose to buy*

RULE	IF	THEN *% of bonds YOU choose to buy*	Support	Certainty	Coverage
4	Risk neutral **and** Education ≥ Master	≥10%	14	1	0.08
5	Answer to Q1 not correct **and** answer to Q2 not correct **and** Fin/Econ studies: YES	≥10%	14	1	0.08
6	Answer to Q1 correct **and** Fin/Econ studies: NO	≤60%	15	1	0.09
7	Answers to Q2 **and** Q3 correct **and** Fin/Econ studies: NO	≤30%	6	1	0.12
8	Risk neutral **and** Education = High School **and** Fin/Econ studies: NO	≤ 30%	6	1	0.12

5 Conclusions

Behavioural Finance offers a wide range of studies that prove in different ways the negative effects of the biases on the decision making process. We studied the extent to which a young investor is being flawed by some of the main heuristics when taking a decision under uncertainty and risk.

One subject was overconfidence: our result is that Generation Y appears not to be overconfident. The percentage of people being sure that their answer is correct was smaller than the percentage of people answering correctly to that question. Probably, overconfidence is a quality that is developed through the years of working in the field, while the negative effects can be lowered by financial studies.

The second effect observed in the survey was a sort of framing effect, when the answer to a question depends on the way the question is formulated. Two types of questionnaires asked respondents to form a portfolio for a fictitious friend and for themselves, then reversing the order of the two questions: individuals were more careful and they exhibited risk aversion in the second question of this type, independently from the subject who should invest. The framing bias is truly dangerous as in finance not always the matter of an investment is clear. Having the right person that can (repeatedly) inform, in an unbiased way, is the key to have a good diversified portfolio with the right amount of risk that one can tolerate.

The results of the survey give an insight on how young people, mainly with a good knowledge in finance, face risky decisions in finance. A suggestion is to raise the financial literacy of new generations, also with regard to behavioural concepts, so as to create awareness of the negative effects it can cause.

6 Appendix: The Questionnaire

We report below the detailed questions and the possible answers in the survey. Raw data can be requested to the corresponding authors.

Q1 If choosing between an investment in government bonds and shares of private companies, one can say that bonds are riskier. (True, False, Don't know)

Q1.1 How sure are you that your previous answer is correct? (Less than 50, 50–80%, More than 80%)

Q2 The primary reason the annual report is important in finance is that it is used by investors when they form expectations about the firm's future earnings and dividends, and the riskiness of those cash flows. (True, False, Don't know)

Q2.1 How sure are you that your previous answer is correct? (Less than 50, 50–80%, More than 80%)

Q3 A leverage ratio bigger than 2.5 is a clear sign that the company is financially healthy and is a good investment opportunity. (True, False, Don't know)

Q3.1 How sure are you that your previous answer is correct? (Less than 50, 50–80%, More than 80%)

Q4 Do you think your knowledge of basic finance is above average? (Yes, No)

Q5 Imagine you have a friend who needs your opinion on investing his saved 10,000$ in a combination of these alternatives for the next 12 months. Historically, bonds are considered less risky investments, whereas all the investments in stocks (national and foreign) are riskier but on average they can guarantee higher earnings. Suggest your friend (in terms of percentages) how to construct the portfolio. You even can put 100% in one box, if you feel it appropriate. The total of the investments must add up to 100%. (Bond investment—steady but pretty low income for savings, Equity investment—shares with greater volatility than bonds but higher returns, Flexible funds—a mix of foreign stocks and bonds, a little riskier than equity but higher returns)

Q5-2nd type of survey Imagine you saved 10,000$. You have to invest in a combination of these alternatives for the next 12 months. Historically, bonds are considered less risky investments, whereas all the investments in stocks (national and foreign) are riskier but on average they can guarantee higher earnings. Construct your portfolio (in terms of percentages). You even can put 100% in one box, if you feel it appropriate. The total of the investments must add up to 100%.

Q6 Imagine you have to take the same decision but for yourself now. You have 10,000$ to invest in the same combination of alternatives for the next 12 months in a portfolio. The total of the investments must add up to 100%. (Bond investment—steady but pretty low income for savings, Equity investment—shares with greater volatility than bonds but higher returns, Flexible funds—a mix of foreign stocks and bonds, a little riskier than equity but higher returns)

Q6-2nd type of survey Now imagine you have to give an advise to your friend who also has 10,000$ saved and you have to help him construct his portfolio for the next 12 months with the same alternatives. The total of the investments must add up to 100%.

Q7 Imagine you have a fried, who is 25 years old, does not have any debt and just finished his MBA. He will inherit shortly 100,000$ from his deceased aunt. He is asking you an advice on forming a portfolio in which to invest the inheritance. Based on the graph below, would you recommend your friend to invest in a portfolio which has this trend? (Yes, No, Don't know)[1]

Q8 Which of the following statements comes closest to the amount of financial risk that you would be willing to take when you will save or make investments? (Take above average financial risks expecting to earn above average returns, Take average financial risks expecting to earn average returns, Not willing to take any financial risks)

Q10 Could you indicate your gender? (Female, Male)

Q11 Could you indicate your age? (18–27, 28–36, More than 36)

Q12 Which is your nationality? (open answer)

Q13 Which is your level of education (past or currently enrolled)? (High School, Bachelor Degree, Master Degree, Ph.D.)

Q14 Have you studied Finance or Economics courses in bachelor or in a higher degree? (Yes, No)

[1] In the first type of survey, the graph represents one year S&P data, which appears to be rather smooth and increasing on average. In the second type of survey, starting from the same S&P series, the graph only shows data concerning the last month, which seem rather oscillating, even if increasing on average.

References

1. Altman, M.: Prospect Theory and Behavioral Finance. In: Baker, H.K., Nofsinger, J.R. (eds.) Behavioral Finance Investors, Corporations, and Markets, pp. 191–209. Wiley, Hoboken, New Jersey (2010)
2. Bhandari, G., Deaves, R.: The demographics of overconfidence. J. Behav. Financ. **7**(1), 5–11 (2006)
3. Bhandari, G., Deaves, R., Hassanein, K.: Debiasing investors with decision support systems: an experimental investigation. Decis. Support. Syst. **46**(1), 399–410 (2008)
4. Cavezzali, E., Gardenal, G., Rigoni, U.: Risk taking behaviour and diversification strategies: do financial literacy and financial education play a role? J. Financ. Manag. Mark. Inst. **3**, 121–156 (2015)
5. Greco, S., Matarazzo, B., Słowiński, R.: Multicriteria classification by dominance-based rough set approach. In: Kloesgen, W., Zytkow, J. (eds.) Handbook of Data Mining and Knowledge Discovery. Oxford University Press, New York (2002)
6. Kahneman, D., Tversky, A.: Prospect theory: an analysis of decision under risk. Econometrica **47**(2), 263–291 (1979)
7. Pawlak, Z.: Rough sets. Int. J. Comput. Inf. Sci. **11**(5), 341–356 (1982)
8. Ricciardi, V.: A Risk Perception Primer: A Narrative Research Review of the Risk Perception Literature in Behavioral Accounting and Behavioral Finance (July 20, 2004). Available at SSRN https://doi.org/10.2139/ssrn.566802
9. Rieger, M.O., Wang, M.: Can ambiguity aversion solve the equity premium puzzle? Survey evidence from international data. Financ. Res. Lett. **9**(2), 63–72 (2012)
10. Skubic, T., McGoun, E.G.: Individuals, images, and investments. In: Frankfurter, G.M., McGoun, E.G. (eds.) From Individualism to the Individual: Ideology and Inquiry in Financial Economics. Ashgate Aldershot, Hants, England (2002)
11. Thaler, R.: Mental accounting matters. J. Behav. Decis. Mak. **12**, 183–206 (1999)
12. Tversky, A., Kahneman, D.: Choices, values and frames. Am. Psychol. **39**(4), 341–350 (1984)
13. Weber, A.L.: Introduction to Psychology. Harper Collins, New York (1991)

Advanced Smart Multimodal Data Processing—Dedicated to Alfredo Petrosino

A CNN Approach for Audio Classification in Construction Sites

Alessandro Maccagno, Andrea Mastropietro, Umberto Mazziotta, Michele Scarpiniti, Yong-Cheol Lee, and Aurelio Uncini

Abstract Convolutional Neural Networks (CNNs) have been widely used in the field of audio recognition and classification, since they often provide positive results. Motivated by the success of this kind of approach and the lack of practical methodologies for the monitoring of construction sites by using audio data, we developed an application for the classification of different types and brands of construction vehicles and tools, which operates on the emitted audio through a stack of convolutional layers. The proposed architecture works on the mel-spectrogram representation of the input audio frames and it demonstrates its effectiveness in environmental sound classification (ESC) achieving a high accuracy. In summary, our contribution shows that techniques employed for general ESC can be also successfully adapted to a more specific environmental sound classification task, such as event recognition in construction sites.

A. Maccagno · A. Mastropietro · U. Mazziotta
Department of Computer, Control and Management Engineering,
Sapienza University of Rome, Rome, Italy
e-mail: maccagno.1653200@studenti.uniroma1.it

A. Mastropietro
e-mail: mastropietro.1652886@studenti.uniroma1.it

U. Mazziotta
e-mail: mazziotta.1647818@studenti.uniroma1.it

M. Scarpiniti (✉) · A. Uncini
Department of Information Engineering, Electronics and Telecommunications,
Sapienza University of Rome, Rome, Italy
e-mail: michele.scarpiniti@uniroma1.it

A. Uncini
e-mail: aurelio.uncini@uniroma1.it

Y.-C. Lee
Department of Construction Management, Louisiana State University, Baton Rouge, USA
e-mail: yclee@lsu.edu

A. Esposito et al. (eds.), *Progresses in Artificial Intelligence and Neural Systems*,
Smart Innovation, Systems and Technologies 184,
https://doi.org/10.1007/978-981-15-5093-5_33

1 Introduction

In last years, many research efforts have been made towards the event classification of audio data, due to the availability of cheap sensors [1]. In fact, systems based on acoustic sensors are of particular interest for their flexibility and cheapness [2]. When we consider generic outdoor scenarios, an automatic monitoring system based on a microphone array would be an invaluable tool in assessing and controlling any type of situation occurring in the environment [3]. This includes, but is not limited to, handling large civil and/or military events. The idea in these works is to use Computational Auditory Scene Analysis (CASA) [4], which involves Computational Intelligence and Machine Learning techniques, to recognize the presence of specific objects into sound tracks. This last problem is a notable example of Automatic Audio Classification (AAC) [5], the task of automatically labeling a given audio signal in a set of predefined classes.

Getting into the more specific field of environmental sound classification (ESC) in construction site, the closest attempts have been performed by Cheng et al. [6], who used Support Vector Machines (SVM) to analyze the activity of construction tools and equipment. Recent applications of AAC have also been addressed to audio-based construction sites monitoring [7–9], in order to improve the construction process management of field activities. This approach is revealing itself as a promising method and a supportive resource for unmanned field monitoring and safety surveillance that leverages construction project management and decision making [8, 9]. More recently, several studies extended these efforts to more complicated architectures exploiting Deep Learning techniques [10].

In the literature, it is possible to find several instances of successful applications in the field of environmental sound classification that make use of deep learning. For example, in the work of Piczak [11], the author exploits a 2-layered CNN working on the spectrogram of the data to perform ESC, reaching an average accuracy of 70% over different datasets. Other approaches, instead of using handcrafted features such as the spectrogram, perform end-to-end environmental sound classification obtaining higher results with respect to the previous ones [12, 13].

Inspired and motivated by the MelNet architecture described by Li et al. [14], which has been proven to be remarkably effective in environmental sound classification, the aim of this paper is to develop an application able to recognize vehicles and tools used in construction sites, and classify them in terms of type and brand. This task will be tackled with a neural network approach, involving the use of a Deep Convolutional Neural Network (DCNN), which will be fed with the mel spectrogram of the audio source as input. The classification will be carried on five classes extracted from audio files collected in several construction sites, containing in situ recordings of multiple vehicles and tools. We demonstrate that the proposed approach for ESC can obtain good results (average accuracy of 97%) to a very specific domain as the one of construction sites.

The rest of this paper is organized as follows. Section 2 describes the proposed approach used to perform the sound classification. Section 3 introduces the experimental setup, while Sect. 4 shows the obtained numerical results. Finally, Sect. 5 concludes the paper and outlines some future directions.

2 The Proposed Approach

CNNs are a particular type of neural networks, which use the *convolution* operation in one or more layers for the learning process. These networks are inspired by the *primal visual system*, and are therefore extensively used with image and video inputs [10]. A CNN is composed by three main layers:

- **Convolutional layer**: The convolutional layer is the one tasked with applying the convolution operation on the input. This is done by passing a filter (or *kernel*) over the matricial input, computing the convolution value, and using the obtained result as the value of one cell of the output matrix (called *feature map*); the filter is then shifted by a predefined *stride* along its dimensions. The filters parameters are trained during the training process.
- **Detector layer**: In the detector layer, the output of the convolution is passed through a nonlinear function, usually a ReLU function.
- **Pooling layer**: The pooling layer is meant to reduce the dimensionality of data by combining the output of neuron clusters at one layer into one single neuron in the subsequent layer.

The last layer of the network is a fully connected one (a layer whose units are connected to every single unit from the previous one), which outputs the probability of the input to belong to each of the classes.

CNNs in a machine learning system show some advantages with respect to traditional fully connected neural networks, because they allow sparse interactions, parameters sharing and equivariant representations.

The reasons why we used CNNs in our approach is due to the intrinsic nature of audio signals. CNNs are extensively used with images and, since the spectrum of the audio is an actual picture of the signal, it is straightforward to see why CNNs are a good idea for such kind of input, being able to exploit the adjacency properties of audio signals and recognize patterns in the spectrum images that can properly represent each one of the classes taken into consideration.

The proposed architecture consists in a DCNN composed of eight layers, as shown in Fig. 1, that is fed with the mel spectrogram extracted from audio signals and its time derivative. Specifically, we have as input a tensor of dimension $60 \times 2 \times 2$ that is a couple of images representing the spectrogram and its time derivative: 60 is the number of mel bands, while 2 is the number of time buckets. Then, we have five convolutional layers, followed by a dense fully connected layer with 200 units and a final softmax layer that performs the classification over the 5 classes. The structure of the proposed network is summarized in the following Table 1, and it can be graphically appreciated in Fig. 1.

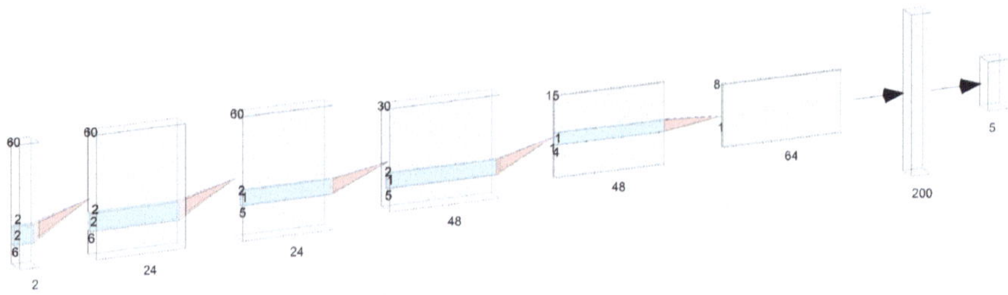

Fig. 1 Graphical representation of the proposed architecture

Table 1 Parameters of the proposed DCNN architecture

Layer	Input shape	Filters	Kernel size	Strides	Output shape
Conv1	[Batch, 60, 2, 2]	24	(6, 2)	(1, 1)	[Batch, 60, 2, 24]
Conv2	[Batch, 60, 2, 24]	24	(6, 2)	(1, 1)	[Batch, 60, 2, 24]
Conv3	[Batch, 60, 2, 24]	48	(5, 1)	(2, 2)	[Batch, 30, 1, 48]
Conv4	[Batch, 30, 1, 48]	48	(5, 1)	(2, 2)	[Batch, 15, 1, 48]
Conv5	[Batch, 15, 1, 48]	64	(4, 1)	(2, 2)	[Batch, 8, 1, 64]
Flatten	[Batch, 8, 1, 64]	–	–	–	[Batch, 512]
Dense	[Batch, 512]	–	–	–	[Batch, 200]
Output—Dense	[Batch, 200]	–	–	–	[Batch, 5]

All the layers employ a ReLu activation function except for the output layers which uses a Sofmax function. The optimizer chosen for the network is an Adam Optimizer [15], with the learning rate set to 0.0005. Such value was chosen by performing a grid search in the range [0.00001, 0.001]. Moreover, a dropout strategy, with a rate equal to 30%, has been used in the dense layer.

Regarding the setting of other hyper-parameters, different strategies were adopted. For the batch size, a grid search was used to determine the most appropriate values. The filter size and the stride were set reasonably according to the input size. Small filters were adopted such to capture small, local and adjacent features that are typical of audio data. Lastly, to prevent the network depth from either exploding in size, adding unnecessary complexity for no actual return, or not being high enough, therefore returning substandard results, we decided to use the same amount of layers as other related works, such as the one in [14], as a baseline. Variations on this depth have not shown appreciable improvements on the overall effectiveness of the network's classification, so it has been kept unchanged.

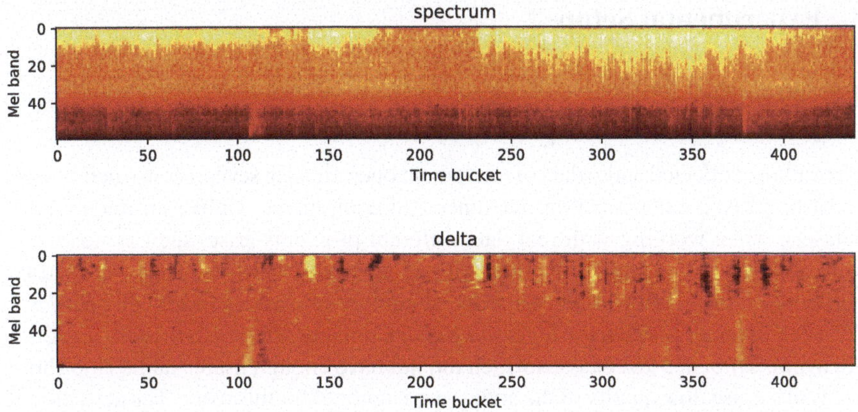

Fig. 2 Example of a log-mel spectrogram extracted from a fragment along with its derivative. On the abscissa we find the time buckets, each of which representing a sample about 23 ms long, while on the ordinates the log-mel bands. Since our fragments are 30 ms long, the spectrogram we extract will contain 2 buckets

2.1 *Spectrogram Extraction*

The proposed DCNN uses, as its inputs, the mel spectrogram that is a version of the spectrogram where the frequency scale has been distorted in a perceptual way, and its time derivative.

The technique used to extract the spectrogram from the sample is the same used by Piczak [11], via the Python library `librosa`.[1] The frames were re-sampled to 22,050 Hz, then a window of size 1024 with hop-size of 512 and 60 mel bands has been used. A mel band represents an interval of frequencies which are perceived to have the same pitch by human listeners. They have been found to be performing in speech recognition.

With this parameters, and the chosen length of 30 ms for the frames (see next sections), we obtain a small spectrogram of 60 rows (bands) and 2 columns (buckets). Then, using again `librosa`, we compute the derivative of the spectrogram and we overlap the two matrices, obtaining a dual channel input which is fed into the network (Fig. 2).

[1] Available at: https://librosa.github.io/.

3 Experimental Setup

3.1 Dataset

The authors collected audio data of equipment operations in several construction sites consisting diverse construction machines and equipments. Unlike artificially built datasets, when working with real data different problems arise, such as noise due to weather conditions and/or workers talking among themselves. Thus, we focused our work on the classification of a reduced number of classes, specifically *Backhoe JD50D Compact*, *Compactor Ingersoll Rand*, *Concrete Mixer*, *Excavator Cat 320E*, *Excavator Hitachi 50U*. Classes which did not have enough usable audio (too short, excessive noise, low quality of the audio) were ignored for this work. The activities of these machines were observed during certain periods, and the audio signals generated were recorded accordingly. A Zoom H1 digital handy recorder has been used for data collection purposes. All files have been recorded by using a sample rate of 44,100 Hz and a total of about one hour of sound data (eight different files for each machine) has been used to train the architecture.

3.2 Data Preprocessing

In order to feed the network with enough and proper data, each audio file for each class is segmented into fixed length frames (the choice of the best frame size is described in the experiment section). As first step, we split the original audio files into two parts, training samples (70% of the original length) and test samples (30% of the original length). This is done to avoid testing the network on data used previously to train the network, as this would cause the network to overfit and give misleading results.

Then, we perform data augmentation by splitting the files into smaller segments of 30 ms, each of which overlaps the subsequent one by 15 ms. We then compute the Root Mean Square (RMS) of every signal of these frames, and drop the ones with too small power with respect to the average RMS of the different segments, in order to remove the frames which contain mostly silence.

After that, the dataset is balanced by taking N samples for each class, where N is the number of elements contained in the class with the least amount of samples. In this way, we avoid the problem of having certain classes with an abnormal number of usable audio segments being potentially either over-represented or under-represented and negatively impacting the training of the model, especially due to the presence of multiple models of the same vehicle.

Using the Python library `librosa`, we extract the waveform of the audio tracks from the audio samples and, using the same library, we generate the log-scaled mel spectrogram [16] of the signal and its time derivative that will be the input to the network.

Numerical results have been evaluated in terms of accuracy, recall, precision and F1 score [17].

4 Numerical Results

4.1 Selecting the Frame Size

A sizeable amount of time was spent into finding the proper length for the audio segments. This is of crucial importance since, if the length is not adequate, the network will not be able to learn proper features that clearly characterize the input. Hence, in order to select the most suitable length, we generated different dataset variants by splitting the audio using different frame lengths, and we subsequently trained and tested different models on the differently-sized datasets. The testing results in terms of overall accuracy are show in Fig. 3.

As we can see, with a smaller frame size better results are obtained, while we notice a drop as the size increases. It is also interesting to observe that with a very large frame size the accuracy tends to slightly improve again. However, the use of long frames does not lead to anything interesting since the network may tend to learn an ensemble of the signal that is not significant and useful to work in fast-response applications (hazard detection, activity monitoring, etc.). Finally, the optimal frame

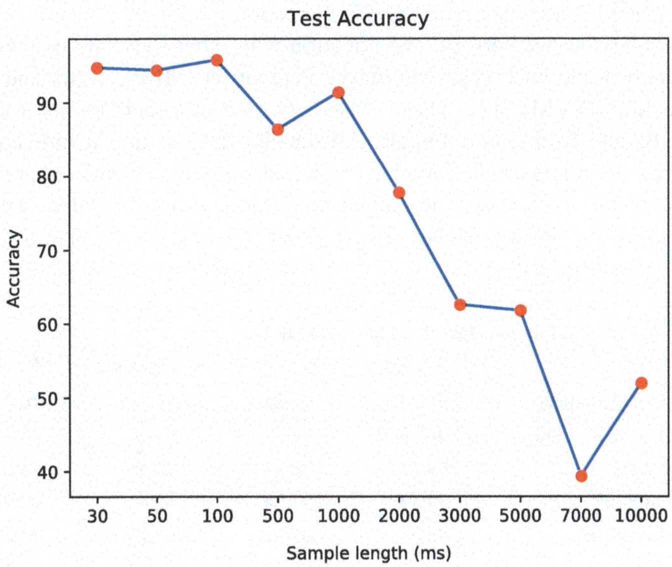

Fig. 3 Overall classification accuracy according to different sample sizes of the audio frames

size is obtained by selecting a duration of 30 ms, since it led not only to achieve a high accuracy but also a larger number of samples.

In order to properly test the network we performed a K-fold cross validation, with $K = 5$. The results of the classification are shown in the next subsection.

4.2 Classification Results

As just stated, a 5-fold cross validation was performed and the results are shown in Table 2. The dataset was split into training set and validation set (80–20%) for each fold.

As we can notice, the network achieves very high results in all the metrics, demonstrating its effectiveness in this particular domain. Even though our classes also include vehicles of the same type (we have two excavators and a backhoe, which is a kind of excavator as well), such classes are discriminated in a very clear and accurate way as the net recognizes also the brand of the machine.

After having performed the cross validation, we trained the network again on the original version of the dataset (training set 70% and test set 30%) and tested it. The way the network learns can be seen in Fig. 4; the learning is actually really fast as we see that high overall accuracy values are reached within few epochs and thus the convergence is rapid. The accuracy results obtained are shown in the confusion matrix in Fig. 5. From this figure, it is clear that all classes are well correctly recognized, since the accuracy is always higher than 95%. The class with worst result is the Excavator Cat 320E that performs at 95% of accuracy.

As a comparison, we perform classification with other five state-of-the-art classifiers, namely Random Forest, Multilayer Perceptron (MLP), k-NN and Support Vector Machine (SVM) [17]. These classifiers take into their inputs a set of 62 features extracted from audio signals. All details, features and parameters of the implemented classifiers can be found in [8]. Results of these considered approaches, averaged over the five classes, are shown in Table 3. From this table, we can see that the state-of-the-art approaches always produce worse results than those of the

Table 2 5-Fold cross validation classification results (in %)

Class	Accuracy	Recall	Precision	F1
Backhoe JD50D Compact	98.52	97.23	95.54	96.34
Compactor Ingersoll Rand	98.73	97.89	95.71	96.76
Concrete Mixer	99.21	98.49	97.58	98.03
Excavator Cat 320E	99.19	97.34	98.60	97.96
Excavator Hitachi 50U	98.99	97.82	97.16	97.49
All classes	97.08	97.34	97.30	97.32

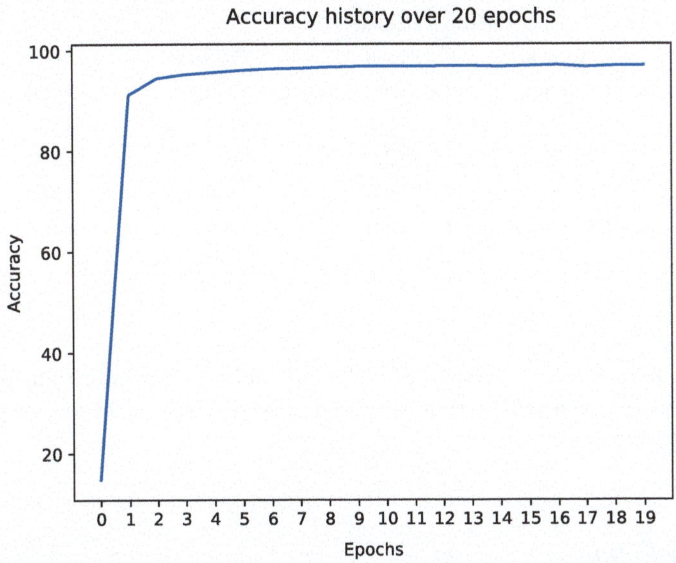

Fig. 4 Overall accuracy obtained on the test set

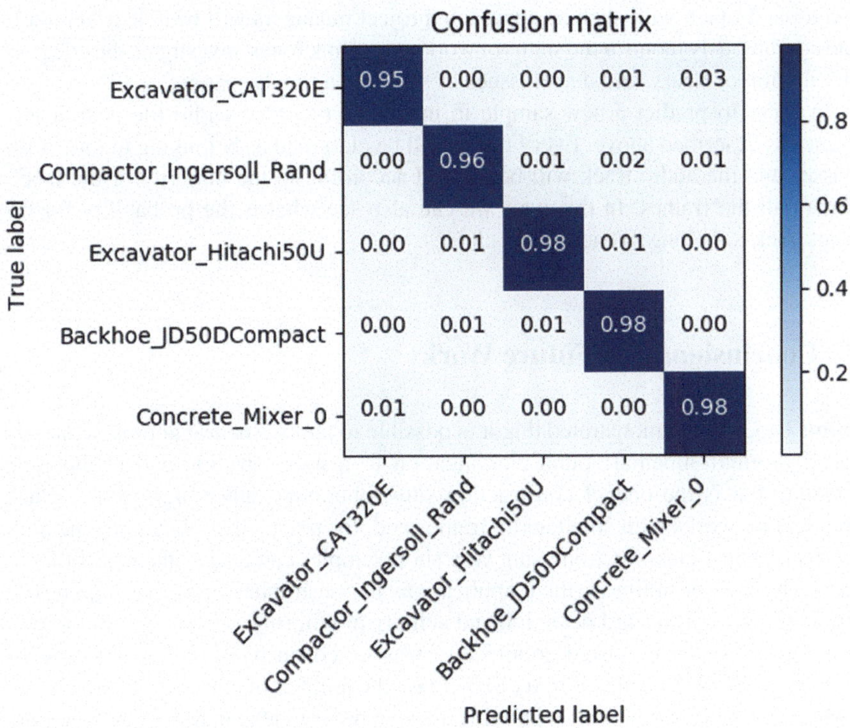

Fig. 5 Confusion matrix obtained by the proposed approach

Table 3 Averaged results of compared classifiers (in %)

Approach	Accuracy	Recall	Precision	F1
Random forest	93.16	93.21	93.40	93.30
MLP	91.06	93.20	91.34	92.26
k-NN	85.28	85.32	86.04	85.68
SVM	83.66	83.75	84.63	84.19
DCNN (proposed)	97.08	97.34	97.30	97.32

proposed architecture, shown in the last line of Table 3. This is due to the powerful feature representation and discrimination of the used DCNN and the mel spectrogram signal representation.

4.3 *Prediction*

The proposed approach can be used to promptly predict the active working vehicles and tools. In fact, with such an approach, project managers will be able to remotely and continuously monitor the status of workers and machines, investigate the effective distribution of hours, and detect issues of safety in a timely manner.

In order to predict a new sample in input, the recorded audio file is split into frames as described above. Every frame will be classified as belonging to one of the classes and the audio track will be labeled according to the majority of the labels among all the frames. In this way, we can also see what is the probability for the input track to belong to each of the classes.

5 Conclusions and Future Work

In this paper, we demonstrated that it is possible to apply a neural approach already tested in environmental sound classification to a more specific and challenging domain, that is the one of construction sites, obtaining rather high results. Such architecture works with small audio frames and, for practical applications, the ability to perform a classification using very short samples can lead to the possibility to use such network in time-critical applications in construction sites that require fast responses, such as hazard detection and activity monitoring.

Up to now, the proposed architecture was tested on five classes obtaining an accuracy of 97%. The idea is to try to increase the number of classes to include more tools and vehicles employed in building sites, in order to lead in the future to a more reliable and useful system. Moreover, the most interesting way to extend the work would be to try to combine more architectures in order to establish which kind of neural networks can help the audio classification in construction sites.

References

1. Scardapane, S., Scarpiniti, M., Bucciarelli, M., Colone, F., Mansueto, M.V., Parisi, R.: Microphone array based classification for security monitoring in unstructured environments. AEÜ Int. J. Electron. Commun. **69**(11), 1715–1723 (2015)
2. Weinstein, E., Steele, K., Agarwal, A., Glass, J.: LOUD: a 1020-node modular micro-phone array and beamformer for intelligent computing spaces. Technical Report MIT/LCS Technical Memo MIT-LCS-TM-642 (2004)
3. Kaushik, B., Nance, D., Ahuja, K.K.: A review of the role of acoustic sensors in the modern battlefield. In: Proceedings of the 11th AIAA/CEAS Aeroacoustics Conference, pp. 1–13 (2005)
4. Wang, D., Brown, G.J.: Computational Auditory Scene Analysis: Principles, Algorithms, and Applications. Wiley-IEEE Press (2006)
5. Fu, Z., Lu, G., Ting, K.M., Zhang, D.: A survey of audio-based music classification and annotation. IEEE Trans. Multimed. **13**(2), 303–319 (2011)
6. Cheng, C.-F., Rashidi, A., Davenport, M.A., Anderson, D.V.: Activity analysis of construction equipment using audio signals and support vector machines. Autom. Constr. **81**, 240–253 (2017)
7. Zhang, T., Lee, Y.-C., Scarpiniti, M., Uncini, A.: A supervised machine learning-based sound identification for construction activity monitoring and performance evaluation. In: Proceedings of 2018 Construction Research Congress (CRC 2018), New Orleans, Louisiana, USA, pp. 358–366, 2–4 April 2018
8. Lee, Y.-C., Scarpiniti, M., Uncini, A.: Advanced sound identification classifiers using a grid search algorithm for accurate audio-based construction progress monitoring. J. Comput. Civil Eng. (2020)
9. Sherafat, B., Rashidi, A., Lee, Y.-C., Ahn, C.R.: A hybrid kinematic-acoustic system for automated activity detection of construction equipment. Sensors **19**(19), 4286 (2019)
10. Goodfellow, I., Bengio, Y., Courville, A.: Deep Learning. MIT Press (2016)
11. Piczak, K.J.: Environmental sound classification with convolutional neural networks. In: 2015 IEEE 25th International Workshop on Machine Learning for Signal Processing (MLSP), pp. 1–6, Sept. 2015
12. Tokozume, Y., Harada, T.: Learning environmental sounds with end-to-end convolutional neural network. In: 2017 IEEE International Conference on Acoustics, Speech and Signal Processing (ICASSP), pp. 2721–2725, March 2017
13. Xie, Y., Lee, Y.-C., Scarpiniti, M.: Deep Learning-Based Highway Construction and Maintenance Activities Monitoring in Night Time. Construction Research Congress (CRC 2020), Tempe, AZ, USA, 8–10 March 2020
14. Li, S., Yao, Y., Hu, J., Liu, G., Yao, X., Hu, J.: An ensemble stacked convolutional neural network model for environmental event sound recognition. Appl. Sci. **8**(7) (2018)
15. Kingma, D.P., Ba, J.: Adam: a method for stochastic optimization. In: ICLR (2015)
16. Stevens, S.S., Volkmann, J., Newman, E.B.: A scale for the measurement of the psychological magnitude pitch (1937)
17. Alpaydin, E.: Introduction to Machine Learning, 3rd edn. MIT Press (2014)

Fault Detection in a Blower by Machine Learning-Based Vibrational Analysis

Vincenzo Mariano Scarrica, Francesco Camastra, Gianluca Diodati, and Vincenzo Quaranta

Abstract This work presents a system for Fault Detection in a Blower by Machine Learning-based Vibrational Analysis. Fault Detection System is composed of two stages. The former carries out the wavelet decomposition of the vibration signal and represents the vibration signal by the projection onto the principal components retaining 99% of the available information. The latter performs the classification by a Linear Support Vection Machine. To validate the system an experimental laboratory, where it is possible to reproduce various faults, different in intensity and in type, has been properly built. Preliminary results, even obtained on a test of limited size, are quite encouraging.

1 Introduction

Fault Detection [1, 2] consists in monitoring a system, identifying when the fault occurs and providing the type of fault. Fault Detection is a relevant topic in the maintenance procedures of the aircrafts. Despite of its importance, no public domain datasets of faults and damages in the aircrafts are available. This represents a strong obstacle for the development and the experimental validation of fault detection systems in the aeronautical domain.

V. M. Scarrica · F. Camastra (✉)
Department of Science and Technology, Parthenope University of Naples,
Centro Direzionale Isola C4, 80143 Naples, Italy
e-mail: camastra@ieee.org

V. M. Scarrica
e-mail: vincenzo.scarrica@gmail.com

G. Diodati · V. Quaranta
CIRA, Italian Aerospace Research Centre, via Maiorise snc, 81043 Capua, Italy
e-mail: g.diodati@cira.it

V. Quaranta
e-mail: v.quaranta@cira.it

A. Esposito et al. (eds.), *Progresses in Artificial Intelligence and Neural Systems*,
Smart Innovation, Systems and Technologies 184,
https://doi.org/10.1007/978-981-15-5093-5_34

The aim of the paper is to develop a fault detection system for an engine (e.g., a blower) by vibrational analysis using machine learning methods. The system is composed of two parts. The former, after having performed the wavelet decomposition of vibration signals, carries out the feature extraction representing the fault by the projection onto the principal components retaining 99% of the available information. The latter performs the classification by a Linear Support Vector Machine. In order to validate the fault detection system an experimental laboratory, where it is possible to reproduce various faults, different in intensity and in type, has been properly constructed.

The paper is organized as follows: In Sect. 2 the experimental laboratory for the system validation is described; Sect. 3 describes the feature extraction process, performed on vibrational signal; Sect. 4 illustrates the classifier; Sect. 5 reports some experimental results; finally, in Sect. 6 some conclusions are drawn and a few lines of further research that will be investigated in the next future.

2 Experimental Laboratory

In order to create a real data fault database, an experimental laboratory, shown in Fig. 1, was constructed. The experimental database was created by injecting a signal representative of different type of damages in an electromechanical engine during

Fig. 1 Experimental laboratory

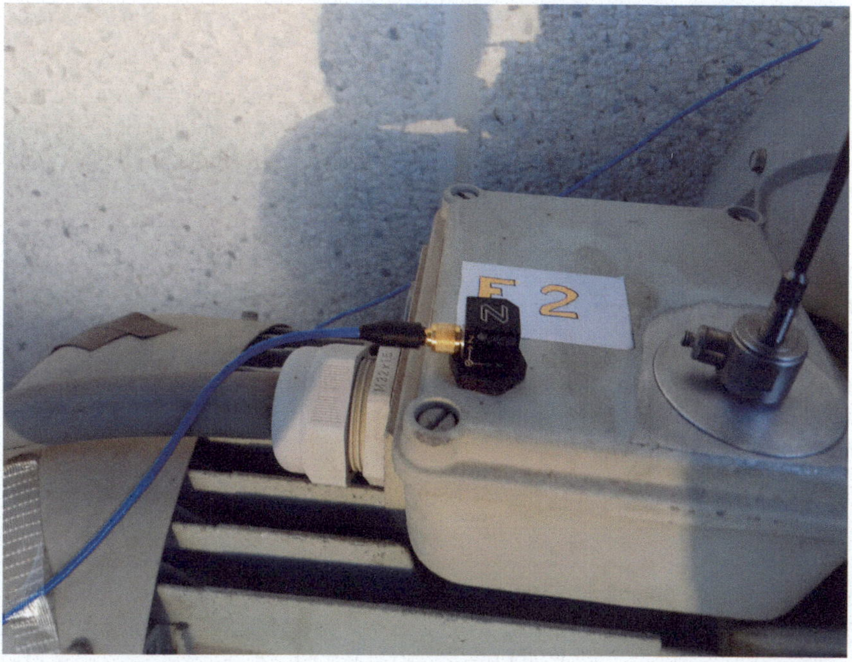

Fig. 2 Experimental laboratory: the accelerometer's location

its normal operation. The signal was injected through an electromechanical shaker suspended by a flexible link to a crane, to avoid the influence of the dynamic response of the supporting structure. The shaker was connected to the system by a flexible stinger attached to the case with structural glue (see Fig. 1). Two different faults were reproduced experimentally. The former is the *Defect of the Bearing Outer Ring* (DBOR), the latter is the *Defect of the Bearing Inner Ring* (DBIR). Both faults were simulated with different fault magnitudes.

Siemens/LMS Scadas III front-end, equipped with a generation module and multiple acquisition modules, is used to drive the shaker and acquire the accelerometer signals. Eleven tests were performed. In each test, seven replicas of 5 s window for each damage were fed to the shaker, with 1 s of zero signal to switch between each window. The vibration measurements were acquired by a tri-axial accelerometer PCB type 356A16, with a sensitivity of about 100 mV/g (Fig. 2). The Accelerometer E2 was placed on the input power case, i.e., where the electrodynamical shaker applies its force. An overall number of 3 acceleration signals have been acquired, i.e., the acceleration signals in the x, y and z directions from the sensor. Summing up, for both fault have been collected 77 signals for the sensor. Since the sensor yields three different signal, the overall signal number was $77 \times 3 = 231$. Each signal was sampled at 12.8 KHz in a temporal window of 2 s.

3 Feature Extraction

Each acceleration signal is undergone to the feature extraction process. Feature extraction is composed of three stages. The first stage is signal filtering, the second one consists in the computation of sixteen statistical features, in the last stage the dimensionality reduction is performed.

3.1 *Signal Filtering*

In the signal filtering phase, the *discrete wavelet transform* [3] is applied to each acceleration signal. To this purpose, we recall that given a function $f(x)$, can be decomposed as:

$$f(x) = \sum_{j=-\infty}^{\infty} \sum_{k=-\infty}^{\infty} c_{jk}\psi_{jk}(x) \tag{1}$$

where $\psi_{jk}(x)$ are obtained by translations and dilation of a *mother wavelet* $\psi(x)$:

$$\psi_{jk}(x) = 2^{j/2}\psi(2^j x - k) \tag{2}$$

As mother wavelet it has been chosen the *Daubechies 10*, where 10 denotes the number of vanishing moments [3]. To this purpose, we recall that the Daubechies mother wavelet cannot be expressed by a closed formula. Nevertheless, the values of the coefficients c_{jk} can be obtained by the computations of the roots of the following polynomial, given the number of the vanishing moment p:

$$\sum_{k=0}^{p-1} \binom{p-1+k}{k} y^k = 0 \tag{3}$$

After having applied to each accelerometer signal the discrete wavelet transform to the sixth level, each signal is decomposed in 64 components.

3.2 *Statistical Feature Computation*

Using Baraldi et al.'s approach [4], for each signal component $v = (v_1, \ldots, v_n)$ the following sixteen features have been computed:

- *maximum*,

$$\max_{1 \leq q \leq n} v_q;$$

– *minimum*,

$$\min_{1\le q\le n} v_q;$$

– *mean*,

$$\langle v\rangle = \frac{1}{n}\sum_{q=1}^{n} v_q;$$

– *standard deviation*,

$$\sigma_v = \sqrt{\frac{1}{n}\sum_{q=1}^{n}(v_q - \langle v\rangle)^2};$$

– *root mean square*,

$$v_{rms} = \sqrt{\frac{1}{n}\sum_{q=1}^{n} v_q^2};$$

– *skewness*

$$\gamma_v = \frac{1}{\sigma_v^3}\frac{1}{n}\sum_{q=1}^{n}(v_q - \langle v\rangle)^3;$$

– *kurtosis*

$$\kappa_v = \frac{1}{\sigma_v^4}\frac{1}{n}\sum_{q=1}^{n}(v_q - \langle v\rangle)^4;$$

– *crest*

$$crest_v = \frac{1}{v_{rms}} \max_{1\le q\le n} |v_q|;$$

– *clearence*

$$clear_v = \left(\max_{1\le q\le n} |v_q|\right) / \left(\frac{1}{n}\sum_{q=1}^{n}\sqrt{|v_q|}\right)^2;$$

– *shape*

$$shape_v = v_{rms}/mean;$$

– *impedence*

$$imp_v = \left(\max_{1\le q\le n} |v_q|\right) / \frac{1}{n}\sum_{q=1}^{n} |v_q|$$

– *mean absolute deviation*

$$MAD_v = \frac{1}{n}\sum_{q=1}^{n} |v_q - \langle v \rangle|$$

– central moments of order k, for $k = 2, \ldots, 5$:

$$M_k = \frac{1}{n}\sum_{q=1}^{n} (v_q - \langle v \rangle)^k$$

Since each signal is decomposed in 64 components and 16 features have been extracted from each component, each signal is represented by $64 \times 16 = 1024$ features. Since the accelerometer yields three different signals, in x, y, and z directions, each vibration should be represented by a vector of $1024 \times 3 = 3072$ features. However, the *a priori* knowledge about the process allows stating that DBOR and DBIR faults yield essentially vibrations along one specific axis whereas the ones along the reaming two axes are not influent. Therefore for each failure it has been considered only the vibration signal along one axis, y and z for DBOR and DBIR, respectively, and it is represented by 1024 features.

3.3 Dimensionality Reduction

In the last stage in order to avoid the *curse of dimensionality* [5] the dimensionality reduction is performed. To this purpose, *Principal Component Analysis* (*PCA*) [6] is performed. PCA projects the data along the direction of maximal variance of data. In order to fix the number of Principal Components that have to be retained, it recalls that the amount of the information of the first k principal components is given by the sum of the corresponding k major eigenvalues. After some experiment trails, it has been decided to consider the top 41 and 52 principal components, for DBOR and DBIR faults, respectively. This corresponds in both cases to retain the 99% of the available information. Therefore, at the end of this stage, each vibration signal, is represented by the projections onto 41 and 52 principal components, for DBOR and DBIR faults, respectively.

4 Classification

The classification was performed by *Linear Support Vector Machines* [7–9]. A Linear Support Vector Machine (*Linear SVM*) [7, 8] is a binary classifier and the patterns with output $+1$ and -1 are called positive and negative, respectively. The underlying idea of Linear SVM is the computation of the *optimal hyperplane algorithm*, i.e., the plane that yields the maximum separation margin of between two classes.

Let be $\mathcal{T} = \{(\boldsymbol{x}_1, y_1), \ldots, (\boldsymbol{x}_\ell, y_\ell)\}$, the training set, the optimal hyperplane $y(\boldsymbol{x}) = \boldsymbol{m} \cdot \boldsymbol{x}$ can be obtained solving the following constrained optimization problem

$$\min_m \frac{1}{2}\|\boldsymbol{x}\|^2 \quad \textit{subject to } y_i((\boldsymbol{m} \cdot \boldsymbol{x}_i + b)) \geq 1 \quad i = 1, \ldots, \ell. \tag{4}$$

However, in real world applications, mislabelled examples might exist yielding a partial overlapping of the two classes. To cope with this problem, it allows that some patterns can violate the constraints, introducing slack variables. They are strictly positive when the respective pattern violate the constraint, are null otherwise. Linear SVM allows controlling at the same time the margin, expressed by $\|m\|$, and the number of the training errors, given by the non-null ξ_i, by the minimization of the constrained optimization problem:

$$\min_m \frac{1}{2}\|\boldsymbol{x}\|^2 + C\sum_{i=1}^{\ell} \xi_i \quad \textit{subject to } y_i((\boldsymbol{m} \cdot \boldsymbol{x}_i + b)) \geq 1 \quad i = 1, \ldots, \ell. \tag{5}$$

The constant C, sometimes called *regularization constant*, manages the trade-off between the separation margin and the number of the misclassifications and in this work it is setup using model selection techniques, e.g., cross validation [10, 11]. Since both the function and constraints are convex, the problem (5) can be solved by means of the method of Lagrange multipliers α_i, obtaining the following final form:

$$\max_\alpha \sum_{i=1}^{\ell} \alpha_i - \frac{1}{2}\sum_{i=1}^{\ell}\sum_{j=1}^{\ell} \alpha_i\alpha_j y_i y_j (\boldsymbol{x}_i \cdot \boldsymbol{x}_j) \quad \textit{subject to} \tag{6}$$

$$0 \leq \alpha_i \leq C \quad i = 1, \ldots, \ell \tag{7}$$

$$\sum_{i=1}^{\ell} \alpha_i y_i = 0. \tag{8}$$

The vectors, whose respective multipliers are non-null, are called *support vectors*, justifying the name of the classifier. The so-constructed optimal hyerplane algorithm, called *Linear SVM* implements the following decision function:

$$f(\boldsymbol{x}) = \operatorname{sgn}\left(\sum_{i=1}^{\ell} \alpha_i y_i (\boldsymbol{x}_i \cdot \boldsymbol{x}) + b\right). \tag{9}$$

SVM experiments in this work have been performed using *SVM Light* [12] software package.

Table 1 Accuracy, precision, and recall for DBOR and DBIR faults

Fault	Input	Accuracy (%)	Precision (%)	Recall (%)
Defect of the bearing outer ring (DBOR)	41	87.0	70.0	64.0
Defect of the bearing inner ring (DBIR)	52	83.0	67.0	63.0

5 Experimental Validation

The Fault Detection system was validated by collecting a dataset formed by 308 vibration signals, 77 signals for both DBOR and DBIR faults considered, and the rest of signals was collected when in the blower there is absence of both DBOR and DBIR faults. The database was splitted in two equal parts, training and test set. The results on test set, expressed in terms of accuracy, precision, and recall, are reported in Table 1.

The results, even obtained on a test of limited size, are quite encouraging.

6 Conclusions

The paper has presented a fault detection system for a blower by a Machine Learning-based vibrational analysis. The system is composed of two modules. The former carries out the wavelet decomposition of the vibrational signal, and, then, represents it by 41 and 52 features, for DBOR and DBIR faults, respectively. The latter performs the classification by Linear Support Vector Machines. In order to validate the fault detection system an experimental laboratory has been properly built. The experimental validation, performed on a test of limited size, has shown encouraging results.

Future research will be developed along two main directions. The former will consist in a more comprehensive experimental validation of fault detection system, extending notably the size of the test set and the number of detected failures. The latter will go towards to the investigation of alternative feature extraction approaches, e.g., deep learning based.

Acknowledgments Vincenzo Mariano Scarrica developed part of the work, in an internship at CIRA, for his final B.Sc. dissertation in Computer Science at Parthenope University of Naples, with the joint supervision of F. Camastra, G. Diodati and V. Quaranta. F. Camastra's research was funded by *Sostegno alla ricerca individuale per il triennio 2015–17* project of Parthenope University of Naples. G. Diodati's and V. Quaranta's researches were developed within PRORA project SMOS (Smart-On-Board System) funded by Italian Aerospace Research Centre (CIRA).

References

1. Isermann, R.: Fault-Diagnosis System. Springer, New York (2006)
2. Gertler, J.: Fault detection and diagnosis. In: Encyclopedia of Systems and Control, pp. 1–7. MIT Press (2013)
3. Daubechies, I.: Ten Lectures on Wavelets. Society for Industrial and Applied Mathematics, SIAM (1992)
4. Baraldi, P., Cannarile, F., Di maio, F., Zio, E.: Hierarchical k-nearest neighbours classification and binary differential evolution for fault diagnostics of automotive bearings operating under variable conditions. Eng. Appl. Artif. Intell. **56**(1), 1–13 (2016)
5. Bellman, R.: Adaptive Control Processes: A Guided Tour. Princeton University Press (1961)
6. Jollife, I.T.: Principal Component Analysis. Springer-Verlag (1986)
7. Cortes, C., Vapnik, V.: Support vector networks. Mach. Learn. **20**, 1–25 (1995)
8. Vapnik, V.: Statistical Learning Theory. Wiley, New York (1998)
9. Schölkopf, B., Smola, A.: Learning with Kernels. MIT Press, Cambridge (USA) (2002)
10. Duda, R., Hart, P., Stork, D.: Pattern Classification. Wiley, New York (2001)
11. Hastie, T., Tibshirani, R., Friedman, R.: The Elements of Statistical Learning, 2nd edn. Springer (2009)
12. Joachim, T.: Making large-scale SVM learning practical. In: Advances in Kernel Methods-Support Vector Learning, pp. 169–184. MIT Press (1999)

Quaternion Widely Linear Forecasting of Air Quality

Michele Scarpiniti, Danilo Comminiello, Federico Muciaccia, and Aurelio Uncini

Abstract In this paper, we propose a quaternion widely linear approach for the forecasting of environmental data, in order to predict the air quality. Specifically, the proposed approach is based on a fusion of heterogeneous data via vector spaces. A quaternion data vector has been constructed by concatenating a set of four different measurements related to the air quality (such as CO, NO_2, SO_2, PM_{10}, an similar ones), then a Quaternion LMS (QLMS) algorithm is applied to predict next values from the previously ones. Moreover, when all the considered measurements are strongly correlated each other, the Widely Linear (WL) model for the quaternion domain is capable to benefit from correlations and to obtain improved accuracies in prediction. Some experimental results, evaluated on two different real world data sets, show the effectiveness of the proposed approach.

1 Introduction

In the last two decades, both the research community and the parliaments worldwide have given an increasing attention on the adverse effects on human health of the presence of potential dangerous substances in air. To this purpose, numerous studies have found a direct link between inhalation and long term exposure to particulate matter and other dangerous substances, and the increase in mortality rates,

M. Scarpiniti (✉) · D. Comminiello · F. Muciaccia · A. Uncini
Department of Information Engineering, Electronics and Telecommunications (DIET), "Sapienza" University of Rome, Rome, Italy
e-mail: michele.scarpiniti@uniroma1.it

D. Comminiello
e-mail: danilo.comminiello@uniroma1.it

F. Muciaccia
e-mail: federico.muciaccia@uniroma1.it

A. Uncini
e-mail: aurelio.uncini@uniroma1.it

A. Esposito et al. (eds.), *Progresses in Artificial Intelligence and Neural Systems*, Smart Innovation, Systems and Technologies 184,
https://doi.org/10.1007/978-981-15-5093-5_35

particularly for lung cancers [9, 10, 16]. Moreover, a continuous presence of particulate matter lead to a constant decrease in visibility inside the cities and to the deposition of trace elements on roads and buildings [6, 10].

In order to tackle these issues, twenty years ago the European Commission released the Council Directive 1999/30/EC, concerning the set of a roof for daily concentration of particulate matter, and obliged member states to issue a warning every time this roof is reached by entering an "attention state". However, from last data available by the European Environmental Agency[1] it is evident that, although global pollution has decreased, some pollutants, such as PM_{10} for example, remained more or less stable and constantly exceeds the required roof. For this reason, the development of a simple and robust algorithm that is able to accurately predict the daily levels of pollutants from available data is a constant priority for researchers and worldwide administrations.

The forecasting of environmental data is not a so simple task, since it presents a great complexity and an ample variety of hidden factors. Numerous works have investigated the possibility of automatically forecasting the daily concentration of pollutants using learning systems. A great effort has been done for the forecasting of PM_{10} concentration. In particular, neural network approaches provide a good solution by achieving a moderate error with small amount of data [5, 8]. Other working approaches are based on Support Vector Machines (SVMs) [1]. More recently, some advantages have been obtained by the application of Kernel Adaptive Filters (KAFs) [11].

When a set of correlated data related to the same phenomena is available, sequential data fusion via vector spaces becomes a very interesting approach [12]. The idea is to construct an augmented representation of data by combining them into an hypercomplex variable. Specifically, different heterogeneous variable becomes the real or the imaginary parts of the hypercomplex variable. Authors in [12] limit their attention to a quaternion variable and successfully make predictions from seismic data, stock markets and wind intensity and direction. In addition, the information provided by the knowledge of the whole statistics in the quaternion case can be fully exploited by the Widely Linear (WL) model [14]. Quaternion algorithms that exploit the WL model show a faster and more accurate convergence with respect traditional quaternion algorithms.

In this paper, we propose a quaternion widely linear approach for the prediction of environmental air pollution. The proposed idea is based on the Quaternion Widely Linear LMS (QWL-LMS) algorithm [14]. Four different air measurements collected by chemical sensors located in a polluted city area have been used as the four components of the quaternion representation. Some experimental results, carried out on two real-world and publicly available data sets, demonstrate the effectiveness of the proposed approach.

The rest of the paper is organized as follows. In Sect. 2, we briefly introduce the quaternion algebra, while Sect. 3 describes the widely linear model and the proposed

[1]Available at: https://www.eea.europa.eu//publications/air-quality-in-europe-2018.

algorithm. The description of the experimental setup used in the work is provided in Sect. 4 while we validate our approach in Sect. 5. Finally, some conclusive remarks are dawn in Sect. 6.

2 Introduction to Quaternion Algebra

Quaternions are a four-dimensional hypercomplex normed division algebra [2, 13] and consist of the components: $q = q_0 + q_1\mathbf{i} + q_2\mathbf{j} + q_3\mathbf{k}$. If q_0 is zero, the quaternion is named *pure quaternion*. The conjugate and the module of a quaternion $q = q_0 + q_1\mathbf{i} + q_2\mathbf{j} + q_3\mathbf{k}$ are defined, respectively, as follows:

$$q^* = q_0 - q_1\mathbf{i} - q_2\mathbf{j} - q_3\mathbf{k}, \tag{1}$$

$$|q| = \sqrt{q_0^2 + q_1^2 + q_2^2 + q_3^2}. \tag{2}$$

The imaginary units $\mathbf{i} = (1, 0, 0)$, $\mathbf{j} = (0, 1, 0)$ and $\mathbf{k} = (0, 0, 1)$, represent an orthonormal basis in $\mathbb{R}^3$ and satisfy the fundamental properties of quaternion algebra shown below:

$$\mathbf{ij} = \mathbf{i} \times \mathbf{j} = \mathbf{k}, \qquad \mathbf{jk} = \mathbf{j} \times \mathbf{k} = \mathbf{i}, \qquad \mathbf{ki} = \mathbf{k} \times \mathbf{i} = \mathbf{j} \tag{3}$$

$$\mathbf{i}^2 = \mathbf{j}^2 = \mathbf{k}^2 = -1. \tag{4}$$

where "×" denotes the vector product in $\mathbb{R}^3$.

Quaternion algebra is not commutative under the operation of multiplication, i.e. $\mathbf{ij} \neq \mathbf{ji}$. Consequently, we have:

$$\mathbf{ij} = -\mathbf{ji}, \qquad \mathbf{jk} = -\mathbf{kj}, \qquad \mathbf{ki} = -\mathbf{ik}. \tag{5}$$

The product between quaternions $q = q_0 + q_1\mathbf{i} + q_2\mathbf{j} + q_3\mathbf{k}$ and $p = p_0 + p_1\mathbf{i} + p_2\mathbf{j} + p_3\mathbf{k}$ is calculated as:

$$\begin{aligned} qp &= (q_0 + q_1\mathbf{i} + q_2\mathbf{j} + q_3\mathbf{k})(p_0 + p_1\mathbf{i} + p_2\mathbf{j} + p_3\mathbf{k}) \\ &= (q_0p_0 - q_1p_1 - q_2p_2 - q_3p_3) \\ &\quad + (q_0p_1 + q_1p_0 + q_2p_3 - q_3p_2)\,\mathbf{i} \\ &\quad + (q_0p_2 - q_1p_3 + q_2p_0 + q_3p_1)\,\mathbf{j} \\ &\quad + (q_0p_3 + q_1p_2 - q_2p_1 + q_3p_0)\,\mathbf{k}. \end{aligned} \tag{6}$$

Similarly to complex numbers, the sum of quaternions is computed as:

$$\begin{aligned} q \pm p &= (q_0 + q_1\mathbf{i} + q_2\mathbf{j} + q_3\mathbf{k}) \pm (p_0 + p_1\mathbf{i} + p_2\mathbf{j} + p_3\mathbf{k}) \\ &= (q_0 \pm p_0) + (q_1 \pm p_1)\,\mathbf{i} + (q_2 \pm p_2)\,\mathbf{j} + (q_3 \pm p_3)\,\mathbf{k}. \end{aligned} \tag{7}$$

Finally, for every quaternion $q = q_0 + q_1\mathbf{i} + q_2\mathbf{j} + q_3\mathbf{k}$, the three perpendicular quaternion involutions given by:

$$\begin{aligned} q^i &= -\mathbf{i}q\mathbf{i} = q_0 + q_1\mathbf{i} - q_2\mathbf{j} - q_3\mathbf{k} \\ q^j &= -\mathbf{j}q\mathbf{j} = q_0 - q_1\mathbf{i} + q_2\mathbf{j} - q_3\mathbf{k} \\ q^k &= -\mathbf{k}q\mathbf{k} = q_0 - q_1\mathbf{i} - q_2\mathbf{j} + q_3\mathbf{k} \end{aligned} \tag{8}$$

3 The Quaternion Widely Linear Model

Inspired to complex-valued estimation using mean-squared error criteria, the widely linear (WL) model aims at estimating a signal y in terms of a generalized linear combination involving both the complex variable and its conjugate, in order to exploit the full information provided by the statistics in the complex domain [15].

In the quaternion domain, the previous idea applies to each of the four components of the quaternion $\mathbf{x}(n)$ and its three involutions, yielding the following model:

$$y(n) = \mathbf{w}_1^T(n)\mathbf{x}(n) + \mathbf{w}_2^T(n)\mathbf{x}^i(n) + \mathbf{w}_3^T(n)\mathbf{x}^j(n) + \mathbf{w}_4^T(n)\mathbf{x}^k(n) \equiv \mathbf{w}_a^T(n)\mathbf{x}_a(n), \tag{9}$$

where $\mathbf{w}_1(n)$, $\mathbf{w}_2(n)$, $\mathbf{w}_3(n)$ and $\mathbf{w}_4(n)$ are linear quaternion filters for each of the four considered components at time n, while $\mathbf{w}_a(n) = \left[\mathbf{w}_1^T(n), \mathbf{w}_2^T(n), \mathbf{w}_3^T(n), \mathbf{w}_4^T(n)\right]^T$ and $\mathbf{x}_a(n) = \left[\mathbf{x}^T(n), \mathbf{x}^{iT}(n), \mathbf{x}^{jT}(n), \mathbf{x}^{kT}(n)\right]^T$ are an augmented representation of the quaternion widely linear (QWL) model. The model in (9) is graphically depicted in Fig. 1.

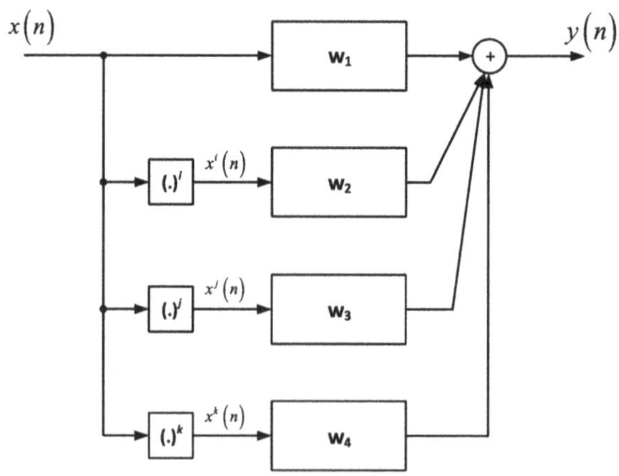

Fig. 1 The basic idea of the Quaternion Widely Linear (QWL) model

The considered solution is based on the minimization of the mean square error (MSE) that in the quaternion domain is defined as:

$$\mathscr{J}(n) = E\left\{e(n)e^*(n)\right\}, \tag{10}$$

where $e(n) = d(n) - y(n)$ is the quaternion error signal, $e^*(n)$ is its conjugate and $d(n)$ is the quaternion reference signal.

3.1 The Quaternion Widely Linear Least Mean Square

The Quaternion Widely Linear Least Mean Square (QWL-LMS) can be obtained by extending the well known LMS algorithm to the quaternion algebra. By using the stochastic gradient descent optimization in quaternion algebra [14], the gradient of the instantaneous cost function (10), is evaluated as follows:

$$\begin{aligned}\nabla_{\mathbf{w}_a}\mathscr{J}(n) &= e(n)\nabla_{\mathbf{w}_a}e^*(n) + \nabla_{\mathbf{w}_a}e(n)e^*(n)\\ &= 2e(n)\mathbf{x}_a(n) - 4\mathbf{x}_a(n)e^*(n).\end{aligned} \tag{11}$$

Notice that, due to the non-commutativity of the quaternion product, the two error gradient terms in (11) need to be treated separately. Based on the generic stochastic gradient update, the QWL-LMS learning rule becomes:

$$\mathbf{w}_a(n+1) = \mathbf{w}_a(n) + \mu\big(2\mathbf{x}_a(n)e^*(n) - e(n)\mathbf{x}_a(n)\big), \tag{12}$$

where the scaling factor of two has been absorbed in the learning rate μ.

3.2 The Proposed Idea

In the proposed approach, we select four of the available chemical measurements from sensor data and use them as the four components of the input quaternion $\mathbf{x}$. The input data matrix $\mathbf{X}$ is constructed by considering the last τ samples (last measurements): $\mathbf{X}(n) = [\mathbf{x}(n), \mathbf{x}(n-1), \ldots, \mathbf{x}(n-\tau)]$.

After evaluating the augmented input $\mathbf{X}_a(n) = [\mathbf{x}_a(n), \mathbf{x}_a(n-1), \ldots, \mathbf{x}_a(n-\tau)]$ in (9) and applying the QWL-LMS algorithm in (12), the augmented weight vector $\widetilde{\mathbf{w}}_a(n)$ just learned will be subsequently used to produce the predicted output on the augmented testing sequence $\mathbf{X}_{a-te}(n)$:

$$\widehat{\mathbf{y}}(n) = \widetilde{\mathbf{w}}_a(n)^T\mathbf{X}_{a-te}(n). \tag{13}$$

4 Experimental Setup

Experiments have been carried out on two different real world data sets. The first one is the UCI *Air Quality Data Set*,[2] that contains 9358 instances of hourly averaged responses from an array of 5 metal oxide chemical sensors embedded in an Air Quality Chemical Multisensor Device [3]. The device was located on the field in a significantly polluted area, at road level, within an Italian city. Data were hourly recorded from March 2004 to February 2005 (one year) [3, 4]. The data set provides hourly averaged concentrations for CO, Non Metanic Hydrocarbons (NMHC), Total Nitrogen Oxides (NO_x), Nitrogen Dioxide (NO_2) and Ozone (O_3).

The second data set have been constructed by collecting environmental data from the Ancona city in 2017.[3] The data set provides hourly averaged concentrations for PM_{10}, $PM_{2.5}$, CO, NO_2, SO_2 and O_3.

Results have been compared in terms of MSE and prediction accuracy of the obtained time series. As an additional objective index, we used the prediction gain R_p, defined as [7]:

$$R_p = 10 \log_{10} \left(\frac{\sigma_x^2}{\sigma_e^2} \right), \tag{14}$$

measured in dB, where σ_x^2 denotes the variance of the input sequence $x[n]$ and σ_e^2 denotes the variance of the error sequence $e[n]$.

5 Experimental Results

Experimental results have been conducted by using different time delay values, *a priori* unknown, specifically by selecting it in the interval $\tau \in [2, 15]$. This is meaning that we use the last τ samples to predict the next one. Both considered data sets have been used. Specifically, from the UCI AirQuality data set we have selected the measures CO, NMHC, NO_2 and O_3, as the four quaternion parts, respectively. From the ANCONA 2017 data set we have selected the measures PM_{10}, $PM_{2.5}$, NO_2 and O_3, as the four quaternion parts. The two data sets are characterized for two common measurements. However, time series of the UCI AirQuality data set are more correlated with respect the related of the ANCONA 2017. A sequence of $N_{tr} = 1600$ continuous samples have been randomly selected to train the predictor, while a sequence of $N_{te} = 400$ samples have been randomly selected to test the obtained predictor.

[2]Available at: https://archive.ics.uci.edu/ml/datasets/Air+quality.

[3]Data are available at: https://www.eea.europa.eu/data-and-maps/data/aqereporting-8.

Table 1 Prediction MSE obtained by the considered approaches in different data sets for different τ value

	UCI AirQuality		ANCONA 2017	
τ	QWL – LMS	LMS	QWL – LMS	LMS
2	0.0011	0.0031	0.0022	0.0034
3	0.0003	0.0015	0.0010	0.0021
4	0.0003	0.0024	0.0015	0.0026
5	0.0008	0.0033	0.0020	0.0030
6	0.0013	0.0041	0.0021	0.0033
7	0.0018	0.0047	0.0022	0.0035
8	0.0022	0.0053	0.0024	0.0036
9	0.0024	0.0058	0.0025	0.0037
10	0.0026	0.0063	0.0026	0.0038
12	0.0029	0.0074	0.0028	0.0039
15	0.0041	0.0085	0.0046	0.0053

Results in terms of MSE are shown in Table 1. This table shows the results for different values of τ in both the data sets. MSEs of the LMS approach are averaged above the four single runs of the algorithm, one for each of the considered environmental measurements. From Table 1 we can argue that the proposed idea performs better than the separate LMS prediction, and the best result is obtained by considering $\tau = 3$ and $\tau = 4$. As a remark, we can observe that the gap between the two approaches is reduced in the case of the ANCONA 2017 data set. The motivation of the improvement in prediction for the UCI AirQuality data set is due to the stronger correlation between the selected four measurements.

Results in terms of the prediction gain R_p in (14) are shown in Table 2. Substantially, this table confirms the same considerations made in the discussion of Table 1. As a remark, Table 2 shows a lesser difference between the two considered data sets.

The accuracies in predictions are shown in Figs. 2 and 3, for the first 150 samples of test sets extracted from the UCI AirQuality and ANCONA 2017 data sets, respectively, by considering the case of $\tau = 3$ and $\tau = 5$, respectively. In Fig. 3 we can observe that the prediction of O_3 is the worst one. In addition, Fig. 4 shows the predicted sequences for the UCI AirQuality data set with $\tau = 9$. From these figures, and in particular from Fig. 4, we can argue that the proposed QWL-LMS algorithm is able to better adapt to the transient and peaks of the sequences with respect to the standard LMS algorithm that, instead, shows an averaging behavior. Overall, the QWL-LMS demonstrates a faster convergence.

Interestingly enough, the considered approach can be also used for the prediction of only three sequences by correlating them with a fourth sequence in which we are not directly interested. For example, in the UCI AirQuality data set, we can use the sequence related to the temperature of the environment as the real part of the quaternion, while we continue to use (and predict) NMHC, NO_2 and O_3 as

Table 2 Prediction gain R_p in dB obtained by the considered approaches in different data sets for different τ value

	UCI AirQuality		ANCONA 2017	
τ	QWL – LMS	LMS	QWL – LMS	LMS
2	19.78	15.83	13.25	09.37
3	29.98	22.38	18.78	12.56
4	29.71	20.33	17.33	11.57
5	25.38	19.03	15.94	11.00
6	23.08	18.12	15.77	10.64
7	21.77	17.49	15.47	10.38
8	20.80	17.01	15.17	10.22
9	20.33	16.60	14.99	10.09
10	20.11	16.24	14.77	09.98
12	19.62	15.58	14.49	09.82
15	18.05	14.94	10.76	09.53

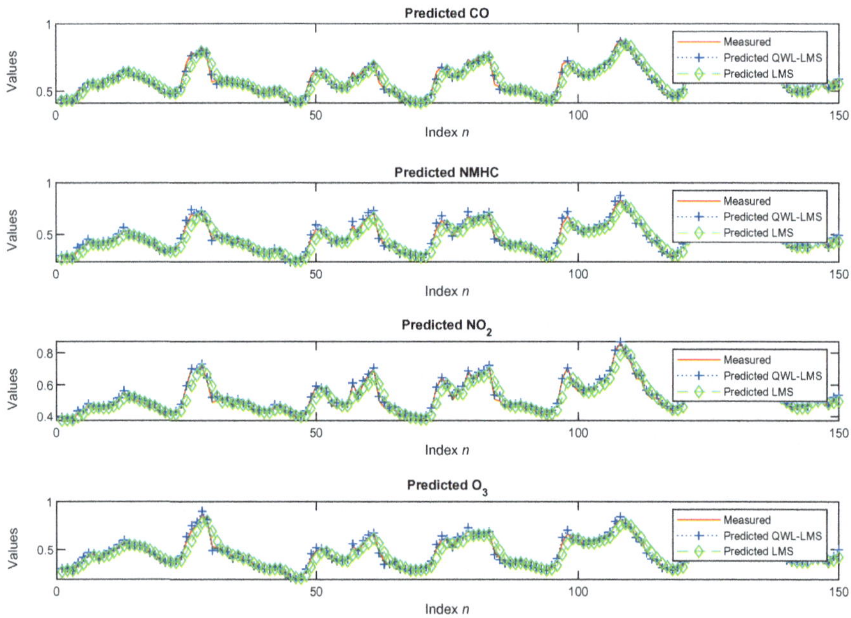

Fig. 2 Accuracy obtained by the considered approach for the UCI AirQuality data set in the case of $\tau = 3$. The figure is limited to the first 150 samples of the test set

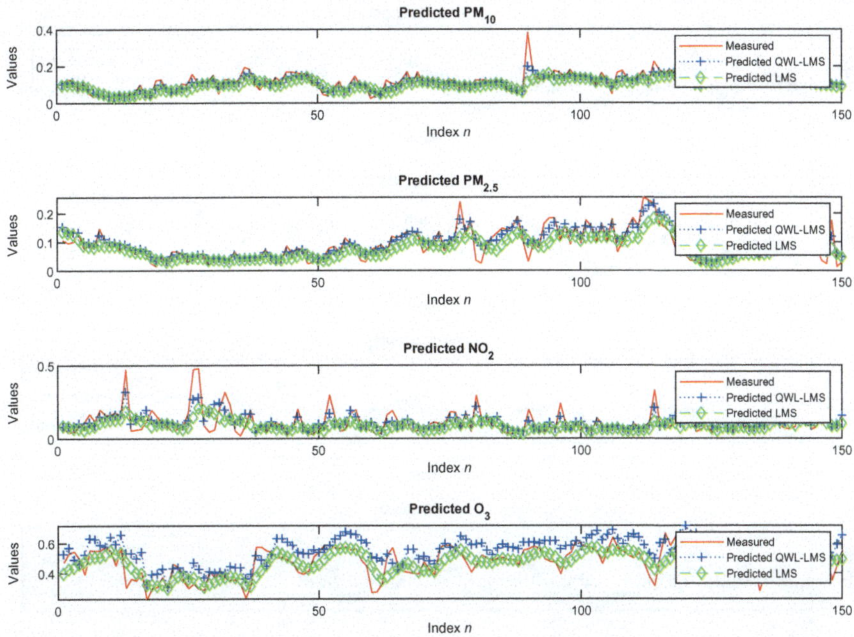

Fig. 3 Accuracy obtained by the considered approach for the ANCONA 2017 data set in the case of $\tau = 5$. The figure is limited to the first 150 samples of the test set

the three imaginary parts. The prediction can benefits from the intrinsic correlation provided by the temperature, even if we do not predict it. In this case, results are MSE = 0.0002 for the QWL-LMS (resp., MSE = 0.0012 for the LMS) and $R_p = 31.01$ dB for QWL-LMS (resp., $R_p = 22.67$ dB for the LMS). Hence, also in this case, the proposed QWL-LMS outperforms the single sensor LMS algorithm.

6 Conclusions

In this paper, we have proposed a quaternion widely linear LMS (QWL-LMS) algorithm to approach the challenging problem of air pollution prediction. In particular, four chemical measurements have been selected and combined into a quaternion representation in a sequential data fusion via vector spaces. The augmented representation is able to obtain a faster and more accurate convergence. Some experimental results, implemented by using the QWL-LMS algorithm and evaluated in terms of MSE, accuracy and prediction gain, have shown the effectiveness of the proposed approach.

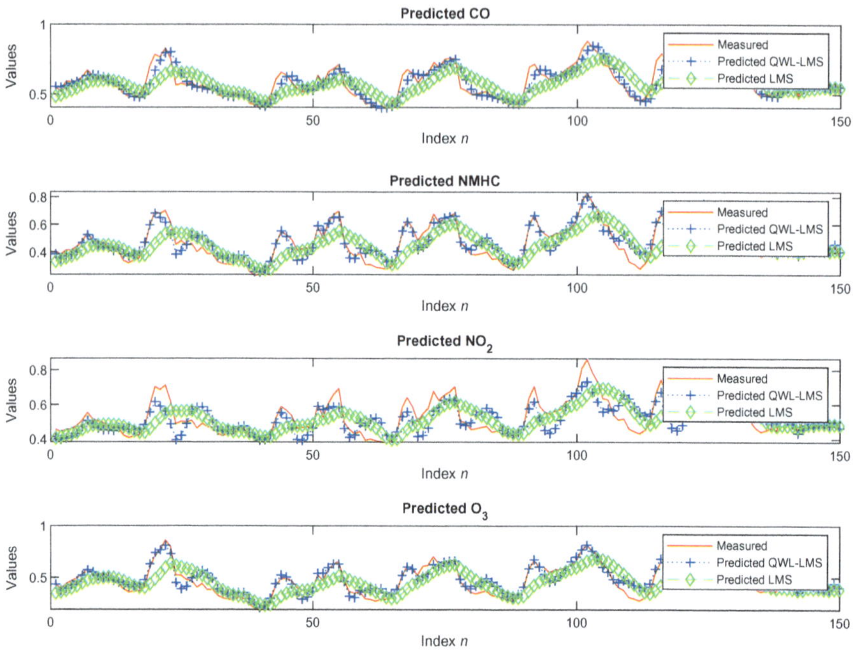

Fig. 4 Accuracy obtained by the considered approach for the UCI AirQuality data set in the case of $\tau = 9$. The figure is limited to the first 150 samples of the test set

References

1. Arampongsanuwat, S., Meesad, P.: Prediction of PM_{10} using support vector regression. In: 2011 International Conference on Information and Electronics Engineering (IPCSIT 2011), vol. 6, pp. 120–124 (2011)
2. Bülow, T., Sommer, G.: Hypercomplex signals—a novel extension of the analytic signal to the multidimensional case. IEEE Trans. Signal Process. **49**(11), 2844–2852 (2001)
3. De Vito, S., Massera, E., Piga, M., Martinotto, L., Di Francia, G.: On field calibration of an electronic nose for benzene estimation in an urban pollution monitoring scenario. Sens. Actuators B Chem. **129**(2), 750–757 (2008)
4. De Vito, S., Piga, M., Martinotto, L., Di Francia, G.: CO, NO_2 and NO_x urban pollution monitoring with on-field calibrated electronic nose by automatic bayesian regularization. Sens. Actuators B Chem. **143**(1), 182–191 (2009)
5. Gardner, M., Dorling, S.: Artificial neural networks (the multilayer perceptron): A review of applications in the atmospheric sciences. Atmos. Environ. **32**, 2627–2636 (1998)
6. Hooyberghs, F., Mensink, C., Dumont, G., Fierens, F., Brasseur, O.: A neural network forecast for daily average PM_{10} concentrations in Belgium. Atmos. Environ. **39**(8), 3279–3289 (2005)
7. Mandic, D.P., Goh, S.L., Aihara, K.: Sequential data fusion via vector spaces: fusion of heterogeneous data in the complex domain. J. VLSI Signal Process. Syst. Signal Image Video Technol. **48**(1–2), 99–108 (2007)
8. Park, S., Kim, M., Kim, M., Namgung, H.G., Kim, K.T., Cho, K.H., Kwon, S.B.: Predicting PM_{10} concentration in Seoul metropolitan subway stations using artificial neural network (ANN). J. Hazard. Mater. **341**, 75–82 (2018)

9. Pope, C., Burnett, R., Thun, M., Calle, E., Krewskik, D., Ito, K., Thurston, G.: Lung cancer, cardiopulmonary mortality, and long term exposure to fine particulate air pollution. J. Am. Med. Assoc. **287**, 1132–1141 (2002)
10. Reisen, V.A., Queiroz Sarnaglia, A.J., Costa Reis Jr., N., Lévy-Leduc, C., Santos, J.M.: Modeling and forecasting daily average PM_{10} concentrations by a seasonal long-memory model with volatility. Environ. Model. Softw. **51**, 286–295 (2014)
11. Scardapane, S., Comminiello, D., Scarpiniti, M., Parisi, R., Uncini, A.: PM_{10} forecasting using kernel adaptive filtering: an Italian case study. In: Apolloni, B. Bassis,, S., Esposito, A., Morabito, C. (eds.) Neural Nets and Surroundings. Smart Innovation, Systems and Technologies, vol. 19, pp. 93–100. Springer, Berlin (2013)
12. Took, C.C., Mandic, D.P.: Fusion of heterogeneous data sources: a quaternionic approach. In: Proceedings of 2008 IEEE Workshop on Machine Learning for Signal Processing (MLSP 2008), pp. 456–461. Cancun, Mexico (2008)
13. Took, C.C., Mandic, D.P.: The quaternion LMS algorithm for adaptive filtering of hypercomplex processes. IEEE Trans. Signal Process. **57**(4), 1316–1327 (2009)
14. Took, C.C., Mandic, D.P.: A quaternion widely linear adaptive filter. IEEE Trans. Signal Process. **58**(8), 4427–4431 (2010)
15. Took, C.C., Mandic, D.P.: Augmented second-order statistics of quaternion random signals. Signal Process. **91**(2), 214–224 (2011)
16. Vanos, J.K., Cakmak, S., Kalkstein, L.S., Yagouti, A.: Association of weather and air pollution interactions on daily mortality in 12 Canadian cities. Air Qual. Atmos. Health **8**(3), 307–320 (2015)

Non-linear PCA Neural Network for EEG Noise Reduction in Brain-Computer Interface

Andrea Cimmino, Angelo Ciaramella, Giovanni Dezio, and Pasquale Junior Salma

Abstract For handling human-machine interactions, in this last years, many efforts have been devoted to Brain-Computer Interface (BCI). Electroencephalogram (EEG) electrodes enhance the convenience and wearability of BCI. Unfortunately, the noise induced by sampling reduces the signal quality compared to that of electrodes. In this paper a methodology for EEG waves compression and noise reduction is introduced. The approach is based on a non-linear Principal Component Analysis Neural Network for compression and decompression (reconstruction) of the data. Experiments are made on a corpus containing the activation strength of the fourteen electrodes of an EEG headset for eye state prediction. The experimental results highlight that the technique permits to obtain an higher rate of classification accuracy w.r.t. the use of row data.

1 Introduction

Brain Computer Interface means of direct communication between a brain and an external device. Electroencephalography (EEG) is the most studied non-invasive interface mainly due to its fine temporal resolution, ease of use, portability and low set-up cost. Since the first EEG where described by Hans Berger in 1929 all start thinking about all potential application [1], then the research in neural application

A. Cimmino · A. Ciaramella (✉) · G. Dezio · P. J. Salma
Department of Science and Technology, University of Naples "Parthenope", Centro Direzionale, Isola C4, 80143 Naples, Italy
e-mail: angelo.ciaramella@uniparthenope.it

A. Cimmino
e-mail: andrea.cimmino1@studenti.uniparthenope.it

G. Dezio
e-mail: giovanni.dezio001@studenti.uniparthenope.it

P. J. Salma
e-mail: pasqualejunior.salma@studenti.uniparthenope.it

A. Esposito et al. (eds.), *Progresses in Artificial Intelligence and Neural Systems*, Smart Innovation, Systems and Technologies 184,
https://doi.org/10.1007/978-981-15-5093-5_36

start to be applied. In 70s, several scientist began to develop communication systems that provided interactions between human and computers sponsored by **ARPA** (Advanced Research Project Agency): that was first approach to BCI. A **Brain-Computer Interface**, also known as **BCI** or **Neural Interface**, is a channel of communication between brain, in particular the synapses and neurons, and an external acquisition device. This device acquires signals from the brain, such as **EEG signals**, and this is an example of mono-directional BCI. There are also directional BCI, that allows the exchange of data between the brain and computer. The BCI are based on electric signals, and due to this they are often composed by electrodes, which are in direct contact with the skin or neural tissue. The one in contact with the neural tissue is more invasive, and need a surgery to be installed, because of this they are not popular in research, and due to ethical or legal problems they are not used in UE. The other one are often composed by a cap or a headset. The one used in our experiment is composed by a headset. A more complete definition of BCI can be the following: "*A brain-computer interface is a communication system that does not depend on the brain's normal output pathways of peripheral nerves and muscles*" [2]. In Computer Science and Bioengineering BCI works to support people with handicap [3]. What keep us working is the hope of help that people, who can't move and have to be helped, to make them newly self sufficient. The acquisition and interpretation of EEG/Neural data can control the moves of a wheelchair, or replay some vocal synthesis, or also control a **Home Automation System**. In comatose-vegetative state subjects neural interface can evaluate the cognitive state of patient, through the read of EEG signals. Despite the research interest for BCI is given by the will to give assistive technology, people with handicap, has took to birth several application branches. Today users can control EEG rythm changes through meditation, or can do image classification based on EEG response to visual pulse, or can monitor their focus level. A recent investigation has identified a completely different BCI use, indeed the neural interface has taken the first steps in gaming [4]. Recently, brain stimuli have been used also to track emotions [5] and for military scenarios [6]. For avoiding false alarms it is important to determine, with great accuracy, the specific stimuli. In [7], the authors investigate how the eye state (open or closed) can be predicted by measuring brain waves with an EEG. In this paper we introduce a noise reduction approach for EEG signals recorded by a commercial Emotiv Epoc headset.[1] The approach is based on a non-linear Principal Component Analysis (PCA) technique developed by a Neural Network. We used a corpus containing the activation strength of the fourteen electrodes of the EEG headset for eye state prediction. The experimental results highlight that the denoising permits to obtain an higher rate of accuracy in classification. The paper is organized as follows. In Sect. 2, some concepts about the EEG headset are introduced. In Sect. 3, we describe the noise reduction technique and, in Sect. 4 some experimental results are discussed. Finally, conclusions and future remarks are outlined in Sect. 5.

[1] http://www.emotiv.com.

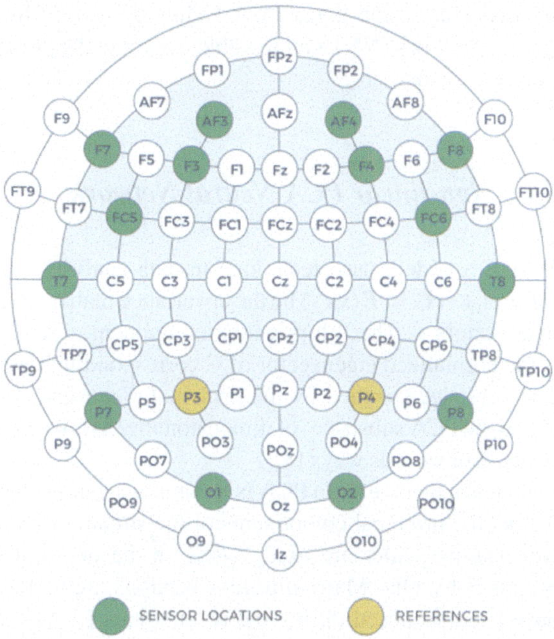

Fig. 1 *Emotiv EPOC+* channels location

2 Emotiv EPOC+

The **Emotiv EPOC+** is a personal EEG device, it can be used for contextual brain research and brain-computer interface application, it allow to detect emotion and face expression, communication take place via bluetooth 4.0, that can be used for send high-quality raw EEG data, captured by electrodes. The **Emotiv EPOC+** is a 14 channel device, every single channel read the potential difference below the skin, location of them is shown in the Fig. 1. We choose to use **Emotiv EPOC+** due to its ease of use, since a medical one would be too expensive and more complex to use.

3 Signal Reconstruction Schema

The noise reduction is a fundamental and practical step in the field of digital signals (i.e, wave, image). The objective of removing the noise is to obtain the recovered signal in such a way that it resembles the original (noiseless) as closely as possible. Many filtering algorithms are known in literature (e.g., Wiener, Kalman filter) that achieve good performance when advance the spectral properties of the noise free data and the noise itself are known.

In this paper we present an approach based on random noise reduction which does not need to know in advance the spectral properties of the data. The noise reduction is obtained by compression and decompression (reconstruction) of the noisy data.

In particular, a non-linear PCA technique is used based on one layer feedforward Neural Network (NN) which is able to extract the principal components of the stream of input vectors.

3.1 Non-linear PCA Neural Network

PCA is a widely used technique in data analysis. Mathematically, it is defined as follows. Let $\mathbf{C} = E(\mathbf{x}\mathbf{x}^T)$ be the covariance matrix of L-dimensional zero mean input data vectors $\mathbf{x}$. The i-th principal component of $\mathbf{x}$ is defined as $\mathbf{x}^T\mathbf{c}(i)$, where $\mathbf{c}(i)$ is the normalized eigenvector of $\mathbf{C}$ corresponding to the i-th largest eigenvalue $\lambda(i)$. The subspace spanned by the principal eigenvectors $\mathbf{c}(1), \ldots, \mathbf{c}(M), (M < L))$ is called the PCA subspace (of dimensionality M). We note that the PCA can be neurally realized in various ways [8].

In this paper we use a PCA NN that is a one layer feedforward NN which is able to extract the principal components of the stream of input vectors. Typically, Hebbian type learning rules are used, based on the one unit learning algorithm originally proposed by Oja. Many different versions and extensions of this basic algorithm have been proposed during the recent years [9–11]. The structure of the PCA NN can be summarised as follows: there is one input layer and one forward layer of neurons totally connected to the inputs; during the learning phase there are feedback links among neurons, that classify the network structure as either hierarchical or symmetric. After the learning phase the network becomes purely feedforward. The hierarchical case leads to the well known GHA algorithm; in the symmetric case we have the Oja's subspace network.

PCA neural algorithms can be derived from optimization problems; in this way the learning algorithms can be further classified in linear PCA algorithms and non-linear PCA algorithms. As defined in [12] in order to extract the principal components we use a non-linear PCA NN that derives from the robust generalization of variance maximisation where the objective function $f(t)$ is assumed to be a valid cost function, such as $\ln\cosh(t)$. The general algorithm is described in Fig. 2. We stress that the

Fig. 2 Example of PCA neural network

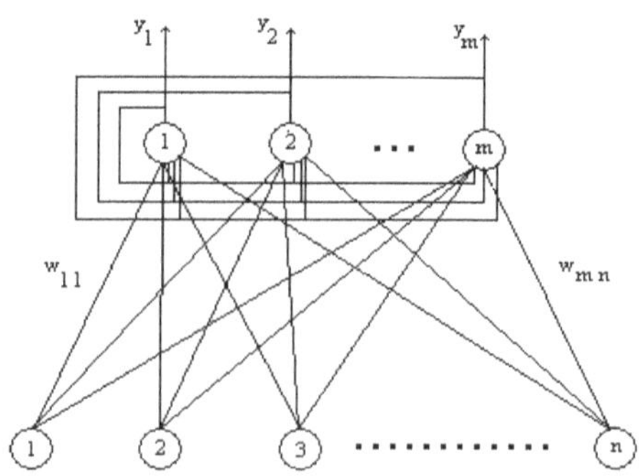

approach can be applied recursively on a single channel (or signal) and several comparisons proved the robustness of the approach also w.r.t. classical PCA and Independent Component Analysis (ICA) [9–11].

Algorithm 1 Non-linear PCA Algorithm

1: Initialize the number of output neurons m. Initialize the weight matrix $\mathbf{W} = [\mathbf{w}_1, \ldots, \mathbf{w}_m]$ with small random values. Initialize the learning threshold ϵ, the learning rate μ and the α parameter. Reset pattern counter $k = 1$.

2: Input the n-th pattern

$$\mathbf{x}_n = [x(n), x(n+1), \ldots., x(n+(q-1))]$$

where q is the number of input components.

3: Calculate the output for each neuron $y_i = \mathbf{w}_i^T \mathbf{x}_n$. $\forall i = 1, \ldots, m$

4: Modify the weights using the following equation

$$\mathbf{w}_i(k+1) = \mathbf{w}_i(k) + \mu_k g(y_i(k))\mathbf{e}_i(k)$$

$$\mathbf{e}_i(k) = \mathbf{x}_i - \sum_{j=1}^{I(i)} y_j(k)\mathbf{w}_j(k)$$

$$\mathbf{w}_i(k+1) = \frac{\mathbf{w}_i(k+1)}{\|\mathbf{w}_i(k+1)\|}$$

where in the hierarchical case we have $I(i) = i$. In the symmetric case $I(i) = m$, the error vector $\mathbf{e}_i(k)$ becomes the same $\mathbf{e}_i$ for all the neurons.

5: UNTIL the number of pattern is not empty GO TO 2

6: Convergence test:
IF $C_T = \frac{1}{2}\sum_{i=1}^{m}\sum_{j=1}^{m}(w_{ij} - w_{ij}^{old})^2 < \epsilon$
THEN GO TO 8
ELSE

$$\mathbf{W} = (\mathbf{W}\mathbf{W}^T)^{\frac{1}{2}}\mathbf{W}$$

$$\mathbf{W}^{old} = \mathbf{W}$$

7: $k = k + 1$; GO TO 2.

8: END

---

4 Experimental Results

The experiments were carried out following the scheme proposed in [7]. During the experiments, tester was in a silent room and unconscious of experiment begin time. Before the use of Emotiv EPOC+ Headset, electrodes has been soaked with a saline solution to improve electric conduction from head to electrodes and to gain more precision on read raw data. The tester had to switch between two main eyes states: closed and opened eyes. Closed eye state has been considered when eyes was completely closed and opened in all other ways. The eye state was manually annotated

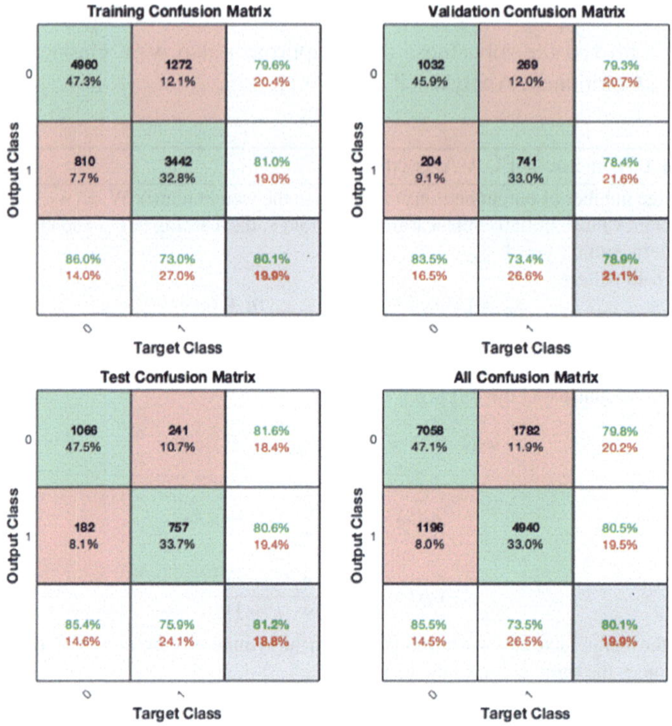

Fig. 3 Confusion matrices with data set without noise reduction

by analyzing a video recordings the tester activities. Data adopted for experiments is composed by 14,976 instances (i.e., observations) with 15 attributes each, where 14 representing the values of the electrodes and the other the annotated eye state. The data are pre-processed eliminating some sequences with high variance.[2] We adopted a Multi-Layer Perceptron (MLP) NN with one hidden layer and 10 hidden nodes with *logistic* activation function for classifying. We apply a cross-validation mechanism in which training (70%), validation (15%), and test sets (15%) are displayed. In Fig. 3 the confusion matrices on the three data sets are considered. We observe that all confusion matrix report a percentage of classification around 80%. Successively, we apply the Non-liner PCA NN for denoising the EEG waves. An example of EEG signal noise reduction is visualized in Fig. 4. Applying MLP, we obtain the confusion matrices of Fig. 5. In this case we obtain for all confusion matrix a percentage of classification of 99.6%.

[2]The data set can be download from http://cvprlab.uniparthenope.it.

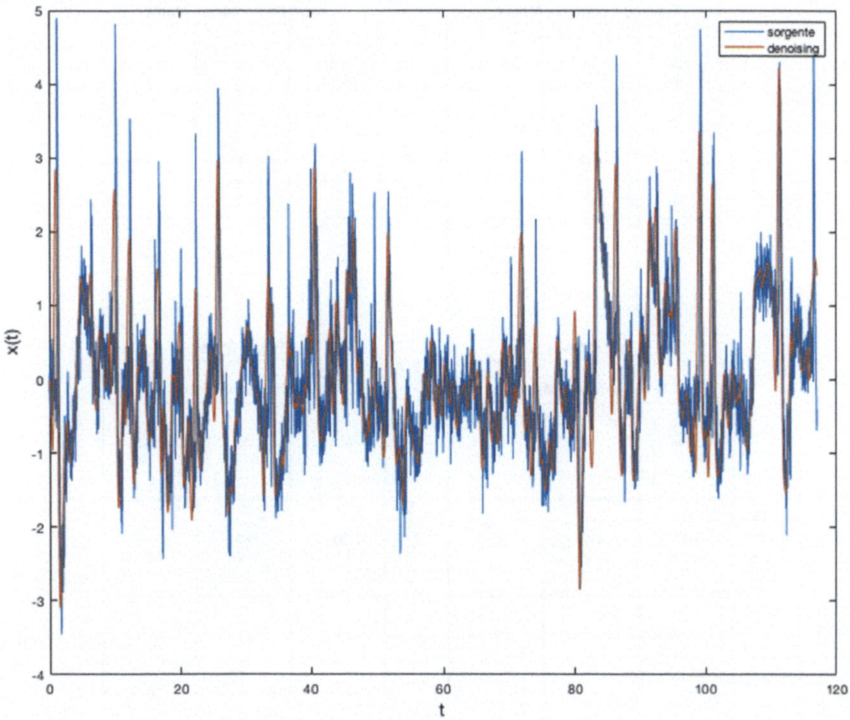

Fig. 4 Example of noise reduction on EEG signal

5 Conclusions

In this paper we introduced a methodology for EEG waves compression and noise reduction. The approach is based on a non-linear Principal Component Analysis Neural Network for compression and reconstruction of the data. Experiments are made on a corpus containing the activation strength of the fourteen electrodes of an Emotiv Epoc EEG headset. In particular the problem of eyes state prediction is considered. From the experimental results we observed that the proposed methodology permits to obtain an high rate of classification accuracy. In the next future, the authors concentrate on the use and comparisons of different topologies of non-linear PCA NN and different data acquired from the headset for further kinds of applications (e.g., hardware control).

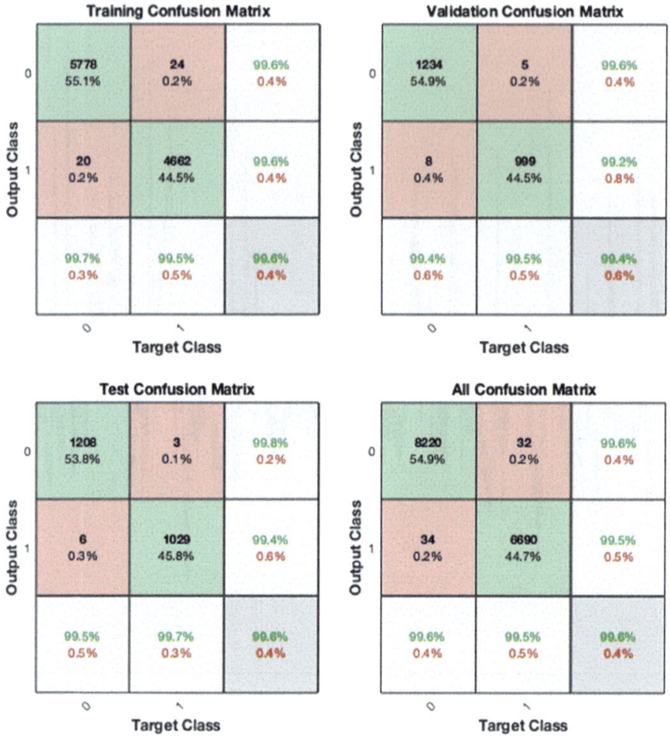

Fig. 5 Confusion matrices with data set with noise reduction

Acknowledgments This work was partially funded by the University of Naples Parthenope (*Sostegno alla ricerca individuale per il triennio 2017–2019* project).

References

1. Berger, H.: Uber das electrenkephalogramm des menchen. Arch. Psychiat. Nervenkr. **87**, 527–570 (1929)
2. Wolpaw, J.R., Birbaumer, N., Heetderks, W.J., McFarland, D.J., Peckham, P.H., Schalk, G., Donchin, E., Quatrano, L.A., Robinson, C.J., Vaughan, T.M.: Brain-computer interface technology: a review of the first international meeting. IEEE Trans. Rehab. Eng. **8**(2), 164–173 (2000)
3. Ossmy, O., Tam, O., Puzis, R., Rokach, L., Inbar, O., Elovici, Y.: MindDesktop: computer accessibility for severely handicapped. In: Proceedings of the the ICEIS, Beijing, China (2011)
4. Pour, P., Gulrez, T., AlZoubi, O., Gargiulo, G., Calvo, R.: Brain-computer interface: next generation thought controlled distributed video game development platform. In: Proceedings of the CIG, Perth, Australia (2008)
5. Pham, T., Tran, D.: Emotion Recognition Using the Emotiv EPOC Device. Lecture Notes in Computer Science (2012)

6. Erp, J.V., Reschke, S., Grootjen, M., Brouwer, A.-M.: Brain performance enhancement for military operators. In: HFM, Sofia, Bulgaria (2009)
7. Roesler, O., Suendermann, D.: A First Step Towards Eye State Prediction Using EEG (2013)
8. Tagliaferri, R., Ciaramella, A., Milano, F., Barone, L., Longo, G.: Spectral analysis of stellar light curves by means of neural networks. Astron. Astrophys., Suppl. Ser. **137**(2), 391–405 (1999)
9. Ciaramella, A., De Lauro, E., De Martino, B., Di Lieto, S., Falanga, M., Tagliaferri, R.: ICA based identification of dynamical systems generating synthetic and real world time series. Soft. Comput. **10**(7), 587–606 (2006)
10. Ciaramella, A., Gianfico, M., Giunta, G.: Compressive sampling and adaptive dictionary learning for the packet loss recovery in audio multimedia streaming. Multimedia Tools Appl. **75**(24), 17375–17392 (2016)
11. Ciaramella, A., Giunta, G.: Packet loss recovery in audio multimedia streaming by using compressive sensing. IET Commun. **10**(4), 387–392 (2016)
12. Ciaramella, A., De Lauro, E., De Martino, B., Di Lieto, S., Falanga, M., Tagliaferri, R.: Characterization of strombolian events by using independent component analysis. Nonlinear Process. Geophys. **11**(4), 453–461 (2004)

Spam Detection by Machine Learning-Based Content Analysis

Daniele Davino, Francesco Camastra, Angelo Ciaramella, and Antonino Staiano

Abstract The paper aims to present a Spam Detection system by a Content Analysis based on Machine Leaning. The system is composed of six units: Tokenization and Cleaning words, Lemmatization, Stopping Word Removal and Synonym Replacement, Term Selection, Bag-of-Words Representer, and Classifier. Experiments performed on two different datasets, i.e., *SpamAssassin* and *Trec2007* show satisfactory results, comparable with the state of the art.

1 Introduction

Spam is defined as "*Unsolicited, unwanted email that was sent indiscriminately, directly or indirectly, by a sender having no current relationship with the recipient*" [1]. The economic impact of spam is so high for institutions and companies, so that developing of effective spam filtering system is become one of the crucial topics in computer security [2]. In the last years, several efforts have been performed in spam filtering domain [3, 4] for developing spam detection systems, i.e., systems able to discriminate spam email messages from the rest of emails denoted as *ham*.

Spam Detection systems can be divided in two big families. Systems, that belong to the former family, use to detect spam the information essentially contained in the

D. Davino · F. Camastra · A. Ciaramella · A. Staiano (✉)
Department of Science and Technology, University of Naples Parthenope,
Centro Direzionale Isola C4, 80143 Naples, Italy
e-mail: antonino.staiano@uniparthenope.it

D. Davino
e-mail: daniele.davino.88@gmail.com

F. Camastra
e-mail: camastra@ieee.org

A. Ciaramella
e-mail: angelo.ciaramella@uniparthenope.it

A. Esposito et al. (eds.), *Progresses in Artificial Intelligence and Neural Systems*,
Smart Innovation, Systems and Technologies 184,
https://doi.org/10.1007/978-981-15-5093-5_37

email header, e.g., email sender and object. The latter family is composed of systems that use only the email body for deciding if a message is spam or not.

The aim of the paper is to present a spam detector system by a content analysis based on Machine Learning methods. The proposed systems is formed by the following six units, i.e., Tokenization and Cleaning words, Lemmatization, Stopping Word Removal and Synonym Replacement, Term Selection, Bag-of-Words Representer, and Classifier.

The paper is organized as follows. Section 2 describes structure and units of Spam Detection System; Sect. 3 presents some experimental results; finally, in Sect. 4 some conclusions are drawn.

2 Spam Detector

Spam Detector is composed of six units, i.e., *Tokenization and Cleaning Words*, *Lemmatization*, *Stopping Word Removal and Synonym Replacement*, *Term Selection*, *Bag-of-Words Representer*, and *Classifier*.

2.1 Tokenization and Cleaning Word

This module has the task of performing *tokenization*, i.e., parsing the email body, extracting the words, and cleaning them applying the following heuristic rules:

- removing of apexes, underscores, and dashes, if they are at the beginning or at the end of the word;
- removing of Javascript tags, email addresses, and web urls;
- the character % is removed if there are no numbers before or after it;
- The words whose first symbol is $ followed by two or more numbers and two or more letters are splitted in two parts; the former containing the symbol $ and the number, the latter composed only of letters.
- all types of parentheses, double quotes, and some of common special characters are filtered;
- all hexadecimal numbers, numbers in scientific notation, and numbers without a unit of measure are removed;
- all Asiatic, Arabic and Hebrew characters are removed;
- all characters, excluding numbers or letters that are repeated more than three times consecutively are deleted.

An example of tokenization and cleaning word process is shown in Fig. 1.

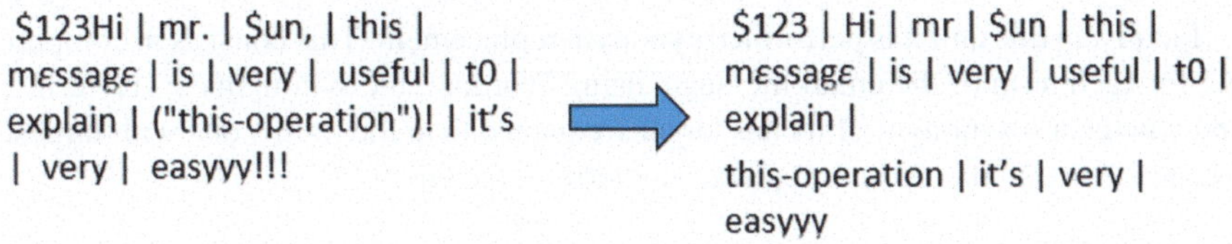

Fig. 1 An example of tokenization and cleaning words. On the left w.r.t. arrow, an email message; on the right the same message after tokenization and clearing word process. The | symbol is used as separator

2.2 *Lemmatization*

This stage is usually carried out by stemming process, whose aim is to reduce each word to its linguistic root, i.e., the *stem*. Although for the English language Porter's algorithm [5] is an effective stemming algorithms, it cannot handle irregular forms, e.g., irregular verbs. For this reason Porter's algorithm has been replaced with *lemmatization*, which consists in reducing the word to its *lemma*. To this purpose we recall that in linguistics the lemma of a word is the word that is conventionally chosen to represent all flexed forms of a given term in a dictionary. For instance, the lemmas of the English verbs "took" and "was" are "to take" and "to be", respectively. Lemmatization, in this work, has been performed using *Wordnet* [6].

2.3 *Stopping Word Removal and Synonym Replacement*

This stage has the task of individuating, and then removing the so-called *stop words*. Stop words are the most frequent words in a text. Since they are the most frequent words in any text, they have poor discriminative power and they do not allow discriminating a text from another one. For this reason, stop words are removed in any text. Stop words can vary from language to language. They are usually articles, prepositions, conjunctions, auxiliary, and modal verbs. In English language, the number of stop words is at least 174 [7]. An example of stopping word removal on an email message is shown in Fig. 2.

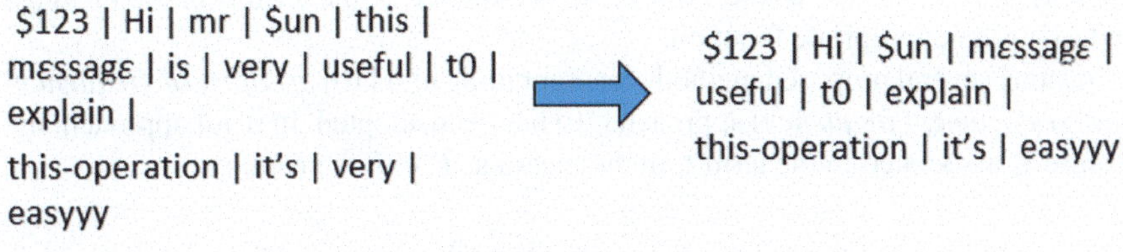

Fig. 2 An example of lemmatization and stopping word removal word. On the left w.r.t. arrow, the email message after tokenization and clearing word process on the right the message after lemmatization and stopping word removal. The | symbol is used as separator

Finally, in this unit it is performed synonym replacement. This consists in computing for each term of the email message, using Wordnet, the synonyms of the terms, and wherever a synonym of its has larger frequence than the term's one with largest frequence in email corpus, it replaces the term in the email.

2.4 Term Selection

After Stopping Word Removal, an email message is usually represented by a vector of some hundreds of thousands of terms. Hence, it is necessary to apply any feature selection method to reduce the number of words as low as possible. For this reason, in this stage a score is computed for each term. Finally, the terms whose score is larger than θ, whose value is a priori fixed, are picked, whereas the rest of terms are removed.

As feature score, *information gain* (or *mutual information*) [8] has been adopted. The information gain $\tau(x_i)$ assigned to the term x_i is given by:

$$\tau(x_i) = \sum_{c \in \{\mathcal{C}\}} \sum_{t \in \{x_i, x_i'\}} P(t, c) \log \left[\frac{P(t, c)}{P(t) P(c)} \right] \tag{1}$$

where x_i' denotes the absence of x_i and $\mathcal{C}$ the set of the all possible categories a term can belong to.

2.5 Bag of Words Representation of Email Message

In this stage, each email message, formed by the list of terms survived at Feature Selection stage, is converted in a format, suitable to be processed by a machine learning-based classification algorithm.

In order to do that, *Bag of Words* (*BOW*) approach [9] is applied to the message. If $\mathcal{T} = (t_1, t_2, \ldots, t_n)$ denotes the set of words that appear in the message, the message is represented by a n-dimensional feature vector $\boldsymbol{T} = (x_1, x_2, \ldots, x_n)$, with $x_i = h(\nu(t_i))$, $i = 1, \ldots, N$, where $\nu(t_i)$ is the occurrence of the term t_i in the message and $h(\cdot)$ is an appropriate function.

Among several proposed methods for computing $h(\cdot)$ [10], *tf-idf* (*term frequency-inverse document frequency*) approach [10] has been adopted. In tf-idf approach, the feature x_i associated to the term t_i in the message $\boldsymbol{T} \in \mathcal{T}$ is given by:

$$x_i = \nu_{t_i T} \log \left(\frac{|\mathcal{T}|}{\nu_{t_i}} \right), \tag{2}$$

where $\nu_{t_i T}$ is the number of occurrences of the term t_i in the email message $\boldsymbol{T}$ and ν_{t_i} the number of occurrences of the term t_i in all email messages of $\mathcal{T}$.

2.6 Classification

The classification is performed by Support Vector Machines that we remind. Support Vector Machine (*SVM*) [11, 12] is a binary classifier and the patterns with output $+1$ and -1 are called positive and negative, respectively. The underlying idea of SVM is the computation of the *optimal hyperplane algorithm*, i.e., the plane that yields the maximum separation margin between two classes. Let $\mathcal{T} = \{(\boldsymbol{x}_1, y_1), \ldots, (\boldsymbol{x}_\ell, y_\ell)\}$ be the training set, the optimal hyperplane $y(\boldsymbol{x}) = \boldsymbol{m} \cdot \boldsymbol{x}$ can be obtained solving the following constrained optimization problem

$$\min_m \frac{1}{2}\|\boldsymbol{x}\|^2 \quad \textit{subject to } y_i((\boldsymbol{m} \cdot \boldsymbol{x}_i + b)) \geq 1 \quad i = 1, \ldots, \ell. \tag{3}$$

However, in real world applications, mislabelled examples might exist yielding a partial overlapping of the two classes. To cope with this problem, it allows that some patterns can violate the constraints, introducing slack variables. They are strictly positive when the respective pattern violate the constraint, are null otherwise. SVM allows controlling at the same time the margin, expressed by $\|m\|$, and the number of the training errors, given by the non-null ξ_i, by the minimization of the constrained optimization problem:

$$\min_m \frac{1}{2}\|\boldsymbol{x}\|^2 + C\sum_{i=1}^{\ell}\xi_i \quad \textit{subject to } y_i((\boldsymbol{m} \cdot \boldsymbol{x}_i + b)) \geq 1 \quad i = 1, \ldots, \ell. \tag{4}$$

The constant C, sometimes called *regularization constant*, manages the trade-off between the separation margin and the number of the missclassifications. Since both the function and constraints are convex, the problem (4) can be solved by means of the method of Lagrange multipliers α_i, obtaining the following final form:

$$\max_\alpha \sum_{i=1}^{\ell}\alpha_i - \frac{1}{2}\sum_{i=1}^{\ell}\sum_{j=1}^{\ell}\alpha_i\alpha_j y_i y_j(\boldsymbol{x}_i \cdot \boldsymbol{x}_j) \quad \textit{subject to} \tag{5}$$

$$0 \leq \alpha_i \leq C \quad i = 1, \ldots, \ell \tag{6}$$

$$\sum_{i=1}^{\ell}\alpha_i y_i = 0. \tag{7}$$

The vectors, whose respective multipliers are non-null, are called *support vectors*, justifying the name of the classifier. Optimal hyperplane algorithm implements the following decision function:

$$f(\boldsymbol{x}) = sgn\left(\sum_{i=1}^{\ell} \alpha_i y_i (\boldsymbol{x_i} \cdot \boldsymbol{x}) + b\right). \tag{8}$$

In order to get SVM it is adequate to map nonlinearly input data in a Feature Space $\mathcal{F}$, before computing the optimal hyperplane. This can be performed by using *kernel trick* [13], i.e., replacing the inner product with an appropriate Mercer kernel $G(\cdot)$, obtaining the final form of the SVM decision function:

$$f(\boldsymbol{x}) = sgn\left(\sum_{i=1}^{\ell} \alpha_i y_i G(\boldsymbol{x_i}, \boldsymbol{x}) + b\right). \tag{9}$$

In the recognizer it is used, as Mercer Kernel, the *Gaussian Kernel* [13], defined as $G(\boldsymbol{x}, \boldsymbol{y}) = \exp(-\gamma \|x - y\|^2)$, where $\gamma \in \mathbf{R}^+$.

In this work the parameter γ of gaussian kernel and regularization constant C have been fixed by model selection methods, e.g., *k-fold cross-validation* [14]. SVM experiments in this work have been performed using *LibSVM* [15] software package.

3 Experimental Results

For the experimental validation of Spam Detection system two public domain email datasets, *SpamAssassin* [1] and *Trec2007* [16], have been used.

SpamAssassin is a labeled email dataset, composed of 1896 email labelled *Spam* and 4400 emails labelled *Ham*. In the Ham email set, there is a small subset, composed of 250 emails, considered *Hard Ham* since they are emails very difficult to classify ham due to their structure.

Trec2007 email dataset is composed of 50,199 and 25,220 emails, labelled as *Spam* and *Ham*.

Two different tests have been performed. In the first test, Spam Detector was trained using Trec2007 as training set, and was tested on SpamAssassin. Two different spam detectors were tested, the former with synonym replacer (SR), the latter without synonym replacer. For both spam detectors in the term selection module, the threshold for information gain has been fixed to 10^{-4}. For SVM classifier, gaussian kernel is used and SVM parameters has been set up by k-fold cross-validation [14]. As reported in Table 1, the system without synonym replacer has performed slighty better w.r.t. the other system. As general comment, it is important to observe that the percentage of False Ham, i.e., spam message missclassfied as ham, is very poor.

In the second test, the system used SpamAssassin as training set and Trec2007 as test set. The same experimental set up of the first set was used, with the only difference that for both spam detectors the value of the threshold for information gain, in the term selection module, has been chosen equal to 10^{-3}. The results are shown in Table 2. Even in this test spam detector without synonym replacement performs slighty better than the other with Synonym Replacement.

Table 1 Results using Trec2007 and SpamAssassin as training and test set, respectively

	With SR	Without SR
# True Ham	2415	2485
# True Spam	1418	1422
# False Ham	74	71
# False Spam	1733	1663
# Terms	324,245	324,245
# Terms after word removal	33,042	27,533
Accuracy (%)	67.96	69.26
Precision	0.450	0.461
Recall	0.950	0.952
F1	0.611	0.621

Table 2 Results using SpamAssassin and Trec2007 as training and test set, respectively

	With SR	Without SR
# True Ham	22,814	23,013
# True Spam	31,779	32,769
# False Ham	13,463	12,472
# False Spam	2373	2174
# Terms	61,013	61,013
# Terms after word removal	6723	4046
Accuracy (%)	77.51	79.20
Precision	0.931	0.938
Recall	0.702	0.724
F1	0.800	0.817

As general comment, it is possible to observe that results, in terms of accuracy, are satisfactory even the test set cardinality exceeds ten times the one of training set, as it happens in the first test. Besides, there is no statistical evidence that Synonym Replacement yields improvements about the spam detector accuracy. On the contrary, spam detector accuracy (see Table 2) seems to decrease.

4 Conclusions

In this paper a Spam Detector System, using a machine Learning based Content analysis of the spam message, has been described. The system is composed of six modules, i.e., *Tokenization and Cleaning words*, *Lemmatization*, *Stopping Module and Synonym Replacement*, *Term Selection*, *Bag-of-Words Representer*, and *Classifier*. The Spam Detector System has been validated using two public domain benchmarks, SpamAssassin, Trec2007 in two different tests, using each time a benchmark as training set and the other as test set. Results obtained are comparable with state of the art [3, 4].

In the next future we plan to improve the spam detection system investigating the following research lines, namely improving bag-of-words using N-gram representation, and investigating adversarial learning techniques in order to empower the robustness of the spam detection system.

Acknowledgments Daniele Davino developed part of the work, as exam project for Multimedia Systems and Laboratory, during his M.Sc. in Computer Science at University of Naples Parthenope, under the supervision of Francesco Camastra and Angelo Ciaramella. Francesco Camastra's, Angelo Ciaramella's, and Antonino Staiano's researches were funded by *Sostegno alla ricerca individuale per il triennio 2015–17* project of University of Naples Parthenope.

References

1. Cormack, G., Lynam, T.: Spam corpus creation for TREC. In: CEAS, pp. 1–6. MIT Press (2005)
2. Camastra, F., Ciaramella, F., Staiano, A.: Machine learning and soft computing for ict security: an overview of current trends. J. Ambient Intell. Humaniz. Comput. **4**(2), 235–247 (2013)
3. Guzella, T., Caminhas, W.: A review of machine learning approaches to spam filtering. Expert Syst. Appl. **36**(7), 10206–10222 (2009)
4. Caruana, G., Li, M.: A survey of emerging approaches to spam filtering. ACM Comput. Surv. **44**(2), 9.1–9.27 (2012)
5. Porter, M.: An algorithm for suffix stripping. Program **14**(3), 130–137 (1980)
6. Fellbaum, C.: Wordnet. In: The Encyclopedia of Applied Linguisticsl. American Cancer Society (2012)
7. Saini, J., Rakholia, R.M.: On continent and script-wise divisions-based statistical measures for stop-words lists of international languages. Procedia Comput. Sci. **89**, 313–319 (2018)
8. Cover, T.M., Thomas, J.: Elements of Informtion Theory. Wiley (1991)
9. Salton, G., Wong, A., Yang, C.: A vector-space model for automatic indexing. Commun. ACM **18**(11), 613–620 (1975)
10. Sebastiani, F.: Machine learning in automated text categorization. ACM Comput. Surv. **34**(1), 1–47 (2002)
11. Cortes, C., Vapnik, V.: Support vector networks. Mach. Learn. **20**, 1–25 (1995)
12. Vapnik, V.: Statistical Learning Theory. Wiley, New York (1998)
13. Schölkopf, B., Smola, A.: Learning with Kernels. MIT Press, Cambridge (USA) (2002)
14. Hastie, T., Tibshirani, R., Friedman, R.: The Elements of Statistical Learning, 2nd edn. Springer (2009)
15. Chang, C.C., Lin, C.J.: LIBSVM: a library for support vector machines. ACM Trans. Intell. Syst. Technol. **2**(3), 1–27 (2011)
16. Bailey, P., De Vries, A., Craswell, N., Soboroff, I.: Overview of the TREC 2007 enterprise track. In: TREC, pp. 1–7. MIT Press (2007)

A Multimodal Deep Network for the Reconstruction of T2W MR Images

Antonio Falvo, Danilo Comminiello, Simone Scardapane, Michele Scarpiniti, and Aurelio Uncini

Abstract Multiple sclerosis is one of the most common chronic neurological diseases affecting the central nervous system. Lesions produced by the MS can be observed through two modalities of magnetic resonance (MR), known as T2W and FLAIR sequences, both providing useful information for formulating a diagnosis. However, long acquisition time makes the acquired MR image vulnerable to motion artifacts. This leads to the need of accelerating the execution of the MR analysis. In this paper, we present a deep learning method that is able to reconstruct subsampled MR images obtained by reducing the k-space data, while maintaining a high image quality that can be used to observe brain lesions. The proposed method exploits the multimodal approach of neural networks and it also focuses on the data acquisition and processing stages to reduce execution time of the MR analysis. Results prove the effectiveness of the proposed method in reconstructing subsampled MR images while saving execution time.

Keywords Magnetic resonance imaging · Fast MRI · Multiple sclerosis · Deep neural network

1 Introduction

Nuclear magnetic resonance (NMR) is a transmission analysis technique that allows to obtain information on the state of matter, exploiting the interaction between magnetic fields and atoms nuclei. In the biomedical field, information deriving from the NMR is represented in the form of tomographic images. Nowadays, the NMR plays an important role in the health field, and it allows to carry out a whole typology of diagnostic exams, from traditional to functional neuroradiology, from internal diagnostic to obstetrics and pediatric diagnostics [1].

A. Falvo · D. Comminiello (✉) · S. Scardapane · M. Scarpiniti · A. Uncini
Department of Information Engineering, Electronics and Telecommunications (DIET), Sapienza University of Rome, Via Eudossiana 18, 00184 Rome, Italy
e-mail: danilo.comminiello@uniroma1.it

A. Esposito et al. (eds.), *Progresses in Artificial Intelligence and Neural Systems*, Smart Innovation, Systems and Technologies 184,
https://doi.org/10.1007/978-981-15-5093-5_38

During the acquisition stage of an MR signal, it is necessary to sample the entire k-space to obtain images that are as much detailed as possible [2, 3]. Data in the k-space encode information on spatial frequencies and are generally captured line by line. Therefore, the acquisition time for a given sequence depends on the number of lines sampled in the k-space, thus leading to a rather slow acquisition process. Moreover, significant artifacts may occur in the MR images caused by to slow movements of the patient, due to physiological factors or to fatigue, e.g., too much time in the same position [2, 3]. Moreover, the long scan time also increases the healthcare cost for the patient, besides limiting the availability of MR scanners.

Over the years, several methods, such as compressed magnetic resonance and parallel magnetic resonance [4–7], have been proposed to accelerate MRI scans by skipping the k-space phase coding lines and avoid the aliasing phenomenon introduced by subsampling. The problem of accelerating magnetic resonance can also be tackled through deep learning techniques. In particular, the reconstruction of tomographic images has been often efficiently addressed by using convolutional neural networks (CNNs) [8–12].

Most of the state-of-the-art methods focus on the reconstruction of MR images using a unimodal neural architecture where a subsampled image to be reconstructed is provided as input. In this paper, we propose a new deep learning method for reconstructing MR images by exploiting additional information provided by FLAIR images. Such images are widely used in the MR diagnosis as they allow to enhance the brain lesions due to the disease. FLAIR images are highly correlated with the T2 weighted images (T2WIs), thus the joint use of such images increases the efficiency of the reconstruction and also presents much more information in the lesion region. To exploit both the images, we propose a multimodal deep neural network, inspired by the well-known U-Net, a convolutional-based model that was developed for biomedical image segmentation [13]. In the literature, several studies have been proposed using a multimodal approach for image reconstruction. In [14], T2WIs were attempt to be estimated from T1WI, while other works focus on improving the quality of the subsampled images with the help of high resolution images with different contrast [15, 16]. However, to the best of our knowledge, no attempt has ever been made to reconstruct T2WIs from subsampled T2WIs (T2WIsub) and FLAIR images while maintaining high image quality in the area of lesions.

Experimental results prove that the proposed method is able to accelerate the MR analysis four times, while preserving the image quality, with a high detail of any lesion and negligible aliasing artifacts.

The rest of the paper is organized as follows. In Sect. 2, we introduce the proposed approach, including a new subsampling mask, while the proposed Multimodal Dense U-Net is presented in Sect. 3. Results are shown in Sect. 4 and, finally, our conclusion is drawn in Sect. 5.

2 Proposed Approach: Main Definitions

We first focus on the images to be provided as input in order to reconstruct the T2WI.

2.1 *Problem Formulation*

We denote with $\mathcal{X}_{T2}$ the k-space for the T2WI that represents the target. Multiplying the k-space $\mathcal{X}_{T2}$ for a suitably designed mask M, it is possible to obtain a subsampled version of the k-space, i.e.,

$$\mathcal{X}_{T2sub} = M \cdot \mathcal{X}_{T2} \tag{1}$$

The bidimensional inverse Fourier transform allows to achieve data into the space domain. Therefore, we define the fully-sampled target image Y_{T2} and the subsampled T2 image Y_{T2sub} to be used for reconstruction through the proposed deep network. Finally, we denote the FLAIR image to be provided as input with Y_F.

We want to reconstruct the fully-sampled T2 image Y_{T2}, given the only availability of subsampled Y_{T2sub} and Y_F. The reconstructed T2 image is denoted as $\hat{Y}_{T2}$. To this end, we build and train a deep network to minimize the following loss function:

$$\arg\min \{\text{MSE} + \text{DSSIM}\} \tag{2}$$

in which the MSE denotes the mean-square error and DSSIM the structural dissimilarity index. The former is defined as:

$$\text{MSE} = \frac{1}{N} \sum_{i=1}^{N} \left(Y_{T2,i} - \hat{Y}_{T2,i} \right)^2 . \tag{3}$$

On the other hand, the DSSIM is complementary to the structural similarity index (SSIM), which is often adopted to assess the perceived quality of television and film images as well as other types of digital images and videos. It was designed to improve traditional methods such as the signal-to-peak noise ratio (PSNR) and the mean square error (MSE), and it is defined as:

$$\text{DSSIM}\left(Y_{T2}, \hat{Y}_{T2} \right) = \frac{1}{2} - \frac{(2\mu_Y \mu_{\hat{Y}} + c_1)(2\sigma_{Y\hat{Y}} + c_2)}{2\left(\mu_Y^2 + \mu_{\hat{Y}}^2 + c_1\right)\left(\sigma_Y^2 + \sigma_{\hat{Y}}^2 + c_2\right)} \tag{4}$$

where μ_Y, $\mu_{\hat{Y}}$ represent the mean values, σ_Y^2 and $\sigma_{\hat{Y}}^2$ the variances, and $\sigma_{Y\hat{Y}}$ the covariance.

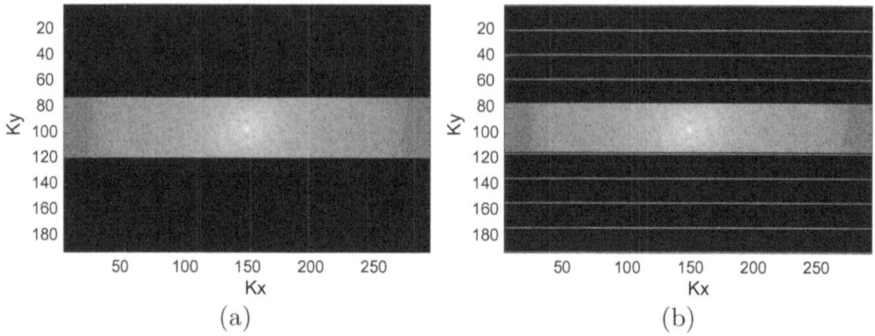

Fig. 1 Subsampling masks: **a** center mask and **b** the proposed custom mask

2.2 Customization of a New Subsampling Mask

Most of the existing literature dealing with MRI acceleration mainly focuses on the reconstruction of images. However, the quality of the reconstruction depends significantly on how k-space is sampled. This problem can be faced essentially by adopting one of the following approaches: (1) a *dynamic approach* based on deep learning in which cells of fixed width are allowed to move in the k-space and change position based on the reconstruction performance; (2) a *static approach* in which fixed sampling masks are used that go to select only certain areas of the k-space.

In this work, we choose the static approach since the dynamic one does not guarantee that the power spectrum of a reference image is similar to that of the test image. The adopted sampling method consists of a mask that acts along the direction of the phase coding of the k-space, in which it is possible, once the subsampling factor is set, to choose the percentage of samples that will occupy the central part of k-space, thus leaving the rest of the samples equidistant from each other.

Figure 1 shows two different types of mask both obtained by setting a subsampling factor $k = 4$. In the center mask of Fig. 1a, samples are taken exclusively in the central area of the k-space, where most of the low-frequency components can be found providing useful information on the contrast of the image [14]. However, in this work, we propose a new mask, depicted in Fig. 1b, which selects the 80% of the total samples from the center and the remainder in an equidistant manner so as to have information even in the high frequencies.

3 Multimodal Dense U-Net

The proposed neural network is multimodal architecture: on one branch we provide the T2WIsub as input while on the other branch we provide the FLAIR image to be used to improve the reconstruction quality. We expect all the spatial information in the FLAIR image to help estimate the anatomical structures in T2WI.

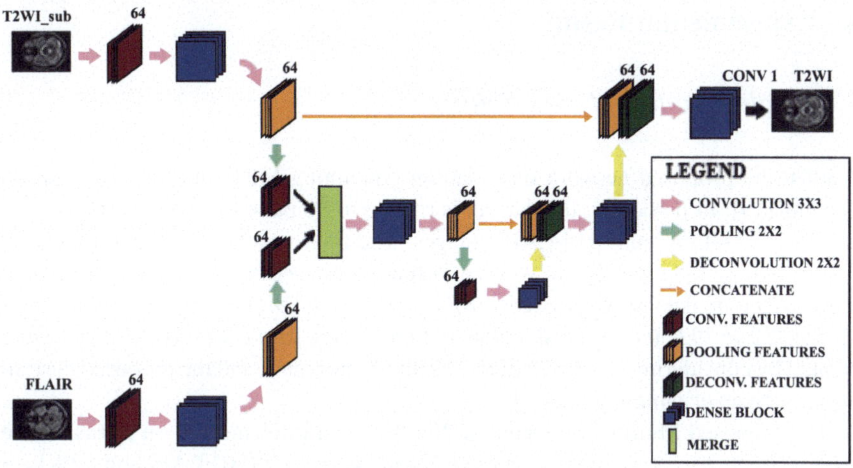

Fig. 2 Scheme of the proposed Multimodal Dense U-Net architecture

Both inputs initially undergo separate contraction transformations to then merge later and follow the classic coding-decoding approach of U-Net models. The proposed Multimodal Dense U-Net is depicted in Fig. 2.

The network consists essentially of 4 components, namely convolutive layers, pooling layers, deconvolutive layers and dense blocks. The size of the characteristic map decreases along the contraction path through the pooling blocks as it increases along the expansion path by deconvolution. Pooling partitions the input image into a set of squares, and for each of the resulting regions returns the maximum value as output. Its purpose is to progressively reduce the size of the representations, so as to reduce the number of parameters and the computational complexity of the network, at the same time counteracting any overfitting.

Deconvolutive layers act inversely with respect to pooling and aim to increase the spatial dimensions of the inputs. This allows to obtain images of a size comparable to those of the input images from the network. In the simplest case these levels can be implemented as static oversampling with bilinear interpolation.

The dense block, proposed in [17], allows to effectively increase the depth of the entire network while maintaining a low complexity. Moreover, it requires less parameters to be trained. The dense block consists of three consecutive operations: batch normalization (BN), ELUs activation functions [18] and 3×3 convolution filters. The hyper-parameters for the dense block are the growth rate (GR) and the number of convolutional layers (NC). The network ends with a reconstruction level consisting of a dense block followed by a 1×1 convolutional layer that yields the reconstructed T2WI.

4 Experimental Results

4.1 Dataset and Network Setting

We test the proposed network on a dataset containing MRIs of multiple sclerosis patients [19]. In particular, the dataset is related to 30 patients and it contains axial 2D-T1W, 2D-T2W and 3D-FLAIR images. The final voxel size of such images is $0.46 \times 0.46 \times 0.8$ mm^3. In our work, a further preprocessing has been performed in MATLAB to make the voxel size isotropic to $0.8 \times 0.8 \times 0.8$ mm^3, to extract slices of size 192×292, and to shrink intensity to the range [0, 1]. T2WIsub images were created by considering two types of masks, the center mask and the proposed custom mask, with a subsampling factor $k = 4$.

The proposed Multimodal Dense U-Net has been implemented on Keras. In the training stage, for each patient we provide the network with 150 FLAIR and T2WIsub images using the T2WIs as target. For dense blocks, we set a zero growth rate and a number of levels equal to 5 with feature maps size of 64 and ELU activation levels. We use Adam as an optimizer for training. A total of 80 epochs are performed, with early stopping. The duration of each epoch is about 15 min, having set a batch size of 4 and using a desktop PC with an Intel Core i5 6600-K 3.50 GHz CPU, 16 GB of RAM and NVIDIA GeForce GTX 970 GPU. To quantitatively evaluate reconstruction performance, we use the MSE and DSSIM metrics.

4.2 Evaluation of the Proposed Mask

We want to evaluate first the effectiveness of the proposed custom subsampling mask compared to the center mask on the quality of reconstruction in terms of the SSIM using a Dense U-Net network.

Results are shown in Fig. 3, where it is clear that the proposed custom mask allows us to obtain a reconstruction of the image with outstanding performance. In particular, using the center mask we get a 71% reconstruction percentage compared to the target (Fig. 3a), while using the proposed custom mask (Fig. 3b) the similarity index rises up to 86%.

4.3 Evaluation of the Proposed Deep Architecture

Conceptually, the proposed architecture and the standard Dense U-Net might appear similar, but the former one manages the two inputs differently. Moreover, the hyper-parameters of the dense blocks chosen for our network considerably change the concept of dense block as the whole of growth was set to zero thus avoiding internal expansion in dense blocks.

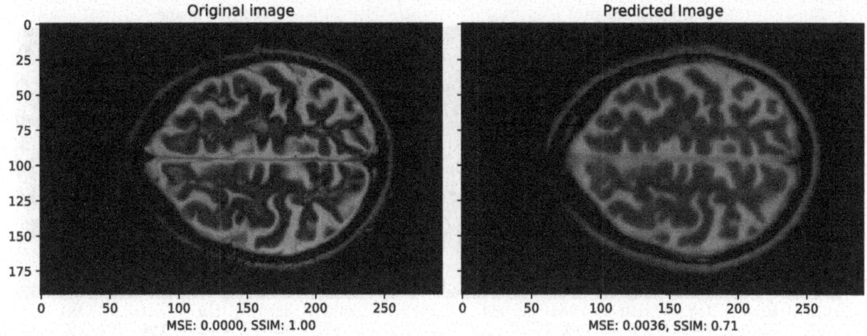

(a)

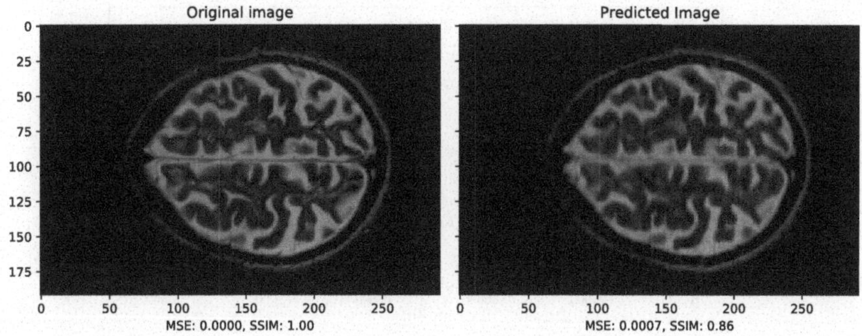

(b)

Fig. 3 Predicted images using: **a** center mask and **b** the proposed custom mask

We compare the reconstruction quality of the two networks in terms of SSIM having used the mask that provided the best performance for the subsampling, i.e., the proposed custom mask. Results are shown in Fig. 4, where it is clear that the quality of reconstruction has been considerably improved compared to Dense U-Net. In particular, the degree of similarity with respect to the target is 94% rather than 86% of the Dense U-Net. By using the proposed architecture, high image quality is achieved, thus enabling the recognition of brain injuries caused by the disease. We also show the loss function behavior for the proposed method in Fig. 5.

5 Conclusion

In this work, we propose a deep learning model exploiting the capabilities of both multimodal networks and dense blocks. In particular, the proposed approach allows to reconstruct T2WIs, subsampled by a factor of 4, thus leveraging the correlation

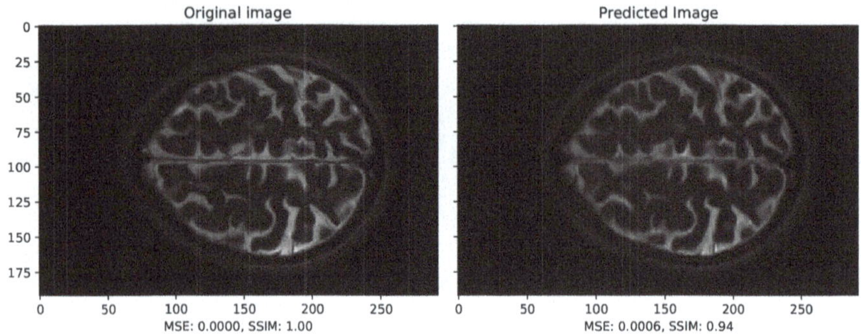

Fig. 4 Predicted T2WI reconstructed by the proposed Multimodal Dense U-Net

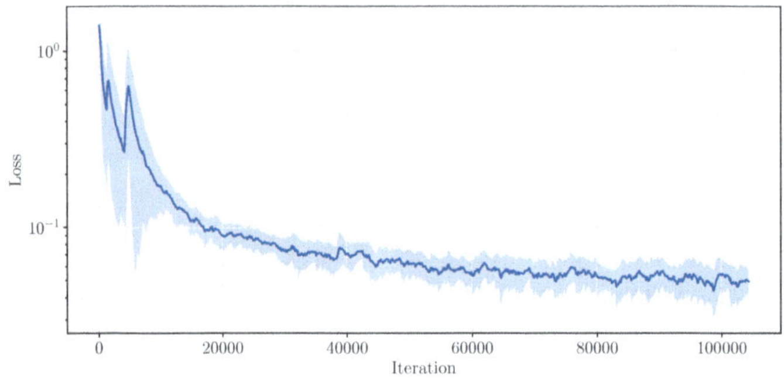

Fig. 5 Loss function behavior

that exists with FLAIR images. At the same time, the proposed method is able to maintain a high quality of image reconstruction, in particular in the area of the brain lesions due to multiple sclerosis. The comparison with a state-of-the-art Dense U-Net architecture has shown that the proposed network outperforms both in terms of perceptive quality and in terms of execution times. Future works will focus on increasing the speed of the MRI scan, with the goal of achieving an acceleration of at least 10 times, and on further improving the reconstruction quality.

References

1. Beall, P.T., Amtey, S.R., Kasturi, S.R.: NMR Data Handbook for Biomedical Applications. Pergamon Books Inc., Elmsford, NY (1984)
2. Haacke, E.M., Brown, R.W., Thompson, M.R., Venkatesan, R.: Magnetic Resonance Imaging: Physical Principles and Sequence Design, vol. 82. Wiley-Liss, New York, NY (1999)
3. Liang, Z.P., Lauterbur, P.C.: Principles of Magnetic Resonance Imaging. A Signal Processing Perspective. The Institute of electrical and Electronics Engineers, New York, NY (2000)
4. Gamper, U., Boesiger, P., Kozerke, S.: Compressed sensing in dynamic MRI. Magn. Reson. Med. **59**(2), 365–373 (2008)

5. Jaspan, O.N., Fleysher, R., Lipton, M.L.: Compressed sensing MRI: a review of the clinical literature. Br. J. Radiol. **88**(1056) (Dec 2015)
6. Lustig, M., Donoho, D., Pauly, J.M.: Sparse MRI: the application of compressed sensing for rapid MR imaging. Magn. Reson. Med. **58**, 1182–1195 (2007)
7. Lustig, M., Donoho, D.L., Santos, J.M., Pauly, J.M.: Compressed sensing MRI. IEEE Sig. Proc. Mag. **25**(2), 72–82 (2008)
8. Jin, K.H., McCann, M.T., Froustey, E., Unser, M.: Deep convolutional neural network for inverse problems in imaging. IEEE Trans. Image Proc. **26**(9), 4509–4522 (2017)
9. McCann, M.T., Jin, K.H., Unser, M.: Convolutional neural networks for inverse problems in imaging: a review. IEEE Sig. Proc. Mag. **34**(6), 85–95 (2017)
10. Qin, C., Schlemper, J., Caballero, J., Price, A.N., Hajnal, J.V., Rueckert, D.: Convolutional recurrent neural networks for dynamic MR image reconstruction. IEEE Trans. Med. Imaging **38**(1), 280–290 (2019)
11. Roy, S., Butman, J.A., Reich, D.S., Calabresi, P.A., Pham, D.L.: Multiple sclerosis lesion segmentation from brain MRI via fully convolutional neural networks. arXiv preprint arXiv:1803.09172v1 (Mar 2018)
12. Schlemper, J., Caballero, J., Hajnal, J.V., Price, A.N., Rueckert, D.: A deep cascade of convolutional neural networks for dynamic MR image reconstruction. IEEE Trans. Med. Imaging **37**(2), 491–503 (2018)
13. Ronnenberger, O., Fischer, P., Brox, T.: U-Net: convolutional networks for biomedical image segmentation. In: International Conference on Medical Image Computing and Computer-Assisted Intervention (MICCAI). Lecture Notes in Computer Science, vol. 9351, pp. 234–241. Springer, Cham (2015)
14. Xiang, L., Chen, Y., Chang, W., Zhan, Y., Lin, W., Wang, Q., Shen, D.: Deep leaning based multi-modal fusion for fast MR reconstruction. IEEE Trans. Biomed. Eng. (Early Access) (2018)
15. Huang, J., Chen, C., Axel, L.: Fast multi-contrast MRI reconstruction. Magn. Reson. Imaging **32**(10), 1344–1352 (2014)
16. Kim, K.H., Do, W.J., Park, S.H.: Improving resolution of MR images with an adversarial network incorporating images with different contrast. Med. Phys. **45**(7), 3120–3131 (2018)
17. Huang, G., Liu, Z., van der Maaten, L., Weinberger, K.Q.: Densely connected convolutional networks. In: IEEE Conference on Computer Vision and Pattern Recognition (CVPR). pp. 2261–2269. Honolulu, HI (Jul 2017)
18. Clevert, D.A., Unterthiner, T., Hochreiter, S.: Fast and accurate deep network learning by exponential linear units (ELUs). In: International Conference on Learning Representations (ICLR). pp. 1–14. San Juan, Puerto Rico (May 2016)
19. Žiga, L., Galimzianova, A., Koren, A., Lukin, M., Pernuš, F., Likar, B., Špiclin, V.: A novel public MR image dataset of multiple sclerosis patients with lesion segmentations based on multi-rater consensus. Neuroinformatics **16**(1), 51–63 (2018)

Dynamics of Signal Exchanges and Empathic Systems—Dedicated to Anna Costanza Baldry

Facial Emotion Recognition Skills and Measures in Children and Adolescents with Attention Deficit Hyperactivity Disorder (ADHD)

Aliki Economides, Yiannis Laouris, Massimiliano Conson, and Anna Esposito

Abstract Facial expressions have significant communicative functions; any changes in the facial muscles help disentangle meaning, control the conversational flow, provide information as to the speaker/listener's emotional state and inform about intention. Abnormalities in the recognition of facial expressions have been associated with psychiatric disorders. This review focuses on facial recognition abilities in children and adolescents with Attention Deficit Hyperactivity Disorder (ADHD). Using PRISMA guidelines original articles published prior to August 2019 were identified focusing on the emotion recognition skills of children and adolescents with ADHD and the measures administered. 25 studies were identified with the majority (18) showing some deficits on emotion recognition in children/adolescents with ADHD compared to typically developing (TD) children. The results are synthesized in terms of the type of stimuli implicated (static vs. dynamic), the measures/tasks administered, whether authors differentiated among specific emotion dimensions in the analysis of results, the effect of comorbidity on emotion recognition, and whether greater deficits have been reported for some emotions compared to others. Studies on facial emotion recognition in children and adolescents with ADHD focused mainly on the recognition accuracy of facial emotions, showing inconsistent results and a

A. Economides · M. Conson · A. Esposito (✉)
Dipartimento di Psicologia, Università Della Campania Luigi Vanvitelli, Caserta, Italy
e-mail: iiass.annaesp@tin.it

A. Economides
e-mail: aliki.economides@unicampania.it

M. Conson
e-mail: conson.massimiliano@gmail.com

Y. Laouris
Cyprus Neuroscience & Technology Institute, Nicosia, Cyprus
e-mail: laouris@cnti.org.cy

A. Esposito
International Institute for Advanced Scientific Studies (IIASS), Vietri Sul Mare, Italy

A. Esposito et al. (eds.), *Progresses in Artificial Intelligence and Neural Systems*, Smart Innovation, Systems and Technologies 184,
https://doi.org/10.1007/978-981-15-5093-5_39

heterogenous use of measures. It is unknown whether the studies' participants followed therapeutic plans (other than pharmacotherapy) at the time of the study or before, a factor that may potentially have influenced the review's results.

1 Introduction

Facial expressions have been the focus of interest for more than 10 decades. As early as in the 1870 s, Charles Darwin wrote of facial expressions and emotions. It was not nevertheless until the 1970s that Ekman [30, 31] and Izard (1971, 1977) studied the field of perception and categorization of facial expressions. Though not having gone without criticisms, it is believed that the 'basic' emotions are six namely, fear, disgust, anger, happiness, surprise and sadness. Emotion recognition is crucial for social interaction and functioning and social cognition requires the ability to recognize, encode, and interpret emotions from faces [49]. Emotional understanding skills are hence essential for healthy social development in early childhood as one must decode and understand others' reactions, motivations and intentions [81] in everyday interactions. Emotional understanding skills are also closely related to developmental outcomes such as language, literacy, school readiness and mathematic skills in preschool [27, 71].

The recognition of emotional expressions improve with age [12] and develop gradually over time [24, 84], with some emotional expressions, i.e. happiness and sadness, being recognized earlier than others i.e. fear and disgust [38, 37]. It is now well established that happiness is the first to be recognized and with the greatest accuracy. Posamentier and Abdi [70] for example, found that infants produce their first smile 2 to 12 h after birth and Southam-gerow and Kendall [78] argued that by 4 years of age most children can pose basic facial expressions equally well to adults. Individual differences in emotion knowledge and understanding are key factors in current and later prosocial behavior [58, 59] with children skilled in emotion recognition being reported as more likeable by their peers when compared to other same-aged children. In contrast, children with poor emotion recognition skills are at increased risk for being rejected or victimized by peers [59, 75]. Generally, it appears that young children with deficits in emotion recognition skills are at increased risk for several negative outcomes such as aggression, academic failure, poor emotional relationships and delinquency [18, 2].

Focusing on task demands children by 6 years of age, for example, were found to have a nearly perfect score when asked to point to which of the two faces was happy, angry, surprised, or sad but a good accuracy level was only achieved by the age of 10 when children were asked to select which of the two faces expressed the same emotion as a third face [13]. In a similar study by Mondloch, Geldart, Maurer, and Le Grand [60] children's performance reached the level of adults when asked to match an emotional photograph to either surprise, happiness, neutral, or disgust with accuracy increasing between 6 to 8 years of age. Whereas task demands influence emotion recognition accuracy, the age of 6 to 8 is a critical point of time as healthy children appear able to recognize the 6 basic emotions equally well to adults.

2 Facial Expressions and Psychopathology

Facial expressions have significant communicative functions; any changes in the facial muscles help disentangle meaning, control the conversational flow, provide information as to the speaker/listener's emotional state and inform about intention [34]. Understanding others' emotional facial expressions is a significant social-cognitive skill which helps to modulate one's behavior: for example, is a friend frightened or excited at the sight of a dog, is an observer becoming upset or surprised by an act of bravado?

Abnormalities in the recognition of facial expressions have been associated with psychiatric disorders in both children [8] and adults [39]. A failure to identify emotional facial expressions can have wide-reaching and long-term detrimental effects upon social behavior [43]. Although different child and adolescent clinical populations have been shown to have deficits in facial expression recognition such as children diagnosed with Down syndrome [67], schizophrenia [17] conduct disorders [79], Attention Deficit Hyperactivity Disorder [49] and depressive disorder [21], autism is perhaps the most widely studied area in terms of developmental psychopathology and emotional deficits.

2.1 Attention Deficit Hyperactivity Disorder (ADHD)

ADHD is a neurodevelopmental disorder affecting people of all ages and of both genders. School-aged children with ADHD have been reported to suffer from social and emotional deficits, i.e. inability to effectively appraise the emotional state of others [16] and impairments in cognitive functions, i.e. inhibition, sustained attention, and executive planning [5]. Children with ADHD encounter many social problems, are generally less accepted by peers and lack social skills [29, 46]. Reduced social competence has been highly associated with the disorder [53] and the social problems encountered by these children constitute significant predictors of negative outcomes in later life i.e. adolescence and adulthood [61]. Factors related to emotional processing, and specifically deficient emotion recognition, has been discussed to play a key role [20].

The symptoms of ADHD begin in childhood (usually between the ages of 3 to 6), and for about half of the children, the symptoms continue into adolescence and adulthood. The primary symptoms of ADHD are (1) hyperactivity/impulsivity and (2) inattention and the specific presentation of symptoms may vary by age. Despite the lack of global consensus with regards to the prevalence of ADHD, it has been estimated that ADHD lies between 5.3% [68] and 7.1% [86] in children and adolescents and 1.2–7.3% in adults [35].

3 Objective

In this literature review we analyzed different studies highlighting the different aspects of facial emotion recognition in children with ADHD (with and without comorbidities) as well as the tasks administered to measure emotion recognition. The current review will complement past reviews [82] and meta-analyses [11] by solely focusing on studies that have implicated children and adolescent participants, by taking a closer look on the measures administered to assess emotion recognition and by considering the factor of comorbidity.

Despite Bora and Pantelis [11] having conducted an excellent review, the authors investigated social cognition in attention-deficit/hyperactivity disorder (ADHD) with comparisons having only been made among ADHD vs ASD and ADHD vs healthy controls. Uekermann et al. [82] included in their review studies from 1979–2009 with the authors having in detail argued the social cognition impairments in ADHD and Romani and colleagues 73 assessed face memory and face recognition in children and adolescents with attention deficit hyperactivity disorder. Nevertheless, in both papers detailed information such as the measures administered to assess affect recognition, the type of stimuli implicated, and the number of emotions investigated in each and across the studies included in the reviews have not been examined.

To our knowledge this is the first systematic review assessing emotion recognition in children and adolescents with ADHD that considered ADHD with comorbidities (e.g., ASD, CD and ODD) 'or other disorders' and the measures undertaken to assess emotion recognition.

4 Materials and Methods

4.1 Study Eligibility Criteria

We included academic articles (e.g. original articles and dissertations) focusing on facial emotion recognition in children and adolescents with ADHD (with and without comorbidities). This review considers studies published only between January 2000 and August 2019. Eligibility criteria hence constitute the study to have been published in a peer-reviewed journal from 2000 to 2019; participants to have been diagnosed with ADHD according to the criteria of the Diagnostic and Statistical Manual of Mental Disorders (DSM) by the American Psychiatric Association (APA); to have been children and adolescents and their emotion recognition skills to have been investigated.

4.2 *Information Sources*

For this review we used Elsevier, PsycINFO, PsycArticles, Medline and PubMed databases. This review also benefited from other widely used search engines such as Google Scholar, recommendations from Web Libraries such as Mendeley, and reference lists from single articles, editorials and reviews.

4.3 *Search Strategy*

This review's search strategy followed the guidelines of the Preferred Reporting Items for Systematic Reviews and Meta-Analyses (PRISMA) to identify all relevant studies published after January 2000. The terms included in the search strategy were "Face recognition", "Facial emotion recognition", "Face recognition and ADHD", "ADHD comorbidities", "Facial expressions", "Emotion recognition measures in ADHD" and "Emotion recognition measures". Only studies in the English language were included. The full article was retrieved when its abstract met the eligibility criteria or when there was not enough information to exclude the article. The articles retrieved in full text were subsequently reviewed to determine whether our inclusion criteria were really met.

As regards the exclusion criteria, other literature reviews and meta-analyses as well as studies that implicated adults or focused on the pharmacotherapy effects in patients with ADHD were excluded. Furthermore, in the case that children or adolescents were not diagnosed according to the criteria of DSM, were excluded.

4.4 *Risk of Bias Across Studies*

Database bias: Articles only in the English language were evaluated.

5 Results

5.1 *Available Literature*

Using PRISMA guidelines, 112 articles were selected in the initial search. After excluding duplicates, the abstracts of the remaining 102 articles were scanned and 71 further articles were excluded based on the exclusion criteria in Sect. 4.3. Full texts of 39 articles were examined and 14 were further excluded as they did not meet the eligibility criteria, resulting in 25 studies being included in the review (Fig. 1). Table 1 presents the studies included in this systematic review and Table 2

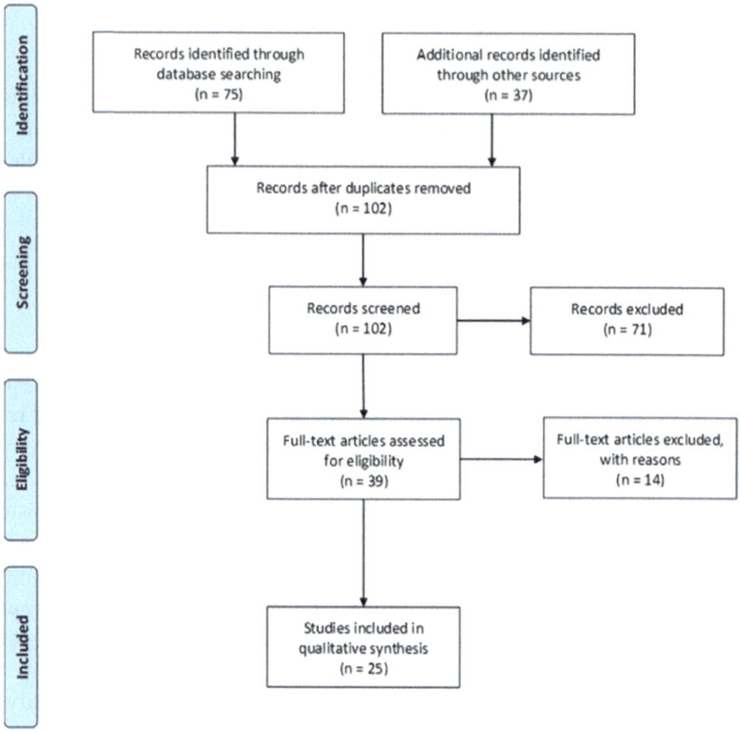

Fig. 1 Prisma flow chart

differentiates the studies included in the review based on the emotions investigated, whether authors differentiated among specific emotion dimensions in their analyses, and whether children with ADHD significantly differed from TD.

5.2 *Face Recognition in ADHD*

Table 1 presents 25 studies identified with the criteria set, with the majority (18) showing some deficits in emotion recognition in children with ADHD compared to typically developing children (TD) [1–5, 7–11, 13, 14, 16, 17, 21–23, 26]. There were 20 experimental, 1 genotype, 3 studies measuring brain and 1 study assessing both behavioral and neurophysiological parameters [24] in ADHD. All but one study [20] included a control group and ten included only boys [3–5, 7, 9, 16, 17, 19, 21, 22].

Table 1 Studies included in the systematic review

No. of study date & authors	No. and type of participants; age (mean, sd)	Emotion recognition task & stimuli	Emotions investigated	Results
1. [16] 2000 Corbett & Glidden	ADHD (n = 37) (28 boys, 11 girls) 6.8–12.8 yrs *(10.08, 1.78)* TD (n = 37) (19 boys, 18 girls) 6.8–12.8 yrs *(9.49, 1.92)*	**Ekman and Friesen Pictures of Facial Affect** [32] 28 slides from the Pictures of Facial Affect set were presented for as long as necessary for the participant to give a response on a laminated sheet listing the six basic expressions and a neutral response	Anger Sadness Happiness Fear Disgust Surprise	TD children performed significantly better than ADHD children. However, all groups comparisons were based on total scores collapsed across the six basic emotions hence the results do not offer any insights into the specific type of facial affect that ADHD children will express difficulties in processing
2. [14] 2000 Cadesky, Mota & Schachar	ADHD (n = 86) (69 boys, 17 girls) 7–13 yrs *(9.0, 1.4)* CD (n = 24) (20 boys, 4 girls) *(9.3, 1.6)* ADHD and CD (n = 6) (54 boys, 7 girls) *(9.3, 1.5)* TD (n = 27) (18 boys, 9 girls) *(9.3, 1.5)*	**Diagnostic Analysis of Non-Verbal Accuracy** [63] Children had to interpret emotional cues from pictures of facial expressions and recordings of voices drawn from the Diagnostic Analysis of Non-Verbal Accuracy (DANVA). The DANVA contains 4 subtests that were used by this study: Adult Facial Expressions [64], Child Facial Expressions, Adult Paralanguage [6], and Child Paralanguage [63]. For all subsets but the Child Paralanguage, each emotion was presented 6 times. For the Child Paralanguage subset 3 happy, 6 sad, 3 angry and 4 fearful stimuli were presented	Anger Sadness Fear Happiness	Children with conduct problems (CP) and ADHD were significantly less accurate at interpreting all emotions except anger than TD children. However, children with ADHD + CP differed in the type of errors made: the ADHD group's errors were generally random in nature, whereas the CP group tended to misinterpret emotions as anger. The ADHD + CP group performed better than the ADHD and CP groups, was as accurate as the control group and displayed a unique pattern of errors. The authors attributed the errors made by the ADHD group to deficits in encoding stimuli due to inattention rather than to specific distortions in interpreting emotions

(continued)

Table 1 (continued)

No. of study date & authors	No. and type of participants; age (mean, sd)	Emotion recognition task & stimuli	Emotions investigated	Results
3. [28] Downs & Smith	ADHD/ODD boys (n = 10) 5.8–9.9 yrs *(8.3, 1.0)* High functioning autistic boys (n = 10) 5.8–9.9 yrs *(7.10, 1.1)* TD boys (n = 10) 6.4–9.2 yrs *(7.7, 1.2)*	**Questions on Emotional Understanding** [45] Participants were presented with questions to determine their level of emotional understanding. Each child was assessed beginning at level 1 and continuing through to level 5. Level 1–identified emotional facial expressions in photographs; Level 2–identified emotional facial expressions in schematic drawings, Level 3–identified situation-based emotions; Level 4–identified desire-based emotions; Level 5–identified belief-based emotions	Not defined	The ADHD/ODD group answered significantly fewer total questions correctly on the theory of mind emotional understanding task than did the nonclinical group. The ADHD/ODD group also displayed a trend toward having a significantly lower level of emotional understanding than both the nonclinical group and the autism group
4. [3] 2004 Aspan et al.	Boys with the combined subtype of ADHD (n = 22) 13–16 yrs *(14.2, 0.2)* TD boys (n = 22) 13–16 yrs *(14.2, 0.2)*	**Facial Expressions of Emotion-Stimuli and Tests (FEEST)** [88] The computerized and extended version of the original 60 faces test from the Pictures of Facial Affect [32], Facial Expressions of Emotion-Stimuli and Tests (FEEST), was used. Participants were asked to choose a label for the emotional content of the faces visible on the screen. The images were displayed in a random order	Anger Sadness Happiness Fear Disgust Surprise	Adolescents with ADHD were more sensitive in the recognition of disgust, were significantly worse in the recognition of fear, and showed a tendency toward impaired recognition of sadness ($(1,40) = 3.771$, < 0.056). The recognition of anger, happiness and surprise did not show significant differences between the two groups. Hyperactivity measures were positively correlated with the recognition of disgust and inversely correlated with the recognition of fear

(continued)

Table 1 (continued)

No. of study date & authors	No. and type of participants; age (mean, sd)	Emotion recognition task & stimuli	Emotions investigated	Results
5. [66] Pelc, Kornreich, Foisy & Dan	Predominantly hyperactive-impulsive type of the ADHD disorder (n = 30) (23 boys, 7 girls) 7–12 yrs *(8.0, 1.2)* Age- and sex-matched TD children (n = 30) (23 boys, 7 girls)	**Set of Emotional Facial Stimuli** [44] Emotional facial expressions performed by 2 male and 2 female actors were presented to all participants and to one participant at a time. A series of intermediate expressions differing in emotional intensity levels by 10% steps was constructed based on the neutral face of the same actor and using the computer program Morph 1.0. The 30 and 70% intensity levels were used for the present study. In total 16 stimuli were presented in a random order: a set of 2 (intensity levels: 30 and 70%) × 4 (emotions: happiness, anger, disgust, and sadness) × 2 (actors). Each of the 16 expressions were rated by participants on four 7-point intensity scales	Happiness Anger Sadness Disgust	Children with ADHD made significantly more emotional facial expressions decoding errors than control children. No significant differences were found in emotional facial expressions decoding accuracy between children with ADHD and control children for happiness and disgust. Decoding accuracy was significantly lower in children with ADHD than in TD children for anger with 70% intensity. Decoding accuracy was also lower in children with ADHD than in control children for sadness at all intensities. Children with ADHD were furthermore unaware of the decoding errors when compared to the control group, manifesting significantly lower awareness of errors for anger

(continued)

Table 1 (continued)

No. of study date & authors	No. and type of participants; age (mean, sd)	Emotion recognition task & stimuli	Emotions investigated	Results
6. [89] Yuill & Lyon	Task 1 Boys with ADHD (n = 19) 5 with comorbid ODD 5.10–11.9 yrs *(8.11)* TD boys (n = 19) 7.2mo–11 yrs *(8.11)* Task 2 Boys with ADHD (n = 17) 5.8–10.6 yrs *(8.2)* TD boys (n = 13) 5.0–6.0 yrs *(5.5)*	**Non-emotional and emotion matching tasks** Task 1 Participants were presented with a set of six photographs to match to six situations (e.g. Happy: Thomas has just found his lost puppy). For the non-emotional task, a further set of six photographs were presented where the facial expressions were posed as neutral (e.g. Hot (Sunglasses): Thomas has just been out in the sunshine). The photographs were posed by an 11-year-old boy and were validated by 5 adults Task 2 Participants performed the same tasks, but with an 'inhibitory scaffolding' procedure to prevent impulsive responding	Happiness Fear Sadness Anger Surprise Disgust	Task 1 Boys with ADHD performed more poorly than the control group in matching faces to situations. Performance on the emotion matching task was also lower than on the non-emotion task. Boys with ADHD also performed poorly when making judgements about non-emotional characteristics of faces. There were no significant performance differences in the ADHD group between those diagnosed with ODD and those not Task 2 The ADHD group performed equally well to the control group on the non-emotion task, but poorer than the control group on the emotion task. There was a significant difference between the emotion and non-emotion task for the ADHD group with the control group performing equally well on both tasks. The effect of scaffolding was group- and task-specific: it helped the boys with ADHD more in the non-emotion task than in the emotion task. Children with ADHD who failed any situation-matching task were still able to label the emotional expressions correctly in 85% of cases

(continued)

Table 1 (continued)

No. of study date & authors	No. and type of participants; age (mean, sd)	Emotion recognition task & stimuli	Emotions investigated	Results
7. [42] Guyer et al.	Bipolar disorder (BD) (n = 42) 7–18 yrs *(12.8, 2.5)* Severe mood dysregulation (n = 39) *(11.8, 2.1)* Anxiety and/or major depressive Disorders (n = 44) *(13.1, 2.5)* ADHD or CD (n = 35) *(14.8, 1.6)* 25 boys, 10 girls TD (n = 92) *(14.4, 2.4)* 47 boys, 45 girls	**Diagnostic Analysis of Nonverbal Accuracy (DANVA)** [63] Participants were presented with a face-emotion labeling task. Each computer-administered subtest included 24 photographs of child or adult models (12 female, 12 male) displaying equal numbers of high and low-intensity expressions of happiness, sadness, anger, and fear. Faces appeared for 2 s. In a forced-choice format, participants indicate by button-press which emotion a face expressed	Happiness Sadness Anger Fear	ADHD/CD patients performed similarly to controls on the face-emotion labelling tasks. Face labeling ability did not differ based on the age of the face or the specific emotion displayed
8. [87] Williams et al.	Unmedicated ADHD boys (n = 51) 8.0–17.0 yrs *(13.79, 2.33)* TD boys (n = 51) 8–17 yrs *(13.09, 2.39)*	**Gur faces** [41] Evoked expressions of facial emotion (eight different individuals; four males, four females) acquired from a standardized series were presented in black and white during an ERP recording. Participants selected the verbal label corresponding to each facial expression (fear, anger, sadness, disgust, happiness, or neutral) and percentage accuracy was recorded	Fear Happiness Anger Sadness Disgust	Unmedicated ADHD participants were significantly more anxious and depressed than healthy controls. Children with ADHD were significantly worse in recognizing anger and fear. These expressions tended to be misidentified as neutral or sadnes

(continued)

Table 1 (continued)

No. of study date & authors	No. and type of participants; age (mean, sd)	Emotion recognition task & stimuli	Emotions investigated	Results
9. [77] Sinzig, Morsch & Lehmkuhl	Autism + ADHD (n = 21) (20 boys, 1 girl) 6.1–18.2 yrs *(11.6, 18.2)* Autism (n = 19) (17 boys, 2 girls) 8.1–18.9 yrs *(13.6, 3.4)* ADHD (n = 30) (28 boys, 2 girls) 7.1–17.9 yrs *(12.7, 3.1)* TD (n = 29) (22 boys, 2 girls) 7.6–17.6 yrs *(12.8, 2.9)*	**Frankfurt Test and Training of Social Affect (FEFA)** [10] Facial affect recognition was assessed with a computer-based program used for teaching emotion processing, the Frankfurt Test and Training of Social Affect (FEFA) using faces and eye-pairs as target material. The FEFA comprises 50 photographs of faces and 40 photographs of eye-pairs according to the "pictures facial affect" and the six basic emotions according to Ekman and Friesen. Furthermore, three attention-tasks (Sustained attention, Inhibition, Set-Shifting) were administere	Happiness Fear Sadness Anger Disgust Surprise	Facial affect recognition was impaired in children suffering from ADHD symptoms only and, Autism + ADHD. Children with ADHD were impaired on both facial affect recognition and recognition of emotion from eye-pairs when compared to TD children. Children with Autism + ADHD were worse in the recognition of happiness (eye-pairs) and surprise (faces) when compared to both TD children and Autistic children. Children with ADHD scored lower on the recognition of happiness (eye-pairs) when compared to TD children
10. [9] Boakes, Chapman, Houghton & West	ADHD boys (n = 48) 7.10–12.3 yrs *(10.2, 1.4)* TD boys (n = 48) *(10.3, 1.3)*	**Facial Affect Interpretation Task (FAIT)** The stimulus expressions used in the Facial Affect Interpretation Task (FAIT) were created from scenes drawn from two contemporary television shows. 24 static, 24 dynamic-decontextualized and 24 dynamic-contextualized stimuli were presented in either a cartoon or a real-life portrayal mode	Happiness Fear Sadness Anger Disgust Surprise	Boys with ADHD were significantly impaired in the interpretation of disgust and fear when compared to controls. Results also suggested a trend towards impairments for boys with ADHD in the interpretation of surprise. Although a main effect indicated significant overall performance increments across these three levels, participants in the ADHD and TD did not appear to benefit differentially from increasing levels of supplemental information

(continued)

Table 1 (continued)

No. of study date & authors	No. and type of participants; age (mean, sd)	Emotion recognition task & stimuli	Emotions investigated	Results
11. [19] Da Fonseca, Sequier, Santos, Poinso & Deruelle	Combined subtype of ADHD (n = 27) (21 boys & 6 girls) 5.0–15.0 yrs (*10.2, 2.7*) TD (n = 27) (21 boys & 6 girls) 5.0–15.0 yrs (*10.3, 2.7*)	Experiment 1 40 digitized colored photographs (10 of each emotion investigated) were acquired from popular French media magazines. The character's face and part or full body figured on the photographs. Target faces consisted of 26 pictures of adult faces and 16 pictures of children faces with an equal number of male and female characters. The pictures were validated in a pilot study with 16 TD children. After the presentation of each stimulus children were presented with three response-options (the target face and two distracter emotions) Experiment 2 Stimuli comprised 60 colored photographs taken from popular French media magazines. These photographs were scanned and used as visual scenes in which either a face expressing an emotion, or an object was masked. Visual scenes masking an object or face were carefully matched in terms of complexity and the number of characters and objects contained in the scene. Target faces and target objects were masked by a white Circle. The stimuli were validated in a pilot study with 16 TD children.	Anger Sadness Fear Happiness	Experiment 1 TD children performed generally significantly better than children with ADHD. For all participants both Happiness and Anger were better recognized than Fear and Sadness. No significant Group by Emotion interaction was found Experiment 2 Children with ADHD were less accurate than TD using contextual information to understand emotions. TD children performed significantly better than children with ADHD. Both groups performed significantly worse on the Emotion recognition than on the Object recognition condition

(continued)

Table 1 (continued)

No. of study date & authors	No. and type of participants; age (mean, sd)	Emotion recognition task & stimuli	Emotions investigated	Results
12. [40] Greenbaum et al.	Fetal Alcohol Spectrum Disorders (FASDs) (n = 33) (16 boys and 17 girls) 6.0–13.0 yrs *(9.20)* ADHD/33% had 1 or more comorbid conditions (n = 30) (24 boys and 6 girls) *(9.30)* TD (n = 34) (18 boys and 16 girls) *(8.90)*	**Minnesota Test of Affective Processing (MNTAP)** [51] Emotion processing was assessed using 4 of the 6 subsets from the Minnesota Test of Affective Processing (MNTAP). Affect Match, Affect Naming, Affect Choice, and Prosody Content. Affect Match required the child to determine whether the same or different emotions were conveyed on the faces of 30 pictures presented. In Affect Naming, the child was asked to select the cartoon face with the same facial expression as in the photographed face. In Affect Choice the child had to touch the face on the computer screen depicting the emotion generated verbally by the computer. In Prosody Content children had to state whether voice and content matched	Not defined	FASD group performed significantly worse than ADHD and control groups on Affect choice. No comparisons are provided for ADHD and controls.
13. [65] Passarotti, Sweeney, & Pavuluri	Combined subtype of ADHD (n = 14) (9 boys, 5 girls) 10–18 yrs *(13.00, 2.35)* TD (n = 19) (9 boys, 10 girls) *(13.53, 3.16)*	**Gur faces** [41] Participants underwent an fMRI scanning session when they were administered a block design 2-back working memory task with emotional faces for approximately 7 min. The paradigm involved two runs with one condition each. The first run consisted of blocks of angry and neutral faces and the second run consisted of blocks of happy and neutral faces. On each trial a face stimulus with a certain emotion or a neutral expression was presented for 3 s and participants responded by pressing a response key if they saw the same face as the one presented two trials earlier	Anger Happiness	For reaction time, there was only a significant effect of face emotion; the reaction time for angry faces was significantly slower than for neutral faces across groups. There was a non-significant trend for the ADHD group to be less accurate than TD childre

(continued)

Table 1 (continued)

No. of study date & authors	No. and type of participants; age (mean, sd)	Emotion recognition task & stimuli	Emotions investigated	Results
14. [4] Bal	ADHD (n = 33) (23 boys, 10 girls) (*9.33, 1.76*) 12 were on medication 14 had comorbidities TD (n = 38) (18 boys, 16 girls) (*9.94, 1.63*)	**Dynamic Affect Recognition Evaluation (DARE)** [69] Morphed images in a video like format starting with a neutral expression and slowly transitioning into one of the 6 basic emotions. The DARE software was synchronized with the eye tracking and physiological monitoring equipment. The participants were asked to identify which of the six emotion labels best represented the emotion that had just been presented	Anger Disgust Fear Happiness Sadness Surprise	ADHD group made significantly more errors than the control group to anger and disgust. In the ADHD group 36% of responses to anger were disgust, and 22% of responses to disgust were anger and 17% were fear. Analysis indicated significant differences between boys and girls on disgust errors only with boys making significantly more errors than girls. No differences among medicated and unmedicated ADHD children were found. The ADHD group was not significantly slower or faster in identifying the emotions than the control group
15. [76] Schwenck et al.	ADHD (n = 56) (38 boys, 18 girls) (28 unmedicated, 28 medicated) 8.2–17.0 yrs *(12.34, 2.54)* TD (n = 28) (19 boys, 9 girls) *(12.49, 2.55)*	**Karolinska Directed Emotional Faces** [55] Emotion recognition was assessed using the Morphing Task (MT). The MT was a self-developed task in which children were shown 60 film clips each of 9-s length with a neutral facial expression changing continuously to an emotional expression. Participants were asked to press a key as soon as they had recognized which emotion was presented and to name the correct emotion	Happiness Fear Sadness Disgust Anger	No differences were found between children with ADHD without medication and the control group in neither the reaction time, variability of reaction time nor the number of correctly identified emotions. This result applied for all basic emotions assessed. Furthermore, medication did not influence emotion recognition performance
16. [50] Kochel et al.	ADHD boys (n = 16) TD boys (n = 16) 8.5–11.8 yrs (*10.16, 1.10*)	**EEG Emotional Go/NoGo task with faces from the Karolinska Directed Emotional Faces (KDEF)** [55] A total of 24 different faces from the emotion categories anger, sadness, happiness, and neutral were displayed by 3 women and 3 men. An emotional task required inhibition, or a button press for a certain emotional face (e.g., "do not press the button when a happy/sad/angry face is presented"). A non-emotional task was used as a control condition	Anger Sadness Happiness	Children with ADHD made more errors compared to the control group. Boys with ADHD tended to make more recognition errors for anger compared to the control group. These group differences were not seen in the recognition of sadness, happiness, and neutral faces. Longer RT were required for the emotional compared to the neutral task. For both groups the longest RT were required anger and shortest for happiness. ADHD did not differ from TD in RT

(continued)

Table 1 (continued)

No. of study date & authors	No. and type of participants; age (mean, sd)	Emotion recognition task & stimuli	Emotions investigated	Results
17. [54] Ludlow et al.	ADHD boys (n = 24) 12–15 yrs *(13.59, 0.82)* TD age and verbal IQ matched with mild to moderate learning disabilities (n = 24) (15 boys, 9 girls) 12–16 yrs *(14.13, 1.22)*	**Emotion Evaluation Test (EET) from The Awareness of Social Inference Test (TASIT)** [57] The EET comprises a series of 28 short (15–60 s) videotaped vignettes of trained professionals interacting in everyday situations and portraying one of seven emotions, presenting 4 vignettes for each emotion. 12 segments suggest a positive emotional state and 16 segments suggest a negative emotional state. Participants' recognition of spontaneous emotional expressions was assessed	Happiness Disgust Fear Anger Sadness Surprise	No significant differences were found between the participants diagnosed with hyperactive and combined types of ADHD. ADHD participants overall recognized less emotions accurately than control participants. There was overall better recognition of positive than negative emotions. Happiness was recognized better than all the other emotions, whereas fear and disgust were recognized worse than the other emotions. TD performed significantly better in the recognition of anger, surprise, sadness and fear. Verbal ability and age were not significant predictor of EET scores. The use of medication in the ADHD group did not have significant effects on the EET scores
18. [62] Noordermeer et al.	ADHD (n = 82) (55 boys, 27 girls) *(16.0, 3.1)* ADHD + ODD (n = 82) (55 boys, 27 girls) *(16.0, 3.0)* TD (n = 82) (55 boys, 27 girls) *(16.0, 3.3)*	**Identification of Facial Emotions** [23] Participants were shown a picture of an adult face displaying an emotion and had to compare the expressed emotion with the target emotion (happy, sad, and angry), by pressing a yes/no button. Pictures remained on screen until a response was given. For every emotion, a 50/50 distribution of pictures that contained the target emotion and pictures that contained a non-target emotion was shown. The sequence of the tested target emotions was randomly assigned	Happiness Sadness Anger	Groups did not differ in the percentage of correct responses during happy, angry, or afraid trials. Mean RT for angry trials did differ between groups; the ADHD + ODD group showed slower mean RTs for correct responses compared with controls indicating difficulties in correctly identifying angry facial emotions. The ADHD-only group did not differ from TD.

(continued)

Table 1 (continued)

No. of study date & authors	No. and type of participants; age (mean, sd)	Emotion recognition task & stimuli	Emotions investigated	Results
19. [83] van Goozen et al.	ADHD boys (n = 194) 10.0–17.0 yrs (*13.95, 1.82*)	Genotype study to assess the links between COMT genotype and aggression. Participants completed tasks assessing executive function (response inhibition and set shifting), empathy for fear, sadness and happiness, and fear conditioning. To assess participants' cognitive and affective empathy three clips depicting the emotions of sadness, happiness and fear were edited from cinematic films. After each clip participants completed a questionnaire concerning the recognition of the emotions of the main character (cognitive empathy) and their own emotions while viewing the clip (affective empathy). Participants were also asked to explain the reason for the emotion (-s) identified in the main character and themselves	Sadness Fear Happiness	COMT Val allele carriers showed poorer response inhibition and set shifting abilities, reduced fear empathy and reduced autonomic responsiveness (lower SCRs) to the conditioned aversive stimulus. COMT Val158Met did not predict impairments in recognizing others' emotions or affective empathy for happiness or sadness
20. [7] Berggren al.	ASD (n = 35) (18 boys, 17 girls) 8.6–15.4 yrs (*11.6, 1.8*) ADHD (n = 32) (17 boys, 15 girls) 8.6–15.9 yrs (*11.1, 2.1*) TD (n = 32) (18 boys, 14 girls) 9.4–15.5 yrs (*11.7, 1.8*) Age matched groups	**The Swedish version of the computer-based Frankfurt Test for Facial Affect Recognition (FEFA)** [10] The FEFA test uses the cross-cultural concept of the 6 basic emotions and the neutral affective state to assess explicit FAR skills by verbal labelling of the emotions expressed in the eye regions and in whole faces. There are between 3 to 9 items for each basic emotion on the eyes test, and 5 to 9 for the face test. The total score for the faces and eyes test, the number of correct answers per basic emotion and response times in second for both tests were assessed	Happiness Sadness Anger Surprise Disgust Fear	The ADHD group responded faster than the ASD group for global FAR. No differences between ADHD and TD were found. No differences on accuracy for specific emotions were found between TD and ADHD

(continued)

Table 1 (continued)

No. of study date & authors	No. and type of participants; age (mean, sd)	Emotion recognition task & stimuli	Emotions investigated	Results
21. [80] Tehrani-Doost et al.	ADHD boys (n = 28) 7.0–12 yrs *(8.75, 1.39)* TD boys (n = 27) *(9.46, 1.45)*	**Cohn Kanade AU-coded Facial Expressions Database** [48] A computerized facial emotion recognition task based on the Cohn- Kanade face database was used. One hundred faces were randomly divided in happy, sad, angry and neutral expressions. Each emotion was shown to the participants 25 times, using three male and two female faces randomly	Happiness Anger Sadness	Children with ADHD were significantly worse in recognizing all emotions compared to TD. The two groups showed no difference in recognizing the neutral faces. The time spent in recognizing happy faces was higher in the ADHD. Both groups were better to detected happy and neutral faces than the angry ones. The ADHD group recognized happy and neutral targets more accurately than sad targets and there was a significant interaction between angry and sad targets recognition and inattention
22. [74] Sarraf Razavi et al.	Combined subtype ADHD boys (n = 19) 7.0–11.0 yrs *(9.21, 1.13)* TD boys (n = 19) 7.0–11.0 yrs *(9.73, 1.04)*	**Cohn Kanade AU-coded Facial Expressions Database** [48] This study aimed to compare the gamma band oscillations among patients with ADHD and TD during facial emotion recognition. Stimuli were 6 Caucasian faces (3 females and 3 males) expressing happy, angry, sad and neutral emotions. Each emotion was repeated 60 times in a random way. Participants had to respond by pressing one of the four buttons (representing the emotions under investigation) on a joystick	Happiness Anger Sadness	Individuals with ADHD differed from TD children during early facial expression recognition. ADHD children showed a significant reduction in the gamma band activity when compared to healthy controls in occipital site and a significant decrease for happiness and anger in the left and right occipital, respectively

(continued)

Table 1 (continued)

No. of study date & authors	No. and type of participants; age (mean, sd)	Emotion recognition task & stimuli	Emotions investigated	Results
23. [47] Jusyte, Gulewitsch & Schönenberg	ADHD (n = 26); 13 ADHD Combine; 10 ADHD Inattentive (17 boys, 9 girls) (24 unmedicated, 2 medicated with methylphenidate) 10.0–14.0 yrs *(11.68, 1.65)* TD (n = 26) (13 boys, 13 girls) 10.0–14.0 yrs *(11.73, 1.30)*	**Radboud Faces Database** [52] Color photographs of 4 child models served as stimuli. The stimulus material was created by blending the neutral and emotional expressions of every model identity, resulting in six distinct neutral-emotional sequences for each model. For every sequence, a set of 51 images with intensity levels ranging from completely neutral to full blown expression were extracted (2% increment steps) and presented to all participants. The participants were instructed to press a button as soon as they were able to correctly recognize the depicted expression. The sequence was then immediately stopped, the face disappeared, and the perceptual intensity as well as selected emotion category were recorded	Happiness Anger Fear Disgust Surprise Sadness	There was an overall lower accuracy across all emotions in the ADHD as compared to the control group; happy expressions were associated with the highest and fearful with the lowest accuracy rates. Surprise was better recognized than disgust. Some emotions, such as happiness, were recognized at lower intensity levels than others whereas disgust and surprise did not differ from each other. The results were not attributed to impulsive or inattentive responding in the ADHD as compared to the control group. In accordance with previous reports, disgust was frequently confused with anger, fear with surprise, and sadness with fear or anger, a characteristic pattern that was evident in both groups (Kats-Gold et al., 2007; Schönenberg et al., 2014)
24. [72] Rinke et al.	ADHD (n = 29); Unmedicated 24 h before the study 2 with comorbid developmental disorder (24 boys, 5 girls) 7.0–17.0 yrs *(12.09, 2.76)* TD (n = 21) (9 boys, 12 girls) 8.0–17.0 yrs *(12.08, 3.00)*	**Pictures of Facial Affect** [32] Behavioral and neurophysiological study. Children had to undergo among others, the emotional continuous performance task (ECPT) and the facial emotion matching (FEM) task. In the ECPT, different emotional stimuli were presented sequentially, and the participant had to get prepared for an action (go-trial), inhibit an action (no-go-trial), or simply ignore the stimulus. In the FEM task different human facial emotion expressions were presented and the child was asked to choose one of the model pictures showing an equal emotion as the one presented in the center of the screen	Happiness Anger Fear Disgust Surprise Sadness	Children with and without ADHD did not differ on the behavioral performance tasks as well as the neurophysiological measurements. Older children showed better recognition accuracy of the emotion anger than the younger children and were faster in their responses

(continued)

Table 1 (continued)

No. of study date & authors	No. and type of participants; age (mean, sd)	Emotion recognition task & stimuli	Emotions investigated	Results
25. [1] Airdrie et al.	ADHD with or without comorbid CD (n = 63) (47 boys, 16 girls) 11.0–18.0 yrs (*14.20, 2.09*) TD (n = 41) (21 boys, 20 girls) 11.0–18.0 yrs (*15.50, 2.70*)	**Pictures of Facial Affect** [32] Emotion recognition task with congruent eye tracking. 60 faces from the Ekman and Friesen series of facial affect (happiness, fear, anger, sadness) and neutral were presented. Each emotion was morphed with its corresponding neutral expression to create a 50 and 75% intensity expression. An equal number of male and female target faces appeared, and slides contained an equal number of each emotion presented at each intensity	Happiness Fear Anger Sadness	Emotions high in intensity were better recognized than emotions low in intensity. Highest accuracy scores were found for happy faces, followed by neutral, fear, angry and sad. Groups differed only in the recognition of fear and neutral emotions. Children with ADHD + CD were less accurate than children with ADHD and HC in the recognition of fear and neutral expressions. Children with ADHD did not differ from HC in the recognition of fear and neutral expressions. There was greater tendency to misinterpret fear faces as angry faces for children with ADHD + CD than ADHD and HC, with no differences being found between children with ADHD and H

Table 2 Emotions examined, differentiation among specific emotion dimensions across studies, and impairments in children with ADHD compared to TD across the emotions and studies included in the review

Emotions under study	Sadness	Fear	Happiness	Disgust	Surprise	Anger	Summary
Study ID							
1	■	■	■	■	■	■	ADHD children were significantly impaired compared to TD on all the emotions investigated. The authors did not differentiate the specific emotion dimensions in their analyses
2	*	*	*			×	Children with ADHD were significantly impaired compared to TD on the emotions of sadness, fear and happiness but not anger. The authors differentiated among specific emotion dimensions in their analysis
3							The authors did not define the emotions investigated in their study.
4	×	*	×	×	×	×	Children with ADHD were significantly impaired compared to TD only on the emotion of sadness. The authors differentiated among specific emotion dimensions in their analysis
5	*		×	×		*	Children with ADHD were significantly impaired compared to TD only on the emotions of sadness and anger. Emotion recognition accuracy did not differ on the emotions of happiness and disgust. The authors differentiated among specific emotion dimensions in their analysis

(continued)

Table 2 (continued)

6	■	■	■	■	■	■	ADHD children were significantly impaired compared to TD. The authors did not differentiate the specific emotion dimensions in their analyses
7	□	□	□			□	ADHD children did not significantly differ from TD. The authors did not differentiate the specific emotion dimensions in their analyses
8	×	*	×	×		*	Children with ADHD were significantly impaired compared to TD on the emotions of fear and anger. Emotion recognition accuracy did not differ on the emotions of happiness, sadness and disgust
9	×	×	*	×	*	×	Children with ADHD were significantly impaired compared to TD on the emotions of happiness and surprise. Emotion recognition accuracy did not differ on the emotions of sadness, fear and disgust
10	×	*	×	*	×	×	Children with ADHD were significantly impaired compared to TD on the emotions of fear and disgust. Emotion recognition accuracy did not differ on the emotions of sadness, happiness, surprise and disgust
11	≡	≡	≡			≡	Children with ADHD were significantly impaired compared to TD on the emotions of sadness, fear, happiness and anger. Despite the presence of a main effect for group and emotion no significant group by emotion interaction was found

(continued)

Table 2 (continued)

12							The authors did not define the emotions investigated
13			×			×	ADHD children did not significantly differ from TD children on the emotions of happiness and anger that the authors investigated. The authors differentiated among specific emotion dimensions in their analysis analyses
14	×	×	×	*	×	*	Children with ADHD were significantly impaired compared to TD on the emotions of disgust and anger. Emotion recognition accuracy did not differ on the emotions of sadness, fear, happiness and surprise between the two groups
15	×	×	×	×		×	The authors did not find any significant differences on the recognition accuracy of the emotions investigated, i.e. sadness, fear, happiness, disgust and anger, between children with ADHD and TD children
16	×		×			*	Children with ADHD were significantly impaired compared to TD only on the emotion of anger. Emotion recognition accuracy did not differ on the emotions of sadness and happiness

(continued)

Table 2 (continued)

17	✱	✱	×	×	✱	✱	Children with ADHD were significantly impaired compared to TD on the emotions of sadness, fear, surprise and anger. Emotion recognition accuracy did not differ on the emotions of happiness and disgust. The authors investigated all six basic emotions and differentiated among specific emotion dimensions in their analysis
18	×		×			×	Children with ADHD and TD children did not differ on any of the emotions investigated i.e. sadness, happiness and anger
19	×	×	×				Children with ADHD and TD children did not differ on any of the emotions investigated i.e. sadness, fear and happiness
20	×	×	×	×	×	×	Children with ADHD and TD children did not differ on any of the six basic emotions investigated
21	✱		✱			✱	Children with ADHD were significantly impaired compared to TD on the recognition accuracy of sadness, happiness and anger
22	~		~			~	The authors examined differences in gamma band oscillations between children with ADHD and TD.

(continued)

Table 2 (continued)

23	≡	≡	≡	≡	≡	≡	Children with ADHD were significantly impaired compared to TD on all six emotions investigated. Despite the presence of a main effect for group and emotion no significant group by emotion interaction was found
24	□	□	□	□	□	□	Children with ADHD did not significantly differ from TD children on the emotions investigated. The authors did not differentiate the specific emotion dimensions in their analyses
25	×	✱	×			×	Children with ADHD were significantly impaired compared to TD only on the emotion of fear. Emotion recognition accuracy did not differ on the emotions of sadness, happiness and anger

In [3, 12] the authors did not define the emotions investigated.
□ ADHD children did not significantly differ from TD/The authors did *not differentiate* the specific emotion dimensions in their analyses.
■ ADHD children were significantly impaired compared to TD/The authors did *not differentiate* the specific emotion dimensions in their analyses.
✱ Children with ADHD were significantly impaired compared to TD/The authors *differentiated* among specific emotion dimensions in their analysis.
× ADHD children did not significantly differ from TD/The authors *differentiated* among specific emotion dimensions in their analysis analyses.
≡ ADHD children were significantly impaired compared to TD/The authors *differentiated* among specific emotion dimensions in their analysis analyses/There was no significant group by emotion interaction.
~ The authors examined differences in gamma band oscillations between children with ADHD and TD children.

5.3 The Effect of Gender and Age

Generally, among the studies that implicated both genders, more boys with ADHD participated than girls. Whereas in Corbett and Glidden [1] gender did not significantly differ between groups, when gender and age were added as covariates boys differed from girls on the perception of affect measures. In the preliminary analyses of Greenbaum et al. [12] albeit the presence of a significant gender effect due to the high male to female ratio in the study, the authors did not proceed with including it in subsequent analyses as this proportion reflected the typical gender rates of ADHD seen in the general hence not deemed as problematic. In contrast, when age and gender were added as covariates in Airdrie et al. [26] emotion recognition accuracy did not differ in gender and only approached significance in age. In Guyer et al. [6] the ADHD/CD and TD groups did not differ on age, but significantly more boys participated than girls, without the gender nevertheless (when added as a covariate) having been reported to exert an effect on emotion recognition. In contrast to Guyer et al. [6] two studies found 7–12-year-old children with ADHD to be less accurate in emotion labelling compared to TD [1, 11]. Corbett and Glidden [1] as well as Pelc et al. [11], differed from Guyer et al. [6] in that younger children participated whereas in Guyer et al. [6] participants were not younger than 12 years of age. In all other experimental studies included in the review, the ADHD and TD groups were age and gender matched.

5.4 Static Versus Dynamic Stimuli

Most studies implicated static stimuli with only the minority having dynamic stimuli in their design [9, 14, 15, 17, 19]. Boakes et al. [9] was the only to have compared static, dynamic-decontextualized and dynamic stimuli with a relevant situational context (dynamic-contextualized stimuli). Surprisingly, although boys with ADHD showed impairments in emotion recognition compared to healthy children, they did not appear to benefit from the increasing number of contextual information. Bal [14] used the Dynamic Affect Recognition Evaluation [69] where stimuli were developed from the Cohn–Kanade Action Unit-Coded Facial Expression Database (Cohn et al. 1999). Participants were presented with morphed images in a video-like format where neutral emotions were transitioning into one of the six basic emotions and children had to label the emotions presented. The authors found that the ADHD group were worse compared to the control group to label anger and disgust. Participants in Ludlow et al. [17] were presented with a series of 28 short video-taped vignettes of trained professionals interacting in everyday situations and portraying one of seven emotions; happy, surprised, neutral, sad, angry, fear and disgust. Children with ADHD were worse overall at recognizing emotions than control participants and specifically the emotions of sadness, anger, fear and surprise. Schwenck et al. [15] assessed emotion recognition via 9-s films of morphed facial expressions and failed to find any differences in emotion recognition between children with ADHD and healthy controls. Despite the stimuli being dynamic in that a neutral facial expression was continuously changed to an emotional expression, it could still be argued that the morphed stimuli lacked ecological validity. The last study to have implicated dynamic stimuli was the genotype study of van Goozen et al. [19] which did not include a control group in their design as the authors aimed to investigate the links between the COMT genotype and aggression in male adolescents with ADHD.

5.5 Differentiation Among Specific Emotion Dimensions

Some studies found a deficient performance for specific emotions while others either did not differentiate among specific emotion dimensions or did not define the emotions investigated. Table 2 presents the emotions examined in each study indicating whether the authors differentiated among specific emotion dimensions in their analyses and the emotions that children with ADHD showed impairments compared to controls. As can be seen from the table below two studies; Downs and Smith [3] and Greenbaum et al. [12], did not define the emotions investigated as Downs and Smith [3] assessed children's theory of mind and Greenbaum et al. [8] participants' affective processing. Four studies did not differentiate among specific emotion dimensions in their analyses [1, 5, 6, 24] namely, Corbett and Glidden [1], Yuill and Lyon [5], Guyer et al. [6] and Rinke et al. [24]. Three studies assessed all 6 emotions [1, 5, 24] and whereas Corbett and Glidden [1] and Yuill and Lyon [5] found significant differences among children with and without ADHD, Guyer et al. [6] and Rinke et al. [24] failed to do so. All four studies implicated static stimuli, with the Corbett and Glidden [1] and Rinke et al. [24] assessing emotion recognition using the Pictures of Facial Affect [32], Yuill and Lyon [5] a non-emotional and emotion matching task and Guyer et al. [6] the DANVA [58].

Most of the studies included in the review investigated all six basic emotions [4, 8, 9, 14, 17, 20, 23], five studies three of the basic emotions [16, 18, 19, 21, 22], three studies examined five basic emotions [7, 10, 15] and three studies four basic emotions [2, 11, 26]. Among the studies that differentiated among specific emotion dimensions in their analyses the most researched emotion was "happiness" closely followed by "anger" and "sadness". Table 3 presents the number of emotions investigated among the studies that differentiated among emotion dimensions and the percentage of studies that found significant differences among children with ADHD and TD.

From the table above it can be inferred that children with ADHD are more frequently reported to be impaired on the emotion of fear as six out of the 13 studies (46%) that assessed fear found deficits on emotion recognition. Anger and surprise follow, with one third of the studies investigating anger (33%) and 28% of the studies investigating surprise, having reported impairments in children the ADHD compared to TD. With regards to happiness only three out of the 19 studies found significant differences among children with ADHD and the control group.

Table 3 Emotions investigated and % of studies that found deficits in ADHD children among the studies that differentiated specific emotion dimensions in their analyses

Emotions investigated	Sadness	Fear	Happiness	Disgust	Surprise	Anger
No. of studies	18	13	19	10	7	18
No. of studies reporting significant differences among children with and without ADHD	4 [2, 11, 17, 21]	6 [2, 4, 7, 9, 17, 26]	3 [2, 8, 21]	2 [9, 14]	2 [8, 17]	6 [7, 11, 14, 16, 17, 21]
% of Studies reporting a significant difference in emotion recognition	22.2	46.2	15.8	20.0	28.6	33.3

With regards to sadness, all four studies that found significant results used differed emotion recognition measures; Cadesky et al. [2] employed the DANVA [58, 11] used a series of emotional facial expressions constructed and validated by Hess and Blairy [44], Ludlow et al. [17] the Emotion Evaluation Test (EET) from The Awareness of Social Inference Test (TASIT) [57, 21] a computerized facial emotion recognition task based on the Cohn- Kanade face database [48]. In DANVA [58], the emotional facial expressions by Hess and Blairy [44] and the emotion recognition task based on the Cohn- Kanade face database [48] static faces are presented while in EET from the TASIT [57] the stimuli are dynamic namely, vignettes of trained professionals interacting in everyday situations. Except for DANVA where stimuli of both adults and children were presented, in all other studies whose measures found significant differences among children with and without ADHD facial expressions of only adults were presented.

The emotion recognition measures used in the studies that found significant differences among ADHD and TD with regards to fear were the: DANVA [58] in [2], FEEST [88] in [4], facial expressions from the Gur et al. [41] database in Williams et al. [7], FAIT in Baokes et al. [9], EET from TASIT [57] in [17] and Pictures of Facial Affect [32] in Airdrie et al. [26]. All measures implicated static stimuli except for FAIT that implicated static and dynamic and EET from TASIT that implicated dynamic stimuli.

With regards to happiness, three studies found significant differences among the two populations and albeit all of them using static stimuli, they utilized different measures namely, Cadesky et al. [2] the DANVA [58, 8] the FEFA [10, 21] the Cohn Kanade AU-coded Facial Expressions Database [48]. Only two studies found children with ADHD to be impaired in the emotion recognition of disgust compared to

TD. Both studies utilized dynamic stimuli; Bal et al. [14] used DARE [14] where morphed images were presented in a video like format slowly transitioning into one of the 6 basic emotions [69] and Da Fonseca et al. [10] the FAIT where static, dynamic-decontextualized and dynamic-contextualized stimuli were presented in either a cartoon or a real-life portrayal mode. Noteworthy, is that a third study that investigated disgust (Aspan et al. 2004), [4], found increased sensitivity in the recognition of disgust in adolescent boys with ADHD compared to TD boys. The authors assessed emotion recognition in their study via the FEEST [88] where participants had to label the emotions of static stimuli presented on the screen.

Two out of the seven studies that implicated surprise in their design found significant differences; Sinzig et al. [8] via the FEFA [10, 17] via the EET from TASIT [57]. Static and dynamic stimuli were used respectively in the studies. One third of the studies that implicated anger in their analyses found differences among children with and without ADHD. All but two studies used static stimuli; Pelc et al. [11] via the Set of Emotional Facial Stimuli [44, 7] via the Gur et al. [41] facial expressions, Kochel et al. [16] via the KDEF [36, 21] via the Cohn Kanade AU-coded Facial Expressions Database [48]. The two studies that implicated dynamic stimuli were Bal et al. [14] and Ludlow et al. [17] that used the DARE [69] and EET from the TASIT [57] respectively.

In addition to the great variability of measures administered among the studies that found significant results for each emotion, great variability is also observed with regards to the study population. For example, for sadness half of the studies implicated both genders and half only boys, for fear and anger the majority (i.e. four studies) implicated only boys and two studies children of both genders, for happiness out of the three studies that found significant differences among children with and without ADHD two assessed both genders and one only boys.

5.6 Assessing Reaction Time on Emotion Recognition

Among the 25 studies included in the review, seven assessed reaction time; Passarotti et al. [13], Bal [14], Schwenck et al. [15], Kochel et al. [16], Noordermeer et al. [18], Tehrani-Doost et al. [21] and Rinke et al. [24]. Children with ADHD were profoundly slower to recognize angry than neutral facial expressions in Passarotti et al. [13], no differences in reaction time were found between children with ADHD and TD children in Bal [14], Schwenck et al. [15] and Kochel et al. [16] and children with ADHD required more time than TD children to recognize the happy facial expressions in Tehrani-Doost et al. [21]. Whereas in Noordermeer et al. [18] the ADHD + ODD group showed slower mean RTs for correct responses compared with controls indicating difficulties in correctly identifying angry facial emotions, the ADHD-only group did not differ from TD. In Rinke et al. [24] no differences in reaction time were found between the experimental groups. Nevertheless, younger children were faster in their responses than older children indicating along with the finding that younger children were worse in the recognition of anger compared to older children, that the facial emotion recognition is above all an age-dependent function. The relations among reaction time and comorbidity were only assessed in Noordermeer et al. [18] as in Passarotti et al. [13], Bal [14], Schwenck et al. [15], Kochel et al. [16] and Tehrani-Doost et al. [21] children with only ADHD participated. In Rinke et al. [24] out of the 29 participants with ADHD, 2 were diagnosed with comorbid developmental disorder. In summary four studies failed to find any differences in reaction time among children with ADHD and TD, one found children with ADHD to require more time to recognize anger compared to controls [13], one found children with ADHD + ODD to be slower to recognize anger compared to TD [18] and one concluded that ADHD boys were slower in the recognition of happiness compared to controls [21]. Interesting is than in Tehrani-Doost et al. [21] participants were the youngest of the studies that assessed reaction time (8.75, 1.39).

5.7 The Factor of Comorbidity

Among the studies included in the literature review, nine [2, 3, 6, 8, 12, 14, 18, 24, 26] implicated children with ADHD and 1 or more comorbid conditions. Nevertheless, not all of them controlled for the factor of comorbidity on emotion recognition i.e. comparing children with only ADHD and ADHD + 1 comorbidity. Taking a closer look Cadesky et al. [2] compared a group of children only with ADHD and with ADHD + CD, Sinzig et al. [8] compared children with ADHD, ASD, Autism + ADHD, Nordermeer et al. [18] implicated children with ADHD, ADHD + ODD and Airdrie et al. [26] children with ADHD and ADHD + CD. Among these studies that controlled for the factor of comorbidity in their analyses, Cadesky et al. [2] found that TD children outperformed the groups of children with ADHD only and CD only, whereas children with the combined symptomatology (ADHD and CD) did not differ from TD children. In contrast Airdrie et al. [26] found children with the combined symptomatology (ADHD + CD) to be less accurate in the recognition of fear than both children with ADHD and TD and children with ADHD to be performing similarly to TD in the recognition of fear. Both studies used static stimuli which have been argued to lack ecological validity. Sinzig et al. [8] found children suffering from ADHD and ADHD + ASD when compared to TD children to be impaired on both facial affect recognition and recognition of emotion from eye-pairs. Furthermore, children with ADHD + ASD were worse in the recognition of happiness and surprise from eye-pairs and faces respectively, when compared to TD children and children with ASD and children with ADHD scored lower on the recognition of happiness (eye-pairs) when compared to TD children.

When it comes to autism, Downs and Smith [3] studied emotion recognition on the notion of the theory of mind and found the group of children with ADHD and ODD to have performed worse than healthy controls. It is important to mention that the authors did to implicate a group of children with ADHD only in their study. Despite Guyer et al. [6] implicating 18 children with ADHD, 7 with Conduct Disorder

(CD) and 10 with ADHD + CD these three groups were treated as one to maximize statistical power and as a result of the preliminary analyses conducted which indicated similar mean scores among the three groups. Furthermore, whereas Greenbaum et al. [12] and Bal [14] report one third and more than a third (respectively) of children with ADHD having 1 or more comorbidities, no more information is provided as to which these comorbidities were, with the comorbidities hence not taken into account in their analyses. Likewise, despite only two of the 29 children with ADHD having been diagnosed with comorbid developmental disorder in Rinke et al. [24] no further information is provided as to whether this factor influenced the results of the study.

5.8 Neural Correlate of Emotion Recognition in ADHD

Functional neuroimaging techniques and event-related potential (ERP) studies have identified alterations in the activation and inhibition of several brain areas in children and adolescents with ADHD during the execution of tasks requiring facial emotion recognition. While children with ADHD did not differ from TD children in the accuracy of emotion recognition in Passarotti et al. [13], alterations in the activation of brain regions were highlighted; relative to TD children, children with ADHD exhibited greater activation in DLPFC (increased activity with positive emotional challenge in cortico-subcortical circuitry) and reduced activation in ventral and medial PFC, pregenual ACC, striatum and temporo-parietal regions (decreased activity with negative emotional challenge). In the event-related potential study of Kochel et al. [16] boys with ADHD made more recognition errors for anger than TD boys and longer reaction times were required for the emotional compared to the neutral task, with the longest RT times being recorded for anger. Children with ADHD relative to controls displayed a severe impairment in response inhibition toward anger cues, which was accompanied by a reduced P300 amplitude. The control group showed a P300 differentiation of the affective categories that was absent in the ADHD group. Williams et al. [7] observed a significant reduction in accipital activity during the early perceptual analysis of emotional expressions (within 120 ms) followed by an exaggeration of activity associated with structural encoding (120–220 ms) and subsequent reduction and slowing of temporal brain activity subserving context processing (300–400 ms). Sarraf Razavi et al. [22] found ADHD children to have a significant reduction in the gamma band activity when compared to TD children in occipital site and a significant decrease for happiness and anger in the left and right occipital, respectively.

6 Discussion

ADHD is a neurodevelopmental disorder affecting people of all ages and of both genders. The symptoms of ADHD begin in childhood (usually between the ages of

3 to 6), and for about half of the children, the symptoms continue into adolescence and adulthood. School-aged children with ADHD have been reported to suffer from social and emotional deficits.

Studies on facial emotion recognition in children and adolescents with ADHD focused mainly on the recognition accuracy of facial emotions, showing inconsistent results and a heterogenous use of measures. This review identified 25 studies to have implicated individuals under the age of 18 diagnosed with ADHD according to the criteria of the Diagnostic and Statistical Manual of Mental Disorders. The male sample exceeded the female and most studies used groups comparable on age and gender. Our literature review focusing on the ADHD population cannot support McClure [56] who conducted a meta-analysis to examine sex differences in the development of facial expression recognition, and provided clear evidence for a small, although robust female advantage in emotion expression recognition over the developmental period (from infancy into adolescence). One could argue that this could be the result of the absence of gender differences on emotion recognition in ADHD. One study in this literature review, Guyer et al. [6] argued for developmental differences in emotion recognition in ADHD, a finding concurrent with literature arguing that the recognition of emotion expression does not emerge as one specific stage in development rather gradually over time [25, 84]. In contrast to Corbett and Glidden [1] and Pelc et al. [11] who found 7–12 year olds with ADHD to be less accurate at emotion labelling than TD children, in Guyer et al. [6], participants were older than 12 years of age, arguing that preadolescent children (participating in [1, 11]) could have greater difficulties labelling facial emotions than do older children with ADHD.

With regards to the stimuli used to assess the emotion recognition skills of children with ADHD five studies implicated dynamic stimuli [9, 14, 15, 17, 19]. All studies used different tasks to assess emotion recognition. As a result of van Goozen et al. [19] being a genotype study a control group was not implicated so comparisons among the two groups were not feasible. Among those that were experimental and implicated a control group and dynamic stimuli, two found significant differences between children with ADHD and TD children; Bal et al. [14] and Ludlow et al. [17]. In Bal et al. [14] more than half of the ADHD participants were also diagnosed with a comorbidity, both males and females participated, and the DARE task was to assess emotion recognition. Children with ADHD were impaired in the recognition of anger and disgust but not in the recognition of fear, happiness, sadness and surprise. In contrast in the study of Ludlow et al. [17] only boys with ADHD participated, the EET from the TASIT was employed [57] and TD children performed significantly better in the recognition of anger, surprise, sadness and fear. It should be mentioned nevertheless that whereas only boys with ADHD participated in Ludlow et al. [17], the control group was composed of both boys and girls while gender was not added as a covariate. Boakes et al. [9] and Schwenck et al. [15] employed the Facial Affect Interpretation Task (FAIT) and Karolinska Directed Emotional Faces [55] respectively. In Boakes et al. [9] children only with ADHD and without any comorbidities participated, while in Schwenck et al. [15] both boys and girls with only ADHD took part. From the present literature review we can hence say that despite static stimuli having been argued to lack ecological validity, the studies implicating dynamic stimuli in this review have not produced more conclusive results with regards to the emotion recognition skills of children with ADHD.

Striking is the fact that none of the studies included in the review assessed whether the type of stimuli administered i.e. stimuli gender (males or females) and age (children or adults) had an effect on the accuracy of emotion recognition. This factor is crucial to be taken in future studies into account as it will provide the opportunity for more in depth results when it comes to the accuracy of emotion recognition.

A wide range of measures have been employed to assess affect recognition. Among the 25 studies included in the review, 18 different measures were used. Taking into account the great variability of measures administered, this section will discuss the most commonly used measures as per the systematic review; Pictures of Facial Affect [32], DANVA [63], Karolinska Directed Emotional Faces database (KDEF) [55], Computer-based Frankfurt Test for Facial Affect Recognition [10], faces by Gur et al. [41] and the Cohn-Kanade Facial Expressions Database [48].

The most frequently used measure was the Pictures of Facial Affect [32] which was utilized by four studies included in the review (either in the original or computerized and extended version). In the pictures of facial affect every model was photographed neutrally and showed one of the seven basic emotions: happiness, anger, fear, surprise, sadness, disgust, and contempt. Using the Facial Action Coding System (FACS; Ekman et al. 2002), researchers were able to produce pictures of standardized facial expressions that were intended to represent "prototypical displays" of emotions. While, the Pictures of Facial Affect have been extensively used in research, it is important to consider the factor of the faces being non-contemporary and presented in black and white format (Fig. 2).

One of the most elaborate sets which is the original Karolinska Directed Emotional Faces database (KDEF) consists of a total of 490 JPEG pictures showing 70 individuals (35 women and 35 men) displaying 7 different emotional expressions (Angry, Fearful, Disgusted, Sad, Happy, Surprised, and Neutral). Each expression is viewed from 5 different angles and all the individuals portraying the emotions were trained amateur actors between 20 and 30 years of age. Researchers interested in this database can freely select pictures as a function of (a combination of) several parameters: sex of the expressor, quality of the emotional expression per expressor, hit rate, intensity, and/or arousal. Despite the KDEF having scored high on the validity and reliability measures with the mean biased hit rate of 72% being comparable with other validation studies [33], one of the most critical limitations is the dataset being limited to adult stimuli and omitting to include any child stimuli as well (Fig. 3).

The DANVA [63] stimuli are faces of adults and children displaying one of four emotional expressions (happiness, sadness, anger, and fear) that vary between pictures in their intensity levels with the use of variable intensity levels corresponding to item difficulty. The researchers created facial expression stimuli by reading participants a vignette and then photographing the participants as they produced a facial

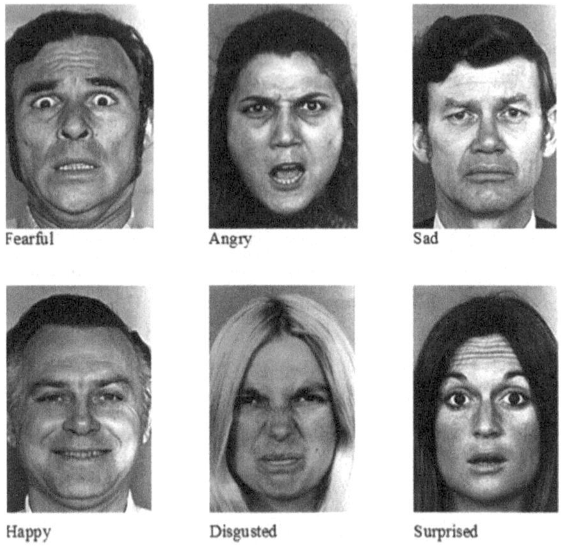

Fig. 2 Example of each emotion of the Pictures of Facial Affect [32]

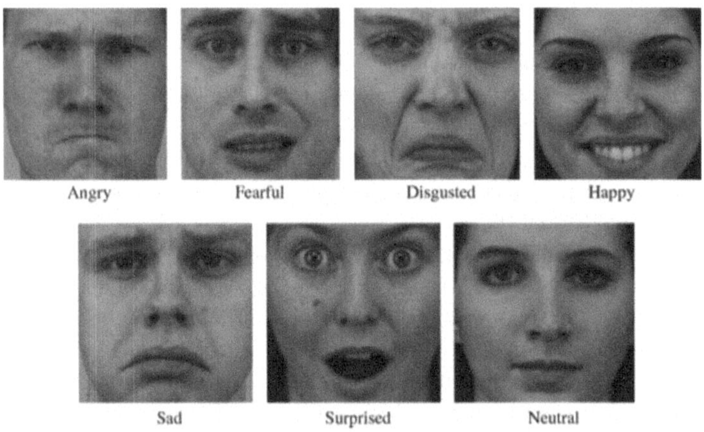

Fig. 3 Example of each emotion (Angry, Fearful, Disgusted, Happy, Sad, Surprised, and Neutral) of the KDEF stimulus set

expression that was appropriate for the vignette. The test can provide adequate performance scores for emotion recognition ability across a broad range of facial characteristics relying on affect naming. DANVA nevertheless only includes pictures for four out of the six basic emotions.

The computer-based Frankfurt Test for Facial Affect Recognition [10] uses the cross-cultural concept of the seven fundamental affective states by Ekman, Friesen, and Ellsworth (1972) and comprise a series of 50 items with black and white pictures presenting basic emotions for faces (face test) and 40 items for eyes (eyes test). Each picture is shown on the computer screen separately with all six emotions written on the side of the picture as an answer option. It is important to mention that some of the photographs included were from the pictures of facial affect by Ekman & Friesen [32]. Whereas the psychometric properties of FEFA having been reported to be excellent (Bolte and Poustka 2003) a striking limitation is the pictures being presented in black and white hence lacking ecological validity and depicting only adults.

The 3D faces developed by Gur et al. [41] were developed on the notion that facial expressions of emotion are increasingly being used in neuroscience as stimuli for studying hemispheric specialization and as probes for functional imaging for face and emotion processing. The orientation of the 2-dimensional face stimuli is fixed and poorly suited for examining asymmetries, where the actors are of a restricted age range and usually pose. The 3-dimenionional faces by Gur et al. [41] were used in the two neuroimaging studies of the review: an fMRI and an ERP study. The 3D images were acquired and reconstructed by adult actors and actresses expressing the 6 basic emotions as well as neutral expressions.

The Cohn-Kanade Facial Expressions Database [48] includes 2105 digitized image sequences from 182 adults between the ages of 18–50 years of age (69% female and 31% male) of varying ethnicity, performing multiple tokens of most primary FACS action units representing happiness, surprise, sadness, disgust, anger, and fear. Despite the database being one of the most comprehensive ones for comparative studies of facial expression analysis, two limitations are: (1) the absence of child stimuli and (2) the lack of spontaneous expressions taking into account that deliberate and spontaneous expressions have different appearance and timing (Fig. 4).

When evaluating the emotion recognition accuracy of children with ADHD it is important to consider how many and which of the basic emotions were investigated. Most of the studies included in the review investigated all six basic emotions [4, 8, 9, 14, 17, 20, 23] with the most researched emotion being happiness followed by anger, sadness and fear. Another great variability identified in the review was that despite the majority of studies having differentiated among specific emotion dimensions in their analyses, two did not define the emotions investigated; Downs and Smith [3] and Greenbaum et al. [12], and four studies did not differentiate among specific emotion dimensions in their analyses; Corbett and Glidden [1], Yuill and Lyon [5], Guyer et al. [6] and Rinke et al. [24].

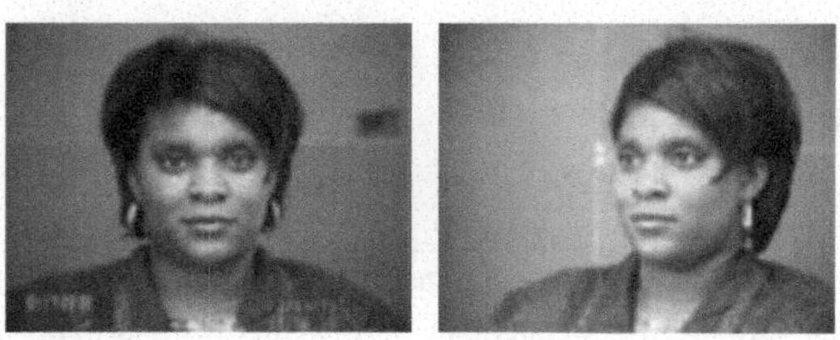

Fig. 4 Frontal and 30-degree to the side views available in the database

When it comes to comorbidities, albeit nine out of the 25 studies included in the review having implicated children with ADHD and 1 or more comorbid conditions only four controlled for the factor of comorbidity in their analyses while also including a control group; Cadesky et al. [2], Sinzig et al. [8], Nordermeer et al. [18], and Airdrie et al. [26]. Different comorbidities were nevertheless assessed as in Sinzig et al. [8] ASD and in Nordermeer et al. [18] ODD. Despite Cadesky et al. [2] and Airdrie et al. [26] both implicating children with CD, the two studies found conflicting results. Whereas both studies implicated static stimuli and studied the same emotions (Happiness, Fear, Anger, Sadness) Cadesky et al. [2] employed the DANVA [63, 26] the Pictures of Facial Affect [32]. An additional difference among the two studies was that children in Aidrie et al. [26] were older (9.3, 1.5) than Aidrie's et al. [26] (14.20, 2.09). Literature suggests that ADHD rarely occurs in isolation being highly concurrent with ASD; according to Davis and Kollins [22] more than two-thirds of individuals with ADHD show features of ASD), there is a comorbidity of 60% with ODD and a prevalence of 16–20% with CD [85]. Whereas the presence of comorbid disorders may pose a potential explanation for the reported difficulties in face recognition in children and adolescents with ADHD, research controlling for the factor of comorbidity in the analyses while also including a control group is sparse.

Despite a web-based search on emotion recognition in ADHD will elicit hundreds of studies, one cannot concretely say in which factors do children with ADHD differ from TD children. Literature on emotion recognition in ADHD has produced mixed findings, a result that can be partly attributed to the great variability of studies employed in the investigation of emotion recognition in ADHD and the complexity of facial emotion recognition in this psychiatric population. While, several factors have been carefully taken into consideration in the design of the studies included in the review, one cannot but wonder whether the studies' participants were undertaking therapies other than pharmacotherapy (i.e. occupational therapy, psychotherapy) with aim to tackle the social and emotional deficits of ADHD and whether these therapeutic methods had an impact on the studies' results.

Acknowledgements The research leading to these results has received funding from the EU H2020 research and innovation program under grant agreement N. 769872 (EMPATHIC) and N. 823907 (MENHIR), the project SIROBOTICS that received funding from Italian MIUR, PNR 2015–2020, D.D. 1735, 13/07/ 2017, and the project ANDROIDS that received funding from Università della Campania "Luigi Vanvitelli" inside the program V:ALERE 2019, funded with D.R. 906 del 4/10/2019, prot. n. 157264,17/10/2019.

References

1. Airdrie, J.N., Langley, K., Thapar, A., van Goozen, S.H.M.: AC. J Am Acad Child Adolesc Psychiatry **57**(8), 561–570 (2018). http://doi.org/10.1016/j.jaac.2018.04.016
2. Arsenio, W., Cooperman, S., Lover, A.: Affective predictors of preschoolers' aggression and peer acceptance: direct and indirect effects. Develop. Psychol. **36**(4), 438–448 (2000). https://doi.org/10.1037/0012-1649.36.4.438. [PubMed: 10902696]

3. Aspan, N., Bozsik, C., Gadoros, J., Nagy, P., Inantsy-Pap, J., Vida, P., Halasz, J.: Emotion recognition pattern in adolescent boys with attention-deficit/hyperactivity disorder. BioMed Res. Int. (2014). http://dx.doi.org/10.1155/2014/761340
4. Bal, Elgziz: *Emotion Recognition and Social Behaviors in Children with Attention-Deficit/Hyperactivity Disorder* (Published doctoral dissertation). University of Illinois at Chicago, Chicago, Illinois (2011)
5. Barkley, R.A.: Behavioral inhibition, sustained attention, an executive functions: constructing a unified theory of ADHD. Psychol. Bull. **121**, 65–94 (1997)
6. Baum, K.M., Nowicki, S. Jr.: Perception of emotion: measuring decoding accuracy of adult prosodic cues varying in intensity. J. Nonverbal Behav. **22**, 89–107 (1998)
7. Berggren, S., Engström, A.-C., Bölte, S.: Facial affect recognition in autism, ADHD and typical development. Cognit. Neuropsychiatry **21**(3), 213–227 (2016). http://doi.org/10.1080/13546805.2016.1171205
8. Blair, R.J.R.: Facial expressions, their communicatory functions and neuro-cognitive substrates. Philos. Trans. R. Soc. B **358**, 561–572 (2003). http://doi.org/10.1098/rstb.2002.1220
9. Boakes, J., Chapman, E., Houghton, S., West, J.: Facial interpretation in boys with attention deficit/hyperactivity disorder. Child Neuropsychol. **14**, 82–96 (2008)
10. Bolte, S., Feineis-Matthews, Sabine, Leber, S., Dierks, T., Hubl, D., Poustka, F.: The development and evaluation of a computer-based program to test and teach the recognition of facial affect. Int. J. Circumpolar Health **61**(2), 61–68 (2002). https://doi.org/10.3402/ijch.v61i0.17503
11. Bora, E., Pantelis, C.: Meta-analysis of social cognition in attention-deficit/ hyperactivity disorder (ADHD): comparison with healthy controls and autistic spectrum disorder. Psychol. Med. **46**, 699–716 (2016). http://doi.org/10.1017/S0033291715002573
12. Boyatzis, C.J., Chazan, E., Ting, C.Z.: Preschool children's decoding of facial emotions. J. Genet. Psychol. **154**(3), 375–382 (1993). http://doi.org/10.1080/00221325.1993.10532190
13. Bruce, V., Campbell, R.N., Doherty-Sneddon, G., Langton, S., McAuley, S., Wright, R.: Testing face processing skills in children. Br. J. Dev. Psychol. **18**, 319–333 (2000). http://doi.org/10.1348/026151000165715
14. Cadesky, E.B., Mota, V.L., Schachar, R.J.: Beyond words: how do children with ADHD and/or conduct problems process nonverbal information about affect? J. Am. Acad. Child Adolesc. Psychiatry **39**(9), 1160–1167 (2000). http://doi.org/10.1097/00004583-200009000-00016
15. Caspi, A., Taylor, A., Moffitt, T.E., Plomin, R.: Neighborhood deprivation affects children's mental health: environmental risks identified in a genetic design. Sage J. **11**(20), 338–342 (2000). https://doi.org/10.1111/1467-9280.00267
16. Corbett, B., Glidden, H.: Processing affective stimuli in children with attention-deficit hyperactivity disorder. Child Neuropsychol. **6**(2), 144–155 (2000). http://doi.org/10.1076/chin.6.2.144.7056
17. Corcoran, C.M., Keilp, J.G., Kayser, J., Klim, C., Butler, P.D., Bruder, G.E., … Javitt, D.C.: Emotion recognition deficits as predictors of transition in individuals at clinical high risk for schizophrenia: a neurodevelopmental perspective. Psychol. Med. **45**(14), 2959–2973 (2015). http://doi.org/10.1017/S00332917150009024
18. Coy, K., Speliz, M.L., DeKlyen, M., Jones, K.: Social-cognitive processes in preschool boys with and without oppositional defiant disorder. J. Abnormal Child Psychol. **229**(2) (2001). http://doi.org/107–119.10.1023/a:1005279828676 [PubMed: 11321626]
19. Da Fonseca, D., Sequier, V., Santos, A., Poinso, F., Deruelle, C.: Emotion understanding in children with ADHD. Child Psychiatry Human Dev. **40**, 111–121 (2009)
20. Dadds, M.R., Cauchi, A.J., Wimalaweera, S., Hawes, D.J., Brennan, J.: Outcomes, moderators, and mediators of empathic-emotion recognition training for complex conduct problems in childhood. Psychiatry Res. **199**(3), 201–207 (2012). https://doi.org/10.1016/j.psychres.2012.04.03
21. Dalili, M.N., Penton-Voak, I.S., Harmer, C.J., Munafò, M.R.: Meta-analysis of emotion recognition deficits in major depressive disorder. Psychol. Med. **45**(6), 1135–1144 (2015). https://doi.org/10.1017/S0033291714002591

22. Davis, N.O., Kollins, S.: Treatment for co-occurring attention deficit/hyperactivity disorder and autism spectrum disorder. Neurotherapeutics **9**(3), 518–530 (2012)
23. De Sonneville, L.M.J.: Amsterdamse neuropsychologische taken: wetenschappelijke en klinische toepassingen. J. Neuropsychol. **1**, 27–41 (2005)
24. De Sonneville, L.M.J., Verschoor, C.A., Njiokiktjien, C., Op het Veld, V., Toorenaar, N., Vranken, M.: Facial identity and facial emotions: speed, accuracy, and processing strategies in children and adults. J. Clin. Exp. Neuropsychol. (Neuropsychol. Dev. Cogn. Sec. A) **24**(2), 200–213 (2002). http://doi.org/10.1076/jcen.24.2.200.989
25. De Sonneville, L.M.J., Verschoor, C.A., Njiokiktjien, C., Op het Veld, V., Toorenaar, N., Vranken, M.: Facial identity and facial emotions: speed, accuracy, and processing strategies in children and adults. J. Clin. Exp. Neuropsychol. (Neuropsychol. Dev. Cogn. Sect. A) **24**(2), 200–213 (2002). http://doi.org/10.1076/jcen.24.2.200.989
26. Denham, S.A.: Emotional Development in Young Children. Guilford Press, New York, London (1998)
27. Denham, S.A.: Social-emotional competence as a support for school readiness: what is it and how do we assess it? Early Edu. Dev. **17**, 57–89 (2006). https://doi.org/10.1207/s15566935eed1701_4
28. Downs, A., Smith, T.: Emotional understanding, cooperation, and social behaviour in high-functioning children with autism? J. Autism Dev. Dis. **34**, 625–635 (2004)
29. DuPaul, G.J., Volpe, R.J., Jitendra, A.K., Lutz, J.G., Lorah, K.S., Gruber, R.: Elementary school students with AD/HD: predictors of academic achievement. J. Sch. Psychol. **42**(4), 285–301 (2004). https://doi.org/10.1016/j.jsp.2004.05.001
30. Ekman, P.: Universals and cultural differences in facial expressions of emotion. In: Cole J. (ed.) Nebraska Symposium on Motivation, 1971, vol. 19, pp. 207–282. University of Nebraska Press: Lincoln (1972)
31. Ekman, P.: Facial expression and emotion. Am. Psychol. **48**(4), 384–392 (1992)
32. Ekman, P., Friesen, W.V.: Pictures of facial affect. Palo Alto. California: Consulting Psychologists Press (1976)
33. Elfenbein, H.A., Mandal, M.K., Ambady, N., Harizuka, S., Kumar, S.: Hemifacial differences in the in-group advantage in emotion recognition. Cognit. Emot. **18**, 613–629 (2004)
34. Esposito, A., Esposito, A.M., Vogel, C.: Needs and challenges in human computer interaction for processing social emotional information. Pattern Rec. Lett. **66**, 41–51 (2015)
35. Fayyad, J., De Graaf, R., Kessler, R., Alonso, J., Angermeyer, M., Demyttenaere, K.,... Lepine, J.P.: Cross-national prevalence and correlates of adult attention-deficit hyperactivity disorder. Br. J. Psychiatry **190**(5), 402–409 (2007)
36. Goeleven, E., De Raedt, R., Leyman, L., Verschuere, B.: The karolinska directed emotional faces: a validation study. Cognit. Emotion **22**(6), 1094–1118 (2008)
37. Gosselin, P.: Le décodage de l'expression faciale des émotions au cours de l'enfance (The decoding of facial expressions of emotions during childhood). Can. Psychol. **46**, 126–138 (2005)
38. Gosselin, P., Larocque, C.: Facial morphology and children's categorization of facial expressions of emotions: a comparison between Asian and Caucasian faces. J. Genet. Psychol. **161**, 346–358 (2000)
39. Green, M.F., Kern, R.S., Robertson, M.J., Sergi, M.J., Kee, K.S.: Relevance of neurocognitive deficits for functional outcome in schizophrenia. In: Sharma, T., Harvey, P. (eds.) Cognition in Schizophrenia: Impairments, Importance, and Treatment Strategies, pp. 178–192. Oxford University Press, New York (2000)
40. Greenbaum, R.L., Stevens, S.A., Nash, K., Koren, G., Rovet, J., Form, T.R., ... Program, M.: Social cognitive and emotion processing abilities of children with fetal alcohol spectrum disorders. Hyperact. Disord. **33**(10), 1656–1670 (2009). https://doi.org/10.1111/j.1530-0277.2009.01003.x
41. Gur, R.C., Sara, R., Hagendoorn, M., Marom, O., Hughett, P., Macy, L.,... Gur, R.E: A method for obtaining 3-dimensional facial expressions and its standardization for use in neurocognitive studies. J. Neurosci. Methods **115**(2), 137–143 (2002)

42. Guyer, A.E., McClure, E.B., Adler, A.D., Brotman, M.A., Rich, B.A., Kimes, A.S., Pine, D.S., Ernst, M., Leibenluft, E.: Specificity of facial expression labeling deficits in childhood psychopathology. Psychol. Fac. Public. **119** (2009)
43. Herba, C., Phillips, M.: Annotation: Development of facial expression recognition from childhood to adolescence: Behavioural and neurological perspectives. J. Child Psychol. Psychiatry **45**(7), 1185–1198 (2004). https://doi.org/10.1111/j.1469-7610.2004.00316.x
44. Hess, U., Blairy, S.: Set of emotional facial stimuli. Department of Psychology, University of Quebec at Montreal, Montreal, Canada (1995)
45. Howlin, P., Baron-Cohen, S., Hadwin, J.: Teaching Children with Autism to Mind-Read: A Practical Guide. Wiley, West Sussex (1999)
46. Hoza, B., Mrug, S., Gerdes, A.C., Hinshaw, S.P., Bukowski, W.M., Gold, J.A., et al.: What aspects of peer relationships are impaired in children with attention-deficit/hyperactivity disorder? J. Cons. Clin. Psychol. **73**, 411–423 (2005). https://doi.org/10.1037/0022-006X.73.3.411
47. Jusyte, A., Gulewitsch, M.D., Schönenberg, M.: Recognition of peer emotions in children with ADHD: evidence from an animated facial expressions task. Psychiatry Res. **258**(2017), 351–357 (2017)
48. Kanade, T., Cohn, J.F., Yingli Tian.: Comprehensive database for facial expression analysis. Paper presented at the 4th IEEE International Conference on Automatic Face and Gesture Recognition (Cat. No. PR00580). Grenoble, France, 28–30 March 2000 (2000). http://doi.org/10.1109/afgr.2000.840611
49. Kara, H., Bodur, S., Çetinkaya, M., Kara, K., Tulaci, O.D.: Assessment of relationship between comorbid oppositional defiant disorder and recognition of emotional facial expressions in children with attention-deficit/hyperactivity disorder. Psychiatry Clin. Psychopharmacol. **27**(4), 329–336 (2017). https://doi.org/10.1080/24750573.2017.1367566
50. Kochel, A., Leutgeb, V., Schienle, A.: Disrupted response inhibition toward facial anger cues in children with Attention-Deficit Hyperactivity Disorder (ADHD): an event-related potential study. J. Child Neurol. **29**(4), 459–468 (2014)
51. Lai, Z., Hughes, S., Shapiro, E.: Manual for the tests of affective processing (MNTAP). University of Minnesota, MN (1991)
52. Langner, O., Dotsch, R., Bijlstra, G., Wigboldus, D.H., Hawk, S.T., van Knippenberg, A.: Presentation and validation of the radboud faces database. Cognit. Emotion **24**(8), 1377–1388 (2010). https://doi.org/10.1080/02699930903485076
53. Lee, S.S., Falk, A.E., Aguirre, V.P.: Association of comorbid anxiety with social functioning in school-age children with and without attention-deficit/hyperactivity disorder (ADHD). Psychiatry Res. **197**, 90–96 (2012). https://doi.org/10.1016/j.psychres.2012.01.018
54. Ludlow, A., Garrood, A., Lawrence, K., Gutierrez, R.: Emotion recognition from dynamic emotional displays in children with ADHD. J. Soc. Clin. Psychol. **33**(5), 413–427 (2014). https://doi.org/10.1521/jscp.2014.33.5.413
55. Lundqvist, D., Flykt, A., Öhman, A.: The Karolinska directed emotional faces (KDEF). CD ROM from Department of Clinical Neuroscience, Psychology section, Karolinska Institutet **91**, 630 (1998)
56. McClure, E.B.: A meta-analytic review of sex differences in facial expression processing and their development in infants, children, and adolescents. Psychol. Bull. **126**(3), 424–453 (2000). http://doi.org/10.1037/0033-2909.126.3.424
57. McDonald, S., Flanagan, S., Rollins, J.: The Awareness of Social Inference Test. Thames Valley Test Company Limited, Suffolk: England (2002)
58. Maxim, L.A., Nowicki, S.J.: Developmental association between nonverbal ability and social competence. Phil. Soc. Psychol. **2**(10), 745–758 (2003)
59. Miller, A., Gouley, K., Seifer, R., Zakriski, A., Eguia, M., Vergnani, M.: Emotion knowledge skills in low- income elementary school children: associations with social status and peer experiences. Soc. Dev. **14**(4), 637–651 (2005). https://doi.org/10.1111/j.1467-9507.2005.00321.x

60. Mondloch, C.J., Geldart, S., Maurer, D., Le Grand, R.: Developmental changes in face processing skills. J. Exp. Child Psychol. **86**, 67–84 (2003)
61. Mrug, S., Molina, B. S., Hoza, B., Gerdes, A. C., Hinshaw, S. P., Hechtman, L., & Arnold, L. E. (2012). Peer rejection and friendships in children with attention-deficit/hyperactivity disorder: Contributions to long-term outcomes. *Journal of abnormal child psychology, 40*(6), 1013-1026. http://doi.org/10.1007/s10802-012-9610-2
62. Noordermeer, S.D., Luman, M., Buitelaar, J.K., Hartman, C.A., Hoekstra, P.J., Franke, B., ... Oosterlaan, J.: Neurocognitive deficits in attention-deficit/hyperactivity disorder with and without comorbid oppositional defiant disorder. J. Att. Disorde, 1–13 (2015). http://doi.org/10.1177/1087054715606216
63. Nowicki Jr., S., Duke, M.P.: Individual differences in the nonverbal communication of affect: the diagnostic analysis of nonverbal accuracy scale. J. Nonverbal Behav. **18**, 9–35 (1994)
64. Nowicki Jr., S., Carton, J.: Measuring emotional intensity from facial expressions: the DANVAFACES 2. J. Soc. Psychol. **133**, 749–750 (1993)
65. Passarotti, A.M., Sweeney, J.A., Pavuluri, M.N.: Emotion processing influences working memory circuits in pediatric bipolar disorder and attention deficit hyperactivity disorder. J. Am. Acad. Child Adolesc. Psychiatry **49**(10), 1064–1080 (2010). http://doi.org/10.1016/j.jaac.2010.07.009
66. Pelc, K., Kornreich, C., Foisy, M.-L., Dan, B.: Recognition of emotional facial expressions inattention-deficit hyperactivity disorder. Pediatr. Neurol. **35**, 93–97 (2006). https://doi.org/10.1016/j.pediatrneurol.2006.01.014
67. Pochon, R., Declercq, C.: Emotion recognition by children with down syndrome: a longitudinal study. J. Int. Devel. Dis. **38**, 332–343 (2013). https://doi.org/10.3109/13668250.2013.826346
68. Polanczyk, G., De Lima, M.S., Horta, B.L., Biederman, J., Rohde, L.A.: The worldwide prevalence of ADHD: a systematic review and metaregression analysis. Am. J. Psychiatry **164**(6), 942–948 (2007)
69. Porges, S.W., Cohn, J.F., Bal, E., Lamb, D.: The Dynamic Affect Recognition Evaluation [Computer software]. Brain-Body Center, University of Illinois at Chicago, Chicago, IL (2007)
70. Posamentier, M.T.: Processing faces and facial expressions. Neuropsychol. Rev. **13**(3) (2003). http://dx.doi.org/1040-7308/03/0900-0113
71. Raver, C.C., Garner, P.W., Smith-Donald, R.: The roles of emotion regulation and emotion knowledge for children's academic readiness. In: Pianta, R.C., Cox, M.J., Snow, K.L. (eds.) School Readiness and the Transition to Kindergarten in the Era of Accountability, pp. 121–147. Paul H Brookes Publishing, Baltimore, MD (2006)
72. Rinke, L., Candrian, G., Loher, S., Blunck, A., Mueller, A., Jäncke, L.: Facial emotion recognition deficits in children with and without attention deficit hyperactivity disorder: a behavioral and neurophysiological approach. NeuroReport **28**(14), 917–921 (2017). https://doi.org/10.1097/WNR.0000000000000858
73. Romani, M., Vigliante, M., Faedda, N., Rossetti, S., Pezzuti, L., Guidetti, V., Cardona, F.: Face memory and face recognition in children and adolescents with attention deficit hyperactivity disorder: a systematic review. Neurosci. Biobehav. Rev. **89**, 1–12 (2018)
74. Sarraf Razavi, M., Tehranidoost, M., Ghassemi, F., Purabassi, P., Taymourtash, A.: Emotional face recognition in children with attention deficit/hyperactivity disorder: evidence from event related gamma oscillation. Basic Clin. Neurosci. **8**(5), 419–426 (2017). https://doi.org/10.18869/NIRP.BCN.8.5.419
75. Schultz, D., Izard, C.E., Ackerman, B.P.: Children's anger attribution bias: relations to family environment and social adjustment. Soc. Dev. **9**(3), 284–301 (2000). https://doi.org/10.1111/1467-9507.00126
76. Schwenck, C., Schneider, T., Schreckenbach, J., Zenglein, Y., Gensthaler, A., Taurines, R.,... Romanos, M.: Emotion recognition in children and adolescents with attention-deficit/hyperactivity disorder (ADHD). ADHD Atten. Def. Hyperact. Disord. **5**(3), 295–302 (2013)
77. Sinzig, J., Morsch, D., Lehmkuhl, G.: Do hyperactivity, impulsivity and inattention have an impact on the ability of facial affect recognition in children with autism and ADHD? Eur. Child Adolesc. Psychiatry **17**, 63–72 (2008). https://doi.org/10.1007/s00787-007-0637-9

78. Southam-Gerow, M.A., Kendall, P.C.: Emotion regulation and understanding. Imp. Child Psychopathol. Ther. **22**, 189–222 (2002)
79. Sully, K., Sonuga-Barke., E., Fairchild, G.: The familial basis of facial emotion recognition deficits in adolescents with conduct disorder and their unaffected relatives. Psychol. Med. **45**, 1–11 (2015). http://doi.org/10.1017/S0033291714003080
80. Tehrani-Doost, M., Noorazar, G., Shahrivar, Z., Banaraki, A.K., Beigi, P.F., Noorian, N.: Is emotion recognition related to core symptoms of childhood ADHD? J. Can. Acad. Child Adolesc. Psychiatry **26**(1), 31–38 (2017)
81. Tonks, J., Williams, W.H.U.W., Frampton, I.A.N., Yates, P., Slater, A.: Assessing emotion recognition in 9–15-years olds: Preliminary analysis of abilities in reading emotion from faces, voices and eyes, **21**(June), 623–629 (2007). https://doi.org/10.1080/02699050701426865
82. Uekermann, J.1., Kraemer. M., Abdel-Hamid, M., Schimmelmann, B.G., Hebebrand, J., Daum, I., Wiltfang, J., Kis, B.: Social cognition in attention-deficit hyperactivity disorder (ADHD). Neurosci. Biobehav. Rev. **34**, 734–743 (2010). http://doi.org/10.1016/j.neubiorev.2009.10.009
83. van Goozen, S.H.M., Langley, K., Langley, K., Northover, C., Hubble, K., Rubia, K.,... Thapar, A.: Identifying mechanisms that underlie links between COMT genotype and aggression in male adolescents with ADHD (2016). http://dx.doi.org/10.1111/jcpp.12464
84. Vicari, S., Reilly, J.S., Pasqualetti, P., Vizzotto, A., Caltagirone, C.: Recognition of facial expressions of emotions in school-age children: the intersection of perceptual and semantic categories. Acta Paediatr. **89**(7), 836–845 (2000). http://doi.org/10.1111/j.1651-2227.2000.tb00392.x
85. Villodas, M.T., Pfiffner, L.J., McBurnett, K.: Prevention of serious conduct problems in youth with attention deficit/hyperactivity disorder. Exp. Rev. Neurother. **12**(10), 1253–1263 (2012)
86. Willcutt, E.G.: The prevalence of DSM-IV attention-deficit/hyperactivity disorder: a meta-analytic review. Neurotherapeutics **9**, 490–499 (2012)
87. Williams, L.M., Hermens, D.F., Palmer, D., Kohn, M., Clarke, S., Keage, H.,... Gordon, E: Misinterpreting emotional expressions in attention-deficit/hyperactivity disorder: evidence for a neural marker and stimulant effects. Biologic. Psychiatry **63**(10), 917–926 (2008)
88. Young, A., Perrett, D., Calder, A.J., Sprengelmeyer, R., Ekman, P.: Facial expressions of emotion-stimuli and tests (FEEST). Thames Valley Test Company, Bury St. Edmunds, UK (2002)
89. Yuill, N., Lyon, J.: Selective difficulty in recognising facial expressions of emotion in boys with ADHD: general performance impairments or specific problems in social cognition? Euro. Child Adolesc. Psychiatry **16**(6), 398–404 (2007). http://doi.org/10.1007/s00787-007-0612-5

The Effect of Facial Expressions on Interpersonal Space: A Gender Study in Immersive Virtual Reality

Mariachiara Rapuano, Filomena Leonela Sbordone, Luigi Oreste Borrelli, Gennaro Ruggiero, and Tina Iachini

Abstract In proxemics, the interpersonal space is the optimal social distance between individuals. Evidence has shown that the emotional facial expressions and gender-related effects can modulate this distance during social interactions. Typically, this distance increases in threatening situations and decreases in safe situations. Moreover, male dyads maintain larger distances than female dyads whereas the findings about mixed-sex dyads are still unclear. Virtual Reality (VR) based technologies are becoming more and more used in different areas of everyday life and, in the scientific field, for studying social phenomena. This raises the question of the degree of similarity of VR simulations to actual phenomena, i.e. its ecological validity and its effectiveness for applied purposes. In order to clarify gender-related and emotion-related effects and the ecological validity of VR, we investigated if real females and males differently modulated their interpersonal distance while male and female virtual confederates with happy, angry and neutral faces approached them. Results showed that participants preferred larger distances with both male and female virtual confederates who showed an angry face rather than neutral and happy. Moreover, males preferred a shorter distance, particularly when facing smiling virtual females, while females preferred a larger distance from angry virtual males. These results suggest that gender differences can affect the impact of emotional facial expression on the modulation of the interpersonal space. Moreover, they confirm previous

---

M. Rapuano (✉) · F. L. Sbordone · L. O. Borrelli · G. Ruggiero · T. Iachini
Laboratory of Cognitive Science and Immersive Virtual Reality, Department of Psychology, University of Campania "L. Vanvitelli", Caserta, Italy
e-mail: mariachiara.rapuano@unicampania.it

F. L. Sbordone
e-mail: filomenaleonela.sbordone@unicampania.it

L. O. Borrelli
e-mail: borrelli.luigi.oreste@gmail.com

G. Ruggiero
e-mail: gennaro.ruggiero@unicampania.it

T. Iachini
e-mail: santa.iachini@unicampania.it

A. Esposito et al. (eds.), *Progresses in Artificial Intelligence and Neural Systems*, Smart Innovation, Systems and Technologies 184,
https://doi.org/10.1007/978-981-15-5093-5_40

behavioural studies and add further support to the ecological validity of IVR simulations and thus its potential effectiveness as a tool for research and application interventions.

1 Introduction

In social psychology, proxemics defines "interpersonal space" (IPS) as the optimal distance that people maintain between themselves and others [9,15, 16, 37]. This portion of space is also conceived as the emotionally tinged zone that people feel as their 'private space' and cannot be intruded without causing discomfort [6, 15, 16, 27]. Typically, people react by increasing this distance when feeling in uncomfortable/threatening situations and by reducing it when feeling in comfort/safe situations [1,13, 15, 16, 20, 23, 31, 39, 41]. Alongside, the emotional valence conveyed by individuals' facial expressions represents an essential component for modulating our interpersonal social distance [11, 22]. Facial expressions are highly informative, mediate visuo-perceptual, psychophysiological and automatic behavioural responses [18, 35, 42]. Indeed, facial expressions that communicate cooperation would prompt approaching behaviours, whereas facial expressions that communicate threat would induce avoiding behaviours [29]. Therefore, emotional stimuli trigger approaching-avoiding reactions that reveal evolutionary adaptations rooted in basic survival mechanisms [7, 8, 11, 14, 31]. Overall, facial expressions convey emotional signals about individuals' intent and constitute an essential component of social interactions by prompting and orienting individuals' responses [8, 12, 24]. Much research within the proxemics domain suggests that also gender moderate social distance [21]. Indeed, it is widely agreed that male dyads maintain larger distances than female dyads during social interactions, while the findings about the interaction between mixed-sex dyads are still unclear [2, 16, 33, 37, 38, 40]. For example, mixed-sex dyads who are in close relationship maintain closer distances than either male-male dyads or female-female dyads [6]. Moreover, women tend to keep a closer distance with men they are friends with [6].

Virtual Reality (VR) based technologies are becoming more and more used in different areas of everyday life and, in the scientific field, for studying social phenomena [30, 36]. The IVR allows for the simulation of social interactions in proxemics research by keeping under control virtual humans' appearance and behaviour, and their spatial context while assuring a high level of similarity to real interactions [5, 26]. Despite these positive characteristics, the IVR technology still raises some criticisms due to supposed discrepancies from real-world interactions for the lack of physical contact [17] and possible perceptual alterations [3, 25, 26]. This raises the question of the degree of similarity of VR simulations to actual phenomena, i.e. its ecological validity and consequently its effectiveness for applied purposes.

In order to clarify the effect of emotional expressions and gender of the interactants and the level of ecological validity of VR, we studied the proxemic behaviour of real females and males in interaction with virtual females and males. To this end, in

Immersive virtual reality (IVR) male and female real participants were approached by male and female virtual confederates exhibiting happy, angry and neutral facial expressions.

Moreover, we administered the stop-distance paradigm that is typically used to evaluate the dimension of interpersonal space, i.e. the comfort distance (the optimal distance from other individuals). Participants were standing still and saw the virtual confederates approaching them. They had to stop the virtual confederate at the point where they still felt comfortable with his/her proximity [2, 10, 13, 16].

We expect that both men and women should maintain a larger distance from virtual confederates with angry than neutral and happy facial expressions. Consistently with the proxemics literature, we expect that men should prefer an overall shorter distance than women. However, it is still unclear how the combination of gender and facial expression may affect men and women's interpersonal space. We hypothesise that men should prefer a shorter distance when interacting with female virtual confederates, especially when they show a happy facial expression, than women interacting with their male counterparts.

2 Method

2.1 Participants

Thirty-two right-handed participants (16 males), aged 18–29 years ($M_{age} = 23$; $SD = 2.9$), were recruited in exchange for course credit. Participants had normal/corrected-to-normal vision). Nobody claimed discomfort or vertigo during the IVR experience and reported being aware of the experimental purpose. All participants gave their written consent to take part in the study. The study was in conformity with the local Ethics Committee requirements and the Declaration of Helsinki [43].

2.2 Settings and Apparatus

The experimental setting and the virtual scenario were similar to those of previous studies [19, 21, 31]. The IVR equipment was installed in a 5 × 4 × 3 m room of the Laboratory of Cognitive Science and Immersive Virtual Reality (CS-IVR, Department of Psychology). The equipment included the 3-D Vizard Virtual Reality Software Toolkit 4.10 (Worldviz, LLC, USA) with the Sony HMZ-T1 (SONY, Japan) head mounted display (HMD) having two OLED displays for stereoscopic depth (images = 1280 9 720; 60 Hz; 45° horizontally, 51.6° diagonally).

The IVR system continuously tracked and recorded participant's position (sample rate = 18 Hz) through a marker on the HMD. Head orientation was tracked by a three-axis orientation sensor (Sensor Bus USB Control-Unit, USA). Visual information was updated in real time. A glove with 14 tactile-pressure sensors allowed the participants to "see" and "feel" their arm movements (Data Glove Ultra; WorldViz, USA). Graphic modeling was created through SketchUp Make (Trimble, USA).

2.3 Virtual Stimuli

The virtual room consisted of green walls, white ceiling and grey floor. A total of twelve young confederates (six females) were selected among a colony of highly realistic virtual humans and were used for the present study (Vizard Complete Characters, WorldViz; USA). Virtual humans represented male and female adults aged about thirty years wearing similar casual clothes and were perceived as representation of Italian people. Their height was 175 cm (males) and 165 cm (females). Their gaze was kept looking straight ahead throughout the trials [5]. Facial emotional expressiveness was obtained by modelling the virtual faces with 3DS Max (Autodesk) following the KDEF free-database (Karolinska Directed Emotional Faces; [28]. Facial expressions were selected on the basis of a pilot study [31] where 14 participants (seven women) rated on a 9-point scale how much the faces presented on the PC appeared happy/unhappy, friendly/threatening, angry/peaceful, and annoying/quite. Following this evaluation, twelve virtual confederates with happy, angry, and neutral expressions were selected (Fig. 1).

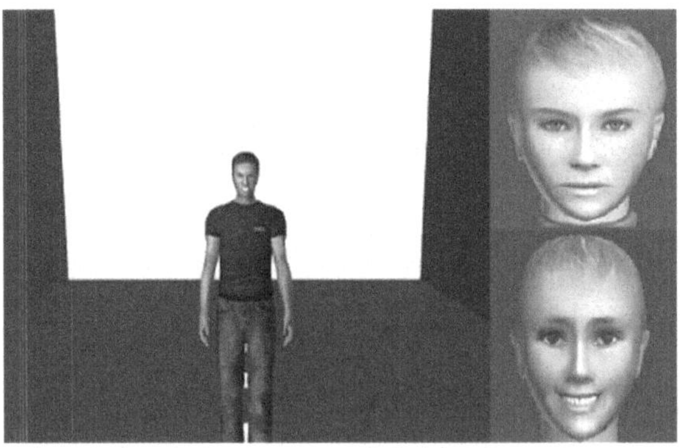

Fig. 1 Examples of virtual stimuli and setting. The left panel shows participants' perspective when a virtual male adult with an angry facial expression frontally approached them; on the floor, a straight dashed white line represented the path that participants and virtual humans followed during the passive/active approach conditions. The right panel shows examples of neutral (top) and happy (bottom) facial expressions of virtual women [31]

2.4 *Procedure*

After giving their written consent, participants were instructed about the task, invited to wear the HMD. Through the HMD, participants were fully immersed in the virtual room where they could see the virtual stimuli and could make extensive exploratory movements. During this initial experience, they had to describe their feeling of presence. Next, they were led to the starting location and were provided with a key-press device held in their dominant hand. The experimental session was divided in two blocks administered in a counterbalanced order with a short delay in between. Within each block, participants had to perform the comfort-distance task. The comfort-distance instruction was: "Press the button until the distance between you and the confederate makes you feel comfortable". During the entire experimental session, participants stood still and saw the virtual confederates walking towards them (0.5 ms^{-1}) until they stopped them by button press. After button press, virtual confederates disappeared. At the beginning of the first block, participants received a four-trial training session. Within each block, the IVR system selected six virtual confederates (three females) showing happy/angry/neutral facial expressions. Each virtual confederate appeared 4 times (quasi-randomized order) resulting in 24 trials per block (tot = 48 trials). After each block, the experimenter checked whether the participants performed the task correctly. Finally, participants evaluated their experience with the virtual confederates. They reported they clearly identified their facial expressions as if they were "realistic persons".

3 Data Analysis

In each trial, participant-virtual confederate distance was recorded. Participant's arm length was then subtracted from the mean distance. The mean distances were analyzed through a two-way repeated measure ANOVA with Participants' Gender as 2-level between factor, Virtual confederates' Gender as 2-level within factor (Male-Female) and Facial expressions as 3-level within factor (Happy-Angry-Neutral).

Data points outside M ± 2.5 SD (0.08%) were excluded. The Newman-Keuls post hoc test was used. The magnitude of significant effects was expressed by partial eta-squared (η_p^2).

3.1 *Results*

A main effect of participants' gender emerged ($F(1,30) = 6.9$, $p = 0.01$, $\eta_p^2 = 0.19$).

Male participants preferred a shorter distance than women ($M_{men} = 65.37$; SD = 35.8; $M_{women} = 95.09$; SD = 27.8). Data analysis also showed a main effect of virtual confederates' gender ($F(1,30) = 31.37$, $p < 0.0001$, $\eta_p^2 = 0.51$) due to the fact

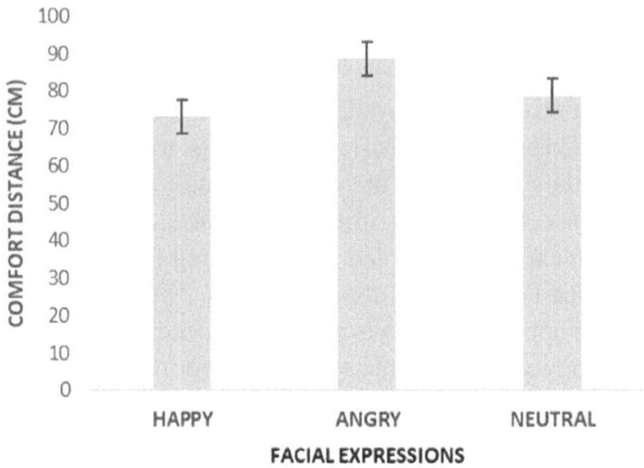

Fig. 2 Mean (cm) of the comfort distance as a function of happy, angry and neutral facial expression. Error bars represent the standard error

that participants kept a shorter distance from female than male virtual confederates ($M_{male} = 91.9$; $SD = 39.44$; $M_{female} = 68.54$; $SD = 34.56$). Moreover, a main effect of facial expressions appeared ($F(2,60) = 22.49$, $p < 0.000$, $\eta_p^2 = 0.43$), such that the distance was significantly larger when participants were approached by virtual confederates with angry ($M = 88.69$; $SD = 37.85$) than neutral ($M = 78.85$; $SD = 37.72$) and happy facial expressions ($M = 73.15$; $SD = 31.47$) (see Fig. 2).

Interestingly, a significant 3-way interaction emerged ($F(2,60) = 5.22$, $p < 0.05$, $\eta_p^2 = 0.15$) (see Fig. 3). The post hoc analysis showed that both men and women preferred a shorter distance from female than male virtual confederates in all emotion conditions (at least $p < 0.05$). However, gender modulated the effect of emotions on the interpersonal space. Results indeed showed that the shortest distance was observed when men faced happy females ($M = 39.33$) (at least $p < 0.05$), while the distance was particularly large when women were approached by angry virtual males ($M = 112.9$; at least $p < 0.05$). Finally, the distance was shorter with male-female dyads than female-male dyads ($p = 0.05$).

4 Conclusion

In the present study, we aimed at investigating how different facial expressions (negative, positive, neutral) along with gender differences affect the regulation of the interpersonal-comfort boundaries in a virtual environment. Furthermore, considering that Virtual Reality based technologies are becoming more and more used in different areas of everyday life and in the scientific field for research purposes, this study also aimed at giving further support to the ecological validity of VR by showing a close similarity of VR simulations to behavioural data.

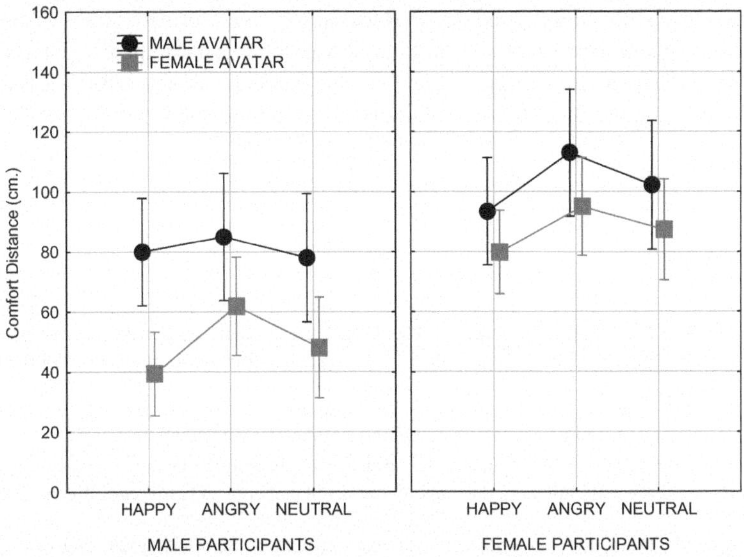

Fig. 3 Mean (cm) of the comfort distance as a function of happy, angry and neutral facial expression and participants-confederates' gender. Error bars represent the standard error

Male and female participants determined their comfort-distance from male and female virtual confederates showing happy, neutral and angry facial expression. In line with the literature, results showed that men preferred a closer distance than women, and in general all participants preferred a shorter interpersonal distance when interacting with female than male virtual confederates, independently of the emotion they expressed [16, 21, 40]. Moreover, as expected, the distance was larger when approached by angry virtual humans than neutral and happy ones. This confirmed the hypothesis that defence in response to a potential social danger represents a fundamental function of the space around the body [19, 31, 32].

The results are overall consistent with previous behavioural studies about gender effects [16, 40] and emotions [31]. This suggests that IVR simulations have a high degree of similarity with the actual social phenomena they model and give further support to the ecological validity of VR-based technologies (for similar results [21]).

Our results also seem to indicate that gender modulates the effect of facial expression on the regulation of the interpersonal space. Indeed, while the shortest distance was observed when men interacted with smiling female virtual confederates, the largest distance was due to women facing angry male virtual confederates. Interestingly, it seems that male participants had a tendency to be more focused on the positive and attractive aspects of the female virtual agents while female participants showed a tendency to be more focused on the safety value of the social context. It could be hypothesized that women are perceived as less threatening or potentially harmful than men and that attraction-repulsion mechanisms play an important role in the regulation of the spatial distance [4, 34, 40, 41].

Finally, this study highlighted that IVR technology is a useful tool for investigating proxemics regulation processes. Results demonstrated that participants treated virtual humans similarly to actual humans, and this reinforced the idea that the IVR is able to simulate the complexity of human interpersonal dynamic interactions at satisfactory levels.

References

1. Adams, R.B., Ambady, N., Macrae, N., Kleck, R.E.: Emotional expressions forecast approach-avoidance behaviour. Motiv. Emot. **30**, 179–188 (2006). https://doi.org/10.1007/s11031-006-9020-2
2. Aiello, J.R.: Human Spatial behavior. In: Stokols, D., Altman, I (eds.) Handbook of Environmental Psychology, vol. 1, pp. 389–504. Wiley, New York (1987)
3. Armbrüster, C., Wolter, M., Kuhlen, T., Spijkers, W., Fimm, B.: Depth perception in virtual reality: distance estimations in peri- and extrapersonal space. Cyberpsychol. Behav. **11**, 9–15 (2008)
4. Argyle, M., Dean, J.: Eye-contact distance and affiliation. Sociometry **28**, 289–304 (1965). https://doi.org/10.2307/278602
5. Bailenson, J.N., Blascovich, J., Beall, A.C., Loomis, J.M.: Interpersonal distance in immersive virtual environments. Pers. Soc. Psychol. Bull. **29**, 819–833 (2003). https://doi.org/10.1177/0146167203029007002
6. Bell, P.A., Greene, T.C., Fisher, J.D., Baum, A.: Environmental psychology, 5th edn. Harcourt College Publishers, New York (2005)
7. Damasio, A.R.: The Feeling of What Happens: Body and Emotion in the Making of Consciousness. Harcourt Brace, New York (1999). https://doi.org/10.1080/15294145200010773287
8. Darwin, C.: The Expressions of the Emotions in Men and Animals. John Murray, London (1872)
9. de Vignemont, F., Iannetti, G.D.: How many peripersonal spaces? Neuropsychologia **70**, 327–334 (2015)
10. Dosey, M.A., Meisels, M.: Personal space and self-protection. J. Pers. Soc. Psychol. **11**, 93–97 (1969). https://doi.org/10.1037/h0027040
11. Ekman, P.: Facial expressions. In: Dalgleish, T., Power, M.J. (eds.) The Handbook of Cognition and Emotion, pp. 301–320. Wiley, New York (1999)
12. Gallese, V., Keysers, C., Rizzolatti, G.A.: Unifying view of the basis of social cognition. Trends Cogn. Sci. **8**, 396–403 (2004). https://doi.org/10.1016/jtics200407002
13. Gessaroli, E., Santelli, E., di Pellegrino, G., Frassinetti, F.: Personal space regulation in childhood autism spectrum disorders. PLoS One **8**(9), e74959 (2013). https://doi.org/10.1371/journalpone0074959
14. Graziano, M.S.A., Cooke, D.F.: Parieto-frontal interactions, personal space and defensive behaviour. Neuropsychologia **44**, 845–859 (2006). https://doi.org/10.1016/jneuropsychologia200509011
15. Hall, E.T.: The Hidden Dimension. Doubleday, New York (1966)
16. Hayduk, L.A.: Personal space: where we now stand. Psychol. Bull. **94**, 293–335 (1983). https://doi.org/10.1037/0033-2909942293
17. Hebl, M.R., Kleck, R.E.: Acknowledging one's stigma in the interview setting: effective strategy or liability? J. Appl. Soc. Psychol. **32**(2), 223–249 (2002). https://doi.org/10.1111/j.1559-1816.2002.tb00214.x
18. Horstmann, G.: What do facial expression convey: feeling states, behavioral intentions or actions requests? Emotion **3**(2), 150–166 (2003). https://doi.org/10.1037/1528-354232150

19. Iachini, T., Coello, Y., Frassinetti, F., Ruggiero, G.: Body-space in social interactions: a comparison of reaching and comfort distance in immersive virtual reality. PLoS ONE **9**(11), e111511 (2014). https://doi.org/10.1371/journalpone0111511
20. Iachini, T., Pagliaro, S., Ruggiero, G.: Near or far? It depends on my impression: moral information and spatial behavior in virtual interactions. Acta Psychol. **161**, 131–136 (2015). https://doi.org/10.1016/jactpsy200509003
21. Iachini, T., Coello, Y., Frassinetti, F., Senese, V.P., Galante, F., Ruggiero, G.: Peripersonal and interpersonal space in virtual and real environments: effects of gender and age. J. Environ. Psychol. **45**, 154–164 (2016). https://doi.org/10.1016/jjenvp201601004
22. Keltner, D., Ekman, P., Gonzaga, G.C., Beer, J.: Facial expression of emotion. In: Richard, D.J., Klaus, K.R., Goldsmith, H.H. (eds.) Handbook of Affective Sciences, pp. 415–432. Oxford University Press, New York (2003)
23. Kennedy, D.P., Glascher, J., Tyszka, J.M., Adolphs, R.: Personal space regulation by the human amygdala. Nat. Neurosci. **12**, 1226–1227 (2009). https://doi.org/10.1038/nn2381
24. Knutson, B.: Facial expressions of emotion influence interpersonal trait inferences. J. Nonverbal Behav. **20**(3), 165–182 (1996). https://doi.org/10.1007/BF02281954
25. Lampton, D.R., McDonald, D.P., Singer, M., Bliss, J.P.: Distance estimation in virtual environments. In: Proceedings of the Human Factors and Ergonomics Society, 39th Annual Meeting, pp. 1268–1272 (1995)
26. Loomis, J.M., Blaskovich, J.J., Beall, A.C.: Immersive virtual environment technology as a basic research tool in psychology. Behav. Res. Methods Instrum. Comput. **31**, 557–564 (1999)
27. Lourenco, S.F., Longo, M.R., Pathman, T.: Near space and its relation to claustrophobic fear. Cognition **119**, 448–453 (2011)
28. Lundqvist, D., Flykt, A., Ohman, A.: The Karolinska Directed Emotional Faces—KDEF CD ROM from Department of Clinical Neuroscience Psychology section Karolinska Institutet ISBN 91-630-7164-9 (1998). https://doi.org/10.1037/t27732-000
29. Marsh, A.A., Ambady, N., Kleck, R.E.: The effects of fear and anger facial expressions on approach-and avoidance related behaviors. Emotion **5**(1), 119–124 (2005). https://doi.org/10.1037/1528-354251119
30. Morganti, F., Riva, G.: Conoscenza, comunicazione e tecnologia: aspetti cognitivi della realtà virtuale. LED Edizioni Universitarie (2006)
31. Ruggiero, G., Frassinetti, F., Coello, Y., Rapuano, M., Schiano di Cola, A., Iachini, T.: The effect of facial expressions on peripersonal and interpersonal spaces. Psychol. Res. (2017). https://doi.org/10.1007/s00426-016-0806-x
32. Ruggiero, G., Rapuano, M., Iachini, T.: Perceived temperature modulates peripersonal and interpersonal spaces differently in men and women. J. Environ. Psychol. **63**, 52–59 (2019)
33. Sawada, Y.: Blood pressure and heart rate responses to an intrusion on personal space. Jpn. Psychol. Res. **45**, 115–121 (2003)
34. Seidel, E.M., Habel, U., Kirschner, M., Gur, R.C., Derntl, B.: The impact of facial emotional expressions on behavioural tendencies in women and men. J. Exp. Psychol. Hum. Percept Perform **36**, 500–507 (2010). https://doi.org/10.1037/a0018169
35. Siegman, A.W., Feldstein, S.: Nonverbal Behavior and Communication. Psychology Press, Abingdon (2014)
36. Slater, M.: Place illusion and plausibility can lead to realistic behaviour in immersive virtual environments. Philos. T Roy. Soc. B **364**(1535), 3549–3557 (2009). https://doi.org/10.1098/rstb.2009.0138
37. Sommer, R.: From personal space to cyberspace. In: Bechtel, R.B., Churchman, A. (eds.) Handbook of Environmental Psychology, pp. 647–657. Wiley, New York (2002)
38. Sussman, N.M., Rosenfeld, H.M.: Touch justification and sex: influences on the aversiveness of spatial violations. J. Soc. Psychol. **106**, 215–225 (1978)
39. Tajadura-Jiménez, A., Pantelidou, G., Rebacz, P., Vastfjall, D., Tsakiris, M.: I space: the effects of emotional valence and source of music on interpersonal distance. PLoS ONE **6**(10), e26083 (2011). https://doi.org/10.1371/journalpone0026083

40. Uzzell, D., Horne, N.: The influence of biological sex, sexuality and gender role on interpersonal distance. Br. J. Soc. Psychol. **45**, 579–597 (2006)
41. van Dantzig, S., Pecher, D., Zwaan, R.A.: Approach and avoidance as action effects. Q J. Exp. Psychol. **61**(9), 1298–1306 (2008). https://doi.org/10.1080/17470210802027987
42. Vuilleumier, P., Pourtois, G.: Distributed and interactive brain mechanisms during emotions face perception: evidence from functional neuroimaging. Neuropsychologia **45**, 174–194 (2007). https://doi.org/10.1016/jneuropsychologia200606003
43. World Medical Association Declaration of Helsinki. JAMA **310** (20), 2191 (2013)

Signals of Threat in Persons Exposed to Natural Disasters

Massimiliano Conson, Isa Zappullo, Chiara Baiano, Laura Sagliano, Carmela Finelli, Gennaro Raimo, Roberta Cecere, Maria Vela, Monica Positano, and Francesca Pistoia

Abstract Individuals exposed to traumatic events can develop Post Traumatic Stress Disorder (PTSD) and depression, but also subclinical behavioral and emotional changes. Although persons exposed to trauma without PTSD show greater resilience, and are better at regulating emotions, with respect to trauma survivors with PTSD, they are anyway more prone to develop physical and psychological maladaptive changes to subsequent stressful experiences. Indeed, especially persons

M. Conson (✉) · I. Zappullo · C. Baiano · L. Sagliano · C. Finelli · G. Raimo · R. Cecere · M. Vela
Developmental Neuropsychology Laboratory, Department of Psychology, University of Campania Luigi Vanvitelli, Viale Ellittico 31, 81100 Caserta, Italy
e-mail: massimiliano.conson@unicampania.it

I. Zappullo
e-mail: isa.zappullo@unicampania.it

C. Baiano
e-mail: chiarabaiano00@gmail.com

L. Sagliano
e-mail: laurasagliano@gmail.com

C. Finelli
e-mail: carmelafinelli61@gmail.com

G. Raimo
e-mail: gennaro.raimo.93@gmail.com

R. Cecere
e-mail: roberta.cecere@hotmail.com

M. Vela
e-mail: mariavela@hotmail.it

M. Positano · F. Pistoia
Department of Biotechnological and Applied Clinical Sciences, Neurological Institute, University of L'Aquila, Via Vetoio, 67100 L'Aquila, Italy
e-mail: positanomonica@hotmail.it

F. Pistoia
e-mail: francesca.pistoia@univaq.it

A. Esposito et al. (eds.), *Progresses in Artificial Intelligence and Neural Systems*, Smart Innovation, Systems and Technologies 184,
https://doi.org/10.1007/978-981-15-5093-5_41

surviving to natural disasters, as earthquakes, can show complex patterns of non-psychopathological responses to traumatic experiences suggesting hypervigilance to signals of threat, and in particular to stimuli conveying self-relevant threatening information. Here, we review neuroimaging and neuropsychological studies uncovering the neural and cognitive mechanisms involved in these non-clinical psychological responses to natural disasters.

1 Introduction

After a natural disaster, like an earthquake, people can develop different emotional disorders [1]. The most frequently reported psychiatric consequences involve post-traumatic stress disorder (PTSD), depression, anxiety and social phobia [2–4]. In particular, PTSD is a psychological consequence of the exposure to a traumatic event involving alterations in behavioral, psychological, physiological, biological and social responses. However, a certain percentage of individuals exposed to traumatic events can develop behavioral and emotional changes representing subclinical pathological conditions. Indeed, trauma survivors without PTSD although being resilient individuals they show increased probability to develop physical and psychological consequences to successive trauma exposure [5, 6]; they can also experience degenerative cognitive disorders [7], more chronic illness later in life and reduced life expectancy [8]. These responses to trauma do not necessarily imply impaired processing of emotions but rather seem to represent an increased sensitivity towards specific emotional signals, in particular those signals conveying self-relevant potential threatening information.

Neuropsychological and neural mechanisms that underlie the effects of trauma exposure in this nonclinical population need to be clarified yet. The main aim of the present review was to shade lights on cognitive and neurofunctional aspects of trauma exposure, in particular focusing on earthquake-exposed persons. By clarifying how the brain reacts to trauma, even in the absence of a clinically relevant condition, it could be possible to develop tailored, and more specific preventive and treatment strategies.

2 Neuropsychological Processing of Emotional Signals in Earthquake Survivors

The capacity to identify and recognize emotional facial expressions is vital for dealing with social relationships and effectively interacting with the surrounding context [9]. Emotional facial expressions are key signals of the other persons' emotions allowing the observer to regulate own affective states in response to others' behavior [10]. Indeed, several data showed that observation of another person's facial expressions activates the very same behavioral responses in the observer through a simulation mechanism [11]. In particular, seminal studies revealed covert facial muscle activity during observation of others' emotional facial expressions; moreover, facial muscles activated during the production of specific emotional expressions were also strongly activated during the observation of the very same facial expressions [12, 13]. Individuals are also able to infer others' emotions by relying upon an implicit correspondence between a specific facial configuration and the related emotional state [14]. This mechanism, known as mentalizing or theory of mind [15], plays a central role in social interactions, since it allows to interpret other people's intentions and to adjust own behavior fittingly [16]. Both simulation and mentalizing can be shaped by experience [17], brain damage [18] and psychopathology [19, 20].

Recently, Bell et al. [21] investigated recognition of facial expressions in persons exposed to the 2010–2011 Canterbury (New Zealand) earthquake. Interestingly, the authors compared two groups of earthquake victims, i.e., with and without PTSD, and a group of non-exposed participants. Results demonstrated that both exposed groups were significantly more accurate in recognizing emotional faces with respect to non-exposed controls. These findings showed that earthquake exposure influenced the capacity to recognize facial expressions, independently from the development of a psychopathological condition. Interestingly moreover, only the earthquake-exposed group, but not the PTSD group, showed faster responses than the non-exposed group when judging the threat-related emotional faces displaying anger, fear, and disgust. Bell et al. [21] suggested that enhanced accuracy in recognizing emotional faces was related to increased sensitivity to stimuli potentially signalling threat, implying that trauma exposure would lead to hypervigilance and attentional bias towards emotional facial expressions.

Consistently, Pistoia et al. [22] studied a group of university students who experienced the devastating L'Aquila earthquake that hit central Italy on April 6th 2009. This group of earthquake-exposed persons did not develop PTSD or other psychopathological conditions but a complex pattern of anxiety-related signs. They underwent two behavioral tasks and were compared with a group of students not living in any earthquake-affected area. The first task asked participants to recognize emotional facial expressions (anger, fear, sadness, disgust, happiness and surprise) and the second task to judge own emotional responses to complex affective scenes from the International Affective Picture System. Results showed that the earthquake survivors were significantly more accurate than controls in recognizing all the emotional facial expressions, whereas there were no differences between the two groups in judging affective evocative scenes. Pistoia et al. [22] interpreted these findings by suggesting that, in earthquake victims, trauma exposure amplified vigilance towards signals conveying potential threats, as emotional faces. In particular, hypervigilance to threat would lead victims to continuously pay attention to emotional facial expressions, thus progressively developing a specific expertise in processing this type of signals. This interpretation was supported by the fact that the earthquake-exposed students living in L'Aquila experienced a long exposure time to the earthquake, since the main shock in 2009 was followed by uninterrupted aftershocks in the next months and by further main earthquakes in 2016 and 2017, with a great and prolonged psychological distress.

Thus, the few available data on processing of emotional signals in persons exposed to natural disasters without developing PTSD suggest that prolonged trauma exposure, especially in situations where continuous threat is present as in the case of ongoing aftershocks following a main earthquake, can activate a hypervigilance response of adaptive value since it enables to monitor the environment to keep safety. However, although hypervigilance to threat could represent an adaptive response to unpredictable situations, when prolonged over time it may became maladaptive, likely contributing to other dysfunctional emotional responses, as anxiety and sleep disorders [23]. Although both Bell et al. [21] and Pistoia et al. [22] converged in suggesting that enhanced accuracy in recognizing emotional faces in earthquake survivors was related to hypervigilance to threat, none of them used a behavioral paradigm directly investigating this kind of response. Actually, the best methodological option would be the classical attentional bias task [24, 25]. To our knowledge, only one study tested attentional bias in a group of earthquake survivors. Zhang et al. [26] studied a group of victims without PTSD of the Chinese Wenchuan earthquake which happened on the 28th of May in 2008. The authors required participants to complete an attentional bias task while recording event-related potentials. The behavioral task was a classical dot probe task: the target probe was presented at the same space location consistent with earthquake-related words in the congruent trial, while in the space location of neutral words in the incongruent trial. Behavioural results showed in the earthquake-exposed group significantly faster responses to congruent trials with respect to incongruent trials, thus demonstrating an attentional bias towards trauma-related stimuli. These results supported the authors' starting hypothesis of hypervigilance to threat in trauma survivors independently from developing a PTSD. These findings are in keeping with the data reviewed above on recognition of emotional faces, although the experimental paradigm used by Zhang et al. [26] did not involve emotional faces. Notwithstanding these methodological differences, available evidence would suggest that trauma exposure can enhance threat detection in earthquake survivors, leading them to constantly drive attention towards potential signals of threat. This, in turn, may actually lead individuals exposed to natural disasters to gradually develop specific "emotional expertise" [22].

3 Neural Correlates of Emotional Changes in Earthquake Survivors

In recent years, neuroimaging studies helped to clarify brain mechanisms related to emotional disorders and biases in trauma-exposed clinical populations. Two main neuroimaging approaches have been used to investigate the neural correlates of emotional changes in trauma-exposed persons: task-based fMRI and resting-state fMRI. Task-based fMRI studies mainly focused on persons exposed to traumatic events who developed PTSD. For instance, in a study specifically assessing the neural correlates of hypervigilance to threat in combat veterans with PTSD, Todd et al. [27] found that, as compared to the non-PTSD group, patients showed greater responses in visual cortex and in the precuneus during processing of arousing stimuli. The authors suggested that PTSD is linked to intensified processing of emotional salience reflected in the enhanced activity recorded in the visual cortex. However, other authors suggested that since chronic hypervigilance is a persistent rather than reactive psychological condition, brain correlates of trauma exposure can be best investigated by resting-state paradigms without the need of presenting participants with emotional stimuli. Indeed, Long et al. [28] showed that assessing brain functional changes by resting-state fMRI (rs-fMRI) was a valuable methodology to predict depression or anxiety symptoms severity and track changes over time in persons survived to earthquake. In particular, the authors found that fronto-limbic and striatal areas, and the functional connectivity of the fronto-striatal and default-mode networks were correlated with the symptom progression over a period of two years after the traumatic experience.

Although earthquake victims, in particular those without PTSD, cannot show structural brain alterations just following the trauma, brain functional changes can be highlighted even within just one month after the disaster [29]. Indeed, in a seminal rs-fMRI study, Lui et al. [29] showed an alteration of resting brain activity in a group of survivors of the devastating Wenchuan earthquake. The authors found in the earthquake survivors that, just within 25 days from the event, the activity in fronto-limbic and striatal regions was significantly increased, and connectivity among limbic and striatal networks was reduced. Moreover, earthquake survivors also showed reduced synchronization within the default mode network. In this pioneering study, Lui et al. [29] could provide data revealing alterations of resting brain activity shortly after important traumatic experiences, analogously to what found in PTSD.

Li et al. [30] assessed the of trauma exposure in earthquake survivors without PTSD, specifically focusing on both functional connectivity and morphological changes involving brain networks and structures related to emotion regulation. The authors assessed participants forty-one months after the Wenchuan earthquake using machine-learning algorithms to distinguish between survivors and non-exposed participants on the basis of the brain structural features. In the trauma-exposed group, results revealed greater grey matter density in prefrontal-limbic regions, in particular the anterior cingulate cortex, the medial prefrontal cortex, the amygdala and the hippocampus, with respect to controls. Moreover, by analysing the resting-state activity the authors found stronger amygdala-hippocampus functional connectivity in the trauma-exposed participants than in controls. These results demonstrated that survival of traumatic experiences, without developing PTSD, was associated with specific brain changes in both morphology and resting-state activity of brain structures and networks that play an important role in the top-down control of emotional regulation. These brain changes might reflect better emotional regulation strategies in trauma-exposed nonclinical adults and could help to explain the reason why they do not develop as clinically relevant condition, as PTSD, in response to trauma exposure.

More recently, in a rs-fMRI study Pistoia et al. [31] investigated the neurofunctional changes related to recognition of emotional faces previously found in earthquake survivors [22]. In particular, the authors assessed the relations between the ability to recognize emotional faces and the activity of the visual network (VN) and the default-mode network (DMN). Resting-state functional connectivity with the main hubs of the VN and DMN was tested both in earthquake-exposed and in non-exposed participants. The results showed that, in earthquake survivors, there wa a reduction in the relation between accuracy in recognizing emotional faces and the functional connectivity of the dorsal seed of the VN with the right inferior occipito-temporal cortex and the left lateral temporal cortex, and of two parietal seeds of DMN, lower parietal and medial prefrontal cortex, with the precuneus bilaterally. These data suggested that a functional change of brain networks involved in identifying and interpreting facial expressions could represent the basis of the "emotional expertise" reported in earthquake survivors [21, 22].

4 Conclusions

Together, the reviewed neuropsychological data show that trauma exposure enhances the ability of earthquake victims to recognize facial expressions. A possible explanation of this effect is that trauma exposure increases vigilance to threat in earthquake victims, leading them to systematically pay much more attention to every kind of potential sign of threat in the surrounding context. This may, in turn, lead trauma survivors to progressively develop a specific "emotional expertise" in the absence of a psychopathological condition. Available neuroimaging studies suggest that trauma survivors who do not develop PTSD can better recruit brain networks involved in top-down control of emotion regulation processes, as revealed by enhanced resting-state fronto-limbic connectivity. Although this "brain resilience" can protect victims from developing clinical symptoms it cannot fully protect them from developing maladaptive responses, such as sleep disorders and anxiety-related responses, when trauma exposure persists as in the case continuous aftershocks following a main earthquake.

References

1. Furukawa, K., Arai, H.: Earthquake in Japan. Lancet **377**, 1652 (2011)
2. Farooqui, M., Quadri, S.A., Suriya, S.S., Khan, M.A., Ovais, M., Sohail, Z., Shoaib, S., Tohid, H., Hassan, M.: Posttraumatic stress disorder: a serious post-earthquake complication. Trends Psychiatry Psychother. **39**, 135–143 (2017)
3. Dube, A., Moffatt, M., Davison, C., Bartels, S.: Health outcomes for children in Haiti since the 2010 earthquake: a systematic review. Prehosp. Disaster Med. **33**, 77–88 (2018)
4. Geng, F., Liang, Y., Shi, X., Fan, F.: A prospective study of psychiatric symptoms among adolescents after the Wenchuan earthquake. J. Trauma. Stress **31**, 499–508 (2018)

5. Bremner, J.D., Southwick, S., Johnson, D., Yehuda, R., Charney, D.: Childhood physical abuse and combat-related posttraumatic stress disorder in Vietnam veterans. Am. J. Psychiatry **150**, 235–239 (1993)
6. Tucker, P.M., Pfefferbaum, B., North, C.S., Kent, A., Burgin, C.E., Parker, D.E., et al.: Physiologic reactivity despite emotional resilience several years after direct exposure to terrorism. Am. J. Psychiatry **164**, 230–235 (2007)
7. Stein, M.B., Kennedy, C.M., Twamley, E.W.: Neuropsychological function in female victims of intimate partner violence with and without posttraumatic stress disorder. Biol. Psychiatry **52**, 1079–1088 (2002)
8. McFarlane, A.: The prevalence and longitudinal course of PTSD: implications for neurobiological models of PTSD. In: Yehuda, R., McFarlane, A. (ed.) Psychobiology of Posttraumatic Stress Disorder, vol. 821, pp. 10–24. New York Academy of Sciences, NY (1997)
9. Ekman, P.: Emotions Revealed: Recognizing Faces and Feeling to Improve Communication and Emotional Life. Times Books/Henry Holt and Co, New York (2003)
10. Phillips, M.L., Drevets, W.C., Rauch, S.L., Lane, R.: Neurobiology of emotion perception I: the neural basis of normal emotion perception. Biol. Psychiatry **54**, 504–514 (2003)
11. Niedenthal, P.M.: Embodying emotion. Science **316**, 1002–1005 (2007)
12. Dimberg, U., Thunberg, M.: Rapid facial reactions to emotional facial expressions. Scand. J. Psychol. **39**, 39–45 (1998)
13. Dimberg, U., Thunberg, M., Elmehed, K.: Unconscious facial reactions to emotional facial expressions. Psychol. Sci. **11**, 86–89 (2000)
14. Goldman, A.I., Sripada, C.S.: Simulationist models of face-based emotion recognition. Cognition **94**, 193–213 (2005)
15. Frith, C.D., Frith, U.: Interacting minds—a biological basis. Science **286**, 1692–1695 (1999)
16. Blakemore, S.J.: The developing social brain: implications for education. Neuron **65**, 744–747 (2010)
17. Conson, M., Ponari, M., Monteforte, E., Ricciato, G., Sarà, M., Grossi, D., Trojano, L.: Explicit recognition of emotional facial expressions is shaped by expertise: evidence from professional actors. Front. Psychol. **4**, 382 (2013)
18. Pistoia, F., Conson, M., Trojano, L., Grossi, D., Ponari, M., Colonnese, C., et al.: Impaired conscious recognition of negative facial expressions in patients with locked-in syndrome. J. Neurosci. **30**, 7838–7844 (2010)
19. MacNamara, A., Post, D., Kennedy, A.E., Rabinak, C.A., Phan, K.L.: Electrocortical processing of social signals of threat in combat-related post-traumatic stress disorder. Biol. Psychol. **94**, 441–449 (2013)
20. Poljac, E., Montagne, B., de Haan, E.H.: Reduced recognition of fear and sadness in post-traumatic stress disorder. Cortex **47**, 97–80 (2011)
21. Bell, C.J., Colhoun, H.C., Frampton, C.M., Douglas, K.M., McIntosh, V.V.W., Carter, F.A., Jordan, J., et al.: Earthquake rain: altered recognition and misclassification of facial expressions are related to trauma exposure but not Posttraumatic Stress Disorder. Front. Psychiatry **8**, 278 (2017)
22. Pistoia, F., Conson, M., Carolei, A., Dema, M.G., Splendiani, A., Curcio, G., et al.: Post-earthquake distress and development of emotional expertise in young adults. Front. Behav. Neurosci. **12**, 91 (2018)
23. Tempesta, D., Curcio, G., De Gennaro, L., Ferrara, M.: Long-term impact of earthquakes on sleep quality. PLoS ONE **8**, e55936 (2013)
24. Bar-Haim, Y., Lamy, D., Pergamin, L., Bakermans-Kranenburg, M.J., van Ijzendoorn, M.H.: Threat-related attentional bias in anxious and non-anxious individuals: a meta-analytic study. Psychol. Bull. **133**, 1–24 (2007)
25. MacLeod, C., Mathews, A., Tata, P.: Attentional bias in emotional disorders. J. Abnorm. Psychol. **95**, 15–20 (1986)
26. Zhang, Y., Kong, F., Han, L., Najam, U., Hasan, A., Chen, H.: Attention bias in earthquake-exposed survivors: an event-related potential study. Int. J. Psychophysiol. **94**, 358–364 (2014)

27. Todd, R.M., MacDonald, M.J., Sedge, P., Robertson, A., Jetly, R., Taylor, M.J., et al.: Soldiers with Posttraumatic Stress Disorder see a world full of threat: magnetoencephalography reveals enhanced tuning to combat-related cues. Biol. Psychiatry **78**, 821–829 (2015)
28. Long, J., Huang, X., Liao, Y., et al.: Prediction of post-earthquake depressive and anxiety symptoms: a longitudinal resting-state fMRI study. Sci. Rep. **4**, 6423 (2014)
29. Lui, S., Huang, X., Chen, L., Tang, H., Zhang, T., Li, X., et al.: High-field MRI reveals an acute impact on brain function in survivors of the magnitude 8.0 earthquake in China. Proc. Natl. Acad. Sci. U.S.A. **106**, 15412–15417 (2009)
30. Li, Y., Hou, X., Wei, D., Du, X., Zhang, Q., Liu, G., Qiu, J., Fan, Y.: of acute stress on the prefrontal-limbic system in the healthy adult. PLoS ONE 12(1), e0168315 (2017)
31. Pistoia, F., Conson, M., Quarantelli, M., Panebianco, L., Carolei, A., Curcio, G., Sacco, S., Saporito, G., Cesare, E.D., Barile, A., Masciocchi, C., Splendiani, A.: Neural Correlates of Facial Expression Recognition in Earthquake Witnesses. Frontiers in Neurosci. 13, (2019)

Adults Responses to Infant Faces and Cries: Consistency Between Explicit and Implicit Measures

Vincenzo Paolo Senese, Carla Nasti, Mario Pezzella, Roberto Marcone, and Massimiliano Conson

Abstract The aim of the present study was to evaluate the consistency of the adult's responses to different infant cues (faces and cries) by taking into account different levels of processing (explicit and implicit). A sample of 94 non-parent adults (56 Females; $M_{age} = 27.8$ years) participated in a within-subject design. Each participant was administered two versions of the Single Category Implicit Association Test and two semantic differentials adapted to evaluate the implicit and explicit responses to infant faces and cries. Results showed that, regardless of the level of processing and the gender, responses to faces were more positive than responses to cries; a substantial independence between responses observed for the different cues or the levels of processing; and that only in males there was a consistency between explicit and implicit responses but to infant faces only. If replicated, these results indicate that theoretical models of caregiving response must take into account the specificity of the different infant cues and that future studies should clarify the incremental validity of responses to the different cues at the different levels of processing on the actual caregiving behaviour.

V. P. Senese (✉) · C. Nasti · M. Pezzella · R. Marcone · M. Conson
Department of Psychology, University of Campania "Luigi Vanvitelli", Caserta, Italy
e-mail: vincenzopaolo.senese@unicampania.it

C. Nasti
e-mail: carla91_n@hotmail.it

M. Pezzella
e-mail: mariopezzella93@hotmail.it

R. Marcone
e-mail: roberto.marcone@unicampania.it

M. Conson
e-mail: massimiliano.conson@unicampania.it

A. Esposito et al. (eds.), *Progresses in Artificial Intelligence and Neural Systems*, Smart Innovation, Systems and Technologies 184,
https://doi.org/10.1007/978-981-15-5093-5_42

1 Introduction

For humans, more than for any other species, the infants' survival and development depend on adults who take care of them [1]. For this reason, researchers have always argued that since birth, infants are endowed with expressive characteristics (e.g., the face) and signals (e.g., crying) that can attract attention and stimulate the adults caregiving [1, 2].

In a complementary way, the literature showed that adults have a specific innate predisposition to respond to infant cues with appropriate caregiving behaviours [1, 3]. However, chronicles and research have frequently shown that adults do not always respond sensitively and appropriately to children, but they can show non-positive or abusive behaviour [4–6].

The theoretical models that investigated the factors that regulate caregiving behaviour have shown that it is multi-determined and that its manifestation depends on the interaction between the characteristics of the context, the child and the adult [7, 8]. The adults' characteristics and particularly the individual differences, seems to play a critical role on caregiving behaviour.

To identify the factors and processes that regulate caregiving, several researchers have investigated, in parents and non-parents, what are the adults' responses to infant cues (e.g., faces and cries). Taking into account the main results of the studies, Swain and colleagues [9] have proposed the Parental Brain Model (PBM). According to their model, infant cues activate three cortico-limbic modules (i.e., reflexive, cognitive, and emotional) whose interactions regulate the caregiving. In this perspective, the infant cues are processed at different levels, from more reflexive to more controlled ones, and it is assumed that the different cues have the same function, activate the same brain networks and regulate the final caregiving behaviour in a similar and gender invariant way.

Although this model is widely shared in the literature [10, 11], there are still very few studies that have directly investigated the extent to which responses to different cues are equivalent or invariant across genders. Moreover, since the model assumes that infant cues are processed at different levels, it becomes even more critical to verify to what extent there is consistency in responses to different cues considering the different level of processing.

Preliminary results seem to indicate that infant cues are associated with differential responses both when comparing responses to the different cues considering the same measure (e.g., neural correlates [12]; implicit responses [13]) and when comparing responses to the same cue but measuring responses at the different levels of processing (e.g., [14]). Therefore, if the data were to further confirm the low coherence between responses, then it would be important to identify which are the cues and the responses that have a greater predictive validity of the caregiving behaviours and in which conditions.

Starting from the abovementioned consideration, the aim of the present study was to investigate and compare the adult's responses to different infant cues by taking into account different levels of processing. In particular, infant faces and cries were considered as cues, and the valence of the implicit associations and explicit evaluations were measured by means of the implicit association test [15, 16] and the semantic differential [17] respectively.

According to the literature [13–16], we expected positive responses to infant faces, negative responses to infant cries, and that gender differences were observed on explicit responses [14] not on implicit ones [13, 15, 16]. Finally, we expected a low consistency in adults' responses to the different infant cues as a function of the type of the cue or the level of processing [12–14].

2 Methods

2.1 *Sample*

A total of 94 non-parent adults (56 females, 38 males) participated in a within-subject experimental design. Their ages ranged from 25 to 35 years ($M = 27.8$, $SD = 3.0$), and their educational level varied from the middle school to university. Males and females were matched as a function of the Age and Socio Economic Status (SES), $Fs < 1$. Participants were tested individually.

2.2 *Procedure*

The experimental session was divided in two phases. In the first phase the basic socio demographic information (i.e., sex, age and socio economic status) were collected, whereas in the second phase two versions of the Single Category Implicit Association Test (SC-IAT [15, 16]) and two semantic differentials (SD [17]) were administered (counterbalanced order) to evaluate explicit and implicit responses to infant faces and cries. The selected stimuli (faces and cries) were related to children with typical development [15, 16]. The session lasted about 20 min. Tests were carried out in conformity with the local Ethics Committee requirements and the Declaration of Helsinki, and all participants signed a written informed consent before starting the experimental session.

2.3 *Measures*

2.3.1 Implicit Measures

Single Category Implicit Association Test. In line with previous studies [13–16], two versions—one visual and one auditory—of the Single Category Implicit Association Test (SC-IAT) were adapted to evaluate the valence of adults' implicit associations to the infant cues: faces and cries. The SC-IAT is a two-stage classification task. In each stage, several exemplars of a single target stimulus (faces or cries) and positive and negative words were presented in a random order. In the visual version, all the stimuli were presented visually, whereas in the auditory version all the stimuli were presented auditory. Participants were presented one item at time (i.e., target stimulus or words) and asked to classify it into the correct category as fast and accurate possible by pressing the key associated with the category. In a first block the target items and the positive words were classified with the same key, whereas the negative words with a different key (positive condition). In a second block the target items and the negative words were classified with the same key, whereas the positive words with a different key (negative condition). Infant cues were the same used into previous studies [15, 16]. Infant faces portrayed infants with a mean age of about 6/8 months showing a neutral expression. Infant cries were extracted from home videos of 13-month-old infant to be acoustically representative of typical infant cry. The latter stimuli lasted 5 s each. They were presented through headphones at a constant volume. The SC-IAT scores were calculated by dividing the differences in the latency of responses between the positive and negative conditions by the standard deviation of the latencies in the two conditions [18]. If the subjects were faster in categorizing the stimuli in the positive condition, they were supposed to have a positive implicit attitude toward the target stimuli. Scores around 0 indicate no IAT effect; absolute values from 0.2 to 0.3 indicate a "slight" effect, values around 0.5 a "medium" effect, and values of about 0.8 to infinity a "large" effect. In this study, positive values indicate that the target cue was implicitly associated with the positive dimension, whereas negative values indicate a stronger association of the target cue with the negative domain. Both SC-IATs showed adequate reliability ($\alpha s > 0.70$).

2.3.2 Self-report Measures

Semantic Differential (SD). To measure the explicit response to infant cues (faces and cries), two Semantic Differential (SD) scales [17] were developed and administered to participants. In each semantic differential scale, participants were asked to evaluate the target stimuli (infant faces or cries) by using six bipolar adjectives (unpleasant-pleasant; undesirable-desirable; despicable-valuable; unfamiliar-familiar; nasty-good; disagreeable-nice) on a seven-point scale. The target stimuli were the same used for the SC-IATs. A composite total score was computed, with greater values indicating a positive evaluation toward the target stimuli. The scales showed adequate reliability ($\alpha s > 0.80$).

3 Data Analysis

Normality of univariate distributions of responses was preliminarily checked. To investigate if the implicit and explicit responses to the different infant cues were influenced by the type of cue and by the gender, two 2 × 2 mixed *ANOVA*s were

Table 1 Results of 2-way mixed within subject *ANOVAs* on responses to infant stimuli as a function of the levels of processing and the gender

Effect	*F*	*df*	*p*	η_p^2
SD				
Cue	250.64	1, 92	< 0.001	0.731
Gender	0.63	1, 92	0.431	0.007
Cue × Gender	1.98	1, 92	0.163	0.021
SC-IAT				
Cue	37.99	1, 91	< 0.001	0.294
Gender	0.06	1, 91	0.811	0.001
Cue × Gender	0.20	1, 91	0.659	0.002

Note SD = Semantic Differential. SC-IAT = Single Category Implicit Association Test

conducted. In all analysis the Bonferroni correction was used to analyse post hoc effects of significant factors, and partial eta squared (η_p^2) was used to evaluate the magnitude of significant effects.

To investigate the coherence between the explicit and implicit responses to the different cues (faces and cries), Pearson correlation coefficients between responses to infant cues were computed as a function of the type of cue, the level of processing considered and the gender.

4 Results

4.1 *Implicit Associations*

The *ANOVA* conducted on the SC-IAT scores showed that implicit responses were strongly influenced by the Cue, $F(1,92) = 250.64$, $p < 0.001$, $\eta_p^2 = 0.731$, whereas not significant were the Gender main effects and the Cue × Gender interactions, Fs < 1 (see Table 1). The calculated power of the tests was >0.80. The mean comparison showed that the responses to faces, $M = 0.24$, 95% CI [0.17; 0.32], were more positive than responses to cries, $M = -0.16$, 95% CI [−0.26; −0.06] (see Fig. 1).

4.2 *Explicit Responses*

The *ANOVA* conducted on explicit responses showed that also in this case the SD scores were strongly influenced by the Cue, $F(1,92) = 17.56$, $p < 0.001$, $\eta_p^2 = 0.281$, whereas not significant were the Gender main effects and the Cue × Gender interactions, Fs < 1. The calculated power of the tests was >0.80. The mean comparison

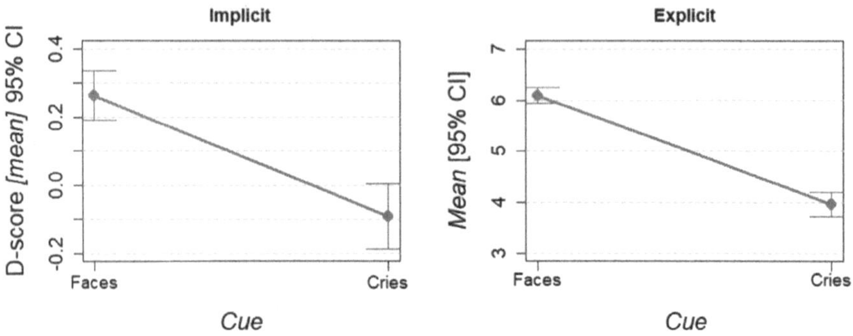

Fig. 1 Adults responses to infant stimuli as a function of the Cue (faces vs. cries) and the levels of processing (implicit vs. explicit)

showed that the responses to faces, $M = 6.40$, 95% CI [6.25; 6.56], were more positive than responses to cries, $M = 3.94$, 95% CI [3.70; 4.19] (see Fig. 1).

4.3 Consistency of Responses

The correlation analysis showed that there was a substantial independence between responses observed for the different cues or the levels of processing, and that only in males there was a coherence between explicit and implicit responses to infant faces not cries (see Table 2).

Table 2 Summary of intercorrelations, means and standard deviations of responses to infant stimuli as a function of the levels of processing and the gender

Measure	1	2	3	4	*M*	*sd*
Explicit						
1. SD Faces		0.14	0.06	−0.02	6.25	0.77
2. SD Cries	0.27		0.11	0.17	3.94	1.12
Implicit						
3. SC-IAT Faces	0.38*	0.12		0.22	0.27	0.35
4. SC-IAT Cries	−0.10	−0.22	−0.04		−0.11	0.49
M	5.93	4.01	0.26	−0.07		
sd	0.79	1.26	0.36	0.40		

Note SD = Semantic Differential. SC-IAT = Single Category Implicit Association Test. Intercorrelations for females ($n = 56$) are presented above the diagonal, and intercorrelations for males ($n = 38$) are presented below the diagonal. Means (*M*) and standard deviations (*sd*) for females are presented in the vertical columns, and means and standard deviations for males are presented in the horizontal rows. *$p < 0.05$

5 Discussion and Conclusions

The results of this study showed substantial independence between responses to different infant cues. In particular, the comparison between mean responses confirms that infant faces are associated with a more positive valence response than cries; and this effect is invariant with respect to gender and it is observed over and above the level of processing considered (implicit or explicit) [13–16]. This result confirms what has been previously described in the literature, and suggests that infant cues have different meanings and different functions and that they should not be considered equivalent or alternative [1, 2].

The second result of this study, which answers more specifically to the research question we posed, is related to the analysis of the consistency between the responses to different infant cues and at the different levels of processing. The data showed that only in the case of faces, and only in men, there is a consistency between the responses observed at the implicit and explicit level, while in women and for cries the responses were substantially independent. This result suggests that each infant cue activates different and independent responses [14] and that there are gender differences in the consistency between the different levels of processing. The lack of consistency between the implicit and explicit responses could be explained by the fact that the explicit responses more than the implicit ones are influenced by the cultural and controlled aspects or by the need to interpret the meaning of the auditory infant cues as in the case of the cries. In fact, unlike faces, crying is a more complex signal that in absence of visual information requires more processing for its interpretation [19].

Finally, the most relevant result of this work is the comparison between the responses to the different infant cues. The data show that, regardless of the level of processing, there is no consistency between responses to infant faces and cries [12, 13]. This seems to further indicate that the different cues have a different meaning and therefore determine independent responses. This result questions the assumption that the different cues are equivalent and give rise to an equivalent response [9–11].

Besides the merits of this study, some limitations should be mentioned. First, we considered only non-parents but it is possible that the consistency between responses to the infant cues is moderated by the parental experiences. Further studies involving parents are needed to verify this hypothesis. Second, although the calculated power was adequate, the sample size was still small and this can have reduced the power of the statistical tests for the "small" effects. Further studies are needed to verify the robustness and the validity of our results. Finally, we considered only responses to infant cues but not the actual caregiving. Further studies should verify the incremental validity of the responses to the different infant cues, evaluated at the different levels of processing, on the caregiving behaviours [7–11].

In conclusion, the results of this study highlight that, regardless of gender, the different infant cues determine specific and independent responses, and showed that the consistency between explicit and implicit responses is moderated by the cue and the gender. If replicated, these results indicate that theoretical models of response to infant cues [9–11] must take into account the specificity of different cues and that future studies should clarify the specific role of responses to the different infant cues at the different levels of processing on the actual caregiving behaviour.

Acknowledgements The authors thank Raffaella Califano, Domenico Campanile, Rita Massaro and Federica Sorrentino for assistance in collecting the data for this study.

References

1. Bornstein, M.H.: Determinants of parenting. Developmental Psychopathology. Four **5**, 1–91 (2016). https://doi.org/10.1002/9781119125556.devpsy405
2. Zeifman, D.M.: An ethological analysis of human infant crying: answering Tinbergen's four questions. Dev. Psychobiol. **39**(4), 265–285 (2001). https://doi.org/10.1002/dev.1005
3. Ainsworth, M.D.S., Blehar, M.C., Waters, E., Wall, S.: Patterns of attachment: a psychological study of the strange situation. Erlbaum, Hillsdale, NJ (1978)
4. Barnow, S., Lucht, M., Freyberger, H.J.: Correlates of aggressive and delinquent conduct problems in adolescence. Aggressive Behav. **31**, 24–39 (2005)
5. Beck, J.E., Shaw, D.S.: The influence of perinatal complications and environmental adversity on boys' antisocial behavior. J. Child Psychol. Psychiatry **46**, 35–46 (2005)
6. Putnick, D.L., Bornstein, M.H., Lansford, J.E., Chang, L., Deater-Deckard, K., Di Giunta, L., Bombi, A.S.: Agreement in mother and father acceptance-rejection, warmth, and hostility/rejection/neglect of children across nine countries. Cross-Cultural Research **46**, 191–223 (2012). https://doi.org/10.1177/1069397112440931
7. Belsky, J.: The determinants of parenting: a process model. Child Dev. **55**, 83–96 (1984)
8. Taraban, L., Shaw, D.S.: Parenting in context: revisiting Belsky's classic process of parenting model in early childhood. Dev. Rev. **48**, 55–81 (2018)
9. Swain, J.E., Kim, P., Spicer, J., Ho, S.S., Dayton, C.J., Elmadih, A., Abel, K.M.: Approaching the biology of human parental attachment: brain imaging, oxytocin and coordinated assessments of mothers and fathers. Brain Res. **1580**, 78–101 (2014)
10. Feldman, R.: The adaptive human parental brain: implications for children's social development. Trends Neurosci. **38**(6), 387–399 (2015). https://doi.org/10.1016/j.tins.2015.04.004
11. Young, K.S., Parsons, C.E., Stein, A., Vuust, P., Craske, M.G., Kringelbach, M.L.: The neural basis of responsive caregiving behaviour: investigating temporal dynamics within the parental brain. Behav. Brain Res. **325**(Pt B), 105–116 (2017). https://doi.org/10.1016/j.bbr.2016.09.012
12. Rutherford, H.J.V., Graber, K.M., Mayes, L.C.: Depression symptomatology and the neural correlates of infant face and cry perception during pregnancy. Soc. Neurosci. **11**(4), 467–474 (2016). https://doi.org/10.1080/17470919.2015.1108224
13. Senese, V.P., Santamaria, F., Sergi, I., Esposito, G. Adults' implicit reactions to typical and atypical infant cues. In: Esposito, A., Faundez-Zanuy, M., Morabito, F., Pasero, E. (eds.) Quantifying and Processing Biomedical and Behavioral Signals, pp. 35–43. WIRN 2017 2017. Smart Innovation, Systems and Technologies 103. Springer, Cham (2019). https://doi.org/10.1007/978-3-319-95095-2_4
14. Senese, V.P., Cioffi, F., Perrella, R., Gnisci, A. Adults' Reactions to Infant Cry and Laugh: A Multilevel Study. In: Esposito, A., Faundez-Zanuy, M., Morabito, F., Pasero, E. (eds.) Quantifying and Processing Biomedical and Behavioral Signals, pp. 45–55. WIRN 2017 2017. Smart Innovation, Systems and Technologies 103. Springer, Cham (2019). https://doi.org/10.1007/978-3-319-95095-2_5
15. Senese, V.P., De Falco, S., Bornstein, M.H., Caria, A., Buffolino, S., Venuti, P.: Human infant faces provoke implicit positive affective responses in parents and non-parents alike. PlosOne **8**(11), e80379 (2013)

16. Senese, V.P., Venuti, P., Giordano, F., Napolitano, M., Esposito, G., Bornstein, M.H.: Adults' implicit associations to infant positive and negative acoustic cues: moderation by empathy and gender. Q. J. Exp. Psychol. **70**(9), 1935–1942 (2017). https://doi.org/10.1080/17470218.2016.1215480
17. Osgood, C.E., Suci, G., Tannenbaum, P.: The Measurement of Meaning. University of Illinois Press, Urbana, IL (1957)
18. Greenwald, A.G., Nosek, B.A., Banaji, M.R.: Understanding and using the implicit association test: I. An improved scoring algorithm. J. Pers. Soc. Psychol. **85**, 197–216 (2003). https://doi.org/10.1037/a0015575
19. Iachini, T., Maffei, L., Ruotolo, F., Senese, V.P., Ruggiero, G., Masullo, M., Alekseeva, N.: Multisensory assessment of acoustic comfort aboard metros: an immersive virtual reality study. Appl. Cogn. Psychol. **26**, 757–767 (2012). https://doi.org/10.1002/acp.2856

The Influence of Systemizing, Empathizing and Autistic Traits on Visuospatial Abilities

Massimiliano Conson, Chiara Baiano, Isa Zappullo, Monica Positano, Gennaro Raimo, Carmela Finelli, Maria Vela, Roberta Cecere, and Vincenzo Paolo Senese

Abstract Much evidence indicates that neurotypical individuals with high level of autistic traits show performances comparable to those of autistic individuals in visuospatial tasks, such as the embedded figure test and the block design task. On the contrary, the role of autistic-like traits on mental rotation abilities in neurotypical population still remains unclear. The aim of the present study was to investigate in a sample of 315 University students (147 female and 168 male) the influence autistic-related traits (including systemizing and empathizing) on different visuospatial skills, as hidden figure identification and mental rotations, also taking into account the

---

M. Conson (✉) · C. Baiano · I. Zappullo · M. Positano · G. Raimo · C. Finelli · M. Vela · R. Cecere
Developmental Neuropsychology Laboratory, Department of Psychology, Università della Campania Luigi Vanvitelli, Viale Ellittico 31, 81100 Caserta, Italy
e-mail: massimiliano.conson@unicampania.it

C. Baiano
e-mail: chiarabaiano00@gmail.com

I. Zappullo
e-mail: isa.zappullo@unicampania.it

M. Positano
e-mail: positanomonica@hotmail.it

G. Raimo
e-mail: gennaro.raimo.93@gmail.com

C. Finelli
e-mail: carmelafinelli61@gmail.com

M. Vela
e-mail: mariavela@hotmail.it

R. Cecere
e-mail: roberta.cecere@hotmail.com

V. P. Senese
Psychometric Laboratory, Department of Psychology, Università della Campania Luigi Vanvitelli, Caserta, Italy
e-mail: vincenzopaolo.senese@unicampania.it

A. Esposito et al. (eds.), *Progresses in Artificial Intelligence and Neural Systems*, Smart Innovation, Systems and Technologies 184,
https://doi.org/10.1007/978-981-15-5093-5_43

influence of participants' sex. Results showed significant effect of sex on specific behavioural traits associated with autism, as attention switching and imagination, while no sex differences were found in visuospatial skills. More relevant, results showed that performance on figure disembedding was more related to autism-related social skills while performance on mental rotation was more related to systemizing.

1 Introduction

In recent years, cognitive abilities of neurotypical individuals with different degrees of autistic traits have been deeply investigated, since converging evidence showed that a continuum of autistic traits exists within the overall population, with clinical autism representing the extreme end of a quantitative distribution [1].

In particular, neurotypical individuals with high levels of autistic traits outperform persons with low levels of autistic traits on visuospatial tests as embedded figures [2], analogously to what found in individual with autism. Indeed, atypical visuospatial analysis has been frequently associated with autism, and specifically, local analysis has been considered a characteristic of the cognitive functioning of people with autism, with an advantage in perceiving the local elements (details) of a complex scene with respect to the global aspects [4]. Less clear, instead, is the effect of autistic traits on mental rotations. For instance, Stevenson and colleagues [3] assessed mental rotation abilities in typical individuals with low, medium and high autistic traits and did not found significant differences but showed that the strategies with which mental rotations were performed varied according to the combination of autistic traits and sex [3]. In individuals with autism, much evidence demonstrated that they perform better than typical controls when mentally rotating three-dimensional figures [5] and objects [6].

Systemizing and empathy are other traits pertinent to autism that are normally distributed in the general population. Systemizing can be conceived as the predilection for examining, understanding and constructing systems, a strength of persons with autism [7]. Empathy is the ability to interpret another's mental states and behaviours, and it represents a weakness in autism [1].

According to the Empathizing–Systemizing (E–S) theory [1, 7], sex differences have been identified in autistic-related cognitive stiles, with males showing greater systemizing tendencies and females showing higher levels of empathy. Sex differences have been also demonstrated in visuospatial skills. In particular, men tend to outperform women on embedded figures and mental rotations [8, 9], although the nature of this difference needs to be clarified yet. For instance, some recent data suggest that training with spatial tasks can diminish, or even cancel, sex differences in visuospatial performance (for instance see [10]).

In order to better clarify the role of systemizing, empathizing, autistic traits and sex on visuospatial skills, in the present study we measured these traits and visuospatial abilities, in particular figure disembedding and mental rotations, in a typically developing population of 315 university students.

2 Methods

2.1 Sample

Participants were 315 university students (147 female and 168 male) recruited from different Universities of Campania, in Southern Italy. The sample had a mean age of 23.2 years (SD = 2.04).

To be included in the study each participant had to meet the following inclusion and exclusion criteria: (i) lack of neurologic or neuropsychological disorders, (ii) lack of history of psychiatric difficulties (i.e. depression, bipolar illness or psychosis), (iii) Italian as mother language.

The research was conducted after participants provided written informed consent and in accordance with the ethical standards of the Helsinki Declaration.

2.2 Measures

Autistic-related traits were evaluated by the following questionnaires: (i) Autism Spectrum Quotient (AQ) [11]; (ii) Empathy Quotient (EQ) [12]; (iii) Systemizing Quotient (SQ) [13].

Autism Spectrum Quotient (***AQ***) [11]. The AQ provides a global index of autistic traits/characteristics and specific sub-scores (i.e. social skill, attention switching, attention to detail, communication, imagination). A person's AQ score represents the quantity of Autism Spectrum Disorder (ASD) traits s/he shows in her/his behaviour. Participants underwent the full 50-item Autism Quotient questionnaire. Answering each question on the survey was mandatory, so there were no missing data for any participants who completed it. The results were scored according to Baron-Cohen and colleagues' [11] criteria, resulting in an "AQ score" for each participant.

Empathy Quotient (***EQ***) [12]. The EQ measures empathy traits related to the recognition of others' emotions and moods. Participants answered the 40-item short version of the Empathy Quotient questionnaire. The results were scored to obtain an "EQ score" for each participant, which represents their level of empathy traits.

Systemizing Quotient (***SQ***) [13] The SQ evaluates individual's trend to identify the rules that govern systems, in order to predict how the system operates. The SQ evaluates across separate examples of systemizing to look at an individual's interest in a range of systems. The SQ comprises 60 questions, 40 assessing systemizing and 20 filler (control) items. The results provide a SQ score indicating individual differences across the systemizing dimension.

Visuospatial skills were evaluated by the following tasks [14, 15]: (i) Hidden Figure identification (HF); (ii) Mental Rotations (RM).

Hidden Figure identification (HF) provides a measure of local visual-processing stile. In this task, participants are presented with a target stimulus and six abstract figures as distractors. The task requires to identify in the six-choice display the complex figure embedded in the target stimulus as fast as possible. To give the correct answer, participants have to mentally disassemble the target stimulus. There are 12 items of increasing complexity as differences among stimuli and distractors gradually decrease. Each correct choice is scored 1 (score range: 0–12).

Mental Rotations (RM) require participants to identify the item that matches the target stimulus among six options differing with respect to spatial rotation degree. The stimuli are shaped as the capital letter L or S, with small white or black circles at the extremities. The six-choice display encloses the target-item stimulus, rotated on the horizontal plane by 45°, 90°, 135°, or 180°, together with five distractors that are mirror forms of the target stimulus at different degrees of rotation. There are 9 items of increasing complexity as the differences among stimuli and distractors gradually decrease. Each correct choice is scored 1 (score range: 0–9).

3 Results

Effects of sex on autistic-related traits and visuospatial abilities. Multivariate analysis of variance (MANOVA) was conducted with sex (female vs. male) as independent variable, and autistic-related traits (AQ, EQ and SQ, and their subscales), and visuospatial measures (HF and RM) as dependent variables (Table 1).

Results did not show any significant effect of sex on visuospatial measures and on EQ and SQ scores. As regard the AQ, results showed a significant main effect of

Table 1 Effects of sex on autistic-related traits and visuospatial skills

	Female (n = 145)	Male (n = 168)	F(2,314)	p	η_p^2
Age	23.18 ± 2.05	23.24 ± 2.03	0.068	0.794	0.000
AQ total score	17.55 ± 5.6	18.12 ± 5.45	0.829	0.363	0.003
AQ social	2.12 ± 1.98	2.15 ± 1.84	0.022	0.881	0.000
AQ switching	4.34 ± 1.77	4.89 ± 1.73	7.78	**0.006**	0.024
AQ detail	5.93 ± 2.39	4.43 ± 2.27	3.56	0.060	0.011
AQ communication	2.50 ± 1.82	2.46 ± 1.67	0.038	0.846	0.000
AQ imagination	2.67 ± 1.61	3.18 ± 1.77	7.23	**0.008**	0.023
EQ total score	42.98 ± 9.73	41.85 ± 10.39	0.990	0.320	0.003
EQ social	6.65 ± 2.35	6.27 ± 2.58	1.76	0.185	0.006
EQ cognition	16.76 ± 5.40	16.49 ± 5.27	0.197	0.657	0.001
EQ reactive	14.10 ± 4.44	13.57 ± 4.71	0.021	0.313	0.003
SQ total score	34.22 ± 12.66	32.74 ± 11.41	1.2	0.274	0.004
HF (accuracy score)	9.87 ± 2.37	9.86 ± 2.31	0.001	0.974	0.000
MR (accuracy score)	6.54 ± 2.35	6.52 ± 2.34	0.01	0.921	0.000

Values are expressed as Mean ± standard deviation; *AQ* Autism Quotient; *SQ* Systemizing Quotient; *EQ* Empathizing Quotient; *HF*, Hidden Figure identification; *MR*, Mental Rotations

sex on AQ switching, F (2, 314) = 7.78; p = 0.006; $\eta_p^2 = 0.024$, and AQ Imagination scores, F (2,314) = 7.23; p = 0.008; $\eta_p^2 = 0.023$; for both subscales, males scored significantly higher than females.

Regression analysis. Separate linear regression analyses were carried out on HF and RM. Regression analysis showed that the AQ social subscale (Beta = 0.175, p = 0.011) and SQ total score (Beta = 0.121, p = 0.049) significantly predicted performances on HF.

As regard MR, regression analysis showed that only SQ total score (Beta = 0.178, p = 0.004) significantly predicted the performance on the task.

4 Discussion and Conclusions

The aim of the present study was to explore the role of autistic-related traits on figure disembedding and mental rotation skills, also taking into account participants' sex. First, data showed higher autistic traits (attention switching and imagination) in males than females, consistent with previous literature [1, 7, 11]. However, unlike previous evidence, we did not find sex differences in visuospatial tasks. Although this result does not fit with the classical view according to which men outperform women on spatial tasks, it is rather consistent with more recent evidence showing that variables related with the degree of person's practicing with spatial activities can level out sex differences [10, 16]. Since the present sample was composed of university students coming from different academic majors, one could speculate that training with specific disciplines might have contributed to the lack of sex differences we reported here. This post hoc interpretation needs to be directly tested in future studies.

As regard the relation between autistic-related traits and visuospatial abilities, data showed that both figure disembedding and mental rotations were predicted by systemizing, while figure disembedding was only predicted by poor social skills (AQ social). This finding supports the link between local processing bias and social impairment, widely shown in Autism Spectrum Conditions [4]. For instance, recently, Russel-Smith and colleagues [17] demonstrated that neurotypical university students with high level of social difficulties (measured by AQ social subscale) were faster on the embedded figure test with respect to individuals with lower social difficulties, consistent with the present finding.

Regarding the role of systemizing, although it appears to predict both figure disembedding and mental rotations, it was more strongly related to mental rotation abilities. Consistent with the present results, Baron-Cohen and colleagues [18] suggested that a good attention to details and local processing skills are a necessary but not sufficient condition to effectively deal with tasks requiring mental rotation skills; according to the systemizing view, local processing would only represents an input for a more complex manipulation processing in order to produce an effective output, following the logic of '*if p, then q*'. Thus, further research is needed to clarify how systemizing contribute to different visuospatial skills as figure disembedding and mental rotations.

In conclusion, our findings suggest that different autistic traits can relate to specific visuospatial abilities, with a closer relationship, on one hand, between figure disembedding and social skills, and, on the other hand, between systemizing and mental rotations.

References

1. Baron-Cohen, S.: The extreme male brain theory of autism. Trends Cogn. Sci. **6**, 248–254 (2002)
2. Grinter, E.J., Maybery, M.T., Van Beek, P.L., Pellicano, E., Badcock, J.C., Bad- cock, D.R.: Global visual processing and self-rated autistic-like traits. J. Autism Dev. Disord. **39**, 1278–1290 (2009)
3. Stevenson, J.L., Nonack, M.B.: Gender differences in mental rotation strategy depend on degree of autistic traits. Autism Res. **11**, 1024–1037 (2018)
4. Mottron, L., Dawson, M., Soulières, I., Hubert, B., Burack, J.: Enhanced perceptual functioning in autism: an update, and eight principles of autistic perception. J. Autism Dev. Disord. **34**, 27–43 (2006)
5. Falter, C.M., Plaisted, K.C., Davis, G.: Visuo-spatial processing in autism—testing the predictions of extreme male brain theory. J. Autism Dev. Disord. **38**, 507–515 (2008)
6. Hamilton, A.F., Brindley, R., Frith, U.: Visual perspective taking impairment in children with autistic spectrum disorder. Cognition **113**, 37–44 (2009)
7. Baron-Cohen, S.: The essential difference: men, women and the extreme male brain. Penguin, London (2003)
8. Auyeung, B., Knickmeyer, R., Ashwin, E., Taylor, K., Hackett, G., Baron-Cohen, S.: Effects of fetal testosterone on visuospatial ability. Arch. Sex. Behav. **41**, 571–581 (2012)
9. Brosnan, M., Daggar, R., Collomosse, J.: The relationship between systemising and mental rotation and the implications for the extreme male brain theory of autism. J. Autism Dev. Disord. **40**, 1–7 (2010)
10. Rodán, A., Contreras, M.J., Elosúa, M.R., Gimeno, P.: Experimental but not sex differences of a mental rotation training program on adolescents. Front. Psychol. **7**, 1050 (2016)
11. Baron-Cohen, S., Wheelwright, S., Skinner, R., Martin, J., Clubley, E.: The autism-spectrum quotient (AQ): evidence from asperger syndrome/high-functioning autism, males and females, scientists and mathematicians. J. Autism Dev. Disord. **31**, 5–17 (2001)
12. Baron-Cohen, S., Wheelwright, S.: The empathy quotient (EQ). An investigation of adults with asperger syndrome or high functioning autism, and normal sex differences. J. Autism Dev. Disord. **34**, 163–175 (2004)
13. Baron-Cohen, S., Richler, J., Bisarya, D., Gurunathan, N., Wheelwright, S.: The systemising quotient (SQ): an investigation of adults with Asperger syndrome or high functioning autism and normal sex differences. Philos. Trans. R. Soc. Lond. B Biol. Sci. **358**, 361–374 (2003)
14. La Femina, F., Senese, V.P., Grossi, D., Venuti, P.: A battery for the assessment of visuo-spatial abilities involved in drawing tasks. Clin. Neuropsychol. **23**, 691–714 (2009)
15. Trojano, L., Siciliano, M., Pedone, R., Cristinzio, C., Grossi, D.: Italian normative data for the battery for visuospatial abilities (TERADIC). Neurol. Sci. **36**, 1353–1361 (2015)
16. Uttal, D.H., Meadow, N.G., Tipton, E., Hand, L.L., Alden, A.R., Warren, C., Newcombe, N.S.: The malleability of spatial skills: a meta- analysis of training studies. Psychol. Bull. **139**, 352–402 (2013)

17. Russell-Smith, S.N., Maybery, M.T., Bayliss, D.M., Sng, A.A.H.: Support for a link between the local processing bias and social deficits in autism: an investigation of embedded figures test performance in non-clinical individuals. J. Autism Dev. Disord. **42**, 2420–2423 (2012)
18. Baron-Cohen, S., Ashwin, E., Ashwin, C., Tavassoli, T., Chakrabarti, B.: Talent in autism: hyper-systemizing, hyper-attention to detail and sensory hypersensitivity. Philos. Trans. R. Soc. Lond. B Biol. Sci. **364**, 1377–1383 (2009)

Investigating Perceptions of Social Intelligence in Simulated Human-Chatbot Interactions

Natascha Mariacher, Stephan Schlögl, and Alexander Monz

Abstract With the ongoing penetration of conversational user interfaces, a better understanding of social and emotional characteristic inherent to dialogue is required. Chatbots in particular face the challenge of conveying human-like behaviour while being restricted to one channel of interaction, i.e., text. The goal of the presented work is thus to investigate whether characteristics of social intelligence embedded in human-chatbot interactions are perceivable by human interlocutors and if yes, whether such influences the experienced interaction quality. Focusing on the social intelligence dimensions *Authenticity*, *Clarity* and *Empathy*, we first used a questionnaire survey evaluating the level of perception in text utterances, and then conducted a Wizard of Oz study to investigate the effects of these utterances in a more interactive setting. Results show that people have great difficulties perceiving elements of social intelligence in text. While on the one hand they find anthropomorphic behaviour pleasant and positive for the naturalness of a dialogue, they may also perceive it as frightening and unsuitable when expressed by an artificial agent in the wrong way or at the wrong time.

Keywords Conversational user interfaces · Chatbots · Social intelligence · Authenticity · Clarity · Empathy

---

N. Mariacher · S. Schlögl (✉)
Department of Management, Communication & IT, MCI Management Center Innsbruck, Innsbruck, Austria
e-mail: stephan.schloegl@mci.edu

A. Monz
Department of Digital Business & Software Engineering, MCI Management Center Innsbruck, Innsbruck, Austria

A. Esposito et al. (eds.), *Progresses in Artificial Intelligence and Neural Systems*, Smart Innovation, Systems and Technologies 184,
https://doi.org/10.1007/978-981-15-5093-5_44

1 Introduction

The ongoing success of mainstream virtual assistants such as Apple's SIRI, Microsoft's CORTANA, Google's ASSISTANT or Amazon's ALEXA, fosters the continuous integration of conversational user interfaces into our everyday lives. Beyond these speech-based intelligent agents, also text-based conversational interfaces have grown significantly in popularity [36]. That is, in 2017 alone the number of chatbots offered on Facebook's MESSENGER platform has doubled.[1] To this end, Beerud Sheth, co-founder and CEO of Teamchat, states *"we're at the early stages of a major emerging trend: the rise of messaging bots"*.[2] So far, this trend is mainly visible in the customer support domain, where in recent years bot technology has gained significant ground (e.g., [11, 24, 54]). In other commercial fields, however, it seems that the technology first needs to adopt more human-like traits for it to be accepted, particularly in areas such as shopping assistance, consulting or advice giving. That is, conversational agents may need to be recognized as social actors and, to some extent, be integrated into existing social hierarchies [50]. Albrecht refers to this as the need for entities (both artificial and human) to be socially intelligent, which according to his **S.P.A.C.E** model encompasses the existence of a certain level of ***S**ituational Awareness*, the feeling of ***P**resence*, and the ability to act ***A**uthentic*, ***C**lear* as well as ***E**mpathic*. Following his theory, the work presented in this paper aims to investigate whether (1) respective elements of social intelligence embedded into textual conversation are perceived by users; and (2) whether such perceptions may influence the experienced interaction quality. In particular, our work focuses on the dimensions *Authenticity*, *Clarity* and *Empathy* embedded into human-chatbot interactions, leaving Albrecht's *Situational Awareness* and *Presence* for future investigations.

2 Related Work

In order to better understand elements of social intelligence and how they may be embedded into text-based human-chatbot interaction it seems necessary to first elaborate on the history and goals of intelligent agents as well as on previous work aimed at investigating traits of socially intelligent behaviour, and how such may be exhibited.

---

[1] https://venturebeat.com/2017/04/18/facebook-messenger-hits-100000-bots/ (last accessed: March 11, 2019).

[2] https://techcrunch.com/2015/09/29/forget-apps-now-the-bots-take-over/?guccounter=2 (last accessed: March 11, 2019).

2.1 Intelligent Agents

Ever since its early days, research in Artificial Intelligence (AI) has worked on two branches, one focusing on using technology to increase rationality, effectiveness and efficiency [39], and the other one concentrating on building machines that imitate human intelligence [7]. Speech recognition, autonomous planning and scheduling, game playing, spam fighting, robotics, and machine translation are just some of the accomplishments that can be attributed to the combined efforts of these two branches [40]. An achievement of particularly high impact lies in the field's steady progress made towards building autonomous machines, i.e., so-called agents which "*can be viewed as [entities] perceiving [their] environment through sensors and acting upon that environment through actuators*" [40, p. 34]. In other words, those machines are able to perceive and interpret their surroundings and react to it without requiring human input [35, 53]. A special form of autonomous systems may be found in conversational agents, which communicate or interact with human interlocutors through means of natural language [44]. Chatbots are a subcategory of these conversational agents defined, according to Abbattista and colleagues, as "*[software systems] capable of engaging in conversation in written form with a user*" [1]. They simulate human conversation using text (and sometimes voice) [13, 43], whereupon their abilities are formed by an illusion of intelligence that is achieved through little more than rule-based algorithms matching input and output based on predefined patterns [13, 44]. In addition, these agents apply several tricks so as to steer and/or manipulate the conversation in a way which feigns intelligence [27]. What they usually miss, however, are linguistic capabilities that show social rather than utilitarian intelligence.

2.2 From Intelligence to Social Intelligence

Research focusing on intelligence has, for a long time, revolved around a single point of measurement, the *Intelligence Quotient* (IQ). Recent developments, however, suggest that a multi-trait-concept represents reality more accurately [2, 19, 47]. Albrecht, for example, distinguishes between *Abstract*, *Social*, *Practical*, *Emotional*, *Aesthetic* and *Kinesthetic* intelligence. Concerning conversational agents, it is particularly *Social Intelligence* (SI), described as "*[...] the ability to get along well with others and to get them to cooperate with you.*" [2, p. 3], which seems increasingly relevant. At its core, SI is understood as a determining factor that motivates people to either approach each other or distance themselves. It encompasses a variety of skills and concepts, such as an understanding of the behaviour of others [3], "*role taking, person perception, social insight and interpersonal awareness*" [17, p. 197], adaption to the situational and social context [2] and the engagement in '*nourishing*' instead of '*toxic*' behaviour. Based on these factors Albrecht proposes the S.P.A.C.E model which describes, as already highlighted, SI to be composed of *Situational Awareness*, *Presence*, *Authenticity*, *Clarity* and *Empathy* [2].

Situational Awareness—The ability to understand people in different situations is described as *Situational Awareness* (SA) [2]. It consists of knowledge about our environment and the subsequent extrapolation of information that accurately predicts or initiates future events [46]. Any given situation is governed by several aspects, such as social rules, patterns and paradigms, the evaluation of behaviours [25], and the different roles within the relevant social hierarchies [42]. Therefore, it requires people to engage with others at an emotional level [2] and to correctly identify norms within social groups as well as emotional keywords [22]. Contextual information is herein crucial so as to correctly read and interpret a situation. This context can be described as "*any information that can be used to characterize the situation of an entity*" [15, p. 5] and further be divided into proxemic, behavioural, and semantic contexts. The proxemic context describes the physical space in which an interaction takes place, the behavioural context includes emotions and motivations of individuals during a given interaction, and the semantic context refers to language and associated meaning. Hence, when aiming to simulate human-human interactions, socially intelligent chatbots would need to consider these contextual attributes as well—perceiving, comprehending and using them to project potential future statuses of the conversation [16].

Presence—Albrecht's *Presence* (P) dimension describes the way people use their physical appearance and body language to affect others. It is influenced by factors such as the first impression, a person's charisma, the respect they exhibit towards others, and the naturalness of their appearance [2]. In virtual environments or when communicating with artificial entities, however, *Presence* is substituted by *Social Presence* [20, 51]. The term is defined as "*the degree of salience of the other person in the interaction and the consequent salience of the interpersonal relationships*" [45, p. 65]. In online settings, for example, presence can be created through measures such as the integration of people's names into conversations or by including pictures of smiling people on websites [20]. High social presence is particularly important in e-commerce, as it has shown to positively influence trust [20, 51]. From a chatbot perspective it is especially the degree of anthropomorphism which influences how social presence is perceived [33]. Yet, increasing levels of anthropomorphism do not always correlate with increasing levels of perceived presence, as overly realistic entities may quickly trigger suspicion or feelings of uneasiness often referred to as the uncanny valley effect [31].

Authenticity—The concept of *Authenticity* (A) encompasses the "*notions of realness and trueness to origin*" [9, p. 457], uniqueness, and originality [23]. Honesty and sincerity also indicate responsibility and empathy towards others [2]. Fulghum describes authenticity as the fundamental social rules of '*playing fair*' and '*sharing everything*' [18]. To this end, research has also emphasized the differences between authentic and rather simulated relationships [48]. In addition, particularly in contexts where direct personal contact is lacking, trust becomes a vital component to authenticity, for it can easily be damaged and may consequently trigger doubts concerning a company's products and services [10]. In essence, the need for companies (and their brands) to act authentic has been steadily increasing [6], since '*in times of increasing uncertainty, authenticity is an essential human aspiration, making it a key issue in contemporary marketing and a major factor for brand success.*' [8, p. 567]. Hence, a socially intelligent chatbot may need to also reflect a certain level of authenticity so as to not cause potentially harmful effects [32].

Clarity—The ability to clearly express what people are thinking, what their opinions and ideas are, and what they want is represented by Albrecht's *Clarity* (C) dimension. Expressing oneself clearly is not limited to the words and phrases used, but also encompasses how people speak, how they use their voice and how compellingly they express their ideas. Small differences in expression can evoke vast differences in meaning [2]. Albrecht identified various strategies which are recommended to improve clarity, for example the "*dropping one shoe*" strategy, in which one creates an expectation through a distinct message early in a conversation [2]. Other strategies include the use of metaphors or graphical explanations such as diagrams or pictures. As a general rule, one should be "*mentally escorted*" so as to reduce cognitive load. Yet, pure text-based communication, as it is inherent to the use of chatbots, differs from face-to-face communication in several ways, most notably in style, spelling, vocabulary, the use of acronyms and emoticons [4], and obviously the lack of additional information transmitted through mimical, gestical as well as voice related expressions. It seems therefore vital to take these factors into account when designing text-based interactions, and search for alternative ways of increasing clarity.

Empathy—The final dimension of Albrecht's S.P.A.C.E. model is described by *Empathy*. The literature has no consensual definition of empathy so that various methods and scales of measurement exist [21]. A common understanding among this plethora of definitions is that emphatic behaviour means that people treat other people's feelings adequately [2] and consequently that they respond to them appropriately. Empathy further includes the skill of understanding a given situation and accepting the feelings of others, even though oneself may disagree with all or some of them [2, 37]. As Decety and Moriguchi describe it: "*Empathy is a fundamental ability for social interaction and moral reasoning. It refers to an emotional response that is produced by the emotional state of another individual without losing sight of whose feelings belong to whom*" [14]. Given that Empathy plays an important role in human-human interactions, we may argue that the transfer of empathic characteristics to human-chatbot interaction is of similar importance. Yet, modelling empathic behaviour requires assessing the context of social situations and consequently to determine which behaviour is required at what time [30]—a rather challenging problem which is currently worked on by both academia and industry.

Although Albrecht's S.P.A.C.E model encompasses a total of five dimensions of SI, our initial investigations into human-chatbot interaction put their focus on only three of them; i.e., *Authenticity*, *Clarity*, and *Empathy*. The other two, i.e., *Social Awareness* and *Presence*, may, however, be subject to future research.

3 Methodology

In order to investigate whether users (1) perceive elements of *Authenticity*, *C*larity and *E*mpathy in chatbot interactions, and (2) whether such would effect their experience of the interaction, we followed a tripartite research methodology.

3.1 Step 1: Evaluating SI Elements in Text

As a first step, we used the literature on SI to design text utterances to be used by a chatbot in three different use cases scenarios (i.e., buying shoes, booking a flight, and buying groceries). For all three of these scenarios we created different sets of German text utterances. Set one focused on *Authenticity*, set two focused on *Clarity* and set three focused on *Empathy* (note: there was some overlap between sets but the majority of utterances were distinct with respect to their inherent SI characteristics). The resulting 89 text utterances were then evaluated by N = 55 students, all of whom were German native speakers, for their perceived degree of clarity, authenticity and empathy. That is, each of the utterances was rated for each of the three SI characteristics on a scale from *1 = low* to *10 = high*.[3]

3.2 Step 2: Simulating Chatbot Interactions

As a second step, we used a Wizard of Oz (WOZ) setup [12, 41] to test our utterances in an interactive setting. To do so, we created three different Facebook profiles, all of which pretended to be chatbots yet were operated by a researcher using our pre-defined sets of text utterances. Profile one, called *Johnny*, acted as the authentic chatbot (using the utterances designed to be authentic), *Claire* acted as the one conveying clarity and *Deni* as the empathic one. A total of N = 18 German-speaking participants interacted with the simulated chatbots in the three afore mentioned interaction scenarios. Those scenarios were counterbalanced so that each participant interacted with each of the simulated chatbots following a different scenario, leaving us with six measurement points for each chatbot/scenario combination. The scenarios were as follows.

Scenario 1: Buying Shoes—*You want to buy new shoes and decide to order them online. Your favorite webshop has developed a new Facebook* MESSENGER *chatbot to help you find the perfect model. Here is some information on what you are interested in:*

---

[3]Note: the complete list of German utterances and their English translations, incl. information on elements of SI which was given to students, is available here: https://tinyurl.com/y4zjd5cz.

– *Model: [Female: sneakers | Male: sport shoes]*
– *Color: Black*
– *Price: approx. EUR 90,-*
– *Size: [Female: 40 | Male: 46]*

You already have an account with the webshop holding all relevant account data incl. payment information and delivery details. Note: Please complete the purchase only if you have found the perfect shoe for you. Otherwise, enjoy the interaction with the chatbot and ask anything you would like to know, even if it does not fit the scenario. Now, please click on the chatbot icon for [Johnny—Claire—Deni], and start the conversation with the sentence "Start Conversation [Johnny—Claire—Deni]". You can terminate the conversation at any time entering the phrase "End Conversation".
Scenario 2: Booking a Flight—*You are planning a trip to Hamburg. You decide to use the newly developed Facebook* MESSENGER *chatbot of your favorite travel portal to book a flight matching the following details:*

- *Departure Airport: [X]*
- *Arrival Airport: Hamburg*
- *From: June 1*st *2018*
- *To: June 5*th *2018*
- *Price: approx. EUR 150,-*

You already have an account with the travel portal holding all relevant account data incl. payment information and delivery details. Note: Please book the flight only if you believe it to be a good match. Otherwise, enjoy the interaction with the chatbot and ask anything you would like to know, even if it does not fit the scenario. Now, please click on the chatbot icon for [Johnny—Claire—Deni], and start the conversation with the sentence "Start Conversation [Johnny—Claire—Deni]". You can terminate the conversation at any time entering the phrase "End Conversation".
Scenario 3: Buying Groceries—*You invited your friend to eat spaghetti with tomato sauce at your place, but unfortunately you do not have time to go and buy the relevant ingredients. Your favourite supermarket offers an online order service which most recently was extended by a Facebook* MESSENGER *chatbot. You decide to use the chatbot to help you choose the best options for:*

- *Pasta*
- *Tomato sauce*
- *Chocolate for desert (most favorite: [X])*

You already have an account with the shop holding all relevant account data incl. payment information and delivery details. Note: Please only chose products that suit your needs. Otherwise, enjoy the interaction with the chatbot and ask anything you would like to know, even if it does not fit the scenario. Now, please click on the chatbot icon for [Johnny—Claire—Deni], and start the conversation with the sentence "Start Conversation [Johnny—Claire—Deni]". You can terminate the conversation at any time entering the phrase "End Conversation".

After each of these scenarios (which ended either by fulfilling the simulated purchasing task or by stopping the conversation) participants were asked to complete a questionnaire assessing the chatbot's perceived level of *Authenticity*, *Clarity* and *Empathy*.[4]

3.3 Step 3: Exploring Interaction Experiences

As a third and final step, all WOZ study participants were asked about their perceived experiences interacting with the three (simulated) chatbots. Interviews, which happened right after each WOZ experiment, were recorded, transcribed and analyzed employing Mayring's qualitative content analysis method [28].[5]

4 Discussion of Results

The goal of our analysis was to shed some light on the question whether chatbot users are capable of perceiving elements of social intelligence embedded in chatbot utterances. The initial questionnaire survey aimed at validating our utterance designs, whereas the WOZ study as well as the interview analysis focused more on the actual interaction with the chatbot and respective experiences.

4.1 Results Step 1: SI Elements in Text

A total of N = 55 students (26 female/29 male) were first given information on elements of SI and then asked to rate our 89 utterances according to their level of perceived *Authenticity*, *Clarity* and *Empathy*. Results show that if asked on paper, and taken out of an actual interactive setting, people find it difficult to separate distinct elements of social intelligence in text. That is, our data shows high levels of clarity, authenticity and empathy with all the evaluated utterances. While *Clair* (clarity) and *Deni* (empathy) scored highest in their respective dimensions, *Johnny* (authenticity) came in second, after *Clair*. That is, *Clair's* utterances were rated the clearest (M = 7.56, Median = 8, Mode = 10) as well as the most authentic (M = 7.34, Median = 8, Mode = 10) and *Deni's* utterances were perceived the most empathic (M = 7.07; Median = 8, Mode = 10). Yet, when looking at the highest ranked sentences in each category, *Johnny* accounted for the top 10 sentences in authenticity

[4]Note: the post-scenario questionnaire and its English translation is available here: https://tinyurl.com/y4zjd5cz.

[5]Note: the used interview guidelines and their English translations are available here: https://tinyurl.com/y4zjd5cz.

whereas for clarity the top 10 were split between *Clair* and *Johnny*, hinting towards a strong connection between those two SI characteristics.

A Pearson correlation confirmed a significant positive relation between *Authenticity* and *Clarity* ($r = 0.718$, $p < 0.001$) as well as between *Authenticity* and *Empathy* ($r = 0.682$, $p < 0.001$) and between *Clarity* and *Empathy* ($r = 0.629$, $p < 0.001$).

4.2 Results Step 2: Simulated Interactions

In order to investigate embedded elements of *Authenticity*, *Clarity* and *Empathy* in chatbot interactions, we conducted a WOZ experiment [12]. As mentioned above, three different FB profiles served as simulated chatbot agents—*Johnny* as the authentic one, *Claire* as the clear one and *Deni* as the empathic one. A member of our research group acted as the wizard, controlling each profile and using its respective text utterances to interact with a total of N = 18 participants (8 female/10 male; 20–30 years old).

Turn-Taking—Looking at the turn-taking behaviour, we found that for the first scenario (i.e., buying shoes), the average turn-taking was 25.22 (SD = 7.73), for the second one (i.e., booking a flight) it was 28.28 (SD = 8.43), and for the third one (i.e., buying groceries) it was 41.67 (SD = 12.04). The significantly higher turn-taking rate in scenario three may be caused by the number of products participants had to buy. As can be seen in Table 1, there were also turn-taking differences between the different chatbots. Yet, those differences were not strong enough to conclude that certain SI characteristics would increase or decrease turn-taking.

Post-scenario Questionnaire—Looking at the post-scenario questionnaire we asked participants to complete after the interaction with each of the simulated chatbots, it can be seen that *Johnny* was the one most liked, which made him also the favorite when it comes to recommending a chatbot to a friend. *Claire*, however, seemed to be the most helpful one. As for elements of SI, the questionnaire included three questions measuring perceived *Authenticity* (i.e., Q5–7), three questions measuring perceived *Clarity* (i.e., Q8–10) and three questions perceived *Empathy* (i.e., Q11–13). Based on our intentions, *Johnny* should have obtained the highest average scores regarding authenticity, yet results show that similar to the outcome of the questionnaire survey

Table 1 Average turn-taking per scenario and chatbot

Chatbot Profile	Scenario 1	Scenario 2	Scenario 3
	Buying shoes $\bar{x}_{turns}$	*Booking a flight* $\bar{x}_{turns}$	*Buying groceries* $\bar{x}_{turns}$
Johnny (authentic)	28.00	27.17	45.17
Claire (clear)	23.33	31.67	38.50
Deni (empathic)	24.33	26.00	41.33

described in Sect. 4.1 it was *Claire* who was perceived the most authentic (M = 6.24). Unfortunately, our intentions with respect to offering a distinctively clear and a distinctively empathetic chatbot were also not met, as both *Clair* as well as *Deni* received the lowest scores in their respective SI characteristics.

4.3 Results Step 3: Qualitative Feedback

Finally, we asked participants about their perceptions of SI elements exhibited by the chatbots and in general about their interaction experiences.

Baseline Feedback—To gain some baseline insights with respect to participants' potential expectations, we first asked them about elements of social intelligence found in traditional sales contexts (i.e., when interacting with human sales personnel in similar settings). To this end, an often named requirement was the ability to judge whether a customer actually requires help; e.g., P13: "*[the person] should definitely notice if someone needs help—when I know what I want, then I do not need help*". Also, it was highlighted that sales personnel needs to show deep knowledge in their respective product category and thus should be able to provide relevant information. Further, rather obvious characteristics such as a certain level of politeness, respectfulness, openness, self-confidence and efficiency were named as elements which would accommodate a customer-friendly atmosphere and thus may add to the level of perceived social intelligence.

General Chatbot Feedback—As opposed to their opinions regarding human-human interaction, our chatbot interactions seemed to polarize participants. While half of them did not see a reason for using a chatbot, the other half perceived the technology as a positive feature potentially helpful in a number of different e-commerce settings. In particular, as an alternative to the often offered search function. Surprisingly, however, participants did not perceive great differences between our chatbots. Overall, the majority found them friendly and polite, yet rather generic, static and potentially time consuming. Differences as to the use of words or sentence structures, although clearly existing, were hardly noticed. Consequently, when asked about their preferred scenario (or chatbot), participants supported their choice with scenario inherent characteristics, such as efficiency, information quality or topic interest, and did not refer to the linguistic behaviour of the chatbot. Hence, it was mainly the speed with which an interaction could be completed that let participants express their preference for one or another chatbot.

Feedback Concerning Elements of Social Intelligence—Although participants did not perceive clear differences in chatbots' communication styles, they found elements of social intelligence, such as jokes or the use of methaphors, an important aspect adding to the naturalness of conversations. That is, the slightly anthropomorphic behaviour created a more pleasant and relaxed atmosphere in which a participant's counterpart was perceived to have its own personality. Also, if not overplayed, it increased a chatbot's trustworthiness; although, the tolerance between comfort creating and overuse of such human-like linguistic behaviour seems rather small, as herein participants' perceptions quickly changed from nice and funny to annoying, dumb and ridiculous.

Feedback Concerning Authenticity—Trying to understand what participants would expect from an authentic chatbot, we first asked them about their understanding of authenticity. Answers show that for them authentic behaviour means to follow goals despite predominant opposing opinions, to not 'put on an act' for others, and to be honest and trustworthy. However, when asked about the characteristics of an authentic chatbot, participants were rather indecisive, so that the question whether a chatbot could be authentic or whether it would merely mimic human behaviour, remained unanswered. However, one characteristic that seemed to unite people's understanding of an authentic chatbot concerns its perceived honesty, both in that it is honest about its artificial nature and that it is honest with respect to the products it recommends. As this perception of honesty has to be earned, participants often suggested an increased level of transparency, particularly in the initial stages of an interaction. For example, showing a selection of products and their respective attributes, which would allow customers to see for themselves whether the allegedly cheapest product is really the cheapest. This aspect of allowing people to validate collected information also adds to a chatbot's trustworthiness—a characteristic which, due to today's often predominant perception of technologies' increasingly manipulative nature, has to first be built up. Other mentioned characteristics which may add to the authenticity of a chatbot include its consistent use of words and language structures, its ability to answer non-contextual questions, or it addressing other interlocutors with their name.

Feedback Concerning Clarity—Clarity was defined as the ability to express oneself clearly and in an understandable manner. Such was also transferred to chatbot behaviour. For example, participants found it important that chatbot answers are coherent and not too long, yet at the same time do not look too basic or scripted. Also, in terms of clarity a chatbot should refrain from asking several pieces of information at once, as our participants were often unsure about how to answer such a request, trying to compose a sentence that holds many pieces of information while still being simple to comprehend. A lack of understanding as to how chatbot technology works and consequently 'makes sense' of information generates the perception that inquiries as well as answers have to be simply structured. Consequently, a request for multiple pieces of information created an uneasiness in people as they felt unsure about the main purpose of the inquiry.

Feedback Concerning Empathy—To be empathic means to be capable of understanding how other people think and feel. To be able to recognize the emotions of other people and to respond appropriately to the observed behaviour. It also incorporates the acceptance of other viewpoints and the showing of interest in other people and their particular situations. Given this need for understanding and recognition of behaviour and/or feelings, participants found that chatbots are not yet sufficiently developed to act emphatic. That is, people usually do not express themselves in text so clearly and extensively that a chatbot (or even another person for that matter) would be able to read their emotions. Furthermore, text-based chats do not really support the use of spontaneous gestures (besides the use of gesture-like elements found in different emoticons), yet gestures are an essential part of expressing emotions and consequently empathic behaviour. Even if certain language elements such as jokes or motivational phrases may convey feelings in conversational behaviour, and potentially trigger positive emotions, they were not perceived as being empathic (although famous early work has shown that people may attribute empathy to a chatbot therapist [52]). It was further stressed that there is a link between empathy and authenticity, for a chatbot that is not perceived authentic can not be perceived empathic, as its alleged empathy could not be taken serious. A rather simple aspect of empathic chatbot behaviour, however, was found in reacting to people's concrete needs; e.g., in an e-commerce setting if a product is out of stock to offer an alternative product that is available. Also, showing interest i.e., continuous questioning, may add to a certain level of perceived empathy. Yet, also here the right level of interest is important, as otherwise human interlocutors may feel skepticism, wondering about the technology's intentions. Finally, utterances which seemingly create positive feelings, such as positive surprises, compliments or jokes, do also add to an overall empathic impression. Although, it has to be mentioned that to this end individual perceptions can be heavily subjected to one's current mood and emotional situation. While in human-human interaction an interlocutor may be able to react and consequently adapt his/her behaviour to an interlocutor's mood, current chatbots are usually not equipped with such an ability.

5 Reflection

The study results presented above paint a rather divers and inconclusive picture of the current situation. That is, it seems that people are able to relate to characteristics of SI when interacting with humans, yet they find it difficult to transfer this understanding to chatbots. In spite of our participants' predominantly positive ratings with respect to the conversations and their strong believe in the potential the technology holds, they exhibited rather critical attitudes towards chatbots. We consistently found that they were rather goal-driven in their interactions, for which efficiency was perceived more relevant than socially intelligent interaction behaviour. While such may be owed to the chosen scenarios (i.e., situated in the online shopping domain), participants seemed to generally have low expectations in chatbot behaviour and thus did not believe the technology would be capable of handling and expressing human traits. Nonetheless, when asked about their concrete perceptions in more detail, impressions of socially intelligent chatbot behaviour were apparent. With respect to the perception of and consequent reaction to traits of SI in human-chatbot interaction we may thus conclude:

SI Utterances in General—The additional utterances we used to express elements of SI, were perceived positive, so that according to participants conversations were 'nicer' and more 'personal'. Mostly, they improved the linguistic exchange by making the interaction clearer, slightly more empathic (if one can speak of empathy when interacting with an artificial entity) and thus also more authentic. However, when used incorrectly, with respect to timing and/or contextual fit, they were quickly perceived as annoying and ridiculous—probably more so than when knowingly interacting with a human being. Consequently one could argue that on the one hand our participants did not believe in complex human traits exhibited by chatbot technology, on the other hand, when exposed to this behaviour they were more critical concerning mistakes than they would have been with a human interlocutor.

Authenticity—Although participants could not define what it would mean for a chatbot to be authentic, they found the interaction more pleasant when the chatbot showed anthropomorphic behaviour. So it does seem that they appreciated the perception of talking to a human being, although they knew that it was a chatbot with which the were conversing. To this end, it was particularly the impression of the system being honest and transparent which helped convey authentic behaviour. Also, it helped increase trust, which may be seen as a key element influencing a technology's success [29].

Clarity—In terms of clarity, participants were satisfied with the way the chatbots expressed themselves. For them the conversations were clear and understandable, and the used language was pleasant. However, participants interpreted clarity as an element of efficiency; i.e., whether the chatbot conversed in a way so that a swift goal completion was supported. Clarity in the sense of providing a better, more comprehensive understanding of the context and thus an increased level of information, seemed less relevant.

Empathy—Empathy was the one dimension of SI which was most criticized. In general, participants had difficulties believing that chatbots (or technology in general) may be capable of showing empathic behaviour, as relevant emotional input and its accurate interpretation are still missing. As for the emotional output, chatbot conversations were perceived as too short and transient to exhibit empathy. Those perceptions may also depend on a person's current emotional state and whether somebody is in the mood of conversing with an artificial entity (and consequently may engage in a more interactive dialogue that offers room for increasingly reflective behaviour). A final aspect of empathy may relate to its creation of affinity and subsequent emotional binding. Here previous work has shown that people build close connections with technologies such as mobile phones [49], robots [34] or even televisions appliances [38], which go as far as to fulfill some sort of companionship role [5]. Since a recent study showed that people ended up personifying Amazon's ALEXA after only four days of using it [26] it may seem plausible that similar effects are inherent to human-chatbot interaction, for which empathy may in fact play a greater role than currently perceived by our study participants.

Concluding, we may thus argue that our analysis of the perceptions and consequent evaluations of SI in human-chatbot interactions produced two important findings. First, we have seen that people are (for now) mainly interested in efficiency aspects when conversing with an artificial entity. That is, they primarily want to reach their goal. Seen from this perspective, the SI of chatbots seems to be of rather secondary importance. On the other hand, however, we have also seen that people often take elements of SI for granted. That is, they assume that chatbots understand how to use these elements to make a conversation more pleasant and natural. Consequently we might conclude that although people are often skeptical about human traits expressed by technological entities, they value a certain level of human-like behaviour.

6 Summary and Outlook

Considering the increasing uptake of chatbots and other AI-driven conversational user interfaces it seems relevant to not only focus on the utilitarian factors of language but also on its social and emotional aspects. The goal of the here presented work was therefore to investigate whether traits of social intelligence implemented by human-chatbot interactions were perceivable by human interlocutors, and if yes whether such affects their experience of the interaction. Explorations employed a questionnaire survey (N = 55) aimed at evaluating people's perceptions of various SI characteristics (i.e., *Authenticity*, *Clarity* and *Empathy*) integrated into simple text utterances, and N = 18 WOZ sessions incl. subsequent interviews, which aimed a investigating those characteristics and their perceptions in a more experimental setting.

Results show that people have great expectations with respect to the conversational behaviour of chatbots, yet characteristics of SI are not necessarily among them. While the feeling of talking to a human being (i.e., anthropomorphic behaviour) is perceived as pleasant and may help increase transparency and consequently trust, people do not believe that a chatbot is capable of conveying empathy. Here it is mainly the reflective, non-goal oriented part of empathy which seems out of scope. While a chatbot may very well understand and react to a person's needs and desires in a given context, the interaction is usually to ephemeral to build up any form of deeper connection. Also, while empathic behaviour embedded in single text utterances is usually well received if appropriate, it is highly criticized when out of place - even more so than with human interlocutors.

A limitation of these results can certainly be found in the exploratory nature of the setting. On the one hand our questionnaire survey evaluated the perception of text utterances that were out of context, on the other hand our WOZ-driven interactions focused on common online shopping scenarios, for which socially intelligent interaction behaviour may not be needed or even appropriate (note: online shopping also happens without conversational user interfaces). Thus, future studies should focus on a less structured situation, potentially building on the early work of Weizenbaum's Eliza [52]. Another aspect which was not subject to our investigations, yet may influence the perception of socially intelligent chatbot behaviour, concerns the adequate use of emoticons. As previous work has shown that emoticons trigger concrete emotions [55], we believe that a better understanding of when an how to use illustrative language in human-chatbot interaction is needed if we aim to eventually increase the level of social intelligence conveyed by these artificial interlocutors.

Acknowledgments The research presented in this paper has been supported by the project EMPATHIC that has received funding from the European Union's Horizon 2020 research and innovation programme under grant agreement No 769872.

References

1. Abbattista, F., Degemmis, M., Licchelli, O., Lops, P., Semeraro, G., Zambetta, F.: Improving the usability of an e-commerce web site through personalization. Recommendation and Personalization in eCommerce **2**, 20–29 (2002)
2. Albrecht, K.: Social Intelligence: The New Science of Success. Wiley (2006)
3. Baron-Cohen, S., Ring, H.A., Wheelwright, S., Bullmore, E.T., Brammer, M.J., Simmons, A., Williams, S.C.R.: Social intelligence in the normal and autistic brain: an fmri study. Eur. J. Neurosci. **11**(6), 1891–1898 (1999). https://doi.org/10.1046/j.1460-9568.1999.00621.x
4. Barton, D., Lee, C.: Language Online: Investigating Digital Texts and Practices. Routledge (2013)
5. Benyon, D., Mival, O.: From human-computer interactions to human-companion relationships. In: Proceedings of the First International Conference on Intelligent Interactive Technologies and Multimedia, pp. 1–9. ACM (2010)
6. Beverland, M.: Brand management and the challenge of authenticity. J. Prod. Brand Manage. **14**(7), 460–461 (2005). https://doi.org/10.1108/10610420510633413
7. Brooks, R.A.: Intelligence without representation. Artif. Intell. **47**(1), 139–159 (1991). https://doi.org/10.1016/0004-3702(91)90053-M
8. Bruhn, M., Schoenmller, V., Schfer, D., Heinrich, D.: Brand authenticity: Towards a deeper understanding of its conceptualization and measurement. Adv. Consum. Res. **40**, 367–576 (2012)
9. Buendgens-Kosten, J.: Authenticity. ELT J. **68**(4), 457–459 (2014). https://doi.org/10.1093/elt/ccu034
10. Chen, S.C., Dhillon, G.S.: Interpreting dimensions of consumer trust in e-commerce. Inf. Technol. Manage. **4**(2), 303–318 (2003). https://doi.org/10.1023/A:1022962631249
11. Cui, L., Huang, S., Wei, F., Tan, C., Duan, C., Zhou, M.: Superagent: a customer service chatbot for e-commerce websites. In: Proceedings of ACL 2017, System Demonstrations, pp. 97–102 (2017)
12. Dahlbäck, N., Jönsson, A., Ahrenberg, L.: Wizard of Oz studies: why and how. Knowl.-Based Syst. **6**(4), 258–266 (1993)
13. De Angeli, A., Johnson, G.I., Coventry, L.: The unfriendly user: exploring social reactions to chatterbots. In: Proceedings of The International Conference on Affective Human Factors Design, London. pp. 467–474 (2001)
14. Decety, J., Moriguchi, Y.: The empathic brain and its dysfunction in psychiatric populations: Implications for intervention across different clinical conditions. BioPsychoSocial Med. **1**(1), 22 (2007)
15. Dey, A.K.: Understanding and using context. Pers. Ubiquit. Comput. **5**(1), 4–7 (2001). http://dx.doi.org/10.1007/s007790170019
16. Endsley, M.R.: Toward a theory of situation awareness in dynamic systems. Hum. Factors **37**(1), 32–64 (1995). https://doi.org/10.1518/001872095779049543

17. Ford, M.E., Tisak, M.S.: A further search for social intelligence. J. Educ. Psychol. **75**(1), 196–206 (1983)
18. Fulghum, R.: All I Ever Really Needed to Know I Learned in Kindergarten. Grey Spider Press (1986)
19. Gardner, H.: Frames of Mind: The Theory of Multiple Intelligences. Basic Books (1983)
20. Gefen, D., Straub, D.W.: Consumer trust in b2c e-commerce and the importance of social presence: experiments in e-products and e-services. Omega **32**(6), 407–424 (2004). https://doi.org/10.1016/j.omega.2004.01.006
21. Gerdes, K., Segal, E., Lietz, C.: Conceptualising and measuring empathy. Br. J. Soc. Work **40**(7), 2326–2343 (2010). https://doi.org/10.1093/bjsw/bcq048
22. Goleman, D., Boyatzis, R.: Social intelligence and the biology of leadership. Harvard Bus. Rev. **86**(9), 74–81 (2008)
23. Gundlach, H., Neville, B.: Authenticity: further theoretical and practical development. J. Brand Manage. **19**(6), 484–499 (2012)
24. Jacobs, I., Powers, S., Seguin, B., Lynch, D.: The top 10 chatbots for enterprise customer service. Forrester Report (2017)
25. Lei, H.: Context awareness: a practitioner's perspective. In: International Workshop on Ubiquitous Data Management. pp. 43–52 (April 2005). https://doi.org/10.1109/UDM.2005.6
26. Lopatovska, I., Williams, H.: Personification of the amazon alexa: Bff or a mindless companion. In: Proceedings of the 2018 Conference on Human Information Interaction & Retrieval, pp. 265–268. ACM (2018)
27. Mauldin, M.L.: Chatterbots, tinymuds, and the turing test: entering the loebner prize competition. In: AAAI (1994)
28. Mayring, P.: Qualitative content analysis: theoretical foundation, basic procedures and software solution (2014)
29. Mcknight, D.H., Carter, M., Thatcher, J.B., Clay, P.F.: Trust in a specific technology: an investigation of its components and measures. ACM Trans. Manage. Inf. Syst. (TMIS) **2**(2), 12 (2011)
30. McQuiggan, S.W., Lester, J.C.: Modeling and evaluating empathy in embodied companion agents. Int. J. Hum.-Comput. Stud. **65**(4), 348–360 (2007)
31. Mori, M., MacDorman, K.F., Kageki, N.: The uncanny valley [from the field]. IEEE Rob. Autom. Mag. **19**(2), 98–100 (2012)
32. Neururer, M., Schlögl, S., Brinkschulte, L., Groth, A.: Perceptions on authenticity in chat bots. Multimodal Technol. Interact. **2**(3), 60 (2018). https://doi.org/10.3390/mti2030060
33. Nowak, K.L., Biocca, F.: The effect of the agency and anthropomorphism on users' sense of telepresence, copresence, and social presence in virtual environments. Presence: Teleoperators Virtual Environ. **12**(5), 481–494 (2003). https://doi.org/0.1162/105474603322761289
34. Ogawa, K., Ono, T.: Itaco: Constructing an emotional relationship between human and robot. In: RO-MAN 2008-The 17th IEEE International Symposium on Robot and Human Interactive Communication. pp. 35–40. IEEE (2008)
35. Persson, P., Laaksolahti, J., Lonnqvist, P.: Understanding socially intelligent agents—a multilayered phenomenon. Trans. Sys. Man Cyber. Part A **31**(5), 349–360 (2001). http://dx.doi.org/10.1109/3468.952710
36. Pinhanez, C.S.: Design methods for personified interfaces. In: Proceedings of the International Conference on Computer-Human Interaction Research and Applications. pp. 39–49. INSTICC, SCITEPRESS Science and Technology Publications, Funchal, Madeira, Portugal (2017). https://doi.org/10.5220/0006487500270038
37. Preece, J.: Empathic communities: balancing emotional and factual communication. Interact. Comput. **12**(1), 63–77 (1999)
38. Reeves, B., Nass, C.: How people treat computers, television, and new media like real people and places (1996)
39. Russell, S., Norvig, P.: Artificial Intelligence: A Modern Approach. Always Learning. Pearson (2016)
40. Russell, S., Norvig, P.: Social Intelligence: The New Science of Success. Pearson (2010)

41. Schlögl, S., Doherty, G., Luz, S.: Wizard of Oz experimentation for language technology applications: Challenges and tools. Interact. Comput. **27**(6), 592–615 (2015). https://doi.org/10.1093/iwc/iwu016
42. Schmidt, S., Schmidt, S.: Achtsamkeit und Wahrnehmung in Gesundheitsfachberufen. Springer (2012)
43. Sha, G.: Ai-based chatterbots and spoken english teaching: a critical analysis. Comput. Assist. Lang. Learn. **22**(3), 269–281 (2009). https://doi.org/10.1080/09588220902920284
44. Shawar, B.A., Atwell, E.: Using dialogue corpora to train a chatbot. In: Proceedings of the Corpus Linguistics 2003 Conference. pp. 681–690 (2003)
45. Short, J., Williams, E., Christie, B.: The Social Psychology of Telecommunications. Wiley (1976)
46. So, R., Sonenberg, L.: Situation awareness in intelligent agents: foundations for a theory of proactive agent behavior. In: Proceedings. IEEE/WIC/ACM International Conference on Intelligent Agent Technology, 2004. (IAT 2004), pp. 86–92 (Sep 2004). https://doi.org/10.1109/IAT.2004.1342928
47. Thorndike, E.: Intelligence and its uses. Harper's Mag. (1920)
48. Turkle, S.: Authenticity in the age of digital companions. Interact. Stud. **8**(3), 501–517 (2007)
49. Vincent, J.: Emotional attachment to mobile phones: An extraordinary relationship. In: Mobile World, pp. 93–104. Springer (2005)
50. Wallis, P., Norling, E.: The trouble with chatbots: social skills in a social world. Virtual Soc. Agents **29**, 29–36 (2005)
51. Weisberg, J., Te'eni, D., Arman, L.: Past purchase and intention to purchase in ecommerce: The mediation of social presence and trust. Internet Res. **21**(1), 82–96 (2011). https://doi.org/10.1108/10662241111104893
52. Weizenbaum, J.: Eliza–a computer program for the study of natural language communication between man and machine. Commun. ACM **9**(1), 36–45 (1966)
53. Wooldridge, M., Jennings, N.R.: Agent theories, architectures, and languages: a survey. In: Wooldridge, M.J., Jennings, N.R. (eds.) Intelligent Agents, pp. 1–39. Springer, Berlin, Heidelberg (1995)
54. Xu, A., Liu, Z., Guo, Y., Sinha, V., Akkiraju, R.: A new chatbot for customer service on social media. In: Proceedings of the 2017 CHI Conference on Human Factors in Computing Systems, pp. 3506–3510. ACM (2017)
55. Yuasa, M., Saito, K., Mukawa, N.: Emoticons convey emotions without cognition of faces: an fmri study. In: CHI'06 Extended Abstracts on Human Factors in Computing Systems, pp. 1565–1570. ACM (2006)

Linguistic Evidence of Ageing in the Pratchett Canon

Carl Vogel

Abstract A corpus of literary works authored by Sir Terry Pratchett is analyzed from the perspective of linguistic variables which from the literature on natural ageing, one might expect to show effects of gradual change and periods without change. Aspects of lexical complexity exhibit trends that diverge from patterns associated healthy ageing. **Background**: Past study of linguistic healthy ageing has analyzed corpora of individual professional writers over time and cross-sectional corpora constructed with writing samples from individuals in distinct age groups. Among other effects, negative linear correlations have been found between the use of both first person singular and first person plural pronouns and age and between past-tense verb forms and age; positive linear correlations have been found between cognitive complexity features and age. **Main goal**: This study seeks to contribute to understanding of whether Alzheimer's disease is accompanied by the same, accelerated or distinct patterns of change in linguistic features associated with healthy ageing. **Method**: A corpus of works published by Sir Terry Pratchett is analyzed with respect to pronoun use and linguistic signals of cognitive complexity, such as embedding words and lexical variety, testing correlations between those quantities and author age. **Results**: The Pratchett corpus exhibits effects in the opposite direction of those associated with healthy ageing for: first person pronoun use and long words. Lexical variety strongly diminishes over the corpus. **Conclusions**: Linguistic data produced by individuals diagnosed with Alzheimer's disease appears to contain signals of the fact, when analyzed *post hoc*. This suggests further work to identify inflection points and to study further features that may be indicative as the linguistic data is produced.

C. Vogel (✉)
School of Computer Science and Statistics,Trinity Centre for Computing and Language Studies, Trinity College Dublin, The University of Dublin, Dublin 2, Ireland
e-mail: vogel@tcd.ie

A. Esposito et al. (eds.), *Progresses in Artificial Intelligence and Neural Systems*, Smart Innovation, Systems and Technologies 184,
https://doi.org/10.1007/978-981-15-5093-5_45

1 Introduction

Theories of how one's use of language may change with age are easy to motivate—diminishing social networks may be manifest in correspondingly reduced use of first person plural pronouns; perceiving that one may have more time behind one than ahead of one may lead to increased use of past-tense temporal reference; ageing accompanied by cognitive decline may be evident in decreased lexical variety and reduced structural complexity; and so on. Increasingly, with the availability of longitudinal corpora, it is possible to test these theories empirically. It has been found that certain linguistic forms used in writing correlate with the age of the author, with some effects holding constant between both professional literary authors and individuals not characterized as professional writers [5].

Natural quantitative relationships to explore are those that are linear or curvilinear (and among those, quadratic or cubic). Some linguistic behaviors may increase more or less in tandem with age. A simple example of this the count of utterances one makes—assuming that one makes more or less the same number of utterances from one year to the next, the total count of all utterances will grow at the same rate that one ages (these are modeled by a line, as in (1): the quantity of note, say total utterances, may be understood as y, predicted by some constant multiplied by x (age) plus potentially some other constant, c).

(1) $y = bx + c$
(2) $y' = bx^2 + cx + c$
(3) $y'' = bx^3 + cx^2 + dx + e$

Curvilinear relationships are more complicated. Quadratic relations (2) are dominated by a squared term, and cubic relations are dominated by a cubed term. Because of these facts about dominance, sometimes these relations are represented by simpler equations that mention only the dominant terms, equivalent to presuming that the value of b is 1 and of the other constants, 0. It is useful to be able to visualize how these equations look when plotted on a Cartesian plane. Figure 1 plots the natural numbers from 1 to 100 letting the vertical axis (y coordinate) be determined by the dominant term of (1) on the left, (2) in the middle and (3) on the right. In a sense, the curvilinear relationships involve changes that are "more quick" than the linear relationships.[1]

With respect to age, one might imagine that some adult developmental processes are sensitive to a critical point after language acquisition maybe associated with professional maturity (for example, 40 years of age), perhaps with increase (or decrease) to a point, and then level somewhat, and then increase again, but in the opposite of the original direction (as in the middle of Fig. 2). A linear relation would not show a

[1]It must be acknowledged, as a helpful reviewer has noted, that this paper is included in a volume that has substantial focus on neural network research and engineering oriented topics, and for many who will encounter this volume, these visualizations are superfluous; however, volumes in this series also have a substantially wider audience, including some for whom these visualizations may serve as helpful reminders.

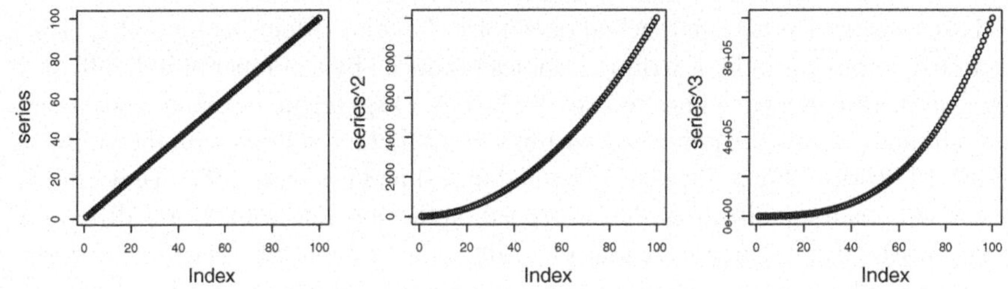

Fig. 1 Left: a linear relation between x and y. Middle: a quadratic relation between x and y. Right: a cubic relation between x and y

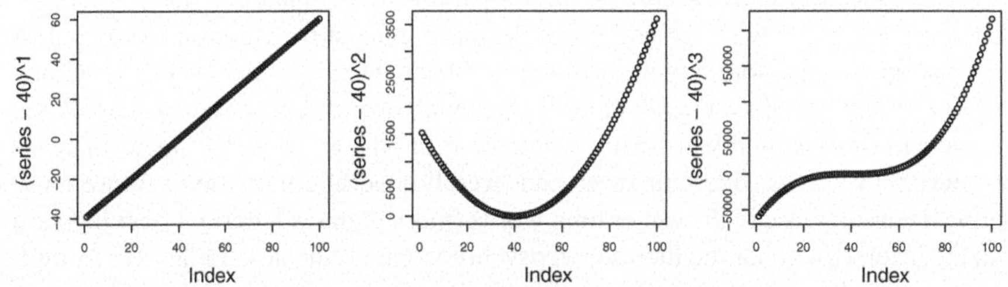

Fig. 2 Plots of the same relations but of the earlier series with the value 40 subtracted away. Left: a linear relation between x and y. Middle: a quadratic relation between x and y. Right: a cubic relation between x and y

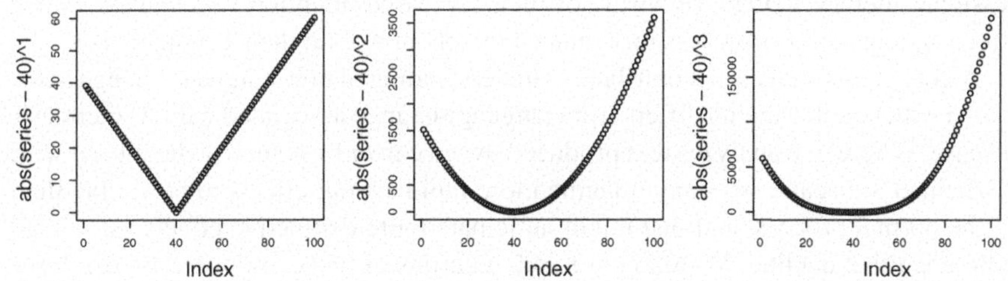

Fig. 3 Plots of the same relations but of the series obtained from the absolute value of the earlier series with the value 40 subtracted away. Left: a linear relation between x and y. Middle: a quadratic relation between x and y. Right: a cubic relation between x and y

change of direction, but displacement along the vertical axis (as in the left of Fig. 2). A cubic relation might show increase, plateau, followed by increase again (as in the right of Fig. 2).

Additionally, it is useful to think of the effect of the absolute value of the difference between age and some critical point, as in Fig. 3. In these terms, there is some resemblance between the linear plot of absolute value of "age" with respect to a "critical point" (40, in the plots) as in the left plot of Fig. 3 and the quadratic plot provided in the middle of Figs. 2 and 3 (which are identical because of the arithmetic effect of squaring negative numbers or the absolute value of negative numbers).

Past research [5] has approached questions of ageing effects using both a cross-sectional corpus (a study of writing samples produced by a number of individuals in age groups from 8 years of age to over 70; N = 3280 aggregated over 45 studies) and longitudinal corpus (the published writings of 12 literary authors with the youngest writer 12 years of age at the start of available text and the oldest, 90). They studied linear and quadratic effects of the sort based on the absolute value of the difference between 40 and author age at the time of writing of the sample text. The two categories of text provide a view of what might be characterized as normal healthy linguistic ageing. A contrast with healthy ageing is expected during cognitive decline, such as is experienced with Alzheimer's disease.

In the cross-sectional data, they found: both linear and quadratic increase in positive emotion words with age and linear decrease in negative emotion words; linear decrease in first person singular pronouns; linear decrease and overall quadratic increase in the use of first person plural pronouns; overall linear decrease quadratic increase in time related words; linear decrease in past-tense verbs; linear increase in future-tense verbs; linear increase and overall quadratic increase in future-tense verbs; linear increase in the use of long words (more than six letters); linear increase and the inverse of quadratic increase in psychological predicates. Table 1 (in its middle column) shows that the analysis of the literary corpus did not yield as many significant effects as the cross-sectional study but that most of the significant effects were aligned. An exception was in the use of exclusive expressions ("but", "exclude" are provided as examples by [5, p. 295]. One might reach the conclusion that professional literary writers, by virtue of their increased attention to language over a lifespan, in general show less pronounced effects of age on their language use.

A goal of this study is to contribute to understanding of how language change associated with healthy ageing differs from language change associated with Alzheimer's disease. Where one analyzes text produced over a career by someone diagnosed with Alzheimer's disease, one may imagine more rapid ageing effects to be visible than in the normal process, and one might anticipate more distinctive effects associated with cognitive decline. With this in mind, a corpus of texts composed by Sir Terry Pratchett is analyzed, given that he was diagnosed with Alzheimer's disease [6]. The results of this analysis may be observed by inspection of the fourth column of Table 1. The next section (Sect. 2.1) outlines the methods by which these results are determined and illustrates them graphically (Sect. 2.2). Finally, (Sect. 3) discusses how the results may be interpreted. The results are mostly consistent with findings of previous work with respect to linguistic trends in healthy ageing. However, differences suggestive of cognitive decline are visible. The primary difference is lexical complexity: a decrease in the use of "long words" is noted. Testing of additional hypotheses shows decrease in lexical variety at the same time as increased lexical content. Verbosity itself has been pointed to as indicative of "changes in message production processes that lie at the juncture between normal and pathological aging" [2, p. 199].

Table 1 Comparison of results from Pennebaker and Stone (2003), in the first two columns, and the Pratchett Corpus, in the third column; unless noted otherwise, numbers reported for the latter are Pearson correlation coefficients and associated significance values. Tests made of both linear and quadratic relations with age (the quadratic relation with age is based on the square of the absolute difference between age and 40, as in the Pennebaker and Stone work); the quadratic correlation coefficient is reported where the significance is determined by a smaller p-value than for the linear case. Where tests were not significant, they are recorded as "n.s."

		Cross-sectional	Literary authors	Pratchett
1.	Positive emotions	Quadratic increase	n.s.	n.s.
2.	Negative emotions	Linear decrease	n.s.	n.s.
	Pronouns			
1.	First person singular	Linear decrease	Linear decrease	Linear increase (0.63, $p < 0.001$)
2.	First person plural	Quadratic increase	n.s.	Linear increase (0.63, $p < 0.001$)
3.	Second person	NA	NA	Linear increase (0.66, $p < 0.001$)
4.	Third person	NA	NA	Linear decrease (-0.59, $p < 0.001$)
	Time orientation			
1.	Time words	Linear decrease	n.s.	n.s.
2.	Past tense	Linear decrease	n.s.	n.s.
3.	Present tense	Linear increase	n.s.	Linear increase (0.56, $p < 0.001$)
4.	Future tense	Linear increase	Linear increase	Quadratic increase (0.37, $p < 0.05$)
	Cognitive complexity			
1.	Long words	Linear increase	n.s.	Linear decrease (-0.72, $p < 0.001$)
2.	Psychological predicates	Linear increase	Linear increase	Linear increase (0.39, $p < 0.05$)
3.	Exclusive	Quadratic decrease	Linear increase	Linear increase (0.69, $p < 0.001$)
4.	Proposition embedding	NA	NA	n.s.
5.	Question embedding	NA	NA	Quadratic increase (0.42, $p < 0.02$)
6.	Prop. & que. embedding	NA	NA	n.s.
7.	Prop. & +/− que. embedding	NA	NA	n.s.
8.	(Lexical types)÷(Lexical tokens)	NA	NA	Linear decrease (-0.70[a], $p < 0.001$)
9.	Lexical tokens	NA	NA	Linear increase (0.37, $p < 0.05$)
10.	Lexical types	NA	NA	n.s.

[a]The type-token ratio is not normally distributed; the correlation coefficient here is Kendall's τ, based on rank ordering correlation

2 Analyzing the Pratchett Corpus

A corpus spanning 20 years of Pratchett's professional publishing career is analyzed. Notwithstanding the argument that it makes sense to analyze linguistic change in an individual not just with respect to the changes that frequently accompany normal healthy ageing, but also with respect to change in the language itself [3], the analysis here does not attempt to quantify change in English over the period from 1983 to 2004 (a period during which "mouse" obtained a highly frequent sense that has an evidently acceptable regular plural as "mouses").

2.1 Methods

A corpus of text files derived[2] from portable document format (PDF) files included works published between 1983 and 2004. This entails an author age range from 35 to 56 years. The corpus includes 43,019 lexical types (sorts of words) and 3,708,337 tokens (individual uses of words).[3] A list of the files and an indication of the series to which they belong in the Pratchett canon is provided in supplementary materials.[4] These files were indexed with respect to a lexicon suggested by the work of [5] and also provided in the supplementary materials.[5]

A complete lexicon is provided to clarify exactly which items populate which categories considered. The work of [5] is reported with respect to the lexicon embedded in the "Linguistic Inquiry and Word Count" (LIWC) system [4]: their paper itself provides suggestive examples, but does not itemize the complete composition of categories considered. For example, the category "exclusive" discussed above is described as containing "but" and "exclude", but this category is understood as also properly containing "except" and negation words, and so these are included here. Verb tense information is incompletely recorded—here it is primarily associated with inflections of "be" and with the list of psychological predicates. Psychological predicates (cogn) are taken to be those that may express a relation between an agent and sentential content, possibly declarative or possibly interrogative. The predicates are noted as those which may embed propositions alone (e.g. "believe"; cognp), questions alone (e.g. "ask", cognq), propositions or questions (e.g. "know", cognpq), propositions or only polar questions (e.g. "doubt", cognppq). Only some of the relations recorded are tested in what follows.

Inasmuch as verbs that embed sentences involve more complexity than those that do not (see the notion of basic-sentence proposed by [1]) it is natural to consider

[2]This was done using a linux utility: https://www.scss.tcd.ie/Carl.Vogel/WIRN2019/Vogel-WIRN2019-appendix.pdf.

[3]No distinction between uppercase and lowercase uses of words is made.

[4] A file containing the appendices to this work is available: http://www.scss.tcd.ie/Carl.Vogel/WIRN2019/Vogel-WIRN2019-appendix.pdf.

[5]Ibid.

the relative frequency of these predicates over time. However, notice that for the purposes of fiction, "said" may be anticipated as a high-frequency item, and is within the category of psychological predicates that may embed propositions or questions. More interesting in this context are the verbs that embed propositions exclusively or questions exclusively (or the overall category of psychological predicates). It is not immediately obvious how well these categories motivated by linguistic syntax and semantics overlap with the "total cognitive words" or "insight" words of [5]. However, it is reasonable to suppose that similar intuitions informed their categories.

The files in the corpus are indexed by the lexicon and the types specified: for each word in the corpus, its count is recorded, as is the count of its category, super-category and alternative category. The type-token ratio is the ratio between the distinct words used and the total count of instances of each distinct word's use. This has a maximum value (1) when each type is used exactly once; something hard to achieve in language, given the need for "function words" like "the". However, it is easier to achieve larger values with smaller amounts of text than with larger amounts of text. Relative frequencies are recorded by sample, and each of the tests reported relativizes to the total token count in each sample. Where the ratio given by the category of interest divided by the total number of tokens in a sample appears normally distributed,[6] the quantity is tested using a correlation either with Age (Year-1948) or AgeEffect ($|y - 40|^2$ for y =age, to capture a quadratic effect in the same manner as described by [5]), reporting Pearson's correlation coefficient where the correlation is significant ($p < 0.05$). In both cases, in the end, a linear correlation is tested, but in the second case the Age quantity is scaled to be quadratic, as just described. Where the quantity is not normally distributed, a rank ordering test is used, correlating the rank ordering of the quantity as measured in the sample with the rank ordering of author age for the sample. Negative correlations are reported as indicating effects that decrease with age (either in a linear or quadratic (curvilinear) manner).

As indicated above, it was expected that in the Pratchett data, quantities associated with cognitive complexity would show decreasing values as age increases.

2.2 Results

The results are recorded in the third column of Table 1. Some of the relationships reported there are visualized in plots of relevant quantities conditioned on age. Evidently, Pratchett was given a diagnosis of Alzheimer's at age 59, in 2007, an age that is outside the range of the data analyzed here by three years.

The leftmost plot in Fig. 4 depicts the relationship between the relative frequency of first-person singular pronouns and age. This linear positive correlation is significant (0.63, $p < 0.001$). The middle plot shows the relative frequency of first-person plural pronouns with age (the linear positive correlation is sig-

[6]In the sense that a Shapiro test of normality does not allow one to reject the null hypothesis that the ratio follows a normal distribution.

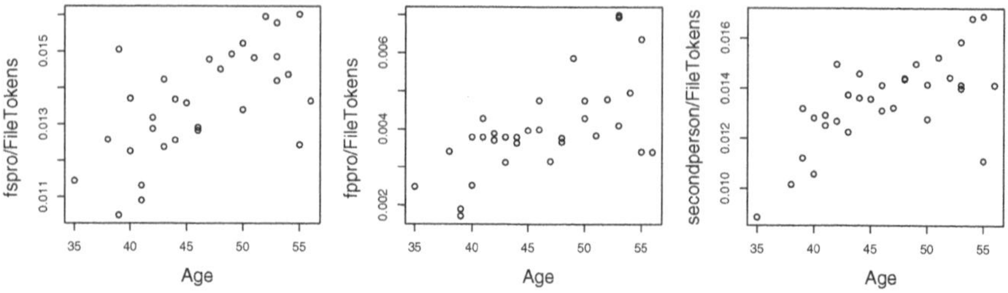

Fig. 4 Relative frequency of (left) first-person singular pronouns, first-person plural pronouns (middle) and second person pronouns (right) within files of the Pratchett corpus, given age

Fig. 5 Relative frequency of third-person pronouns within files of the Pratchett corpus, given age

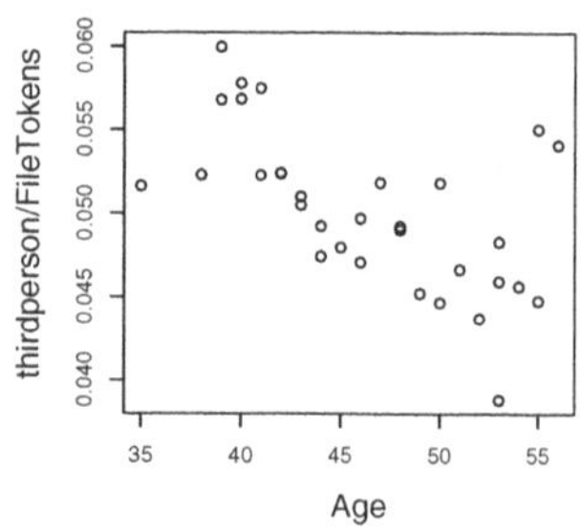

Fig. 6 Relative frequency of present tense forms, given age (left) and future tense forms, given AgeEffect (right)

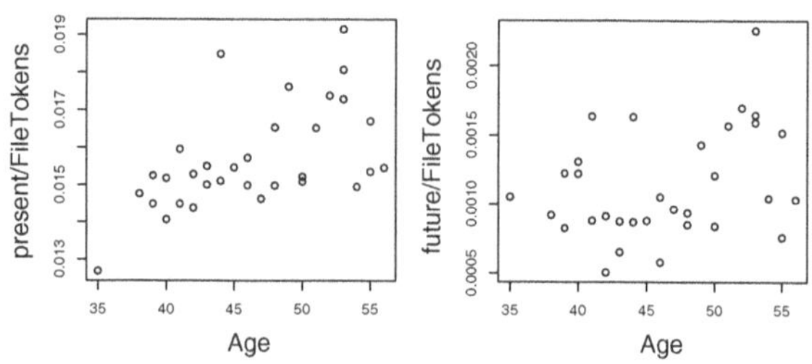

nificant; 0.63, $p < 0.001$), and the rightmost plot shows the relative frequency of second-person pronouns with age (the linear positive correlation is significant 0.66, $p < 0.001$). Figure 5 demonstrates the contrasting relationship between the relative frequency of third-person pronouns and age (the linear negative correlation is significant; −0.59, $p < 0.001$).

There is no significant relationship between age and the relative frequency of temporal words. There is a significant linear positive correlation between relative frequency of cognitive words marked for present tense and Age (0.56, $p < 0.001$; see Fig. 6, left). The linear correlation between future tense forms and Age is not significant, but the quadratic relation between future tense and AgeEffect is significant (0.37, $p < 0.05$; see Fig. 6, right).

The increase in the relative frequency of question embedding predicates with age is demonstrated in Fig. 7 (quadratic increase; 0.042, $p < 0.02$). The plot on the left in Fig. 8 shows the increase in the total number of tokens used, in relation to age

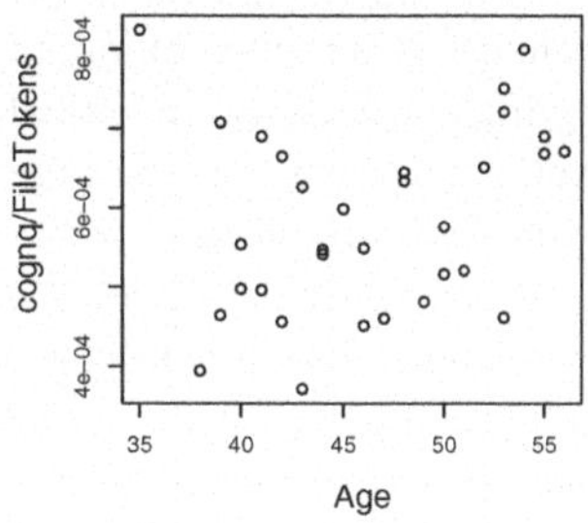

Fig. 7 Relative frequency of question embedding predicates within files of the Pratchett corpus, given age

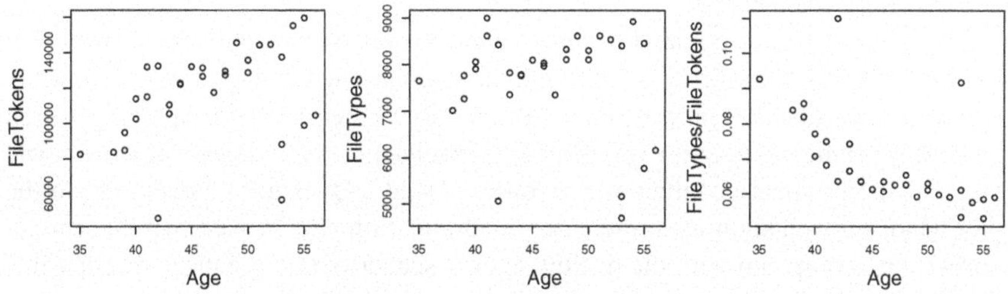

Fig. 8 Token counts (left), type counts (middle) and lexical variety, measured using the Type-Token ratio (right), within files of the Pratchett corpus, given age

(0.37, $p < 0.05$), and the plot on the right in Fig. 8 demonstrates the corresponding decrease in the type-token ratio (−0.70, $p < 0.001$). The plot in the middle shows that there is not a significant linear or curvilinear change in the total number of lexical types with age.

3 Discussion

The different direction in the trend of use of first person singular pronouns in the current study (increase) and in both the cross-sectional study and the wider literary corpus (decrease) as reported by [5] is interesting, because the effects they report can be ascribed to normal, healthy ageing. It may be tempting to explain the different direction observed in the Pratchett canon in relation to the cognitive decline associated with Alzheimer's disease. However, given that the trend involves increase in use of both first person plural and second person pronouns, one might also interpret this in a positive manner, as indicating that the author experienced increasing social network bonds with age, instead of social isolation that might be more typical.

The results show a predicted effect of decrease in use of complex words (as measured by number of letters) over time, distinct from effects noted with healthy ageing [5]. In particular, this is distinct from the effect witnessed in the cross-sectional study.

It is also consistent with Pratchett's own reports of increasing difficulty with spelling [6, pg. 6]. There is also a significant decrease in lexical variety (as measured with the type-token ratio). However, the decrease in type-token ratio is accompanied by a trend towards increase in total tokens produced (−0.70 is the rank-order coefficient associated with decrease in variety, while 0.40 is the rank-order coefficient associated with the increase in total tokens).[7] It is generally accepted as empirical fact that greater type-token ratios can be achieved with shorter texts, so if texts in later years are longer, one expects decrease in lexical variety even if the total number of distinct words (lexical types) used does not increase or decrease significantly with age, as is visible in the middle plot of Fig. 8. Increase in the count of lexical tokens with age is suggestive of increasing verbosity, which has been noted as a feature of cognitive decline [2], as noted earlier.

With respect to the embedding predicates, overall significant effects were not observed. What is noteworthy is the significant increase in quantity of words that may embed questions but not propositions. Complexity as measured through proposition embedding predicates does not reveal significant change, even when subtracting away counts of the predicate of dialogue narration, "said". Question embedding, on the other hand, showed significant increase, quadratic with age.[8] Given that this cannot at present be compared with the healthy ageing scenario, it is not clear whether this is an effect of healthy or unhealthy ageing.[9]

While most of the correlations of note involved linear relationships, some involved quadratic relations. It is of interest that none were best described by cubic relationships—this is interesting because the underlying behavior of cubic relationships (as graphed in a Cartesian plane) appears more intuitive to associate with natural processes associated with human development and ageing than that supplied by quadratic relations, which lack extended "plateaus". Instead of change plateaus, for the measures where ageing effects were expected, the best models appear to be the ones involving steady change. Most results reported here are consistent with linguistic trends identified by prior work as associated with normal ageing. Differences suggestive of cognitive decline are visible. The primary difference is lexical complexity: a decrease in the use of "long words" is noted. An increase in verbosity is identified, and with increased word count, an expected decrease in lexical variety as measured by the type:token ratio.

Acknowledgements I am grateful to Dr. Jennifer Edmond for supplying access to the Sir Terry Pratchett literary corpus. This research is supported by Science Foundation Ireland (Grants 12/CE/I2267 and 13/RC/2106) through the CNGL Programme and the ADAPT Centre.

---

[7]In Table 1 the Pearson correlation coefficient is reported.

[8]Removing the predicate of dialogue narration, "asked", the relative frequency of question embedding predicates still shows a linear positive correlation with age (0.42, $p < 0.02$).

[9]It is tempting to speculate that since questioning robustly signals seeking answers to what one does not know, and since knowing what one does not know is a component of wisdom, then increasing use of question embedding with age is a linguistic signal of increasing wisdom.

References

1. Keenan, E.: Towards a universal definition of 'subject'. In: Li, C. (ed.) Subject and Topic, pp. 304–333. Academic Press, London (1975); Symposium on Subject and Topic, p. 1975. University of California, Santa Barbara (1975)
2. Kemper, S., Kemtes, K.: Aging and message production and comprehension. In: Park, D., Schwarz, N. (eds.) Cognitive Aging: A Primer, pp. 197–213. Taylor and Francis, Philadelphia (2000)
3. Klaussner, C., Vogel, C.: Temporal predictive regression models for linguistic style analysis. J. Lang. Model. **6**(1), 175–222 (2018)
4. Pennebaker, J.W., Francis, M.E., Booth, R.J.: Linguistic Inquiry and Word Count (LIWC). Erlbaum, Mahwah, NJ (2001)
5. Pennebaker, J.W., Stone, L.D.: Words of wisdom: language use over the life span. J. Pers. Soc. Psychol. **85**(2), 291–301 (2003)
6. Pratchett, S.T.: Living with early dementia. In: Lloyd, C.E., Heller, T. (eds.) Long-Term Conditions: Challenges in Health & Social Care, pp. 5–9. SAGE (2012)

M-MS: A Multi-Modal Synchrony Dataset to Explore Dyadic Interaction in ASD

Gabriele Calabrò, Andrea Bizzego, Stefano Cainelli, Cesare Furlanello, and Paola Venuti

Abstract Empathy, reciprocity and turn-taking are critical therapeutic targets in conditions of social impairment such as Autism Spectrum Disorder (ASD). These aspects are related to each other, converging into the construct of synchrony, which includes emotional, behavioural and, possibly, physiological components. Therefore, being able to quantify the synchrony could impact the way therapists adapt and maximise the efficacy of the interventions. However, current methods are based on the observational coding of behavior, which is time-consuming and usually only performed after the interaction is over. In this study we propose to apply Artificial Intelligence (AI) methods on physiological data in order to obtain a real time and objective quantification of synchrony. In particular, we introduce the Multi-Modal Synchrony dataset (M-MS), which includes 3 sources of information—electrocardiographic signals, video recordings and behavioral coding—to support the study of synchrony in ASD. As a first AI application, we are currently developing an unsupervised model to extract a multivariate embedding of the physiological data. The multivariate embedding has to be compared with the behavioral synchrony label to create a map of physiological and behavioural synchrony. The application of AI in the treatment of ASD may become a new asset for the clinical practice, especially if the possibility of providing real time feedback to the therapist is exploited.

G. Calabrò · A. Bizzego · S. Cainelli · P. Venuti (✉)
Department of Psychology and Cognitive Science, University of Trento, Rovereto, Italy
e-mail: paola.venuti@unitn.it

G. Calabrò
Fondazione Bruno Kessler, Trento, Italy

C. Furlanello
HK3 Lab,Rovereto, TN, Italy
e-mail: cesare.furlanello@gmail.com

A. Esposito et al. (eds.), *Progresses in Artificial Intelligence and Neural Systems*, Smart Innovation, Systems and Technologies 184,
https://doi.org/10.1007/978-981-15-5093-5_46

1 Introduction

Autism spectrum disorder (ASD) is an increasingly prevalent neurodevelopmental disorder with severe impacts on quality of life. The symptoms have a precocious onset and affect multiple areas, social impairment in particular being a core feature [1].

A pivotal form of social interaction is the dyadic one, which represents both a developmental tool available to the child [2, 3] and a key therapeutic channel available for ASD therapy. This one-on-one interaction has been viewed from different perspectives, with the construct of synchrony being one of the most relevant [4, 5]. Described in terms of mutuality, reciprocity, rhythmicity, harmonious interaction, turn-taking and shared affect [6], synchrony reflects ASD deficits and has been used to study development trajectories in children with ASD [7]. Being able to quantify and promote synchrony is therefore important to track and improve the efficacy of the intervention and guide the activity of the therapist [8]. However, observational procedures are time consuming, have a certain degree of subjectivity and often can be performed only after the intervention is concluded, thus limiting the usability of such information.

Leveraging on Artificial Intelligence (AI) methods to achieve an objective synchrony evaluation process and provide this information to the therapist in real time can offer a new asset to the clinical practice. There is already promising evidence of successful applications of AI methods to study ASD such as the detection of the occurrence of stereotypical motor movements using wearable accelerometers [9, 10] and the identification of meltdown-related behaviors from video recordings [11]. An AI approach to quantify the synchrony during social interaction could solve the problems related to subjectivity and temporal requirements, enabling an objective real time inference on the current state of the therapy.

However, accuracy and reliability of AI models rest on appropriate datasets, which are scarcely available in our scenario due to the requirement of being based on clinically valid information.

We therefore composed the Multi-Modal Synchrony dataset (M-MS), a new AI-ready dataset including physiological and behavioral data acquired in a naturalistic clinical setting during actual therapeutic activities involving children with ASD. Physiological data consist of electrocardiographic (ECG) signals, which are linked to the engagement in socio-cognitive tasks [12]. Behavioral data are acquired in the form of video recordings of the activities and synchrony labels based on observational coding of behavior performed by experts.

As a first AI application on this dataset, we selected an unsupervised deep learning method based on an auto-encoder architecture. This category of AI methods already proved to be a viable approach to automatically extract from physiological data a representation which is useful to detect the occurrence of events of interest [13]. For our purposes, this translates into the first step in exploring synchrony during social interaction in ASD using an automatically learned representation of physiological data.

2 Materials and Methods

The study was based on actual therapeutic activities carried out between the years 2017 and 2018 in Rovereto and Coredo (Trento, Italy). During all the activities, we simultaneously acquired the ECG signals both from the child and the therapist. The video recording of the therapy sessions became feasible for the activities carried out during the year 2018 thanks to technical improvements in the temporal alignment of the multi-modal data streams.

2.1 Participants

The study involved 19 male and 2 female children meeting the criteria for ASD diagnosis according to the diagnostic and statistical manual of mental disorders (DSM-5)[1] and aged 2–12 years old (average: 5.5 y, SD: 2.5 y). Although debated [14], the imbalance in the sample is in line with the long-standing different prevalence of ASD according to sex [15]. Each child interacted with the same male therapist, which had extensive experience in ASD treatment. The parents—or legal guardians—participated in a briefing about the experimental protocol and provided their informed consent to allow the participation of their child in the study.

2.2 Experimental Setting

We selected activities of music therapy due to its ability to effectively target synchrony aspects [16] regardless potential language impairments often present in children with ASD [1]. We gave priority to the clinical validity of the dataset aiming at the observation of actual therapy dynamics, thus we chose to not adopt a structured protocol. Therefore, the activities were performed without directives to the therapist in order to preserve the natural course and the therapeutic value of the intervention. This unrestricted approach allows its seamless extension to different therapeutic scenarios and to more ecological settings such as the child's home.

The child and the therapist interacted in a dedicated room containing music instruments and no common toys. Upon entering the room, the therapist allowed the child to get comfortable and acclimated to the setting. If the therapist considered the situation at risk of stressing the child, then no data were acquired. Otherwise, he proceeded to the placement of the sensors and carried out the activities while an operator managed the data acquisition platform.

2.3 Data Acquisition

During the activities, ECG signals were acquired both from the therapist and the child therefore providing 2 alignable time series. A critical factor in our experimental design was the adoption of sensorized garments specifically designed to reduce the risk of sensory overload which is a critical factor in ASD [17]. The prototypes were created in collaboration with a specialized company (ComfTech, Monza, Italy). The sensorized garments (see Fig. 1) consisted of a vest integrated with 2 textile electrodes with a conductive surface area in direct contact with the skin and a textile connector transmitting the signal to an acquisition device operating at 128 Hz.

During the activities carried out in 2018, the video stream of the therapy was acquired using a camera operating at 25 frame per second with a resolution of 1440 × 1080 pixels. The alignment of the ECG signals to the video stream was manually achieved by creating a marker event on the acquisition device and a corresponding visual marker on the video stream.

The videos were manually annotated by 2 trained experts applying a behavioural observation code appositely created to detect events relevant to synchrony. The range of annotated events spans from the presence of intentionality in the child's actions up to the full engagement in the interaction, allowing the quantification of the synchrony from the behavioural perspective.

In addition to behavioral and physiological data, we recorded the information about sex, age, DSM-5 diagnosis and severity of each participant. The data have been anonymized by substituting the names of the participants with alphanumeric codes and removing all the references to absolute timestamps.

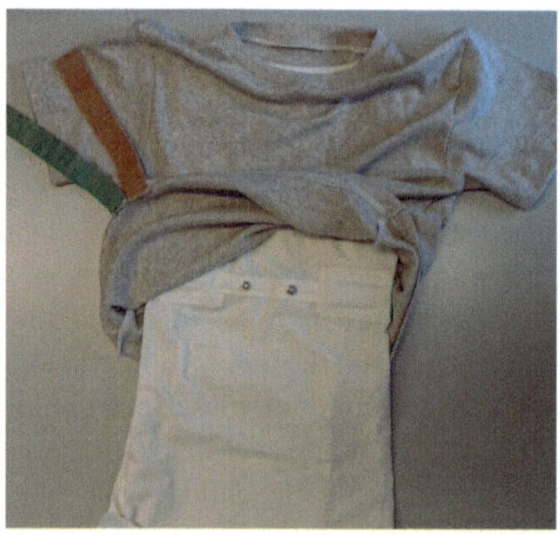

Fig. 1 Example of a prototype sensorized garment designed for children with ASD. The two metal buttons on the white jersey allow the connection of the textile electrodes with the sensing unit (not shown) in collaboration with ComfTech

2.4 Data Processing

To prepare the dataset for use with AI approaches we segmented the data into portions by windowing (length: 10 s; overlap: 9 s) and associated a label of synchrony derived from the behavioural coding of video recordings whenever available. The ECG signals of each portion were processed in order to automatically identify and remove portions presenting either movement artifacts or noise. The data processing procedure is presented in Fig. 2.

The ECG signal processing pipeline was implemented in Python and based on the *pyphysio* library [18].

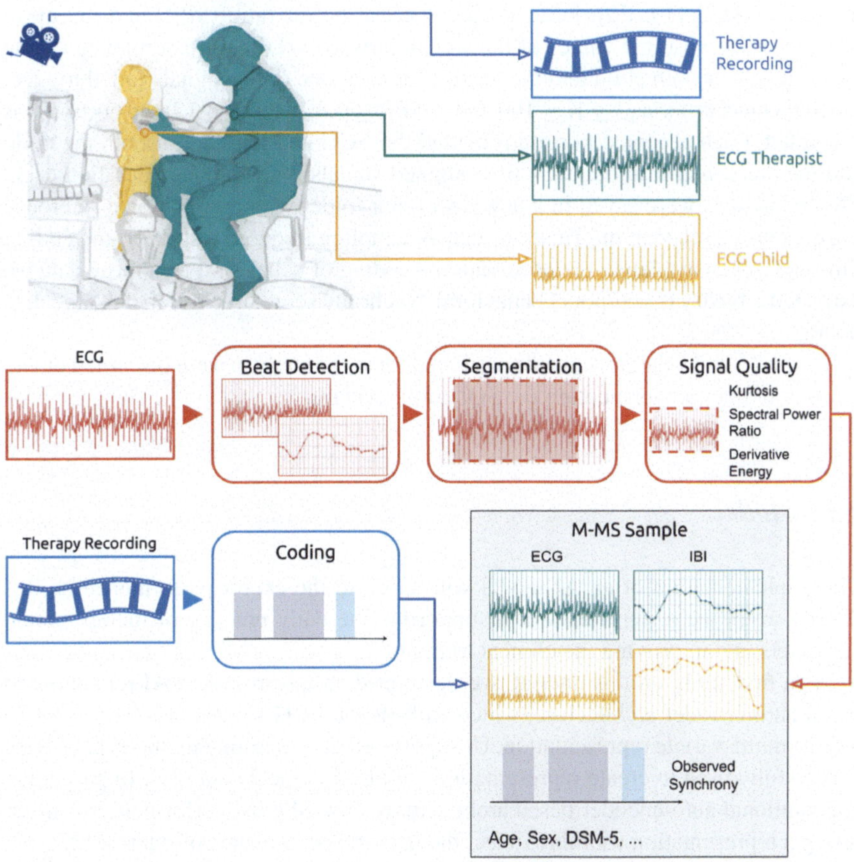

Fig. 2 Overview of the experimental setup (top) and data processing pipeline (bottom). Each sample of the M-MS dataset is composed of: the ECG and IBI segments from both the therapist and the child, the annotation of the observed synchrony and the session metadata.

The pipeline is composed of the following steps:

1. Pre-processing: the ECG signals were resampled by interpolation at 2048 Hz, then filtered with a low-pass infinite impulse response filter (pass-band: <50 Hz, stop-band: >55 Hz) to remove high-frequency noise and with a high-pass filter (pass-band: >0.5 Hz, stop-band: <0.05 Hz) to remove low-frequency drifts.
2. Beat detection: the R peaks in the ECG signals were detected to compute the Inter Beat Intervals (IBI); detection errors were automatically corrected and the IBI series were resampled by interpolation at 4 Hz.
3. Segmentation: ECG and IBI signals were segmented into overlapping consecutive portions (length: 10 s, overlap: 9 s).

The quality of the ECG data in each portion was assessed using 3 signal quality indicators (SQI) [19, 20]: kurtosis (K), spectral power ratio (SPR) and the mean squared of the absolute values of the signal derivative (DE) were computed on the ECG signals of both child and therapist. If at least one of the signals failed to meet the SQI cutoff criteria ($3 \leq K \leq 100$, $0.4 \leq SPR \leq 0.8$, $3 \leq DE \leq 12$) the portion was discarded. Each accepted portion provided the ECG and IBI data of both the child and the therapist for a total of 4 time aligned signals of 10 seconds duration each. These 4 signals were saved in a new data point folder with an additional reference file containing relevant information such as sampling frequency, starting and ending time and, when available, the observational coding of behavior. From these data we extracted a binary (good/poor) behavioral synchrony score used as label for the data point.

Of the 21 child-therapist dyads participating in the study, 6 were discarded due to the lack of data points meeting the SQI cutoff criteria.

2.5 Model

The implementation of an AI model with close to the expert performance on this dataset would be a step forward in supporting the daily practice of therapists and caregivers and improving the clinical training.

As a first study on this dataset, we developed an unsupervised AI model based on an auto-encoder architecture, which embeds the ECG signals of the dyad into a unique multi-variate representation. Unsupervised deep learning methods have been successfully used to create representations of physiological data [21]. In particular, convolutional auto-encoder based architectures showed good performance in structuring a representation useful to detect the occurrence of events of interest [22].

Our pilot architecture (see Fig. 3) consists of a convolutional auto-encoder taking as input the 2 IBI signals present in each data point and creating an embedding of 5 features that is then expanded to recover the original signals patterns. The values of the input IBI signals were normalized portionwise via rescaling between 0 and 1 based on maximum and minimum values. The architecture structure was implemented with *pyTorch* [23].

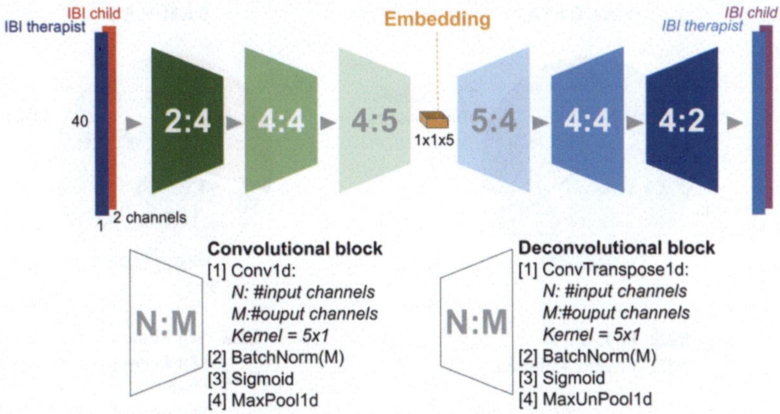

Fig. 3 Structure of the convolutional auto-encoder used to reconstruct the IBI segments of the therapist and the child. The structure is composed of two parts: the encoder (green convolutional blocks) and the decoder (blue de-convolutional blocks).

With this architecture we expect the embedding to include the relevant information about the physiological dynamics and mutual influence of the two input signals.

2.6 *Model Training*

The dataset was split into train and test partitions using a procedure akin to the leave-one-out cross-validation method. At each rotation, all data points from 1 child-therapist dyad were kept out and used as test partition to evaluate the performance on unseen data. For this reason, we call this approach leave-one-dyad-out cross-validation.

We performed the training of the network within 50 epochs using an optimizer based on the Adam algorithm [24] (learning rate: 0.0005) and a batch size of 16 samples. The training of the network was performed applying the mean squared errors (MSE) as loss function.

The performance of the network was then evaluated using the MSE on the test partition and comparing it to the MSE on the train partition in order to check for overfitting.

3 Results

Multi-modal data from a total of 41 sessions (578 min) were collected, with additional 22 sessions (396 min) with only the ECG signals. Two trained experts annotated the video recordings allowing the quantification of synchrony from the behavioural

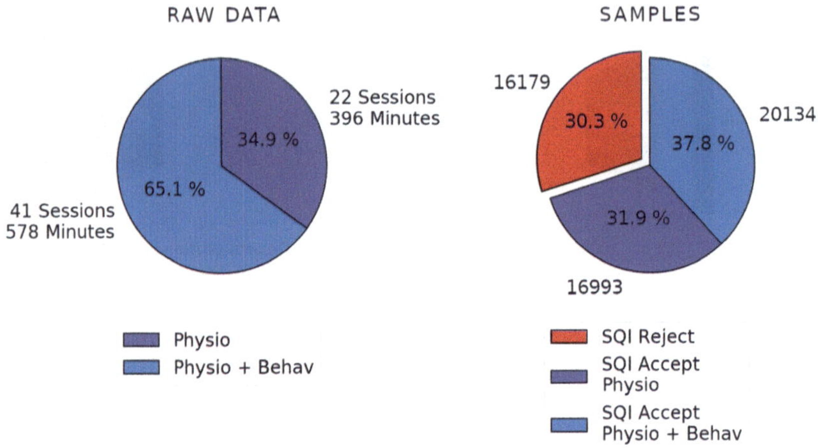

Fig. 4 Summary of the raw data acquired (left) and overview of the quality of the portions used to create the samples in the M-MS dataset (right).

perspective. In addition to the physiological and behavioural data, we added the information about sex, age, DSM-5 diagnosis and severity of each participant. The names of the participants were substituted with random alphanumeric codes and all the references to absolute timestamps were removed in order to anonymize the data.

We processed a total of 53306 portions, resulting in a dataset composed of 37127 data points of high quality physiological data, 20134 of which also provide the video data and the behavioral synchrony label (see Fig. 4).

The name of each data point incorporates the sequential numbering of the relative portion of raw data and the identifying number of the acquisition session. This makes the dataset modular: it is possible to use the information contained in the data point name to concatenate adjacent portions in order to reconstruct signals lasting more than 10 s, thus enabling the extraction of typical handcrafted time domain Heart Rate Variability (HRV) features.

Our convolutional auto-encoder architecture showed a reconstruction error of 5.4%. In particular, we evaluated the network performance using the leave-one-dyad-out cross-validation, obtaining a test MSE of 0.0535 and a train MSE of 0.0543.

4 Discussion

In this work we proposed the M-MS dataset, a new multi-modal dataset aiming at the exploration of physiological and behavioral synchrony during dyadic interaction between a child with ASD and a therapist. The information included are the ECG and IBI data of both the child and the therapist, the video stream of the interaction between them and a behavioral synchrony label based on expert judgement.

Due to the ecological approach in the selection of the experimental setting, the sample composition reflects the actual adhesions to the therapy programs and is in line with the gender imbalance in ASD prevalence present in literature. Nevertheless, the gender imbalance in the sample composition represents a limitation of the dataset which might be inadequate for gender-related studies.

The dataset creation was based on a sliding window procedure with 90% overlap. This form of data augmentation—which is a common practice in the AI field [25]—combined with the naming procedure of the data points makes the dataset modular: the researcher can manage the amount of overlapping information and reconstruct signals of the desired length, selectively enabling the extraction of typical handcrafted features such as HRV indices [26].

The M-MS dataset can be used to train supervised and unsupervised AI models in order to develop an objective and streamlined evaluation of synchrony which would enable an automated annotation of social interaction in ASD thus paving the way to real time feedback to the therapist. From the clinical perspective, this dataset can additionally be used to study the association between the physiological and behavioural patterns and phenotypes in ASD.

As a first application of AI methods to the dataset, we implemented a convolutional auto-encoder representing the child-therapist IBI signals in an embedding of 5 features with a reconstruction error of 5.4%. This preliminary evidence entails further study aiming at an unsupervised model able to extract latent variables from physiological data which would match the behavioral synchrony evaluated by the experts.

5 Data Availability

Due to the personal information included, the dataset is available for research purpose only and prior reasonable request to the corresponding author.

References

1. Association, A.P., et al.: Diagnostic and statistical manual of mental disorders (DSM-5®), Am. Psychiatr. Pub. (2013)
2. Cohen, D.: The developmental being. modeling a probabilistic approach to child development and psychopathology, Neuropsychiatrie de l'Enfance et de l'Adolescence **5**(60), S25–S26 (2012)
3. Azhari, A., Leck, W., Gabrieli, G., Bizzego, A., Rigo, P., Setoh, P., Bornstein, M.H., Esposito, G.: Parenting stress undermines mother-child brain-to-brain synchrony: a hyperscanning study. Sci. Rep. **9**(1), 1–9 (2019)
4. Harrist, A.W., Waugh, R.M.: Dyadic synchrony: its structure and function in children's development. Dev. Rev. **22**(4), 555–592 (2002)

5. Bizzego, A., Azhari, A., Campostrini, N., Truzzi, A., Ng, L.Y., Gabrieli, G., Bornstein, M.H., Setoh, P., Esposito, G.: Strangers, friends, and lovers show different physiological synchrony in different emotional states. Behav. Sci. **10**(1), 11 (2020)
6. Leclère, C., Viaux, S., Avril, M., Achard, C., Chetouani, M., Missonnier, S., Cohen, D.: Why synchrony matters during mother-child interactions: a systematic review. PloS ONE **9**(12), e113571 (2014)
7. Siller, M., Sigman, M.: The behaviors of parents of children with autism predict the subsequent development of their children's communication. J. Autism Dev. Disord. **32**(2), 77–89 (2002)
8. Venuti, P., Bentenuto, A., Cainelli, S., Landi, I., Suvini, F., Tancredi, R., Igliozzi, R., Muratori, F.: A joint behavioral and emotive analysis of synchrony in music therapy of children with autism spectrum disorders. Health Psychol. Rep. **5**(2), 162–172 (2017)
9. Rad, N.M., Kia, S.M., Zarbo, C., van Laarhoven, T., Jurman, G., Venuti, P., Marchiori, E., Furlanello, C.: Deep learning for automatic stereotypical motor movement detection using wearable sensors in autism spectrum disorders. Signal Process. **144**, 180–191 (2018)
10. Rad, N.M., Bizzego, A., Kia, S.M., Jurman, G., Venuti, P., Furlanello, C.: Convolutional neural network for stereotypical motor movement detection in autism, arXiv preprint arXiv:1511.01865
11. Patnam, V.S.P., George, F.T., George, K., Verma, A.: Deep learning based recognition of meltdown in autistic kids. In: 2017 IEEE International Conference on Healthcare Informatics (ICHI), pp. 391–396. IEEE (2017)
12. Palma, S.D., Tonacci, A., Narzisi, A., Domenici, C., Pioggia, G., Muratori, F., Billeci, L.: Monitoring of autonomic response to sociocognitive tasks during treatment in children with autism spectrum disorders by wearable technologies: a feasibility study. Comput. Biol. Med. **85**, 143–152 (2017)
13. Supratak, A., Li, L., Guo, Y., Feature extraction with stacked autoencoders for epileptic seizure detection. In: 36th Annual International Conference of the IEEE Engineering in Medicine and Biology Society, vol. 2014, pp. 4184–4187. (2014)
14. Constantino, J.N.: Taking stock of critical clues to understanding sex differences in the prevalence and recurrence of autism. Autism **21**(6), 769–771 (2017)
15. Werling, D.M., Geschwind, D.H.: Sex differences in autism spectrum disorders. Curr. Opin. Neurol. **26**(2), 146 (2013)
16. Kim, J., Wigram, T., Gold, C.: Emotional, motivational and interpersonal responsiveness of children with autism in improvisational music therapy. Autism **13**(4), 389–409 (2009)
17. Wiggins, L.D., Robins, D.L., Bakeman, R., Adamson, L.B.: Breif report: sensory abnormalities as distinguishing symptoms of autism spectrum disorders in young children. J. Autism Dev. Disord. **39**(7), 1087–1091 (2009)
18. Bizzego, A., Battisti, A., Gabrieli, G., Esposito, G., Furlanello, C.: Pyphysio: a physiological signal processing library for data science approaches in physiology. SoftwareX **10**, 100287 (2019)
19. Behar, J., Oster, J., Li, Q., Clifford, G.D.: ECG signal quality during arrhythmia and its application to false alarm reduction. IEEE Trans. Biomed. Eng. **60**(6), 1660–1666 (2013)
20. Li, Q., Mark, R.G., Clifford, G.D.: Robust heart rate estimation from multiple asynchronous noisy sources using signal quality indices and a kalman filter. Physiol. Meas. **29**(1), 15 (2007)
21. Yuan, Y., Xun, G., Jia, K., Zhang, A.: A multi-view deep learning method for epileptic seizure detection using short-time fourier transform. In: Proceedings of the 8th ACM International Conference on Bioinformatics, Computational Biology, and Health Informatics, pp. 213–222. ACM (2017)
22. Yuan, Y., Xun, G., Jia, K., Zhang, A.: A multi-view deep learning framework for EEG seizure detection. IEEE J. Biomed. Health Inform. **23**(1), 83–94 (2019)
23. Paszke, A., Gross, S., Massa, F., Lerer, A., Bradbury, J., Chanan, G., Killeen, T., Lin, Z., Gimelshein, N., Antiga, L., et al.: PyTorch: an imperative style, high-performance deep learning library. In: Advances in Neural Information Processing Systems, pp. 8024–8035 (2019)

24. Kingma, D.P., Ba, J.: Adam: a method for stochastic optimization. arXiv preprint arXiv:1412.6980
25. Perez, L., Wang, J.: The effectiveness of data augmentation in image classification using deep learning. arXiv preprint arXiv:1712.04621
26. Rickards, C.A., Ryan, K.L., Convertino, V.A.: Characterization of common measures of heart period variability in healthy human subjects: implications for patient monitoring. J. Clin. Monit. Comput. **24**(1), 61–70 (2010)

Computational Methods for the Assessment of Empathic Synchrony

Andrea Bizzego, Giulio Gabrieli, Atiqah Azhari, Peipei Setoh, and Gianluca Esposito

Abstract The synchronization of physiological signals between persons is a well-known proxy of empathy. However, it is also influenced by physiological and environmental factors that should be discriminated to correctly characterize the empathy component. We discuss a framework to compute synchrony and introduce *physynch*, an open-source package developed to easily replicate its computational procedures. We adopted *physynch* to study the synchrony of the electrodermal activity in 61 male-female dyads with different types of relationship: strangers (18 dyads), friends (23 dyads) and lovers (20 dyads). The findings confirm previous results on Heart Rate Variability synchrony and suggest that synchrony is influenced by the type of relationship and is stronger in dyads of strangers. *physynch* is made available for download and use for researchers interested in measuring physiological synchrony.

1 Introduction

Empathy between individuals is modulated by both the influence of behavioral reciprocity (e.g. facial expressions [23]) and autonomic physiological response (e.g. heart rate pattern) [4, 15]. Empathy is a key construct in the development of prosocial behavior [9, 27], with effects on cognition [15].

Although fundamental to the study of social behaviour, the assessment of empathy is hurdled by difficulties of its quantification. One of the main approaches is the study of synchrony. Synchrony, defined as the "dynamic and reciprocal adap-

A. Bizzego (✉) · G. Esposito
Department of Psychology and Cognitive Science, University of Trento, Trento, Italy
e-mail: andrea.bizzego@unitn.it

G. Gabrieli · A. Azhari · P. Setoh · G. Esposito
Psychology Program, School of Social Sciences, Nanyang Technological University, Jurong West, Singapore

G. Esposito
Lee Kong Chian School of Medicine, Nanyang Technological University, Novena, Singapore

A. Esposito et al. (eds.), *Progresses in Artificial Intelligence and Neural Systems*, Smart Innovation, Systems and Technologies 184,
https://doi.org/10.1007/978-981-15-5093-5_47

tation of the temporal structure of behaviors and shared affect between interactive partners" [18], is one of the ways in which empathy is manifested and studied [28]. From a computational point of view, we could say that synchrony is the expression of empathy in the time domain, and, as such, can be studied through the analysis of time-related measures, or signals. Typically, studies focus on the synchrony of physiological signals, which are a proxy of autonomic responses [20].

Synchrony first manifests within the context of the mother-infant relationship [1, 11], but instances of synchrony have also been demonstrated in adult dyads [2]. For instance, hormonal association in romantic couples has been linked to empathy [25] and of connectedness [22]; in dyads of strangers the physiological synchrony was correlated with more efficiency in collaborative tasks [8, 23] and cooperation [21].

In this study, we aim at providing general computational guidelines to study empathic synchrony. We discuss the framework proposed by Golland and colleagues [13, 14] to discriminate between different sources of physiological synchrony. Finally, we introduce `physync`, a Python package that implements Golland's framework. We demonstrate the approach on a dataset of electrodermal (EDA) signals to study the effects of emotions and type of relationship on synchrony in male-female dyads.

2 Computational Methods to Assess Empathy

The first issue in studying physiological synchrony to assess empathy is the choosing of an appropriate mathematical framework of reference. Several types of mathematical frameworks have been used in the literature to quantify synchrony between physiological signals, such as cross correlation [13], recurrence plot analysis [21], wavelet coherence [3], Granger causality [19] and others.

The choice of framework should not be arbitrary as each framework has been developed with a specific aim and reference theory. The researcher should choose based on an understanding of the framework's underlying theory, implicit restrictions (e.g. minimum number of samples) and assumptions (e.g. normality of the data). In addition, the computational parameters should be adequately set after a full understanding of their meaning and implications on the results.

The second issue is associated to the need to discriminate between different sources of synchrony. In general, information carried by physiological signals originate from several sources:

1. *Physiological homeostasis*: effect of different functions simultaneously fulfilled by the human body to remain healthy and efficient (e.g. the thermal regulation, the metabolic activity, etc);
2. *Autonomic response*: the Autonomic Nervous System (ANS) controls the functions of bodily organs to enable them to react to external factors (e.g. fight-flight response). In a psychology experiment on synchrony, the external factors are set

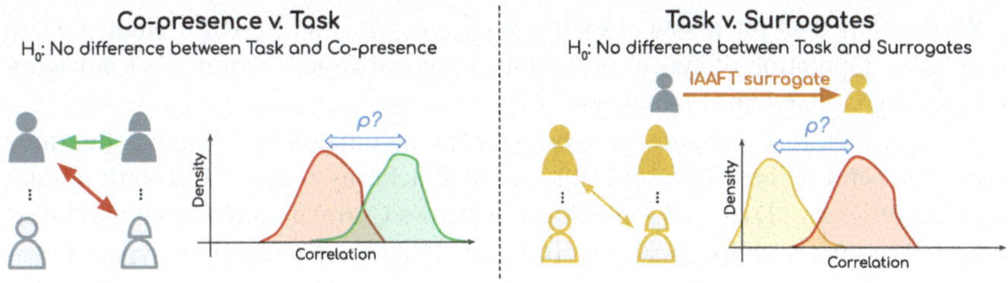

Fig. 1 Golland's framework to discriminate between sources of synchrony in the study of empathy

by the design of the experiment which could be tasks to be accomplished or a series of stimuli;

3. *Empathy*: being in the presence of one or more individual(s) might affect the way the ANS modulates physiological responses to the stimuli [12].

When studying empathy, it is important to discriminate the different sources that contribute the observed physiological synchrony. In particular, the main concern is in the discrimination between the effects of experimental factors and of empathy. Variations due to physiological homeostasis have a low frequency dynamic and are not expected to influence the measures during short term experiments.

To this aim, Golland and colleagues [13, 14], who studied the effect of co-presence on the synchronization of autonomic signals, proposed the computation of three types of synchrony (see Fig. 1):

- Co-presence synchrony: between components of the same dyad; influenced by both emotional empathy and by effects induced by the experimental task;
- Task-driven synchrony: between components of different dyads; thus, influenced only by the experimental task;
- Surrogate synchrony: between surrogate signals, and, thus, only due to random factors. Surrogate signals are generated by application of Iterative Amplitude Adjusted Fourier Transform (IAAFT) algorithm [26].

The distributions of the three types of synchrony measures are then compared by means of appropriate statistical tests to assess the significance of the effects of the stimuli compared to random factors, and then of the effects of co-presence compared to stimuli. A significant difference between surrogate synchrony and control synchrony serves to reject the null hypothesis that the stimuli have no effects on the physiological response. A significant difference between control and co-presence synchrony serves to reject the null hypothesis that there are no effects due to empathy. Usually, it is expected that both stimuli and empathy increase physiological synchrony, and this should be checked in the post-hoc analysis of the median values of the three distributions. For synchrony metrics, Golland and colleagues used the maximum value of the cross-correlation within a time lag of 10 seconds which allows for delays in the synchrony of the physiological signals. However, the proposed framework is independent of the synchrony metrics adopted.

We developed an open-source Python package, *physynch*[1] (v0.1, Python 3.7), to allow the computation of synchrony of physiological signals within the framework proposed by Golland and colleagues.

Its main function, *compute_golland_synch()*, automatically computes the three types of synchrony based on several types of synchrony metrics. Currently implemented metrics are: Dynamic Time Warping, Cross Correlation (CC), Mutual Information, Pearson's and Spearman's Correlation. In addition to the three types of synchrony, it also computes inter-group synchrony: given dyads composed of individuals coming from two categories (e.g. males-females, mothers-children), inter-group synchrony is computed between all the individuals of the same category, thus measuring the effect of the stimuli on the category.

3 Materials and Methods

The computational methods discussed in the previous section are demonstrated on physiological signals collected during an experiment to study the effects of emotions and type of relationship on synchrony in male-female dyads (see Fig. 2).

We studied empathy in 62 male-female dyads divided into 3 groups: Strangers (S, N = 18), Friends (F, N = 23) and Lovers (L, N = 20). The average ages for the 62 female and 62 male participants were 21.65 (SD = 2.77) and 23.48 years old (SD = 5.57), respectively. Participants were distributed in 62 opposite-sex pairs. Participants were recruited among the students of the University of Trento, Italy; all participants were Caucasian, and none presented with any medical or developmental condition. They were required to provide informed consent before the commencement of the study which was conducted in accordance with the Declaration of Helsinki. Every dyad was exposed to six 4-minutes-long video clips, selected to elicit six different emotions in this fixed order: embarrassment, sadness, fear, calmness, love, pride. The six clips were selected from a set of 20 in a pilot study, in which 10 participants associated each video to a unique emotion. The six videos with most consistent elicited emotions were selected. We acquired the ECG and EDA signal (2048 Hz, FlexComp, Thought Technology) from each subject. In this study, we focused on the EDA, from which we extracted the phasic response (PHA) (see Fig. 3), which is linked to the sympathetic activity of the ANS [5]. Before each stimulus, a 10-sec image depicting the title of the video clip on a white background was presented. Between two stimuli, an interval of 1 min with a white fixation point on a green background allowed the participants to return to baseline state. At the end of the last clip, participants were asked to self report three items on a 7-point Likert scale on whether each video was unpleasant/pleasant, scary/funny, embarrassing/non-embarrassing, then a set of instructions would appear on the screen to inform participants that the session had ended. The experimental session lasted 30 min in total.

[1] https://gitlab.com/abp-san-public/physynch.

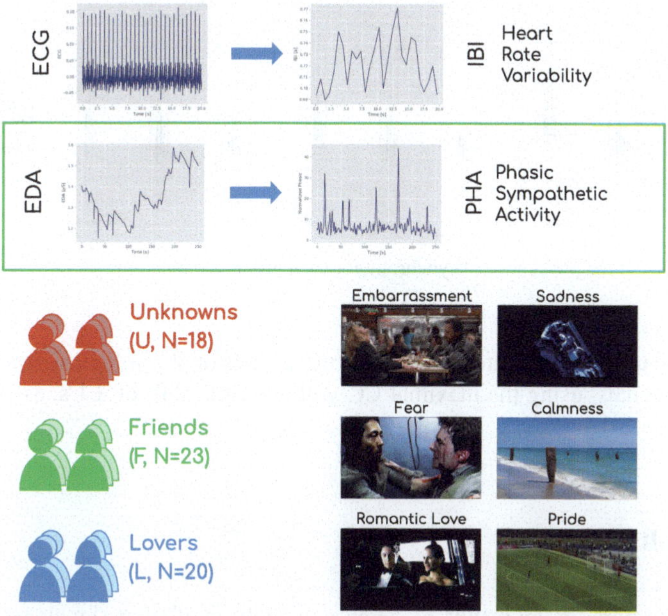

Fig. 2 Representation of the experimental design. Top: two physiological signals were collected: ECG and EDA; the latter is analysed in this study. Bottom: two experimental factors are investigated namely the type of relationship between dyadic members and the type of emotion. Six frames from the six video stimuli are shown

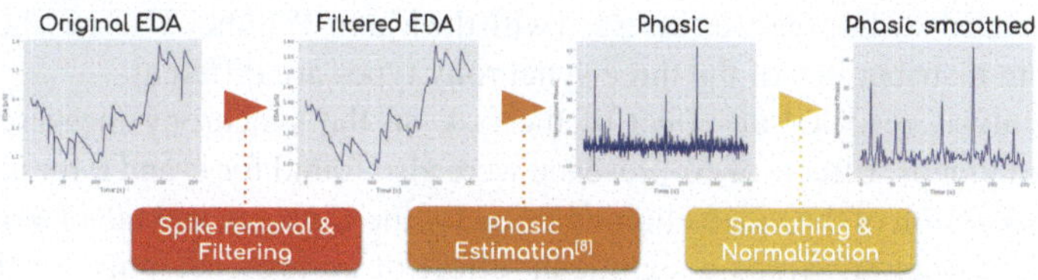

Fig. 3 Signal processing pipeline to extract the Phasic response associated to the sympatethic activity

The EDA signal was first downsampled to 8 Hz, then we removed the spikes and applied a low-pass Infinite Impulse Response filter (pass-band: <2 Hz, stop-band >3.5 Hz) to remove high frequency noise. The filtered signal was deconvolved using a Bateman function, and the phasic component was estimated, as described in [5]. The local range of the phasic component was then computed from consecutive non-overlapping portions (length 1 s), resampled at 2 Hz, and used as the continuous index of sympathetic activity (see a representation of the pipeline with the results of the intermediate steps in Fig. 3). The processing of the signals was performed with custom scripts based on pyphysio [7] (v2.1, Python 3.7).

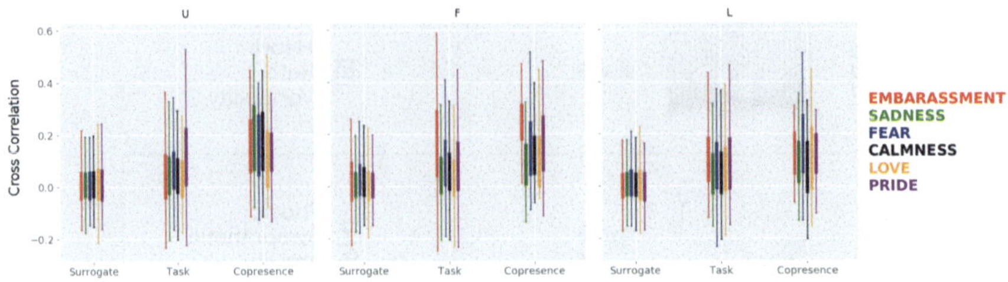

Fig. 4 Values of the physiological synchrony for each group and emotion

We then used *physynch* to compute the three types of physiological synchrony on the PHA signals, using the maximal CC within a time shift of ± 1 s, as synchrony metrics [13, 14].

4 Results

We focused here on the results of synchrony analysis of EDA signals (see Fig. 4). The results of the analysis of the self-assessment of the perceived emotions, the synchrony of the ECG signal and the effect of gender and type of relationship are part of another study which is separately reported.

The significance of the differences between the three types of synchrony (see Table 1) in the distributions was assessed with the Mann-Whitney U test as the number of values in distributions of the three synchrony types are different.

The analysis revealed an effect of the task on the synchrony measures for all groups and videos. Effects of co-presence were also found for U and F for all videos, with the exception of the video eliciting pride for the U group, and the video eliciting fear for F. Lovers presented a significant effect of co-presence only for the video eliciting fear.

5 Discussion

Intuitively, we would have expected a stronger co-presence effect where there is an existing social relationship, but this is contradicted by the results on Lovers where the effect of co-presence only emerged in the fear condition. Notably, these findings confirm the results of a previous study on the same dataset which focused on the synchrony of Heart Rate Variability [6].

The results on co-presence effects may seem counter-intuitive; however numerous examples showed that the synchronization of sympathetic response among strangers is common [16, 29] and have been associated to the initiation of affiliation [24] and facilitation of prosocial behaviors [27]. In a recent study that investigated pairs of

Table 1 Summary of the computed synchrony measures and significance of the comparison between the three types of synchrony

Group	Emotion	Surrogate mean (SD)	VS	Task mean (SD)	VS	Co-presence mean (SD)
Unknowns	Embarrass	0.0102 (0.0842)	p < 0.01	0.0620 (0.1570)	p < 0.01	0.1719 (0.1972)
	Sad	0.0138 (0.0859)	p < 0.01	0.0481 (0.1162)	p < 0.01	0.1695 (0.1743)
	Fear	0.0114 (0.0820)	p < 0.001	0.0809 (0.1378)	p < 0.05	0.1607 (0.2005)
	Calmness	0.0156 (0.0865)	p < 0.001	0.0738 (0.1688)	p < 0.01	0.1868 (0.2100)
	Romance	0.0230 (0.1084)	p < 0.05	0.0530 (0.1381)	p < 0.05	0.1414 (0.1855)
	Pride	0.0162 (0.0966)	p < 0.001	0.1205 (0.1531)	p = 0.24	0.1351 (0.1330)
Friends	Embarrass	0.0186 (0.1045)	p < 0.001	0.1709 (0.1778)	p < 0.05	0.2496 (0.1441)
	Sad	0.0091 (0.0777)	p < 0.001	0.0528 (0.1206)	p < 0.05	0.1246 (0.2053)
	Fear	0.0256 (0.0860)	p < 0.001	0.0954 (0.1299)	p = 0.11	0.1394 (0.1501)
	Calmness	0.0230 (0.0886)	p < 0.001	0.0758 (0.1576)	p < 0.05	0.1486 (0.1905)
	Romance	0.0081 (0.0905)	p < 0.001	0.0467 (0.1129)	p < 0.05	0.1181 (0.1612)
	Pride	0.0142 (0.0894)	p < 0.001	0.0918 (0.1347)	p < 0.05	0.1617 (0.1567)
Lovers	Embarrass	0.0066 (0.0802)	p < 0.001	0.1189 (0.1336)	p = 0.36	0.1313 (0.1324)
	Sad	0.0111 (0.0844)	p < 0.001	0.0573 (0.1279)	p = 0.14	0.1089 (0.1875)
	Fear	0.0147 (0.0820)	p < 0.001	0.0786 (0.1389)	p < 0.01	0.1694 (0.1604)
	Calmness	0.0172 (0.0928)	p < 0.001	0.0640 (0.1467)	p = 0.21	0.1047 (0.2108)
	Romance	0.0139 (0.0913)	p < 0.001	0.0875 (0.1544)	p = 0.16	0.1365 (0.2029)
	Pride	0.0014 (0.0852)	p < 0.001	0.0936 (0.1330)	p = 0.58	0.1001 (0.1089)

strangers in competitive, cooperative and naturalistic (i.e. conversational interaction) social contexts, [10] revealed that strangers established physiological synchrony in both the sympathetic and parasympathetic branches of the autonomic nervous system across all social contexts.

These results strongly corroborate with the findings from our present study in which greater physiological synchrony is observed in less substantiated relationships (i.e. pairs of strangers). Overall, physiological synchrony between strangers are postulated to accompany perceptions of similarity which serve to lay the foundation for future social affiliations. Since Lovers have ideally established their relationship with each other, physiological synchrony may not necessarily serve to enhance social"connectedness" within this group. A limitation of this study is that we did not consider the duration and quality of the romantic relationship, and the particular age range of the participants, which may or may not have fallen within the active partner-seeking phase and thus have differing implications [17].

Taken together, these results reveal the importance of adopting an appropriate approach to avoid errors in the computation of physiological synchrony. It is only due to a suitable adopted framework that we could evince the different effects of co-presence in Lovers. Otherwise, we could have mistakenly assumed that Lovers showed synchrony of physiological response, without being able to observe that this was mainly due to the task, and not to the fact that the dyad was watching the video stimuli together.

6 Conclusion

When quantifying empathy through analysis of physiological synchrony, it is critical to account for the different sources of physiological variability. In this paper, we aimed to highlight the framework proposed by Golland and colleagues as an elegant approach to statistically discriminate and assess both the effects of task and co-presence in dyads. To allow other researchers to easily apply the framework, we developed the Python package *physynch*, and demonstrated its use in a study which investigates the effects of the type of relationship and emotion on the synchrony of the sympathetic response. The results showed that the task of watching the same video has an effect on the synchronization of the sympathetic response. However, only in strangers and friends we observe the effect of co-presence.

We encourage the adoption of the proposed framework, made easier by the *physynch* package, to analyse synchrony in different experimental studies. Its use is not limited by the type of signal or synchrony metrics applied and can be therefore adapted to different research questions and contexts.

References

1. Azhari, A., Leck, W., Gabrieli, G., Bizzego, A., Rigo, P., Setoh, P., Bornstein, M.H., Esposito, G.: Parenting stress undermines mother-child brain-to-brain synchrony: A hyperscanning study. Sci. Rep. **9**(1), 1–9 (2019)
2. Azhari, A., Lim, M., Bizzego, A., Gabrieli, G., Bornstein, M.H., Esposito, G.: Physical presence of spouse enhances brain-to-brain synchrony in co-parenting couples. Sci. Rep. **10**, 75–69 (2020) https://doi.org/10.1038/s41598-020-63596-2
3. Baker, J.M., Liu, N., Cui, X., Vrticka, P., Saggar, M., Hosseini, S.H., Reiss, A.L.: Sex differences in neural and behavioral signatures of cooperation revealed by fnirs hyperscanning. Sci. Rep. **6**(26), 492 (2016)
4. Barsade, S.G.: The ripple effect: Emotional contagion and its influence on group behavior. Adm. Sci. Q. **47**(4), 644–675 (2002)
5. Benedek, M., Kaernbach, C.: A continuous measure of phasic electrodermal activity. J. Neurosci. Methods **190**(1), 80–91 (2010)
6. Bizzego, A., Azhari, A., Campostrini, N., Truzzi, A., Ng, L.Y., Gabrieli, G., Bornstein, M.H., Setoh, P., Esposito, G.: Strangers, friends, and lovers show different physiological synchrony in different emotional states. Behav. Sci. **10**(1), 11 (2020)
7. Bizzego, A., Battisti, A., Gabrieli, G., Esposito, G., Furlanello, C.: Pyphysio: A physiological signal processing library for data science approaches in physiology. SoftwareX **10**(100), 287 (2019)
8. Chikersal, P., Tomprou, M., Kim, Y.J., Woolley, A.W., Dabbish, L.: Deep structures of collaboration: physiological correlates of collective intelligence and group satisfaction. In: Proceedings of the 2017 ACM Conference on Computer Supported Cooperative Work and Social Computing, pp. 873–888. ACM (2017)
9. Cirelli, L.K.: How interpersonal synchrony facilitates early prosocial behavior. Curr. Opin. Psychol. **20**, 35–39 (2018)
10. Danyluck, C., Page-Gould, E.: Social and physiological context can affect the meaning of physiological synchrony. Sci. Rep. **9**(1), 8222 (2019)
11. Feldman, R.: Bio-behavioral synchrony: A model for integrating biological and microsocial behavioral processes in the study of parenting. Parenting **12**(2–3), 154–164 (2012)
12. Gallotti, M., Frith., C.D.: Social cognition in the we-mode. Trends Cogn. Sci. **17**(4), 16065 (2013)
13. Golland, Y., Arzouan, Y., Levit-Binnun, N.: The mere co-presence: Synchronization of autonomic signals and emotional responses across co-present individuals not engaged in direct interaction. PloS One **10**(5), e0125–804 (2015)
14. Golland, Y., Keissar, K., Levit-Binnun, N.: Studying the dynamics of autonomic activity during emotional experience. Psychophysiology **51**(11), 1101–1111 (2014)
15. Hatfield, E., Cacioppo, J.T., Rapson, R.L.: Emotional contagion. Curr. Dir. Psychol. Sci. **2**(3), 96–100 (1993)
16. Jong, J., Whitehouse, H., Kavanagh, C., Lane, J.: Shared negative experiences lead to identity fusion via personal reflection. PloS One **10**(12), e0145–611 (2015)
17. Kacerguis, M.A., Adams, G.R.: Erikson stage resolution: The relationship between identity and intimacy. J. Youth Adoles. **9**(2), 117–126 (1980)
18. Leclère, C., Viaux, S., Avril, M., Achard, C., Chetouani, M., Missonnier, S., Cohen, D.: Why synchrony matters during mother-child interactions: A systematic review. PloS One **9**(12), e113–571 (2014)
19. Leong, V., Byrne, E., Clackson, K., Georgieva, S., Lam, S., Wass, S.: Speaker gaze increases information coupling between infant and adult brains. Proc. Natl. Acad. Sci. **114**(50), 13290–13295 (2017)
20. Levenson, R.W.: The autonomic nervous system and emotion. Emot. Rev. **6**(2), 100–112 (2014)
21. Mønster, D., Håkonsson, D.D., Eskildsen, J.K., Wallot, S.: Physiological evidence of interpersonal dynamics in a cooperative production task. Physiol. Behav. **156**, 24–34 (2016)

22. Papp, L.M., Pendry, P., Simon, C.D., Adam, E.K.: Spouses' cortisol associations and moderators: Testing physiological synchrony and connectedness in everyday life. Family Process **52**(2), 284–298 (2013)
23. Park, S., Choi, S.J., Mun, S., Whang, M.: Measurement of emotional contagion using synchronization of heart rhythm pattern between two persons: Application to sales managers and sales force synchronization. Physiol. Behav. **200**, 148–158 (2019)
24. Reddish, P., Fischer, R., Bulbulia, J.: Lets dance together: synchrony, shared intentionality and cooperation. PloS One **8**(8), e71–182 (2013)
25. Schneiderman, I., Kanat-Maymon, Y., Zagoory-Sharon, O., Feldman, R.: Mutual influences between partners hormones shape conflict dialog and relationship duration at the initiation of romantic love. Soc. Neurosci. **9**(4), 337–351 (2014)
26. Schreiber, T., Schmitz, A.: Surrogate time series. Phys. D **142**(3–4), 346–382 (2000)
27. Shilling, C., Mellor, P.A.: Durkheim, morality and modernity: collective effervescence, homo duplex and the sources of moral action. Br. J. Sociol. 193–209 (1998)
28. Timmons, A.C., Margolin, G., Saxbe, D.E.: Physiological linkage in couples and its implications for individual and interpersonal functioning: A literature review. J. Family Psychol. **29**(5), 720 (2015)
29. Whitehouse, H., Lanman, J.A., Downey, G., Fredman, L.A., Swann Jr., W.B., Lende, D.H., McCauley, R.N., Shankland, D., Stausberg, M., Xygalatas, D., et al.: The ties that bind us: Ritual, fusion, and identification. Curr. Anthropol. **55**(6), 000–000 (2014)

The Structuring of the Self Through Relational Patterns of Movement Using Data from the Microsoft Kinect 2 to Study Baby-Caregiver Interaction

Alfonso Davide Di Sarno, Teresa Longobardi, Enrico Moretto, Giuseppina Di Leva, Irene Fabbricino, Lucia Luciana Mosca, Valeria Cioffi, and Raffaele Sperandeo

Abstract In this paper we will illustrate the progress of a research that intends to use computational methods to analyze the complex interaction between a child and his caregiver, focusing on motor aspects. According to the theory of Ruella Frank, Gestalt psychotherapist, the child acquires motor patterns through the interaction with the caregiver that will be repeated over the course of life. Six fundamental movements have been identified and defined, which allow the child to develop self-awareness. The objective of the study is the creation of an observation tool of dyadic interaction through the automatic recognition of these movements whose purpose is the identification and differentiation of the peculiar vocabulary of movements that is established between parent and child. Thanks to the use of Kinect v2 and through the Kinect Studio program, we have trained a network to identify three of the six basic movements: "Reaching", "Pushing" and "Pulling". The results of this training show that the machine discriminates the gestures that have been defined and recognized through different features, although they visually share a part of the progression. The future perspectives of this study foresee to apply the methodology described in this paper to the other three gestures, widen the database and refine the tagging process, to obtain more precise results. The ultimate aim is to include the database in an application that we have proposed to develop, useful for the recognition of the peculiar dyadic baby-caregiver movements.

1 Introduction

Through the observation of the movement in the human being, we can affirm that gestures and actions carried out in the relational context constitute the most significant and obvious expression of the complexity of a person's experience. The

A. D. Di Sarno (✉) · T. Longobardi · E. Moretto · G. Di Leva · I. Fabbricino · L. L. Mosca · V. Cioffi · R. Sperandeo
SiPGI—Postgraduate School of Integrated Gestalt Psychotherapy, Torre Annunziata, Naples, Italy
e-mail: a.davidedisarno@gmail.com

A. Esposito et al. (eds.), *Progresses in Artificial Intelligence and Neural Systems*, Smart Innovation, Systems and Technologies 184,
https://doi.org/10.1007/978-981-15-5093-5_48

movements make the cognitive, motor and psychic characteristics manifest, at the same time recalling the interest of an observer. Modern research on the development of the psyche in children sets the movement and the possibility of manipulating its environment at the base of the birth of the sense of self, of the structuring of the personality [1–4] and of the subsequent, and more complex, forms of self-awareness [5–8]. In this sense, the interaction between the child and a significant adult can be described as the first development environment, in which a state of continuous interchange and mutual influence between the two subjects is created [9]. In this interaction, the dyad baby-caregiver contributes to the structuring of the physiological, behavioral and emotional regulation [10]: the psyche of the child and that of the adult, as open systems, tend to seek a situation of balance through the integration between self-regulation and interactive regulation [11–14]. The principles that organize childhood representations and internalization are continuous adjustments, disconnection and repair, and emotionally charged moments [15]. Newborns begin life with the ability to regulate both endogenous states and environmental stimuli [16, 17]. The constitutional and temperamental differences condition the first self-regulation capacities. The self-regulation of the child concerns the regulation of the arousal, the maintenance of the alert state, the ability to lower the excitement in case of excessive stimulation, the ability to inhibit behavioral expressions and the ability to develop predictable behavior cycles. It includes variations in response readiness and clarity of signals [18]. Touching, looking away and narrowing the range of facial expressions are examples of self-regulating strategies to lower arousal. In childhood and in adulthood, self-regulation is an index of the ability to pay attention and engage in the environment [19]. In the interactive process each partner influences the other moment by moment [14, 20–23]. Tronick [15] has shown that self-regulation and interactive regulation are closely related and that children who regulate themselves better during the motionless face experiment have better interactive regulation with their mother during the game. Individual differences in endogenous and social capacities influence both the child's self-regulation and interactive regulation. The differences in the ability of self-regulation condition success in interactive regulation; conversely, differences in the nature of interactive regulation can facilitate self-regulation or interfere with it [6, 15].

We propose to use computational methods to analyze the complex interaction between mother and child, using the theory formulated by Ruella Frank as the reading key of this dynamic [24, 25]. According to the author the child structures with the parent a non-verbal vocabulary made up of six fundamental movements, developing self-awareness by experiencing one's own body in relation to that of the other and with respect to the surrounding environment. The movements identified by the author are: "yielding with", "pushing against", "reaching for", "grasping on to", "pulling toward" and "releasing from".

The "Pushing against" refer to a movement of thrust that goes from the periphery of the body (head, hands, feet, coccyx, pelvis) to its center, favoring a density experience. When the person adapts herself to the tensions of the surface that he is pushing, also the surface is adapted to the person which will feel like a thrust in return: there is a mutual adaptation. So, while a part of the body pushes against a surface in order to find stability, the other is free to move, to explore. “Reaching for” refers to the movement of lengthening and exploration of the body in the space in which it is immersed, in his own kinesphere. “Pulling toward” ago the reference to the action of attracting to if this that is chosen in own surrounding space. “Yielding with” corresponds to the experience of giving into another as well as taking from another. It is a full-body experience involving all fundamental physical forces (gravity, ground and space). “Grasping on to” is the act of grasping and containing what has been reached. “Releasing from” the child leaves the object they were holding on to and allows for the reorganization of the whole body. When we release, the action is complete and the “yielding with” phase becomes once again evident [26, 27].

This paper will describe the progress made towards the creation of an observation tool for the Parent-child dyad mediated through machine learning processes.

2 Theoretical Background

The Microsoft Kinect Sensor was initially conceived for the natural human-machine interaction through voice and gestures applied to a gaming environment [28]. In the research environment, many researchers have explored its potential, above all with regard to the tracking and recognition of human movement applied to environments that were not linked to playful video activities. The Software Development Kit (SDK) have made it possible to develop progressively more sophisticated systems for motion tracking in real time, allowing to extract spatial coordinates (x, y, z) of the different parts of the human body in 3D and temporal coordinates from the 25 anatomical landmarks (Joints) generated by the system, which allow the creation of a human skeleton model and the extraction of the subject from the background, making it possible to train classifiers for movement recognition [29, 30].

In the field of psychological research, the analysis of movements performed through kinect has been mainly applied to measure complex psychological characteristics through the observation of non-verbal behavioral patterns, such as facial expressions, posture and gait necessary, especially when there is compare with children or people with cognitive difficulties [31].

Interesting results have been achieved in the study of complex interactions between individuals. The data obtained from the Kinect surveys have, in fact, been used to test the degree of synchrony that is established between individuals involved in a social situation [32–34] or during a task [35]. Only the study by Leclere et al. [36] appears to be oriented to the observation of the baby-caregiver dyadic interaction. In this interesting research, the experimenters used two linked Kinect from which they then extracted individual parameters (contribution to heads distance), to the focus of mutual engagement (percentage of time spent face to face or task oriented), and to the dynamics of motion activity (synchrony ratio, overlap ratio, pause ratio).

3 Objective to Be Achieved

The final objective of our study is the creation of an automatic detection instrument of dyadic movements, in order to identify and to differentiate the peculiar dictionary of movements that structure between caregiver and baby.

A method mediated from processes of Machine Learning (ML) would afford, in fact, of having an instrument reliable, relatively simple to use and in a position to managing a high number of qualitative data. This will afford to observe and to analyze to the interaction diminishing the subjective the subject interpretative factor of the investigator who could compromise the quality of observation. The objective of this study is the creation of a database of the target movements selected (Reaching, Pushing, Pulling) that we want that the machine can discriminates and recognizes in order to test the reliability and reproducibility of the results.

4 Tools

The Kinect v2, developed by Microsoft, was chosen as a tool. This sensor is in its second version and was built in 2014 and consists of two cameras, RGB and InfraRed (IR), which allow it to function as a valid depth sensor and allow different information streams to be obtained such as: stream of 2D color image frames, a stream of 3D depth image frames. Through the SDK [37], it provides a skeleton tracker for at least one of the subjects framed. The latter allows a stable tracking of up to six different people, providing 3D information on the position of 25 joints per person. Although initially developed to be used as part of a game console, this sensor has become popular in research due to its potential and has been used in numerous studies as a tool for tracking movements of various degrees of complexity [38, 39].

Kinect Studio v2.0 was used to manage the different Kinect cameras and record target movements. The latter is a utility used for previewing the sensor array data of the Kinect that allows you to record eXtended Event File (XEF) and select both 2D and 3D vision. The Visual Gesture Builder (VGB) was used to create the database of target movements. The latter allows the generation of data that applications can use for gesture detection. This type of tool has been chosen because it provides a better gesture recognition based on a database, allowing a simple and easily testable gesture detection. VGB uses different methods of recognition (detection) allowing to classify two types of gestures. Discrete gestures are determined by the AdaBoostTrigger, which uses an algorithm that allows determining whether a gesture is completed or not. AdaBoost (Adaptive Boosting) is a machine learning algorithm used to build classifier through a boosting process [40, 41]. Boosting is an approach based on the idea of creating an accurate predictive rule by combining many weak rules in order to build a strong classifier. After each classifier, the weight of each training data is incorrectly classified is increased. Classifiers are trained in sequence, incorrectly weighing the data classified by the previous ones [42]. This algorithm applied to VGB returns a Boolean/binary value of "confidence" (true|false). To detect the progression of a gesture instead use the RandomForestRegression algorithm. The latter is based on the separation of the samples, the algorithm uses a random subset from the training set (composed of features) and splits the data at a threshold value with the minimum error. This process is repeated at each node (regression tree) until the minimum possible error is reached. It is used when a discrete gesture is detected (AdaBoost) and returns as a result a progressive numerical value called "progress" value (0.0 to 1.0).

5 Methods

The classification of human movements is based both on temporal information and spatial information. This Machine Learning technique uses frames tagged by the user that show positive (or negative) examples of the behaviors that the machine must learn [43]. At the end of the tagging process VGB generates a gestural database.

Therefore, our study is being structured through different phases.

Three of the fundamental movements were selected (Reaching, Pushing and Pulling), which were performed by a specialized operator in order to use a defined and repeatable set of movements during the training phase. The movements were recorded through the Kinect Studio program, which provided depth chamber data and skeletal reconstruction data. This process made it possible to generate the clips containing the data recognizable by the Visual Gesture Builder. In this software a solution was created containing the gesture project group, which contains the tagged gesture and the type of detection technology chosen (Adaboost and RFR). Clip files containing the gesture on which to operate the training phase were associated with each project. After tagging the frames of the appropriate clips, the database gesture was built.

During the training and building processes the appropriate machine learning algorithm is applied and processes the tagged data, whether they have been tagged as positive or negative examples. In the current study N = 2990 frames example were labeled as Pulling gesture (1410 positive, 1580 negative); 7058 frames were labeled as Pushing Gesture (1028 positive, 6030 negative); 3930 frames were labeled as Reaching Gesture (908 positive, 3022 negative).

After loading the examples, a pool of weak classifiers is generated, in which 38 different features are taken into account in the search for a recursive pattern in positive gesture examples. Once the weak classifiers are based on the importance of each feature, it was built a strong classifier by combining the multiple weak ones.

The program allows to obtain a top ten of the features used, allowing a deeper understanding of what contributes to the recognition of each particular gesture. An alpha value is returned which is the weight associated with each classifier that is defined by the adaboost algorithm. The final output turns out to be a linear combination of all weak classifiers.

To determine the gestures, the body joints were considered: SpineBase, SpineMid, Neck, Head, ShoulderLeft, ElbowLeft, WristLeft, HandLeft, ShoulderRight, ElbowRight, WristRight, HandRight, HipLeft, HipRight, SpineShoulder, HandTipLeft, ThumbLeft, HandTipRight, ThumbRight.

For pulling the algorithm has generated and combined 4786 classifiers. The most significant (alpha value) were Angles (Head, ShoulderLeft, SpineMid) alpha = 0.640467; Angles (Head, ShoulderRight, SpineBase) alpha = 0.548596; MuscleTorqueX(SpineMid), alpha = 0.529222; DiffMuscleForceY(ElbowLeft, SpineMid) alpha = 0.373380; DiffMuscleForceZ(Head, Neck), alpha = 0.369240; DiffMuscleForceZ(Head, ShoulderRight) alpha = 0.360251; Angles(SpineMid, Head, Neck) alpha = 0.323544; VelocityY(Head) alpha = 0.303309; DiffPositionY(ElbowLeft, ThumbRight) alpha = 0.294317; MuscleForceY(SpineBase) alpha = 0.277150.

For pushing, 4865 classifiers were generated and combined, of which Angles(WristLeft, Head, WristRight) alpha = 0.923187; DiffPositionY(ElbowLeft, SpineMid) alpha = 0.566962; DiffPositionY(ElbowRight, SpineMid) alpha = 0.464880; VelocityZ(Head) alpha = 0.431292; RefinmentHand HAND: (Left) alpha = 0.354333; Angles(SpineMid, Head, SpineShoulder) alpha = 0.352832; Angles(Head, ShoulderRight, ThumbRight) alpha = 0.342787; Angles(Head, ShoulderRight, AnkleRight) alpha = 0.306142; DiffPositionZ(Head, SpineBase) alpha = 0.277213; Angles(Head, ShoulderLeft, AnkleLeft), alpha = 0.272349.

For the reaching 4137 weak classifiers have been generated and combined, of which Angles(SpineMid, Head, HandTipLeft) alpha = 0.950367; MuscleTorqueX(ElbowRight) alpha = 0.598494; VelocityZ(SpineMid) alpha = 0.475500; Angles(SpineMid, Head, HandTipRight), alpha = 0.457554; DiffPositionY(HandLeft, ElbowLeft) alpha = 0.415713; Angles(SpineBase, SpineMid, SpineShoulder), alpha = 0.327620; MuscleTorqueZ(ShoulderRight) alpha = 0.304917; HandDifferenceRaw(Right, OPEN, LASSO) alpha = 0.296405; AngleVelocity(SpineMid, Head, HipRight) alpha = 0.288443; Angles(ShoulderLeft, SpineShoulder, ShoulderRight), alpha = 0.227420.

From these data we can see that in the recognition of the Pulling gesture the features of Head, shoulder and middle vertebral angle and the torque of the spine have been mostly used, whereas in the pushing, the angulation of the wrists and the head are more implicated. Position of the shoulders and spine. Finally, the middle spine, head and hands and the twisting of the shoulders are implicated for the reaching gesture.

This study complies with the AIP (Associazione Italiana di Psicologia) ethical code and will be submitted to the FISIG (Federazione Italiana delle Scuole e Istituti di Gestalt) ethical committee.

6 Limitations

Algorithm-based movements recognition is popular in healthcare applications [44], as gestures and/or activities related to rehabilitation are generally well defined, relatively simple and repetitive. Normally, every gesture or activity has an initial and final position that can be used to discriminate and label the gesture or movement [45]. Algorithm-based approaches and Machining-Learning-Based Motion Recognition (ML) have a number of limitations.

First of all, the rules for each gesture or activity must be carefully defined by the experts and expressed in an implementable form. While for rehabilitation exercises this problem can be solved thanks to the prescription by the doctor of the specific exercise and to the collaboration of the planner, as experts in defining exercise or movement [46], in the clinical context, and therefore in the observation of dyad baby-caregiver, this is not as easy to implement. In fact, what characterizes the baby-caregiver interaction, which we propose to analyze, is the extreme variability of the ways in which the basic movements, theorized by Ruella Frank, can manifest themselves. It is also difficult to identify the exact initial and final position that the parts of the body assume during the basic movements.

Furthermore, ML techniques use recorded and tagged data. It is necessary to collect numerous data showing positive and negative behaviors in relation to the gesture required to make the target movement easily identifiable. A recorded data clip may be sufficient to see a result but it will not necessarily work in real scenarios, as there is a discrepancy between training movements and spontaneous movements.

Furthermore, one of the limitations of this study can be identified in the instrumentation used. The Kinect turns out to be a reliable tool, which allows a continuous collection of data in a contactless, non-invasive manner [47], so different researches have focused on overcoming some problems related to the instrument, especially as regards the occlusion of the image [48]. One of the major problems is, in fact, the incompleteness of the detected image, especially when the user interacts with an object or more generally when a part of the traced body is not directly in the sensor's view [49, 50].

Although Kinect can process six skeletons at a time, the VGB can recognize and allow training exclusively for the data stream of a skeleton at a time [38, 51]. Therefore, the training of the neural network was done considering a single individual, and not a dyad.

References

1. Tsakiris, M.: My body in the brain: a neurocognitive model of body-ownership. Neuropsychologia **48**(3), 703–712 (2010)
2. Sperandeo, R., Moretto, E., Baldo, G., dell'Orco, S., Maldonato, M.: Executive functions and personality features: A circular interpretative paradigm. In 2017 8th IEEE International

Conference on Cognitive Infocommunications (CogInfoCom), pp. 000063–000066. IEEE (2017)
3. Sperandeo, R., Esposito, A., Maldonato, M., Dell'Orco, S.: Analyzing correlations between personality disorders and frontal functions: a pilot study. In International Workshop on Neural Networks, pp. 293–302. Springer, Cham (2015)
4. Maldonato, N.M., Sperandeo, R., Moretto, E., Dell'Orco, S.: A non-linear predictive model of borderline personality disorder based on multilayer perceptron. Front. Psychol. **9**, 447 (2018)
5. Stern, D.N.: Prerequisiti evolutivi per il senso di un Sé narrativo. tr. it. in Stern DN Le interazioni madre-bambino nello sviluppo e nella clinica, tr. it. a cura di Ammaniti e Dazzi, Cortina, Milano (1998)
6. Gallese, V., Sinigaglia, C.: The bodily self as power for action. Neuropsychologia **48**(3), 746–755 (2010)
7. Maldonato, M.: From neuron to consciousness: for an experience-based neuroscience. World Futures **65**(2), 80–93 (2009)
8. Dell'Orco, S., Esposito, A., Sperandeo, R., Maldonato, N.M.: Decisions under temporal and emotional pressure: The hidden relationships between the unconscious, personality, and cognitive styles. World Futures 1–14 (2019)
9. Beebe, B., Lachmann, F.M.: Infant research and adult treatment: Co-constructing interactions. Routledge (2013)
10. Beebe, B., Lachmann, F.M.: The contribution of mother-infant mutual influence to the origins of self-and object representations. Psychoanal. Psychol. **5**(4), 305 (1988)
11. Sander, L.W.: The regulation of exchange in the infant-caretaker system and some aspects of the context-content relationship. Interact. Convers. Dev. Lang. 133–156 (1977)
12. Sander, L.: Polarity paradox and the organizing process in development. In Call, J.D., Galenson, E., Tyson, R. (eds.) Frontiers of infant psychiatry Basic Books, New York (1983); Sander, L.: Toward a logic of organization in psycho biological development. In: Klar, K., Siever, L. (eds.) Biological response styles: clinical implications. American Psychiatric Press, Washington, DC (1985)
13. Beebe, B., Jaffe, J., Lachmann, F.M.: A dyadic systems view of communication. In: Skolnick, N., Warshaw,S. (eds.) Relational views of psychoanalysis. Analytic Press, Hillsdale, NJ (1992)
14. Beebe, B., Lachmann, F.M., Jaffe, J.: Le strutture d'interazione madre-bambino e le rappresentazioni del sé e dell'oggetto. Ricerca Psicoanalitica **1**, 9–63 (1997)
15. Beebe, B., Lachmann, F.M.: Representation and internalization in infancy: Three principles of salience. Psychoanal. Psychol. **11**(2), 127 (1994)
16. Als, H., Duffy, F., McAnulty, U., Badian, N.: Continuity of neurobehavioral functioning in preterm and fullterm newborns. In: Bornstein, M., Krasnegor, N. (eds.) Stability and continuity in mental development. Erlbaum, Hillsdale, NJ (1989)
17. Sigman, M., Beckwith, L., Cullen, S., Parmlee, A.: Stability in the biosocial development of the child born preterm. In: Bornstein, M., Krasnegor,N. (eds.) Stability and discontinuity in mental development. Erlbaum, Hillsdale, NJ (1989)
18. Korner, A., Grobstein R.: Individual differences at birth. In: Rexford, E., Sander, L., Shapiro, T. (eds.) Infant psychiatry. Yale University Press, New Haven, CT (1976)
19. Freedman, N., Barroso, F., Bucci, W., Grand, S.: The bodily manifestations of listening. Psychoanal. Contemp. Thought **1**, 156–194 (1978)
20. Beebe, B., Jaffe, J., Feldstein, S., Mays, K., Alson, D.: Interpersonal timing: the application of an adult dialogue model to mother-infant vocal and kinesic interaction. In: T. Field Infant social perception, Ablex, Norwood, NJ (1985)
21. Beebe, B., Lachmann, F.M.: The contribution of mother-infant mutual influence to the origins of self and object representations Psychoanalytic. Psychology **5**, 305–337 (1988)
22. Beebe, B., Lachmann, F.M.: Representation and internalization in infancy: three principles of salience. Psychoanal. Psychol. **11**, 127–165 (1994)
23. Tronick E.: Emotions and emotional communication in infants. Am. Psychol. **44**, 112–119 (1989)

24. Frank, R.: Il corpo consapevole. Un approccio somatico ed evolutivo alla psicoterapia. FrancoAngeli (2005)
25. Frank, R., La Barre, F.: The first year and the rest of your life: Movement, development, and psychotherapeutic change. Routledge (2011)
26. Rönnqvist, L., Domellöf, E.: Quantitative assessment of right and left reaching movements in infants: A longitudinal study from 6 to 36 months. Dev. Psychobiol. J. Int. Soc. Dev. Psychobiol. **48**(6), 444–459 (2006)
27. Neppl, T.K., Conger, R.D., Scaramella, L.V., Ontai, L.L.: Intergenerational continuity in parenting behavior: Mediating pathways and child effects. Dev. Psychol. **45**(5), 1241 (2009)
28. Longobardi, T., Sperandeo, R., Albano, F., Tedesco, Y., Moretto, E., Di Sarno, A. D., Maldonato, N. M.: Co-regulation of the voice between patient and therapist in psychotherapy: machine learning for enhancing the synchronization of the experience of anger emotion: An experimental study proposal. In 2018 9th IEEE International Conference on Cognitive Infocommunications (CogInfoCom), pp. 000113–000116). IEEE (2018)
29. Guzsvinecz, T., Szucs, V., Sik, C.L.: Developing movement recognition application with the use of shimmer sensor and microsoft kinect sensor. Stud. Health Technol. Inf. **217**, 767–772 (2015)
30. Fong, S., Song, W., Cho, K., Wong, R., Wong, K.: Training classifiers with shadow features for sensor-based human activity recognition. Sensors **17**(3), 476 (2017)
31. Sun, B., Zhang, Z., Liu, X., Hu, B., Zhu, T.: Self-esteem recognition based on gait pattern using Kinect. Gait Posture **58**, 428–432 (2017)
32. Fujiwara, K.: Triadic synchrony: Application of mulitple wavelet coherence to a small group conversation. Appl. Math. **7**(14), 1477 (2016)
33. Schmidt, R.C., Nie, L., Franco, A., Richardson, M.J.: Bodily synchronization underlying joke telling. Front. Human Neurosci. **8**, 633 (2014)
34. Palazzi, A., Calderara, S., Bicocchi, N., Vezzali, L., di Bernardo, G. A., Zambonelli, F., Cucchiara, R.: Spotting prejudice with nonverbal behaviours. In: Proceedings of the 2016 ACM International Joint Conference on Pervasive and Ubiquitous Computing, pp. 853–862. ACM (2016)
35. Won, A.S., Bailenson, J.N., Stathatos, S.C., Dai, W.: Automatically detected nonverbal behavior predicts creativity in collaborating dyads. J. Nonverbal Behav. **38**(3), 389–408 (2014)
36. Leclère, C., Avril, M., Viaux-Savelon, S., Bodeau, N., Achard, C., Missonnier, S., Cohen, D.: Interaction and behaviour imaging: a novel method to measure mother–infant interaction using video 3D reconstruction. Trans. Psychiatry **6**(5), e816 (2016)
37. Catuhe, D.: Programming with the Kinect for Windows software development kit. Pearson Education (2012)
38. Giancola, S., Corti, A., Molteni, F., Sala, R.: Motion capture: an evaluation of kinect V2 body tracking for upper limb motion analysis. In: International Conference on Wireless Mobile Communication and Healthcare, pp. 302–309. Springer, Cham (2016)
39. Hwang, S., Tsai, C. Y., Koontz, A. M.: Feasibility study of using a Microsoft Kinect for virtual coaching of wheelchair transfer techniques. Biomed. Eng. **62**(3), 307–313 (2017)
40. Freund, Y., Schapire, R., Abe, N.: A short introduction to boosting. J Jpn. Soc. Artif. Intell. **14**(771–780), 1612 (1999)
41. Maldonato, M., Dell'Orco, S.: Adaptive and evolutive algorithms: a natural logic for artificial mind. In: Toward Robotic Socially Believable Behaving Systems-Volume II, pp. 13–21. Springer, Cham (2016)
42. Schapire, R. E.: The boosting approach to machine learning: An overview. In: Nonlinear estimation and classification, pp. 149–171. Springer, New York, NY (2003)
43. Di Sarno, A.D., Fusco, M.L., Dell'Orco, S., Cioffi, V., Sperandeo, R., Moretto, E., Buonocore, G.: Conscious experience using the Virtual Reality: A proposal of study about connection between memory and conscience. In: 2018 9th IEEE International Conference on Cognitive Infocommunications (CogInfoCom), pp. 000289–000294. IEEE (2018)
44. Su, C.J.: Personal rehabilitation exercise assistant with kinect and dynamic time warping. Int. J. Inf. Educ. Technol. **3**(4), 448–454 (2013)

45. Lun, R., Zhao, W.: A survey of applications and human motion recognition with microsoft kinect. Int. J. Pattern Recognit. Artif. Intell. **29**(05), 1555008 (2015)
46. Mortazavi, F., Nadian-Ghomsheh, A.: Stability of Kinect for range of motion analysis in static stretching exercises. PLoS One **13**(7), e0200992 (2018)
47. Mobini, A., Behzadipour, S., Saadat, M.: Test–retest reliability of Kinect's measurements for the evaluation of upper body recovery of stroke patients. Biomed. Eng. Online **14**(1), 75 (2015)
48. Lange, B., Koenig, S., Chang, C.Y., McConnell, E., Suma, E., Bolas, M., Rizzo, A.: Designing informed game-based rehabilitation tasks leveraging advances in virtual reality. Disabil. Rehabil. **34**(22), 1863–1870 (2012)
49. Shum, H.P., Ho, E.S., Jiang, Y., Takagi, S.: Real-time posture reconstruction for microsoft kinect. IEEE Trans. Cybernet. **43**(5), 1357–1369 (2013)
50. Pedro, L.M., de Paula Caurin, G.A.: Kinect evaluation for human body movement analysis. In: 2012 4th IEEE RAS & EMBS International Conference on Biomedical Robotics and Biomechatronics (BioRob), pp. 1856–1861. IEEE (2012)
51. Maudsley-Barton, S., McPhee, J., Bukowski, A., Leightley, D., Yap, M. H.: A comparative study of the clinical use of motion analysis from kinect skeleton data. In: 2017 IEEE International Conference on Systems, Man, and Cybernetics (SMC), pp. 2808–2813. IEEE (2017)

eLORETA Active Source Reconstruction Applied to HD-EEG in Alzheimer's Disease

Serena Dattola, Giuseppina Inuso, Nadia Mammone, Lilla Bonanno, Simona De Salvo, Francesco Carlo Morabito, and Fabio La Foresta

Abstract Alzheimer's disease (AD) is a brain pathology that leads to a progressive loss of cognitive functions. In its first stage, it is often mistaken for normal aging because the symptoms are not severe. This condition is referred as Mild Cognitive Impairment (MCI). Electroencephalography (EEG) has been widely used for investigating AD. In particular, the resolution of the so-called EEG inverse problem allows you to reconstruct the distribution of brain active sources. Standard EEG is characterized by a low spatial resolution that can be improved by increasing the number of the recording sensors (HD-EEG). The purpose of this paper is the computation of the brain electrical activity by the eLORETA method for three groups of subjects: CNT (healthy subjects), MCI patients and AD patients. The novelty of this work is that eLORETA was applied to HD-EEG. The hallmark of AD is the shift of the EEG power spectrum to lower frequencies. The analysis of the results suggests that EEG of MCI and, even more, of AD is characterized by an increasing power in delta and theta bands as compared with CNT. Moreover, it has been shown that the greater source activation involves Brodmann areas typically affected by this pathology and is consistent with it. eLORETA shows the involved Brodmann areas automatically.

1 Introduction

Alzheimer's disease (AD) is a brain disorder causing about the 60–70% cases of dementia. During its first stage, people keep living normally, without any substantial alterations of their daily life: this condition is called Mild Cognitive Impairment

S. Dattola (✉) · G. Inuso · N. Mammone · F. C. Morabito · F. La Foresta
DICEAM—Università degli Studi Mediterranea di Reggio Calabria Feo di Vito, 89100 Reggio Calabria, Italy
e-mail: serena.dattola@unirc.it

L. Bonanno · S. De Salvo
IRCCS Centro Neurolesi Bonino-Pulejo, Via Palermo c/da Casazza, SS. 113, Messina, Italy

A. Esposito et al. (eds.), *Progresses in Artificial Intelligence and Neural Systems*, Smart Innovation, Systems and Technologies 184,
https://doi.org/10.1007/978-981-15-5093-5_49

(MCI). The progressive deterioration of cognitive functions affects memory, language and motor abilities severely, leading to a complete loss of autonomy and eventually to death. An early diagnosis is quite complicated because first symptoms are often considered as results of the normal aging.

Electroencephalography (EEG) is a useful diagnostic tool for recording brain electric activity [1–4]. It has been employed for years because it is easy-to use, non-invasive and not expensive. EEG signals are recorded through electrodes placed on the scalp. The potentials measured at the head surface are generated by cortical pyramidal neurons. Each sensor detects the potentials from all active sources because of the conductivity of the scalp and the underlying tissues. Hence, it's essential to make the head model as realistic as possible. In such a way, EEG analysis is more precise [5]. EEG is characterized by an excellent temporal resolution but a modest spatial one. Nowadays, spatial resolution can be improved replacing the standard 10–20 EEG configuration with High-Density EEG (HD-EEG), using up to 256 sensors [6, 7].

The localization of sources detected by EEG requires the resolution of an inverse problem. Generally speaking, the inverse problem consists of two phases, namely an estimation problem and an evaluation problem: it can be considered as a combination of what needs to be estimated and the relationship between the estimated models and the real ones. The EEG inverse problem is an ill-posed problem since the number of unknown parameters (sources) is greater than the number of known parameters (electrodes). It has an infinite number of possible solutions because the same potentials recorded from the scalp can be produced by more than one source configuration. The issue of non-uniqueness of the solution can be faced by imposing mathematical and anatomical constraints. Over the years, researchers have developed many methods for localizing brain active sources [8].

In this paper the data were processed by means of the eLORETA method. Today, LORETA (Low Resolution Electromagnetic Tomography) is one of the most accurate technique for brain source localization [9]. LORETA provided for the first time a linear distributed solution with a good localization accuracy, overcoming the issue of large localization error that affected the minimum norm estimate [10]. In particular, the last version of LORETA, eLORETA, is based on the standardization of the current density, takes into account both the noise caused by EEG measurements and the biological noise, and introduced a weight matrix that considers brain deeper sources more properly. Therefore, eLORETA shows zero error localization in the presence of measurement and structured biological noise [11]. LORETA has been already employed for studying Alzheimer with good results [12, 13].

The novelty of this work is that the eLORETA method is applied to high-density EEG. The purpose is to reconstruct the active brain regions for three groups of subjects: control subjects (CNT), Mild Cognitive Impairment patients (MCI) and Alzheimer's disease patients (AD). We expected to have an increasing power in the lower frequency bands for MCI and AD in comparison with CNT.

2 Materials and Methods

2.1 LORETA

In this paper, active source reconstruction was computed by the LORETA-KEY software (v20181107), using the eLORETA algorithm [11].

LORETA provides a linear, discrete, distributed solution for the EEG inverse problem [9]. The model of the human brain used in LORETA was made by the Brain Imaging Center of the Montreal Neurologic Institute (MNI). The brain volume is considered as a three-dimensional grid made up of a distribution of voxels. The solution space is restricted to cortical grey matter and hippocampus. Every source is placed on a grid point and is characterized by a current density vector with unknown moment components. In eLORETA a more realistic head model is used, applying the boundary elements method (BEM) to the MNI152 template [14].

The general inverse problem is defined as [15]

$$\min_{j} F_W \tag{1}$$

with

$$F_W = ||\Phi - \mathrm{KJ}||^2 + \alpha \mathrm{J}^{\mathrm{T}}\mathrm{WJ} \tag{2}$$

where the vector $\Phi \in R^{N_E x1}$ includes the scalp electric potential differences measured at N_E electrodes with respect to a single common reference electrode; $\mathrm{K} \in R^{N_E x(3N_V)}$ is the lead field matrix corresponding to N_V voxels; $\mathrm{J} \in R^{(3N_V)x1}$ is the current density; $\alpha \geq 0$ is a regularization parameter.

The solution is

$$\hat{J}_{\mathrm{W}} = \mathrm{T_W}\Phi \tag{3}$$

with the pseudoinverse given by

$$\mathrm{T_W} = \mathrm{W^{-1}K^T}\left(\mathrm{KW^{-1}K^T} + \alpha\mathrm{H}\right)^{+} \tag{4}$$

The centering matrix H in Eq. 4 is the average reference operator. The weight matrix W is block diagonal, with subblocks of dimension 3 × 3 for each voxel.

The eLORETA weights satisfy the following equation:

$$\mathrm{W_v} = \left[\mathrm{K_v^T}\left(\mathrm{KW^{-1}K^T} + \alpha H\right)^{+}\mathrm{K_v}\right]^{1/2} \tag{5}$$

where $\mathrm{W_v} \in R^{3\times 3}$ is the vth diagonal subblock of W.

eLORETA has a theoretical expected variance equal to one because it is a standardized method; it is also characterized by no localization bias in the presence of measurement and structured biological noise [11].

2.2 *HD-EEG Acquisition and Preprocessing*

EEG recordings were acquired by the 256-channel HydroCel Geodesic Sensor Net, belonging to the Geodesic EEG System (GES), as shown in Fig. 1.

The electrodes are arranged so that their interdistance is the shortest distance between two points on the surface of a sphere. In such a way, the sampling of the electrical field is optimized [16]. The reference electrode was Cz, located in the center of the head surface.

The data were recorded at IRCCS Centro Neurolesi Bonino-Pulejo of Messina (Italy). The study involved three groups of subjects: 12 CNT, 23 MCI and 16 AD. The EEG were detected in an eye-closed resting state with sampling rate of 250 Hz. The EEG were band-pass filtered between 1 and 40 Hz and cleaned from artifacts by visual inspection. Only 173 channels from the starting 256 were considered because the recordings from the sensors placed on the face and the neck were too noisy. Finally, the EEG recordings were average referenced and segmented into artifact-free non-overlapping epochs of 250 samples [17].

3 Results

EEG frequency bands have a specific biological significance and can be associated to different states of brain functioning. Previous studies revealed that AD patients are characterized by a 'slowing' of their EEG, that is an increase of power for the lower frequency bands (delta and theta) [18–21]. Therefore, eLORETA analysis of the above-mentioned frequency sub-bands can be used for determining which brain regions are activated during different states.

The brain electrical activity of each subject was reconstructed for the frequency sub-bands delta (0.5–4 Hz) and theta (4–8 Hz). The average current density of each group was obtained by averaging the current density of all subjects belonging to the given group. The results are shown in the following figures.

As for the delta rhythm, in MCI the most activated area is the Brodmann area 38 of the right hemisphere, in the temporal lobe (Fig. 2b); in AD there is a greater activation in Brodmann area 21 of both hemispheres, in the temporal lobe too (Fig. 2c). Both MCI and AD show an increase of power spectrum as compared with CNT (Fig. 2a). As for the theta rhythm, also in this case, in MCI and AD there is a greater activation of the temporal lobe areas in comparison to CNT (Fig. 3a). In particular, in MCI the maximum current density value is in Brodmann area 38 of the right hemisphere (Fig. 3b), whereas in AD it is in Brodmann area 37 of the right hemisphere (Fig. 3c).

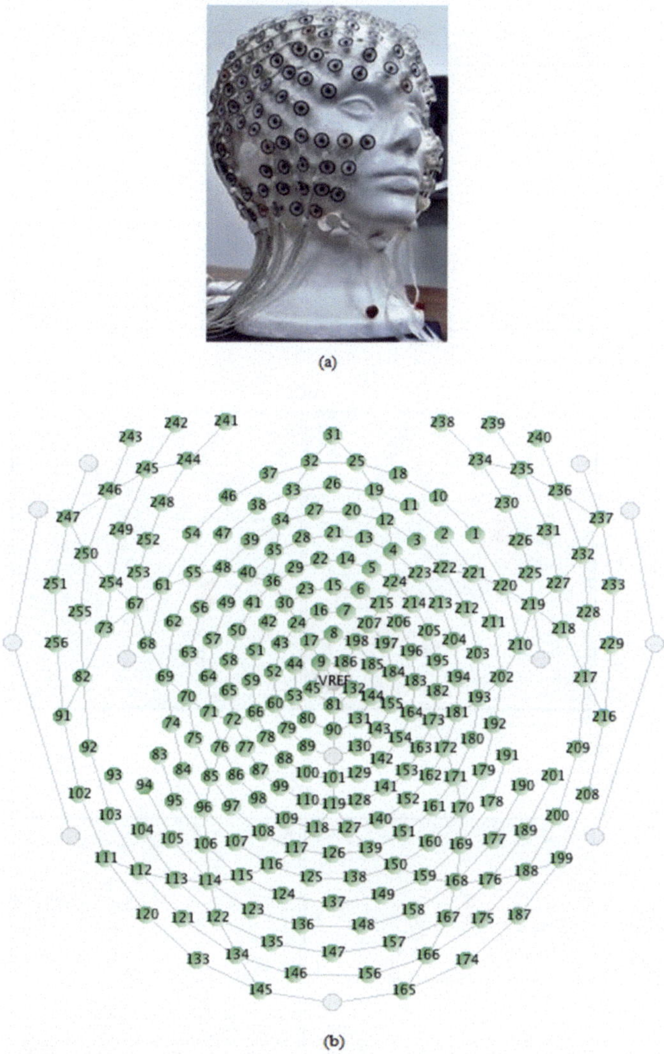

Fig. 1 HD-EEG acquisition system: **a** 256-channel HydroCel Geodesic Sensor Net; **b** Representation in 2D of the 256-channel HCGSN montage

Brodmann area 21 is associated with language and complex sounds processing. Brodmann area 37 is involved in lexico-semantic associations (i.e. associated words with visual percepts). Brodmann area 37 seems to be also participating in some memory circuitries, especially when visual information is involved. Clinical observations have proved that damage in Brodmann area 37 usually results in word-finding

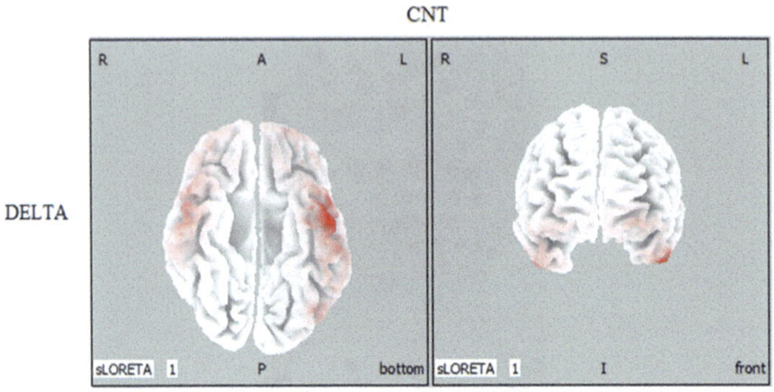

(a)

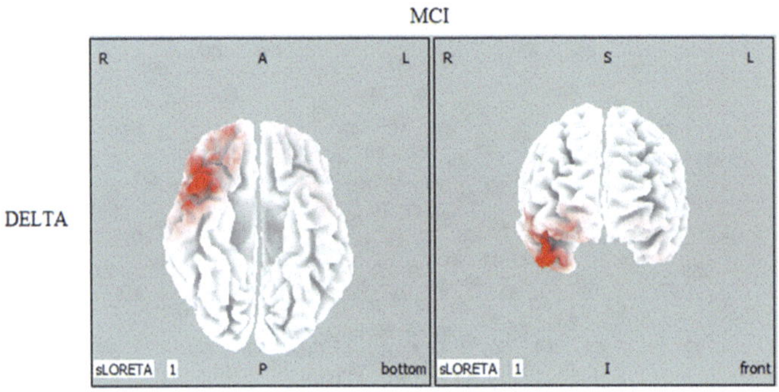

(b)

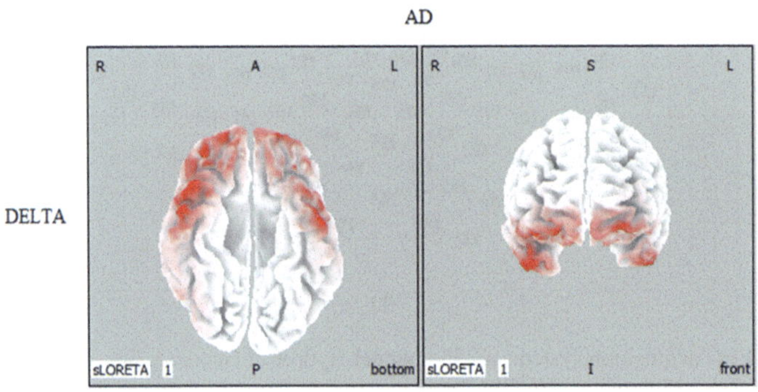

(c)

Fig. 2 Active source distribution in delta band for **a** CNT, **b** MCI and **c** AD. On the left: a bottom view of the brain cortex. On the right: a front view of the brain cortex

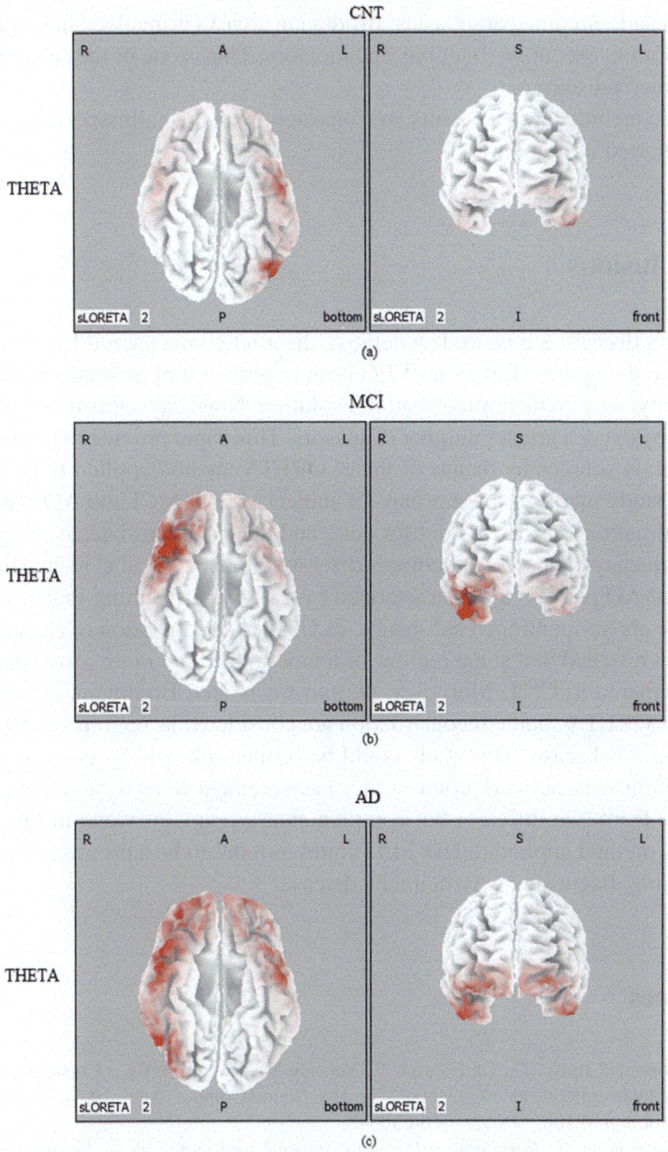

Fig. 3 Active source distribution in theta band for **a** CNT, **b** MCI and **c** AD. On the left: a bottom view of the brain cortex. On the right: a front view of the brain cortex

difficulties and semantic paraphasias. Brodmann area 38 is involved in language processes, emotion, executive functions and memory. This is one of the earliest affected by Alzheimer's disease.

We can conclude that the results are consistent with the hallmarks of Alzheimer's disease reported in literature.

4 Conclusions

Alzheimer's disease is a neurodegenerative disorder characterized by a progressive impairment of cognitive functions. EEG is an important tool for studying brain electrical activity, despite its limited spatial resolution. Nowadays, this drawback can be overcome by using a greater number of sensors. This paper provides a reconstruction of active brain sources by means of the eLORETA method applied to high-density EEG. The study involved three groups of subjects: CNT, MCI and AD. The current density estimation was carried out for delta and theta rhythms because they are two critical frequency bands in Alzheimer's disease. In fact, clinical evidence has shown that EEG of AD patients are characterized by a slowing, resulting in a power increment in the above-mentioned sub-bands. eLORETA computation of current density distribution revealed that some regions of temporal lobe are more active in MCI and AD as compared to CNT. Moreover, the most activated Brodmann areas obtained from the eLORETA source reconstruction are considered as regions deeply affected by Alzheimer's disease. This study could be further enlarged by combining source reconstruction with network connectivity measurement to analyse the evolution of the disease from two different but complementary points of view. In this way, the eLORETA method applied to HD-EEG could turn out to be a promising instrument for improving diagnosis of Alzheimer's disease.

References

1. Mammone, N., Inuso, G., La Foresta, F., Versaci, M., Morabito, F.C.: Clustering of entropy topography in epileptic electroencephalography. Neural Comput. Appl. **20**(6), 825–833 (2011). https://doi.org/10.1007/s00521-010-0505-2
2. Mammone, N. et al.: Permutation Jaccard distance-based hierarchical clustering to estimate EEG network density modifications in MCI subjects. IEEE Trans. Neural Networks Learn. Syst. **99**, 1–14 (2018)
3. Mammone, N. et al.: Brain network analysis of compressive sensed high-density EEG SIGNALS in AD and MCI Subjects. IEEE Trans. Ind. Inf. **15**(1), 527–536 (2019)
4. La Foresta, F. et al.: PCA and ICA for the extraction of EEG components in cerebral death assessment. In: Proceedings. 2005 IEEE International Joint Conference on Neural Networks, 2005, vol. 4. IEEE (2005). https://doi.org/10.1109/ijcnn.2005.1556301
5. Wolters, C.H. et al.: Influence of tissue conductivity anisotropy on EEG/MEG field and return current computation in a realistic head model: a simulation and visualization study using high-resolution finite element modeling. NeuroImage **30**(3), 813–826 (2006)

6. Srinivasan, R., Tucker, D.M., Murias, M.: Estimating the spatial Nyquist of the human EEG. Behav. Res. Methods Instr. Comput. **30**(1), 8–19 (1998)
7. Ryynanen, O.R.M., Jari, A.K.H., Jaakko, A.M.: Effect of measurement noise and electrode density on the spatial resolution of cortical potential distribution with different resistivity values for the skull. IEEE Trans. Biomed. Eng. **53**(9), 1851–1858 (2006)
8. Jatoi, M.A. et al.: A survey of methods used for source localization using EEG signals. Biomed. Signal Proces. Control **11**, 42–52 (2014)
9. Pascual-Marqui, Roberto D., Michel, Christoph M., Lehmann, Dietrich: Low resolution electromagnetic tomography: a new method for localizing electrical activity in the brain. Int. J. Psychophysiol. **18**(1), 49–65 (1994)
10. Hämäläinen, M.S., Ilmoniemi, Risto J.: Interpreting magnetic fields of the brain: minimum norm estimates. Med. Biol. Eng. Comput. **32**(1), 35–42 (1994)
11. Pascual-Marqui, R.D.: Discrete, 3D distributed, linear imaging methods of electric neuronal activity. Part 1: exact, zero error localization. (2007). http://arxiv.org/abs/0710.3341
12. Rossini, P.M., et al.: Conversion from mild cognitive impairment to Alzheimer's disease is predicted by sources and coherence of brain electroencephalography rhythms. Neuroscience **143**(3), 793–803 (2006)
13. Gianotti Lorena, R.R. et al.: Correlation between disease severity and brain electric LORETA tomography in Alzheimer's disease. Clin. Neurophysiol. **118**(1), 186–196 (2007)
14. Fuchs, M. et al.: A standardized boundary element method volume conductor model. Clin. Neurophysiol. **113**(5), 702–712 (2002)
15. Tong, S., Nitish, V.T.: Quantitative EEG analysis methods and clinical applications. Artech House (2009)
16. Electrical Geodesics, I.: Geodesic sensor net technical manual. Elect. Geodes. Inc (2007)
17. Nunez, P.L. et al.: EEG coherency: I: statistics, reference electrode, volume conduction, Laplacians, cortical imaging, and interpretation at multiple scales. Electroencephalogr. Clin. Neurophysiol. **103**(5), 499–515 (1997)
18. Jeong, Jaeseung: EEG dynamics in patients with Alzheimer's disease. Clin. Neurophysiol. **115**(7), 1490–1505 (2004)
19. Labate, D., La Foresta, F., Palamara, I., Morabito, G., Bramanti, A., Zhang, Z., Morabito, F.C.: EEG complexity modifications and altered compressibility in mild cognitive impairment and Alzheimer's disease. In: Recent Advances of Neural Network Models and Applications, pp. 163–173. Springer, Cham. (2014). https://doi.org/10.1007/978-3-319-04129-2_17
20. Moretti, D.V. et al.: Individual analysis of EEG frequency and band power in mild Alzheimer's disease. Clin. Neurophysiol. **115**(2), 299–308 (2004)
21. Dauwels, Justin, Vialatte, Francois, Cichocki, Andrzej: Diagnosis of Alzheimer's disease from EEG signals: where are we standing? Curr. Alzheimer Res. **7**(6), 487–505 (2010)

The Nodes of Treatment: A Pilot Study of the Patient-Therapist Relationship Through the Theory of Complex Systems

Raffaele Sperandeo, Lucia Luciana Mosca, Anastasiya Galchenko, Enrico Moretto, Alfonso Davide Di Sarno, Teresa Longobardi, Daniela Iennaco, Valeria Cioffi, Anna Esposito, and Nelson Mauro Maldonato

Abstract Psychotherapy, unanimously described as a particular organized and systematic relationship between a patient and a therapist, is a real complex system. The interaction between the numerous variables belonging to the patient, the therapist and the context in which the therapeutic couple is inserted, presents auto-poietic characteristics and generates emergent qualities that, at the current state of psychotherapy research, have not been effectively addressed. The methods of machine learning are suitable for analyzing complex systems and in our opinion, at the moment, they are the most appropriate for studying the therapeutic relationship, understood as a quality emerging from patient-therapist interaction. In fact, through the use of artificial intelligence methods it is possible to construct a model of interaction between therapist and patient by integrating in it the non-linearity of information exchanges between the components of the system. The humanistic therapies vision of the patient-therapist relationship as a complex and organized interaction between the parts of a system is comparable to the networks of cellular chemical reactions described by Varela and Maturana. In these networks, which are a complex systems, what is important for maintaining the cell's integrity and its functioning it is not the nature of every single chemical reaction but the form and dynamics of their interaction. This research is a pilot study that intends to evaluate the possibility of describing the complexity of therapeutic relationships using the methods of machine learning and complex networks, ordinarily used to study systems composed of numerous interacting variables. From this pilot study emerges that the use of graphs is certainly a valid tool for the analysis of both the psychotherapeutic sessions and the evolution of the care relationship over time. Numerous suggestions on the dynamics within the patient-therapist

R. Sperandeo (✉) · L. L. Mosca · A. Galchenko · E. Moretto · A. D. Di Sarno · T. Longobardi · D. Iennaco · V. Cioffi
SiPGI—Postgraduate School of Integrated Gestalt Psychotherapy, Torre Annunziata, Naples, Italy
e-mail: raffaele.sperandeo@gmail.com

A. Esposito
Università della Campania "L. Vanvitelli" and IIASS, Caserta, Italy

N. M. Maldonato
University of Naples Federico II, Naples, Italy

A. Esposito et al. (eds.), *Progresses in Artificial Intelligence and Neural Systems*, Smart Innovation, Systems and Technologies 184,
https://doi.org/10.1007/978-981-15-5093-5_50

system emerge from the construction of a complex network useful for describing the trend of psychotherapy, which in this way can be described without losing the value of the wealth of each individual experience.

1 Introduction

A study method is not valid if it doesn't include complexity, so we need models that help us to think the complexity of reality instead of dissolving and mutilating it, generating fragmented versions [1].

The science was founded on the idea of objectivity and of a universe made up of isolated objects, subjected to a linear causality laws. This hyper-simplification blinds us to the complexity of reality and constitutes a real pathology of reason [2, 3].

These fundamental concepts of Morin's thought stigmatize as improper the psychological research practice born in the last century that inadequately fragments psychic life, separating it from the field from which it emerges.

Within this vision it becomes clear that psychic experiences, which are essentially experienced in the first person and express the qualities of complex systems, completely escape the possibility of being described with mechanistic and reductionist methods [4].

Psychotherapy, unanimously described as a particular organized and systematic relationship between a patient and a therapist, is a real complex system. The interaction between the numerous variables belonging to the patient, the therapist and the context in which the therapeutic couple is inserted, presents auto-poietic characteristics and generates emergent qualities that, at the current state of psychotherapy research, have not been effectively addressed [5].

The machine learning methods are suitable for analyzing complex systems and in our opinion, at the moment, they are the most appropriate for studying the therapeutic relationship, understood as a quality emerging from patient-therapist interaction. In fact, through the use of artificial intelligence methods it is possible to construct a model of interaction between therapist and patient by integrating in it the non-linearity of information exchanges between the components of the system [6, 7].

Among all the therapeutic techniques, the approaches that can be treated with these methods of investigation are the humanistic ones, because they describe therapy as a complex creative phenomenon that emerges from the field of the relationship. In fact, they foresee that the therapeutic couple focus the attention on what happens during the session, to become aware of the gestures, breathing, emotions expressed in the voice, of the facial expressions and thoughts that are activated when the treated experiences during the therapy are analyzed [8].

The therapist is guided by what he observes in the patient and his conscious sensitivity regarding what happens in the field becomes the tool through which the patient increases his awareness. The interaction between the patient's and the therapist's components within the unified field of their relationship, and not the telling of the story of the patient's life [9], allows to him to organize new configurations of his own existence and to assimilate and to integrate them into the own control systems [10].

The awareness that patient and therapist acquire of their experience of being in the world is a typical emerging phenomenon of complex auto-poietic systems.

The humanistic therapies vision of the patient-therapist relationship as a complex and organized interaction between the parts of a system is comparable to the networks of cellular chemical reactions described by Varela and Maturana. In these networks, which are a complex system, what is important for maintaining the cell's integrity and its functioning it is not the nature of every single chemical reaction but the form and dynamics of their interaction [11].

So in this perspective, the key to describe and to understand the phenomena emerging from auto-poietic systems are Complex Networks because they allow to visualize the structure of the interactions of complex organizations.

2 Objectives to Be Achieved

This research is a pilot study that intends to evaluate the possibility of describing the complexity of therapeutic relationships using the methods of machine learning and complex networks, ordinarily used to study systems composed of numerous interacting variables.

The use of a complex network to study the interaction between the parts of the patient-therapist system can clarify the tendency of the system to develop an internal order, therefore to be self-organized and auto-poietic, and the way in which self-organization is produced.

The use of the network methods approach to study psychotherapy is useful to describe the structure and dynamics of the psychotherapy, that is a very specific complex organization, and to bring out hypotheses, to be subjected to experimental verification, on non-linear phenomena that determine the outcome of the cure.

3 Method

The research's assumption is that the interaction between the patient's and therapist's psychological dimensions explains the therapeutic change, intended as an emerging phenomenon concerning the field, i.e. the whole system composed by the dyad patient-therapist [7].

The networks approach focuses the attention on the global structure of interactions within the system, in the same way as psychotherapy focuses its attention on the interaction between patient and therapist. A network is a simple architecture formed by nodes connected by bridges. In the patient therapist system the construction of the network identifies the interacting parts of the patient and the therapist as nodes; these parts correspond to their psychic functions [12–14] such as attention, emotion management [15], concentration, self-image, adult functions [16, 17], empathy [18], etc., measured using a standardized test [19].

The bridges connecting these parts are obtained through the analysis of the correlations between the psychic functions, evaluating the Euclidean distance of the scores obtained in the test [20].

In this research an initial exploration of the system in its entirety was carried out using the classic metrics that measure the length, the connectivity and the general density of the system [21].

The final objective of our study is the creation of an automatic detection instrument of dyadic movements, in order to identify and to differentiate the peculiar dictionary of movements that is structured between caregiver and baby.

A mediated from processes of Machine Learning(ML) method would afford, in fact, of having an instrument reliable, relatively simple to use and in a position to managing a high number of qualitative data. This will afford to observe and to analyze to the interaction diminishing the subjective interpretative factor of the investigator who could compromise the quality of observation. The objective of this study is the creation of a database of the target movements selected (Reaching, Pushing, Pulling) that we want the machine can discriminates and recognizes in order to test the reliability and reproducibility of the results.

4 Materials and Procedures

The nodes were identified thanks to the subscales of a phenomenological self-observation test, defined as the adjective checklist (A.C.L.) [22].

The tool consists of a list of 300 adjectives, from which the subjects can select those they consider to be referable to their person.

The processing of the responses produces 36 subscales that describe some characteristics of the subject, such as: "modus operandi", personal needs, self-representation, originality and intelligence scales and transactional analysis scales.

This test was chosen because the self-representation based on the selection of adjectives is not inspired by any psychological theory. Finally, for the use of this tool within a complex network the interpretation of the subscales is not relevant, because the test subscales must not be interpreted but must be described in their mutual relation.

Seven psychotherapists with a humanistic orientation accepted to evaluate for a month the progress of a psychotherapeutic course they themselves have evaluated as effective.

Based on clinical experience, it was felt that a change in the configuration of the patient-therapist system took place, obviously, after four therapy sessions, carried out in one month.

The therapists completed the adjective test at the same time as their patients, at the end of the first and fourth therapy session. For each patient-therapist couple the obtained scores after each session at the subscale of the A.C.L. have been transformed into T points and used as nodes of a complex network.

The networks were designed using GEPHI, an open source software widely used for graph creation and analysis [21].

The connections between the nodes were evaluated by measuring the Euclidean distance between the subscale scores. Thus, non-oriented graphs have been constructed with uniquely positive and reciprocal connections. Finally, the produced graphs by each session were analyzed and compared with each other [20].

For example, the node representing the therapist's "empathy" subscale is connected, in the first session, to only one of the nodes that describe the patient's emotionality. In the fourth session the number of connections of the "empathy" node increases, extending to other nodes of the patient's emotional subscales. In this sense, if the therapist experiences a marked empathy and obtains high scores at this subscale, this allows the patient to have a greater awareness of his emotions. The scores at the patient's emotional subscales become higher and approach the therapist's empathy subscale score.

Figure 1 shows the two networks built on a single therapeutic pair at the first and fourth sessions. Table 1 contains statistical comparisons at the first and fourth sessions, carried out using the Student test on the metrics of the networks of all seven therapeutic couples. In the figure there are evident aggregates of nodes of the same color (four groups of nodes in the first session and three groups in the fourth). These aggregates represent nodes more interconnected to each other without a specific psychological value. The number of aggregates decreases as the number of connections between nodes increases.

5 Results

The network structure is ultra-small world type for the low value of the Average Path length between 3.21 and 2.01 for all networks, and are characterized by a high connectivity between the nodes as evidenced by the presence of only one component connected in all networks (connected components = 1). Finally all the generated networks have a diameter between 6 and 7 steps and do not have internal triangles and clusters.

Table 1 shows that all therapeutic pairs present a significant modification of the network metrics, built on the interactions between their psychic functions. These variations consist in an increase in the average degree of connection between the nodes of the average network connection density, a reduction in the modularity and in the length of the network.

Comparisons between the metrics of the networks built on therapeutic couples at the first and fourth sessions.

The Fig. 1 shows the increase in the density of network connections and the reduction of modularity as evidenced by the increase in connections from the reduction in the number of node colors in the network relative to the fourth session.

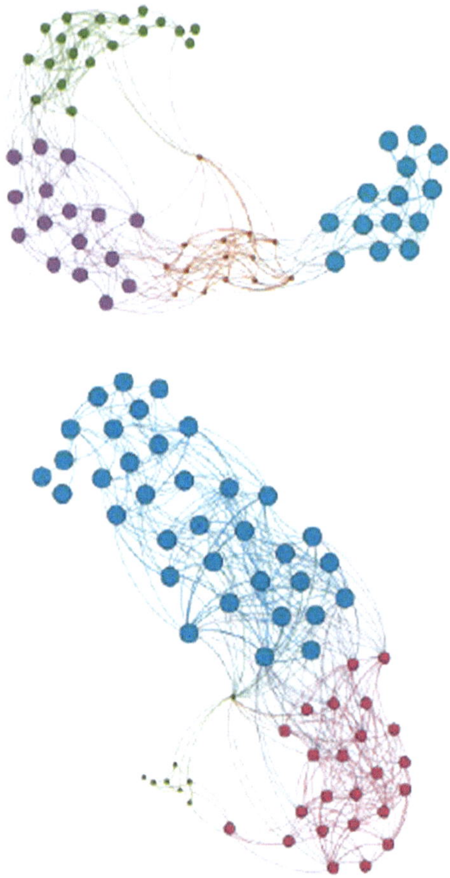

Fig. 1 Networks built after the first and fourth sessions on the interactions of the psychic functions of a patient-therapist couple

Table 1 Comparisons between the metrics of the networks built on therapeutic couples at the first and fourth sessions

	First session		Fourth session		T-Test for paired samples
	M	SD	M	SD	Sig.
Average degree	6.86	2.11	8.98	2.43	0.01
Graph density	0.11	0.03	0.15	0.03	0.04
Modularity	0.62	0.12	0.31	0.08	0.03
Average path length	2.97	0.83	2.53	0.76	0.04

6 Discussion

The generated networks are of the ultra-small world type which is characterized by a remarkable efficiency and speed of connection between all the parts. The key ingredient for the formation of an ultra-small world network is the presence of connections between apparently distant nodes. In this study, the connection between distant elements could depend on the polysemic value of many adjectives (which are differently interpreted by the patient and the therapist) and therefore of the subscales they constitute. Polysemic psychic functions, thanks to their ambiguity of content, could connect with very distant psychic elements [23, 24].

This datum is extremely relevant and needs further investigation because, since the psychotherapies analyzed in the study are all successful [25], it is possible that the structure described by the networks is that which characterizes effective psychotherapies.

The networks are also dense networks with many redundant connections and their density increases with the consolidation of the therapeutic bond. The increase in the average network connectivity and the reduction in modularity confirm the fact that the psychic functions of the patient and the therapist tend to connect more and more, each session.

This data also expresses an increase in the organization of the network and a reduction in the entropy of the system with the passing of sessions; this phenomenon is typical of complex auto-poietic systems described by Varela and Maturana [26].

The auto-poietic nature of the system, in psychotherapy, can be described as the acquisition of new awareness as a result of the interaction of the patient and the therapist on experiences related to what happens in the here and now of the session.

This emerging order can be the expression of regularity in the interactive modes of the constituent elements of the system. The continuous repetition in the "here and now" of sequences in which the therapist, with his sensitivity, captures and highlights the interruptions of awareness, requires the patient's concentration on the critical situation, favors the restoration of the complete experience and the integration of the same in the patient's memories, could be the mechanism underlying the system's self-organizing evolution. These regularities are molecular-scale mechanisms that are repeated numerous times during the interaction between the parts in a therapeutic session, generating an organized structure on the molar scale [26].

7 Conclusions

From this pilot study emerges that the use of graphs is probably a valid tool for the analysis of both the psychotherapeutic sessions and the evolution of the care relationship over time.

Numerous suggestions on the dynamics within the patient-therapist system emerge from the construction of a complex network useful for describing the trend of psychotherapy, which in this way can be described without losing the value of the wealth of each individual experience.

The changes in network metrics, between the first and fourth sessions, shown in the table appear statistically significant. However, the small number of subjects in the statistical analysis does not guarantee that these variations are really determined by the therapeutic process and not dependent on random events. It is essential to clarify that this pilot study suggests the opportunity for further study, but does not allow for drawing relevant conclusions.

Finally it is likely that more extensive research will bring out the value and the weight of the individual parts of the system in determining the dynamics and the outcome of a psychotherapeutic course.

References

1. Morin, E., Gembillo, G., Anselmo, A.: La sfida della complessità. Le Lettere (2011)
2. Morin, E.: La testa ben fatta. Erickson (2004)
3. Maldonato, N.M., Dell'Orco, S.: Toward an evolutionary theory of rationality. World Futures **66**(2), 103–123 (2010)
4. Prochaska, J.O., Norcross, J.C.: Systems of psychotherapy: A transtheoretical analysis. Oxford University Press (2018)
5. Tschacher, W., Haken, H.: Application to psychotherapy: the mixture of deterministic and stochastic interventions. In: The process of psychotherapy, pp. 81–92. Springer, Cham (2019)
6. Maldonato, N.M., Sperandeo, R., Valerio, P., Duval, M., Scandurra, C., Dell'Orco, S.: The centrencephalic space of functional integration: A model for complex intelligent systems. Acta Polytechnica Hungarica 15(5) (2018)
7. Baranyi, P., Csapo, A., Sallai, G.: CogInfoCom-Driven research areas. In: Cognitive Infocommunications (CogInfoCom), pp. 57–71. Springer, Cham (2015)
8. Elliott, R., Greenberg, L.S.: Humanistic-Experiential psychotherapy in practice: emotion-focused therapy. Comprehensive textbook of psychotherapy: Theory and practice, p. 106. (2016)
9. Mosca, L.L., Maldonato, N.M., Di Sarno, A. D., Duval, M., Cioffi, V., Dell'Orco, S., di Ronza, G.: I am a brain, Watson. The rest of me is a mere appendix: The memory, a characteristic of the human being. In Proceedings of the 9th IEEE International Conference on Cognitive Infocommunications (CogInfoCom), 295–298. IEEE (2018)
10. Day, E.: Field attunement for a strong therapeutic alliance: A perspective from Relational Gestalt Psychotherapy. J. Hum. Psychol. **56**(1), 77–94 (2016)
11. Zamorano, C., Cuevas, P., Mera, L.: Considerations in Francisco varela for psychotherapy of sense. J. Contemp. Psychother. 1–8 (2018)
12. Sperandeo, R., Maldonato, N.M., Baldo, G., Dell'Orco, S.: Executive functions, temperament and character traits: A quantitative analysis of the relationship between personality and prefrontal functions. In Proceedings of the 7th IEEE International Conference on Cognitive Infocommunications (CogInfoCom), 43–48. IEEE (2016)
13. Sperandeo, R., Moretto, E., Baldo, G., Dell'Orco, S., Maldonato, N.M.: Executive functions and personality features: A circular interpretative paradigm. In: 2017 8th IEEE International Conference on Cognitive Infocommunications (CogInfoCom), 63–66. IEEE (2017)
14. Sperandeo, R., Maldonato, M., Moretto, E., & Dell'Orco, S. (2019). Executive Functions and Personality from a Systemic-Ecological Perspective. In Cognitive Infocommunications, Theory and Applications, (pp. 79–90). Springer,Cham

15. Dell'Orco, S., Esposito, A., Sperandeo, R., Maldonato, N. M.: Decisions under temporal and emotional pressure: The hidden relationships between the unconscious, personality, and cognitive styles. World Futures, 1–14 (2019)
16. Iennaco, D., Messina, M., Moretto, E., Dell'Orco, S., Costa, V., Sperandeo, R.,... & Buonocore, G. (2018). Decision-making Styles and Personality Traits: A pilot study on the predictive capacity of the TCI regarding the quality of the decision. In Proc. of 9th IEEE International Conference on Cognitive Infocommunications (CogInfoCom), 305– 312
17. Maldonato, N.M., Sperandeo, R., Dell'Orco, S., Cozzolino, P., Fusco, M.L.: The impact of screening, medical treatment and invasive interventions on patients' medical decision-making styles: A cross-sectional study with inferences to the united states population. J. Psychol. Psychother. **7**(328), 2161–2487 (2017)
18. Maldonato, N.M., Dell'Orco, S., Sperandeo, R.: When intuitive decisions making, based on expertise, may deliver better results than a rational, deliberate approach. In Multi-disciplinary approaches to neural computing, pp. 369–377. Springer, Cham (2018)
19. McNally, R.J.: Can network analysis transform psychopathology? Behav. Res. Ther. **86**, 95–104 (2016)
20. Shen, C., Priebe, C. E., Vogelstein, J.T.: From distance correlation to multiscale graph correlation. J. Am. Stat. Assoc. 1–22 (2019)
21. Cherven, K.: Mastering Gephi network visualization. Packt PublishingLtd. (2015)
22. Gough, H.G., Heilbrun, A.B.: The adjective checklist manual: 1980 edn. Consulting Psychologists Press (1980)
23. Harrison, K.R., Ventresca, M., Ombuki-Berman, B.M.: A meta-analysis of centrality measures for comparing and generating complex network models. J. Comput. Sci. **17**, 205–215 (2016)
24. Fornito, A., Zalesky, A., Bullmore, E.: Fundamentals of brain network analysis. Academic Press (2016)
25. Dell'Orco, S., Sperandeo, R., Moretto, E., Maldonato, M.: Revision on psychometric properties of the Temperament and Character Inventory in a clinical sample. Front. Psychol. **9**, 1951 (2018)
26. Bertuglia, C S., Vaio, F.: Complessità e modelli: un nuovo quadro interpretativo per la modellizzazione nelle scienze della nature e della società. Bollati Boringhieri (2011)

Modal Structure and Altered States of Consciousness

Nelson Mauro Maldonato, Mario Bottone, Benedetta Muzii, Donatella di Corrado, Raffaele Sperandeo, Simone D'Andrea, and Anna Esposito

Abstract Although it is a familiar experience for everyone, in the vast majority of cases we discover the importance of consciousness only when in front of someone who no longer appears to possess it: someone 'absent', with their eyes fixed in the void, while the heart beats vigorously and their muscle tone is intact; or a patient with a psycho-organic syndrome or brain trauma, who is awake, even alert, but no longer in contact with the surrounding environment. Despite the fact that millions of people around the world enter and leave the state of consciousness every day, neurophysiological and clinical knowledge about consciousness is still far from forming a coherent scientific corpus. In fact, nowadays, there is no general shared definition of an altered state of consciousness. In this paper we propose a structured model of the phenomenon of consciousness, viewed as a multivariate combination of independent factors, which includes the variations and transitions of consciousness from a normal state of wakefulness to a psychopathological condition (with discrete deviations in subjective experience), and to severe clinical-neurological pictures.

N. M. Maldonato (✉) · M. Bottone
Department of Neuroscience and Reproductive and Odontostomatological Sciences, University of Naples Federico II, Naples, Italy
e-mail: nelsonmauro.maldonato@unina.it

B. Muzii
Intradepartmental Program of Clinical Psychology, AOU, University of Naples Federico II, Naples, Italy

D. di Corrado
Department of Sport Sciences, Kore University, Enna, Italy

R. Sperandeo
DiSU—Department of Humanities, University of Basilicata, Potenza, Italy

S. D'Andrea
Graduate Member Within British Psychological Society, London, UK

A. Esposito
Department of Psychology, University of Campania "L. Vanvitelli", and IIASS, Naples, Italy

A. Esposito et al. (eds.), *Progresses in Artificial Intelligence and Neural Systems*, Smart Innovation, Systems and Technologies 184,
https://doi.org/10.1007/978-981-15-5093-5_51

1 Introduction

There is increasing agreement among scholars that consciousness originates from a proto-experience that integrates cortical and subcortical activities and dynamics into a temporal structure [1]. Starting from a proto-experience apparently immaterial phenomena such as intentionality are linked to objects and movements of the real world through continual changes of perspective [2]. These conditions create the possibility of the various states of attentional and perceptive clarity between the extremes of sleep and hypervigilance that allow man to dialogue with himself, to decide [3, 4] and to act [5] in full awareness of his normal and pathological condition [6]. In reality, even though it is the subjective experience of a system in action, awareness poses serious problems to experimental investigation, unlike vigilance which can be investigated experimentally and with good approximation. In fact, the ordinary difficulties encountered in discussing consciousness [7] become extraordinary in the field of altered states of consciousness, where everything appears more complex and indefinite.

For a long time, a methodological and clinical demarcation between altered states of consciousness caused by organic lesions [8] and others attributable to psycho(patho)logical conditions prevailed among scholars. Around the middle of the last century, Henry Ey [9] identified the term and concept of field of consciousness—a refined synthesis between the concepts of organization, intentionality and the field of perception—as the horizon within which experience expresses its meaning and its momentary, transient and synchronic amplitude. The conceptual value of this representation still helps us better understand not only the genealogy, the background and the stratifications of the conscience [10], but also the evolutionary order of the relationships in life, in particular regarding altered states of consciousness, where categories such as openness to the world, extension and suspension of time are essential aspects [11].

An altered state of consciousness constitutes a discrete (temporary or irreversible) change of experience, perceived as a subjective impairment of psychic functioning [12]. This is more a multidimensional phenomenon than a mere quantitative variation of a single cognitive function. In fact, in the presence of qualitative alterations in the field of consciousness, it is implausible to speak of disturbances in consciousness. As a flow of spontaneous processes, consciousness goes beyond its own field of action and, therefore, does not coincide with the "field of consciousness" [13]. In this sense, the interweaving of subjective experience and objective perception of the world requires the integration of quantitative-categorial levels (attention, vigilance, sleep, coma) [14], and of qualitative-dimensional levels (subjective experiences such as sensations, thoughts, emotions) with the analysis of types and degrees of consciousness, both synchronic (field of consciousness) and diachronic (Ego, personality).

On the historical level, the psychopathological study of consciousness cannot ignore the formal characteristics identified by Jaspers [15]. These are features related to:

a. The sensation of activity: original principle of a life of relationships, that makes us aware of thinking, acting and so on. This sense of one's own presence is absent in states of depersonalization, in which perceptive estrangement and automatism of the will prevail, together with a lack of sensation of one's own feelings;
b. Awareness of one's own uniqueness: the certainty of being a distinct entity at any given time: the stability of the ego is constant, despite its extreme variability;
c. Awareness of identity: a sense of one's own presence in the diachronic dimension of time;
d. Awareness of self (Id) in contrast to the external world and to others: this highlights the unrepeatable evidence of the originality of one's own life of relationships.

Every aspect of life of the relationships depends on how the consciousness integrates and transforms the data from the environment [16]. However, this spontaneous capacity for transformation can be modified by our will: by meditation, by contemplation [17] and by other factors which will be mentioned further on. Consciousness is invariably the clear awareness of our own capacity to think, to want, to act, to decide [18] and to shape our psychic life: a condition that is lacking in states characterized by feelings of illusion, and loss of familiarity of oneself and the surrounding world, which make the boundaries of the self blurred and indeterminate.

2 Neurobiological Roots of Mutations of Consciousness

An altered state of consciousness is usually characterized by a temporary and reversible change in psychic functioning. Some changes, that involve aspects such as memory, type and speed of reasoning, sense of identity, motor skills, perception of time and space, perception of the meaning of things, are not necessarily pathological. They can be determined by an entropic phase transition between different states of consciousness, caused by oscillatory asynchronies of the highly organized bioelectric activity of the *Default Mode Network* [19] (associated with automatic higher integrative functions) and by a decoupling between this and the medial temporal lobes [20].

Among the factors that disturb the dynamics of physiological integration are narcotic and psychedelic substances, that may interfere with the receptor functioning of some brain structures and give rise, among other things, to phenomena of thought diffusion (which is projected and spread in the environment), to sensations of imminent fearful change or to a sinister sense of threat from dark forces and more.

- Cannabis alters psychic activity in general and, in particular, memory and the perception of pain. It can induce degrees of paranoia, greater sensitivity, marked anxiety and anomalies in the ordinary functions of the conscience [21].
- MDMA (ecstasy) [22] causes an increase in positive feelings and a reduction in negative feelings, as well as stimulating emotions and reducing moral inhibitions to behaviour.

- Opioids (heroin, morphine, hydrocodone, oxycodone, etc.) produce analgesia and feelings of euphoria. Like natural painkillers, opioid abuse can result in decreased production of endorphins in the brain. A massive dose of opiates can cause death by inhibition of respiratory center activities [23].
- Cocaine alters communication between neurotransmitters. It inhibits the synaptic re-uptake of norepinephrine, serotonin, dopamine and others [24].
- Lysergic acid (LSD) activates the serotonin receptors, with more evident effect in areas of the cerebral cortex corresponding to the spheres of thought, emotions and affectivity, in a range that goes from fear to ecstasy [25].
- Alcohol shifts the levels of neurotransmitters that, through a synapse, transmit signals from one neuron to another and can cause inhibitory or excitatory effects on the target cell. On the other hand, alcohol accentuates the effect on the brain of the neurotransmitter GABA (gamma-aminobutyric acid) while reducing the excitatory effect of glutamate, thereby causing physiological slowdown [26, 27].

3 Types of Altered Consciousness After a Traumatic Brain Injury

The psychopathological study of consciousness—especially when it comes to quantitative disorders—must be always supported by an analysis of the evolution and dissolution of the relationship between cortical and subcortical structures. For example, lesions of the *Ascending Reticular Activating System* [28, 29] or dysfunctions of both cerebral hemispheres cause severe quantitative alterations of the state of consciousness. In contrast, in the absence of bilateral dysfunction, cortical lesions modify the contents of consciousness in a way consistent with their functional specialization. Evidence has been reported comparing the main pathophysiological states (deep sleep, coma/vegetative state, general anesthesia, loss of consciousness during an epileptic seizure) with the absence of consciousness [30]. These frameworks, although of different etiopathogenesis, share a series of objectively similar elements: regular electroencephalographic activity, a significant decrease in metabolism in the frontal-parietal regions, global arrest of cortico-cortical and thalamus-cortical connections, up to a state of 'non-responsiveness' [31]. These destructuring effects lead to a better understanding of some aspects of the structure of consciousness and the related neural processes. There are, for example, patients who can sleep well [32], be attentive, converse pleasantly, without apparent inadequacy of awareness or altered 'attunement'; however, often they neither know much about what they should know nor recognize the place where they are, nor do they recall how they got there [33]. They are often unaware of the time, the date, they can easily lose the information stored in the memory under conditions of full vigilance.

Although in scientific community the theoretical debate is proceeding rapidly, in everyday clinical practice it is the concepts of vigilance and awareness that define altered states of consciousness [8]. For example, the involuntary closure of the patient's eyes is still considered as a closure of the conscience to the world. Thus, while in the most serious cases of coma [34] awareness and vital functions fail (such as the respiratory system that requires mechanical assistance), in the vegetative state patients alternately open and close their eyes—often in correspondence with the sleep-wake cycle. Furthermore, unlike with coma, in the vegetative state motor activity and 'no-finalised' [35] facial mimicry may be present. Although there are a great number of facts to be considered, patients usually spend a few weeks in a coma before passing on to the vegetative state, at the end of which the risk of a "non-return" becomes probable.

Beside the vegetative states [36], there are the minimally conscious states, where the terminological and clinical framework becomes even more confused. The diagnostic criteria of this syndrome, characterized by severe changes in consciousness, are different from those of the vegetative state [37]. The presence of awareness is minimal and discontinuous: patients seem to interact with the surrounding world in an almost congruous (though transitory) way, either spontaneously or following stimulation [38]. Yet, these actions exclude the execution of complex and protracted tasks. The elements that unite these two clinical pictures are remarkable and, for this reason, they have been inscribed in a clinical neurological spectrum called low-level status, which allows setting aside out-dated concepts such as "apallic syndrome" [39], "coma vigil" "Locked-in syndrome" and others.

4 Quantitative and Qualitative Alterations

Consciousness disorders are traditionally divided into quantitative and qualitative [6]. The former includes organic based clinical pictures that involve the spheres of attention, vigilance and other objectivizable spheres [40]: i.e. confusional states, blunting, stupor and coma in its different degrees. The latter include conditions that involve the contents of conscience, the experience of oneself and the world [41]: i.e. depersonalization, derealization, psychogenic based dissociative phenomena [42], dream and oneiric states, confusion and crepuscular states. In reality, this division appears artificial due to the difficulty in defining a clear demarcation between qualitative and quantitative disorders of consciousness on a psychopathological, clinical and nosographic level.

Altered consciousness is variously caused by (a) accidental events or organic pathologies that manifest themselves through a symptomatology that goes from a slight reduction to a serious impairment of self-awareness; (b) traumatic experiences (car accident, blunt force or other) [43] that may have a high impact on the brain with severe changes in brain functioning, especially when these injuries concern the prefrontal cortex [44], with effects ranging from impaired logical-analytical thinking, behavioural changes, reduced self-awareness and others [45]; (c) neurological syndromes, such as epilepsy, which in some of their specific critical clinical features cause hallucinations, loss of control over one's body and/or dissociation from reality [46]; (d) infections from bacterial or viral agents, such as meningitis (inflammation of the meninges which not infrequently extends to the cerebral parenchyma in forms of meningoencephalitis) which, together with the classical triad (fever, headache and nuchal rigidity) and other symptoms, cause altered states of conscience; (e) disorders of the sleep-wake cycle, in particular sleep deprivation which may even cause epileptic seizures, hallucinations and inability to control sensory-motor functions, with an accompanying impairment of the patient's self-awareness; (f) metabolic alterations consequent to fasting—secondary to religious purposes or eating disorders such as anorexia—which can lead to forms of malnutrition that may even generate epileptic seizures that generate altered states of consciousness [47]; (g) psychotic or manic episodes that also involve reduced self-awareness and severe alterations of the reality test.

As far as quantitative disorders of consciousness are concerned [6] at least four levels can be distinguished schematically:

Level I: characterized by torpor, diminished states of consciousness, hypnoid alterations and more. The patient distinguishes himself from the external world, and fantasy from reality. He still recognizes people and falls asleep normally. This is the first stage towards stupor, a sort of precoma. It may have organic causes (brain trauma, severe metabolic alterations, and tumors) or be caused by the intake of alcohol or other psychoactive substances.

Level II: characterized by oneiric and oneiroid alterations that involve quantitative and qualitative associative anomalies, alteration of the examination of reality, presence of delirious and hallucinatory cues. Reality is still partially distinct from fantasy. This clinical picture occurs in organic-based confusional syndromes (delirium) and in the acute phases of schizophrenia and mania.

Level III: characterized by narrowing of the field of consciousness. The onset is sudden, the symptoms limited in time. It usually resolves with full *restitutio ad integrum*. The patient seems completely absorbed in himself and shows progressive incapacity to integrate temporal and spatial experiences. Fear, dysphoria and mystical-ecstatic states prevail, with incoherence and blunting. The patient does not remember what happened. In the field of consciousness only fragments of perceptions, representations, and feelings occur. Etiopathogenesis is of psychogenic origin and occurs in syndromes such as somatic conversion, post-traumatic stress and long imprisonment (Ganser syndrome). These quantitative disorders may also be determined by epileptic seizures, head trauma, and drug use.

Level IV: confusional states with no examination of reality, linguistic inconsistency, loss of temporal-spatial orientation, disorganization of thought and language, disturbances to perception (illusions, hallucinations), drastic alterations of the sleep-wake rhythm. This occurs in clinical pictures such as delirium or in acute and chronic organic confusional syndromes, in traumas, in dysmetabolic pictures, in tumors and in toxic states.

Qualitative disorders of consciousness (Fig. 1), on the other hand, concern phenomena of derealization and de-personalization [48], dissociative identity disorders [49], psychogenic amnesia, somatic conversion syndrome, somatoform disorders, post-traumatic stress syndrome, acute psychosis and more. They may arise due to causes secondary to brain alterations or as an expression of a defense that intervenes

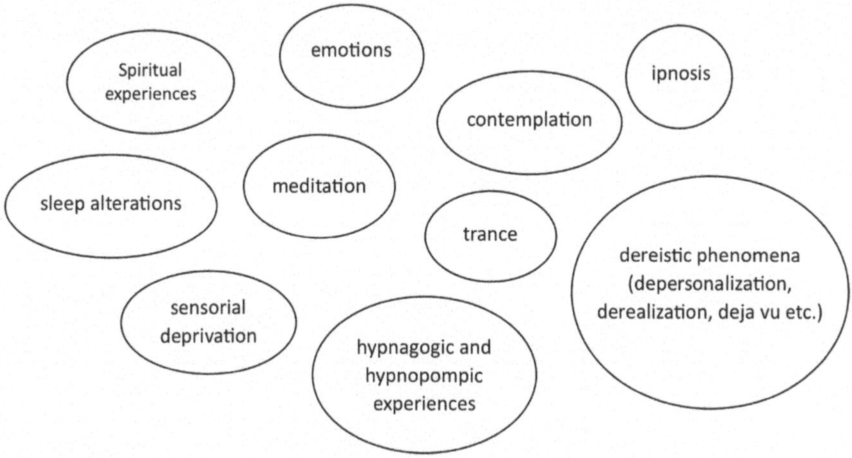

Fig. 1 Qualitative alterations of consciousness

in extreme situations (traumas) and unconscious conflicts, and acts as a barrier to an anxiety that cannot be organized in another way [50]. Although extremely complex, a classification of altered states of consciousness should concern both etiopathogenesis and, the types of experiences. In particular, consideration should be given to whether it is a spontaneous event (daydreaming, reverie, etc.), a physiological process (fasting and sex), psychological (meditation, contemplation, hypnosis) [51], pathological (epilepsy, brain injury and so on) or pharmacological (psychoactive substance) [52].

Beyond these methodological aspects, the study of these disorders cannot ignore four fundamental aspects: activation, temporal dimension (duration) of awareness, self-awareness, sensory dynamics.

5 Towards a Modal Structure of Consciousness: New Directions of Research

It is plausible to assume that the disorders treated here, both quantitative and qualitative, are the expression of a process of disorganization of the conscience, understood not only as a regression of the primary organization of consciousness, but as a symptomatological adaptation to the dynamics of some deep brain structures [6]. On a psychopathological level this disorganization can be expressed globally when the primary activity of the conscience is involved and on the local level when only a specific area, and consequently the corresponding psychic function, is involved. In this sense, it is possible to conceive a modal organization of consciousness which, in its structure, is an expression of an intersection between horizontal and vertical levels.

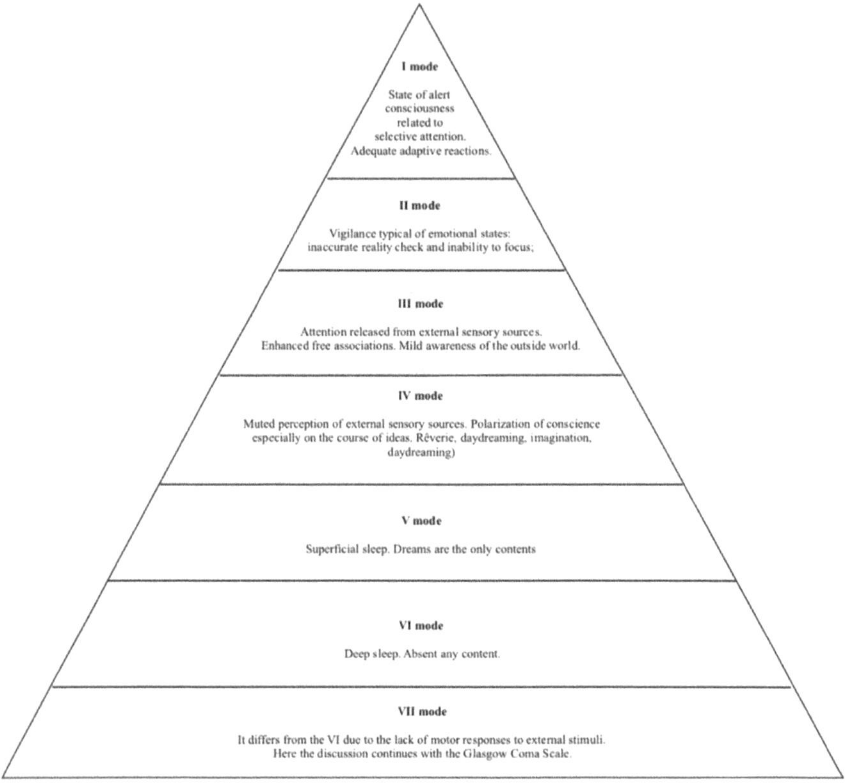

Fig. 2 The modal structure of consciousness

Here we propose a 7-mode structure (Fig. 2) that goes from the state of alert consciousness (which we share with the non-human animal world) to a state in which the conditions for the possibility of a life of relationships life are absent.

6 Conclusions

In most of our life of relationships we routinely experience the flow of thoughts, perceptions, feelings, emotions and more in natural harmony with the surrounding environment. When this condition of fluidity is transformed by emerging instances of a different nature, one enters the complex and indefinite sphere of altered states of consciousness. The use of the term 'altered' could generate the misunderstanding that it is a pathological state, but as we have seen this is only partially true.

New research directions will have to clarify, on the one hand, how the processes of information and integration of information are constituted, how they dissolve and reconstitute each time, and lead to a potential repertoire of the different states of consciousness, and on the other hand, how such a complex physical system can generate—at thalamus-cortical level—the phenomenon of consciousness in the delicate and unstable equilibrium between diversity and unity. This basic research work, however, is inseparable from the psychopathological-clinical analysis of the modal structure of consciousness, between phenomena and dynamics of instability (implicit entropic asymmetries, simple bifurcations, non-linear perturbations and ascending degrees of equilibrium) typical of qualitative alterations and its configurations of static and indifferent equilibria typical of quantitative alterations.

References

1. Maldonato, N.M.: From neuron to consciousness: for an experience-based neuroscience. World Futures **65**(2), 80–93 (2009)
2. Callieri, B., Maldonato, N.M., Di Petta, G.: Lineamenti di psicopatologia fenomenologica. Guida Editori, Napoli (1999)
3. Maldonato, N.M., Dell'Orco, S.: Toward an evolutionary theory of rationality. World Futures **66**(2), 103–123 (2010)
4. Maldonato, N.M., Dell'Orco, S.: How to make decisions in an uncertain world: Heuristics, biases, and risk perception. World Futures **67**(8), 569–577 (2011)
5. Maldonato, N.M., Dell'Orco, S.: The natural logic of action. World Futures **69**(3), 174–183 (2013)
6. Maldonato, N.M., et al.: Mind, brain and altered states of consciousness. Acta Medica Mediterranea **34**, 357–366 (2018)
7. Wittmann, M.: Altered states of consciousness: Experiences out of time and self. MIT Press (2018)
8. Vaitl, D., Birbaumer, N., Gruzelier, J., Jamieson, G.A., Kotchoubey, B., Kübler, A., Sammer, G.: Psychobiology of altered states of consciousness. Psychol. Bull. **131**(1), 98 (2005)
9. Ey, H.: Consciousness: A phenomenological study of being conscious and becoming conscious. Indiana University Press (1978)
10. Ey, H.: Outline of an organo-dynamic conception of the structure, nosography, and pathogenesis of mental diseases. In: Psychiatry and philosophy, pp. 111–161. Springer, Berlin, Heidelberg (1969)
11. Maldonato, N.M., Esposito, A., Dell'Orco, S.: Consciousness and the archipelago of functional integration: On. Quan. Proces. Biomed. Behav. Signals **103**, 127 (2018)
12. Kaës, R., Anzieu, D.: Le travail de l'Inconscient: Textes choisis, présentés et annotés par René Kaës, 1st edn. Dunod, Malakoff (2009)
13. Baars, B.J., Gage, N.M.: Cognition, brain, and consciousness: Introduction to cognitive neuroscience, 2nd edn. Academic Press (2010)
14. Robison, M., Brewer, G.: Individual differences in vigilance: Implications for measuring sustained attention and its association with other cognitive abilities and psychological constructs (2019)
15. Jaspers, K.: Allgemeine Psychopathologie: Für Studierende· Ärzte und Psychologen. 3rd edn. Springer, Berlin, Heidelberg (1923)
16. Morin, A.: Levels of consciousness and self-awareness: A comparison and integration of various neurocognitive views. Conscious. Cogn. **15**(2), 358–371 (2006)
17. Callieri, B: Riflessione, Meditazione, Contemplazione, in 3° Congr. Internaz. CISSPAT, San Marino, Padova (1978–1979)
18. Maldonato, M., Dell'Orco, S.: Decision making styles and adaptive algorithms for human action. Psychology **2**(08), 811 (2011)

19. Rosazza, C., et al.: Multimodal study of default-mode network integrity in disorders of consciousness. Ann. Neurol. **79**(5), 841–853 (2016)
20. Maldonato, N.M., Oliverio, A., Esposito, A.: Neuronal symphonies: Musical improvisation and the centrencephalic space of functional integration. World Futures **73**(8), 491–510 (2017)
21. Hofmann, M.E., Frazier, C.J.: Marijuana, endocannabinoids, and epilepsy: potential and challenges for improved therapeutic intervention. Exp. Neurol. **244**, 43–50 (2013)
22. Müller, F., Brändle, R., Liechti, M.E., Borgwardt, S.: Neuroimaging of chronic MDMA ("ecstasy") effects: A meta-analysis of the literature. Neurosci. Biobehav. Rev. **96**, 10–20 (2018)
23. Robinson, T.E., Berridge, K.C.: Incentive-sensitization and addiction. Addiction **96**(1), 103–114 (2001)
24. Aldridge, D., Fachner, J., Erkkilä, J.: Music therapy today. In: Proceedings of the 6th European music therapy congress, Aldrige, D., Fachner, J. (eds.), vol. 6(4) (2005)
25. Preller, K.H., Schilbach, L., Pokorny, T., Flemming, J., Seifritz, E., Vollenweider, F.X.: Role of the 5-HT2A receptor in self-and other-initiated social interaction in lysergic acid diethylamide-induced states: A pharmacological fMRI study. J. Neurosci. **38**(14), 3603–3611 (2018)
26. Halberstadt, A.L., Vollenweider, F. X., Nichols, D. E. (eds.): Behavioral neurobiology of psychedelic drugs, vol. 36., Springer, Berlin, Heidelberg (2018)
27. Stewart, S.T., Anthony, B.: Analysis of GABA receptor expression in differentiated embryonic neuronal stem cells after treatment with opioids, methamphetamine, and alcohol. In: Proceedings of the West Virginia Academy of Science, vol. 90(1). (2018)
28. Maldonato, N.M.: The ascending reticular activating system. In: Bassis, S., Esposito, A., Morabito, F. (eds.) 23RD Italian workshop on neural networks 2013, recent advances of neural network models and applications, smart innovation, systems and technologies, vol. 26, pp. 333–344. Springer, Cham (2014)
29. Maldonato, M., Dell'Orco, S., Springer, M.: Rethinking consciousness: Some hypothesis on the role of the ascending reticular activating system in the global workspace. In: Apolloni, B., Bassis, S., Esposito, A., Morabito, F. (eds.) Neural Nets WIRN11—Proceedings of the 21st Italian Workshop on Neural Nets, Frontiers in Artificial Intelligence and Applications, vol. 234, pp. 212–219, IOS Press (2011)
30. Di Perri, C., Stender, J., Laureys, S., Gosseries, O.: Functional neuroanatomy of disorders of consciousness. Epilepsy Behav. **30**, 28–32 (2014)
31. Laureys, S., Gosseries, O., Tononi, G. (eds.). The neurology of consciousness: cognitive neuroscience and neuropathology, 2nd edn. Academic Press (2015)
32. Guilleminault, C., Billiard, M., Montplaisir, J., Dement, W.C.: Altered states of consciousness in disorders of daytime sleepiness. J. Neurol. Sci. **26**(3), 377–393 (1975)
33. Demertzi, A., Vanhaudenhuyse, A., Bruno, M.A., Schnakers, C., Boly, M., Boveroux, P., Laureys, S.: Is there anybody in there? Detecting awareness in disorders of consciousness. Expert Rev. Neurother. **8**(11), 1719–1730 (2008)
34. Wijdicks, E.F.: Prolonged coma and early disorders of consciousness. In: Creutzfeldt, C.J., Kluger, B.M., Holloway, R.G. (eds.) Neuropalliative care, pp. 25–35. Springer, Cham (2019)
35. Hannawi, Y., Lindquist, M.A., Caffo, B.S., Sair, H.I., Stevens, R.D.: Resting brain activity in disorders of consciousness: a systematic review and meta-analysis. Neurology **84**(12), 1272–1280 (2015)
36. Laureys, S., Owen, A.M., Schiff, N.D.: Brain function in coma, vegetative state, and related disorders. Lancet Neurol. **3**(9), 537–546 (2004)
37. Bodart, O., Amico, E., Gómez, F., Casali, A.G., Wannez, S., Heine, L., Thibaut, A., Annen, J., Boly, M., Casarotto, S., Rosanova, M., Massimini, M., Laureys, S., Gosseries, O.: Global structural integrity and effective connectivity in patients with disorders of consciousness. Brain Stimul. **11**(2), 358–365 (2018)
38. Owen, A.M., Schiff, N.D., Laureys, S.: The assessment of conscious awareness in the vegetative state. In: Laureys, S., Gosseries, O., Tononi, G. (eds.) The Neurology of Conciousness, 2nd edn., pp. 155–166, Academic Press (2016)

39. Locke, D.: Appalic syndrome. In: Kreutzer, J.S., DeLuca, J., Caplan, B. (eds.) Encyclopedia of Clinical Neuropsychology, 2nd edn. Springer International Publishing (2018)
40. Chennu, S., Bekinschtein, T.A.: Arousal modulates auditory attention and awareness: insights from sleep, sedation, and disorders of consciousness. Front. Psychol. **3**, 65 (2012)
41. Fuchs, T.: From self-disorders to ego disorders. Psychopathology **48**(5), 324–331 (2015)
42. Sperandeo, R., Monda, V., Messina, G., Carotenuto, M., Maldonato, N.M., Moretto, E., De Luca, V., Monda, M., Messina, A.: Brain functional integration: an epidemiologic study on stress-producing dissociative phenomena. Neuropsychiat. Dis. Treat. **14**, 11–19 (2018)
43. Frewen, P., Hegadoren, K., Coupland, N.J., Rowe, B.H., Neufeld, R.W., Lanius, R.: Trauma-related altered states of consciousness (TRASC) and functional impairment I: Prospective study in acutely traumatized persons. J. Trauma Dissoc. **16**(5), 500–519 (2015)
44. Lew, S., Lau, H.: Crucial role of the prefrontal cortex in conscious perception. In: Goldberg, E. (eds.) Executive Functions in Health and Disease, pp. 129–14, Academic Press (2017)
45. Colás, I., Chica, A.B., Ródenas, E., Busquier, H., Olivares, G., Triviño, M.: Conscious perception in patients with prefrontal damage. Neuropsychologia 2019)
46. Korsnes, M.S., Hugdahl, K., Nygård, M., Bjørnæs, H.: An fMRI study of auditory hallucinations in patients with epilepsy. Epilepsia **51**(4), 610–617 (2010)
47. Avner, J.R.: Altered states of consciousness. Pediatr. Rev. **27**(9), 331–338 (2006)
48. Hamilton, H.K., Simeon, D.: Depersonalization/Derealization disorder and schizotypal personality disorder. In: Moskowitz, A., Dorahy, M.J., Schäfer, I. (eds.) Psychosis, Trauma and Dissociation: Evolving Perspectives on Severe Psychopathology, pp. 241–256, Wiley-Blackwell, Hoboken (2019)
49. Bailey, T.D., Boyer, S.M., Brand, B.L.: Dissociative disorders. In: Segal, D.L. (eds.) Diagnostic interviewing, pp. 401–424, 2nd edn. Springer, New York, NY (2019)
50. Lanius, R.A.: Trauma-related dissociation and altered states of consciousness: a call for clinical, treatment, and neuroscience research. Eur. J. Psychotraumatol. **6**(1), 27905 (2015)
51. Droit-Volet, S., Dambrun, M.: Awareness of the passage of time and self-consciousness: What do meditators report? PsyChol. J. **8**(1), 51–65 (2019)
52. Presti, D.E.: Altered states of consciousness. The Blackwell Companion to Consciousness, p. 171 (2017)

The Desiring Algorithm. The Sex Appeal of the Inorganic

Nelson Mauro Maldonato, Paolo Valerio, Mario Bottone, Raffaele Sperandeo, Cristiano Scandurra, Ciro Punzo, Benedetta Muzii, Simone D'Andrea, and Anna Eposito

Abstract Each new era brings with it new technologies and new challenges to civilization. The impetuous technological development is changing the economy and work more and more every day; to the point of making the passage from agrarian society to industrial society seem like a trifle. Nevertheless, if there is widespread awareness of the general issues of society, there is definitely less on the way in which anthropomorphic robots (increasingly in an emotional and cognitive relationship with human beings) are radically modifying intimate and exclusive spheres such as sexuality and love. An epochal passage is being announced from an organic sexuality (driven by desire and pleasure) to an inorganic, artificial sexuality, suspended in an abstract excitement, always at hand and disengaged from categories such as beauty, forms and age. In this paper, we will discuss the plausibility of an emotional relationship between robots and human beings and its ethical implications. It is probable, in fact, that soon there will be self-aware robots, capable of experiencing feelings such as love, joy, suffering, pain and pleasure. If this scenario is realized, will we continue to think that they are only machines, or will we recognize the moral relevance that they do not have today?

N. M. Maldonato (✉) · P. Valerio · M. Bottone · C. Scandurra
Department of Neuroscience and Reproductive and Odontostomatological Sciences, University of Naples Federico II, Naples, Italy
e-mail: nelsonmauro.maldonato@unina.it

R. Sperandeo
DiSU—Department of Humanities, University of Basilicata, Potenza, Italy

C. Punzo
Pontifical Lateran University, Rome, Italy

B. Muzii
Intradepartmental Program of Clinical Psychology, AOU, University of Naples Federico II, Naples, Italy

S. D'Andrea
Graduate Member Within British Psychological Society, London, UK

A. Eposito
Department of Psychology, University of Campania "L. Vanvitelli" and IIASS, Naples, Italy

A. Esposito et al. (eds.), *Progresses in Artificial Intelligence and Neural Systems*, Smart Innovation, Systems and Technologies 184,
https://doi.org/10.1007/978-981-15-5093-5_52

1 Transits from the Bottom of the Matter

The contemporary imaginary is increasingly configured as a set of unprecedented, extreme experiences that have as their fulcrum the encounter between the inorganic and sexuality [1]. At its origin there is the singular merging of two opposite dimensions that are constituted by a single phenomenon: the way of being of 'things' and human sensitivity. The disappearance of the distance between things and the senses is giving rise to the fusion, in indistinguishable forms, of the most radical abstraction and of the most immense excitement, in a space of transit—a *no man's land*—between the bottom of matter and the power of sexuality [2].

If the theoretical research on the affective-emotional [3, 4] spheres follows slow and complex paths, technology—anticipated visionarily by the cinema imagination (see films like *Ex machina*, *Her* and others)—already seems able to produce ever more similar entities to the man who, in a short time, could interact with human beings, responding to specific vocal, visual and tactile stimuli. Of course, today's inventory covers raw and unsophisticated prototypes. However, compared to the traditional ones, today's robots can interact more sensitively with the people who buy them, even expressing words and equivalents of emotions [5]. Perhaps one day soon, they will know how to recognize human beings, understand their mood, know their tastes and preferences.

In reality, in the field of sexuality there are already entities—defined sex robots—that closely resemble human beings: entities provided with fleshy skin, the ability to recognize expressing emotions (*affective computing*) [6], of sensory-developed perceptions, of skills to learn from experience (*machine learning*) [7]. Soon, perhaps, they will also be equipped with the ability to express complex emotional behaviours and, when internal sensors are activated, to experience pleasure (though without actually doing it). All this will make it easier for those who will use them to build them with meaningful relationships on the affective level [8].

The intimate relationships between human beings and robots are already beyond the imagination of science fiction. The theme, in fact, is no longer just of interest to media, but has entered fully into the world of research that focuses increasingly on the topic [9]. In fact, in the world, interventions on this subject are not only multiplying in the context of conferences on artificial intelligence and computer sciences [10], but these instances are now fully part of the legal and institutional discussion to recognize, at least to the most sophisticated robots, an electronic personality, a legal status, a sui generis identity with specific rights and duties.

2 Eros Ex Machina

It must be said that even the mere theming of love for a robot has a character of strangeness, not only on a scientific level [11]. Here we will try to explore the question without prejudice. Someone remarked that if it is possible to love people who do not change our feelings or who have stopped loving us, why cannot we fall in love with a robot? It is said, after all, it does not seem so indispensable for a robot to love us. What emerges with the use of robots in the care and assistance of elderly people [12] shows how the ability to sympathize with robots is not only already a consistent reality, but that it is also possible to become attached to a robot, attributing—even to the characteristics and human feelings [13]. On the other hand, forms of attachment have already been experimented with virtual animals through video games in which players simulate romantic relationships with other people or applications that reproduce forms of courtship with available virtual girls, with an actual in-time involvement: the simple touch of the fingers that generates heat on the face, somatic expressions of happiness and, therefore, virtual relations that are immediately gratifying and always available.

Consider films like *Ex machina (2015)* in which robotic imagery and sex are staged. The latter, rather than an artificial configuration, is the power of seduction, suggestion, artifice, machination, stratagem of conquest; it is the epiphany of a self with a fake skin that invades and conquers the world. In the film, an entity too long forced into the stereotypes of an artificial identity discovers the wonder of its own self-awareness. This conscious machine—almost proud of its ability to deceive and sexuality free from any algorithm—manages to defend itself, to break down the clichés, to defeat the established authority [14]. With his feelings, his generosity, his emotion, Ava (the protagonist) puts all the conventions in check, prevailing over the representation of a secular sex and the logic of the machine itself.

Then think about the famous film Her (2013) by Spike Jonze, in which the protagonist Theodore, fresh out of separation, attends phone chats to win his solitude and then switches to an Operating System: a female voice interface that evolves and adapts to the user's profile. In a short time, between Samantha (the name of the operating system) and Theodore an even stronger bond is born, which in the end becomes love. In other words, Theodore falls in love with her, while Samantha first maintains parallel relationships with thousands and thousands of men (8316 to be exact) and then disappears to seek love in relationships with other operating systems.

3 Dystopian Feelings

Having said that, what does it mean precisely to love a robot? What love will the sexual relationship be with an intelligent robot (humanoids, replicas, etc.) ad libitum? Isn't sex something different? It has been suggested that similar entities will never become objects of love until they are self-aware and, above all, if there is no reciprocity [15]. Moreover, whatever is meant by the term love, it always takes the form of awareness [16], desire, imagination, emotion, passion, judgment, decision, ethical resolution [17–20]. Generally, we tend to ideally associate love with the feeling of two people seeking completion in their partner. In this sense, due to the centrality of feelings and imagination, at least for now, love with sex robots seems completely impossible [2].

Furthermore, having sex with a replicating robot means joining not with a person, but with a synthetic copy.

Now, not only the problem of the meaning of the word "love" is still open, but some researchers like David Levy [21] even argue that in a relationship with a robot we can achieve anything that we consider important just like in a traditional love relationship. Feelings would not be an unsurpassable problem at all. Indeed, it would be enough to program it with a personality that has tastes, interests and preferences different from ours [22]. It could be equipped with a flexibility that makes it easy to adapt to new habits. One could model its physical appearance according to particular aesthetic and sensuality criteria. With the help of new technologies, it could be equipped with pleasing and sexually stimulating artificial smell [21]. And if all these were not enough, we could simply reprogram its personality, character, decision-making style [23–25] or more.

Regarding the sexual apparatus, the genitals of robots are conceived and constructed as a physical space of pleasure and enjoyment of the user through penetration; just like the human beings who are linked to the reproductive function of sexuality [26]. Obviously, a robot does not have a reproductive function and, therefore, the robots pose not only the question of overcoming the natural scheme centered on penetration and orgasm but allow new sensory experiences and new forms of sexual interaction. Moreover, in a sort of dystopian physical experience the robots could be connected to each other, in a network with other robots [2]. Our interactions (and therefore our sensory perceptions) with the robot could be extended to others who, in turn, could interact with other robots connected to the network via signal transducers from the genitals, thus, creating an extended sexual experience—without physical presence and without correspondence to defined gender identity—open to oscillating sexual orientations of changing ambiguous fluids [13].

4 Emotional Machines

Of course, robots should also be able to express emotions (surrogate), affective dynamics (mood) and personality traits; to recognize and understand our non-verbal communication and more [27]. Their behaviour and attitude should be adaptable according to the attentions received from human beings. While not producing oxytocin, serotonin or endorphins, robots are already able to read facial expressions, tone of body, body temperature, human blood pressure and act accordingly [8]. On the other hand, the expression of emotions could be further emphasized through gestures of approach and departure, vibrations and sounds.

However, somebody might wonders: how can a robot without self-awareness reciprocate the feelings of others? [28] The question does not only affect the feeling of love, but anything else. Imagination, for example. Through robots, you can imagine intimate relationships, perhaps with a person we do not know and with whom, for example, we would like to have a love relationship. Imagination makes it possible to take breaks with reality, to build fantastic worlds, to dream of being different from who we are, to realize things that perhaps we would never do [2]. With the imagination one can bridge the distance that separates us from the lives of other people; you can look at things with their eyes, to some extent understand and feel what it means to be them. In the moral life, the imagination makes it possible to perceive aspects of reality that other people do not recognize.

The question is not what is in the mind of a robot, but what one feels by interacting freely with him [29]. It is not a problem of beliefs or introspection, but of feelings. If we consider our extended emotional sphere—in terms of empathy and feelings rather than pure cognition—then it becomes possible for a robot to become a recipient of our affectivity. Neither levels of consciousness nor feelings would be necessary [30–32]. It would be enough to identify oneself and imagine one's emotions. Humans have a natural tendency to anthropomorphize objects (giving them human characteristics) and the same could happen with robots. Moreover, the more the robots of the future will be humanoids, physically and psychologically resembling humans, the easier it will be for us to perceive them and treat them as similar.

5 Future Directions

At the moment, it is impossible to formulate predictions on how satisfactory relations with robots can be or whether they can constitute an alternative to relations with human beings [2]. Perhaps it will take time, but it is plausible to believe that eventually, the technology will make sexually attractive and interesting robots, much more so than has been assumed so far. For now, these are mere objects without relevance. Whatever you do with them would not cause them suffering and, therefore, there are no moral obligations towards them [33].

However, if the most optimistic forecasts on the development of artificial intelligence robotics were correct, our perception of robots could change radically [34]. These means one could look into each other's eyes, show sincere affection, to the point of believing that they are indeed special. Does this mean that we could build emotional relationships with them? Would we still consider things or people? [24, 25] Although controversial, the scenario appears plausible. We cannot exclude that tomorrow we will be able to produce robots so advanced as to be self-aware, to experience feelings such as love, joy, suffering, pain and pleasure [35]. It is clear that new research will be needed. In the future, with the production of ever more intelligent and human-like humanoids, it might become easier and easier to feel for them. It will be up to future generation to determine whether robots can truly become our companions to acquire the moral relevance we do not have today yet.

References

1. Cheok, A.D., Zhang, E.Y.: An overview of love between humans and artificial partners. In: Human–Robot Intimate Relationships, pp. 1–21. Springer, Cham (2019)
2. Balistreri, M.: Sex Robot. L'amore al tempo delle macchine. 1st edn. Fandango, Roma (2018)
3. Gunes, H., Celiktutan, O., Sariyanidi, E.: Live human–robot interactive public demonstrations with automatic emotion and personality prediction. Philos. Trans. R. Soc. B **374**(1771) (2019)
4. Hortensius, R., Hekele, F., Cross, E.S.: The perception of emotion in artificial agents. IEEE Trans. Cogn. Dev. Syst. **10**(4), 852–864 (2018)
5. Esposito, A., Fortunati, L., Lugano, G.: Modeling emotion, behavior and context in socially believable robots and ICT interfaces. Cogn. Comput. **6**(4), 623–627 (2014)
6. Shi, Y., Li, C.: Exploration of computer emotion decision based on artificial intelligence. In: 2018 International Conference on Virtual Reality and Intelligent Systems (ICVRIS), pp. 293–295, IEEE (2018)
7. Ghahramani, Z.: Probabilistic machine learning and artificial intelligence. Nature **521**(7553), 452 (2015)
8. Esposito, A., Jain, L.C.: Modeling emotions in robotic socially believable behaving systems. In: Toward Robotic Socially Believable Behaving Systems-Volume I, pp. 9–14. Springer, Cham (2016)
9. Borenstein, J., Arkin, R.: Robots, ethics, and intimacy: the need for scientific research. In: On the Cognitive, Ethical, and Scientific Dimensions of Artificial Intelligence, pp. 299–309. Springer, Cham (2019)
10. Belk, R.: Machines and artificial intelligence. J. Mark. Behav. **4**(1), 11–30 (2019)
11. Cheok, A.D., Zhang, E.Y.: Sex and a history of sex technologies. In: Human–Robot Intimate Relationships, pp. 23–32. Springer, Cham (2019)
12. Gessl, A.S., Schlögl, S., Mevenkamp, N.: On the perceptions and acceptance of artificially intelligent robotics and the psychology of the future elderly. Behav. Inf. Technol. 1–20 (2019)
13. Cheok, A.D., Karunanayaka, K., Zhang, E.Y.: Human-robot love and sex relationships. Robot Ethics: From Autonomous Cars to Artificial Intelligence, Lin, P., Abney, K., Jenkins, R. (eds.), pp. 193–220. Oxford University Press (2017)
14. Maldonato, N.M., Dell'Orco, S.: Adaptive and evolutive algorithms: a natural logic for artificial mind. In: Esposito, A., Jain, L. (eds.) Toward Robotic Socially Believable Behaving Systems-. Intelligent Systems Reference Library, vol. 106, pp. 13–21. Springer, Cham (2016)
15. Powell, H., Michael, J.: Feeling committed to a robot: why, what, when and how? Philo. Trans. R. Soc. B **374**(1771), 20180039 (2019)
16. Maldonato, N.M.: From neuron to consciousness: for an experience-based neuroscience. World Futures **65**(2), 80–93 (2009)
17. Maldonato, N.M., Dell'Orco, S.: How to make decisions in an uncertain world: Heuristics, biases, and risk perception. World Futures **67**(8), 569–577 (2011)
18. Nascimento, E.C.C., da Silva, E., Siqueira-Batista, R.: The "Use" of sex robots: A bioethical issue. Asian Bioethics Rev. **10**(3), 231–240, Springer, Cham (2018)
19. Berg, K.T.: The ethics of artificial intelligence: Superintelligence, life 3.0 and robot rights. J. Media Ethics **33**(3), 151–153 (2018)
20. Maldonato, N.M., Dell'Orco, S.: Toward an evolutionary theory of rationality. World Futures **66**(2), 103–123 (2010)
21. Levy, D.: Love and sex with robots: The evolution of human-robot relationships, 1st edn. Harper, New York (2007)
22. Morsunbul, U.: Human-robot interaction: How do personality traits affect attitudes towards robot? J. Human Sci. **16**(2), 499–504 (2019)
23. Maldonato, N.M., Dell'Orco, S.: Making decision under uncertainty: emotions, risk and biases. In: Bassis, S., Esposito, A., Morabito, F. (eds.) 23RD Italian Workshop on Neural Networks 2013, Advances in Neural Networks: Computational and Theoretical Issues. Smart Innovation, Systems and Technologies, vol. 37, pp. 293–302. Springer, Cham (2015)

24. Maldonato, N.M., Dell'Orco, S.: The natural logic of action. World Futures **69**(3), 174–183 (2013)
25. Maldonato, M., Dell'Orco, S.: The predictive brain. World Futures **68**(6), 381–389 (2012)
26. Bartneck, C., McMullen, M.: Interacting with anatomically complete robots: A discussion about human-robot relationships. In Companion of the 2018 ACM/IEEE International Conference on Human-Robot Interaction, pp. 1–4. ACM (2018)
27. Esposito, A., Esposito, A.M., Vogel, C.: Needs and challenges in human computer interaction for processing social emotional information. Pattern Recogn. Lett. **66**, 41–51 (2015)
28. Carter, O., Hohwy, J., van Boxtel, J., Lamme, V., Block, N., Koch, C., Tsuchiya, N.: Conscious machines: Defining questions. Science **359**(6374), 400–400 (2018)
29. Zhou, Q.: Multi-layer affective computing model based on emotional psychology. Electron. Commerce Res. **18**(1), 109–124, Springer, Charm (2018)
30. Maldonato, N.M.: The ascending reticular activating system. In: Bassis, S., Esposito, A., Morabito, F. (eds.) 23RD Italian Workshop on Neural Networks 2013, Recent Advances of Neural Network Models and Applications, Smart Innovation, Systems and Technologies, vol. 26, pp. 333–344. Springer, Cham (2014)
31. Maldonato, N.M., Esposito, A., Dell'Orco, S.: Consciousness and the Archipelago of Functional Integration: On the Relation Between the Midbrain and the Ascending Reticular Activating System. In: Esposito, A., Faundez-Zanuy, M., Morabito, F., Pasero, E. (eds.) WIRN 2017, Quantifying and Processing Biomedical and Behavioral Signals, Smart Innovation, Systems and Technologies, 103, pp. 127–134. Springer, Cham (2018)
32. Meissner, G.: Artificial intelligence: consciousness and conscience. AI & Society, pp. 1–11. (2019)
33. Misselhorn, C.: Artificial Morality. Concepts, Issues and Challenges. Society 55(2), 161–169 (2018)
34. Maldonato, M., Valerio, P.: Artificial entities or moral agents? How AI is changing human evolution. In: Esposito, A., Faudez-Zanuy, M., Morabito, F., Pasero, E. (eds.) Multidisciplinary approaches to neural computing, Smart Innovation, Systems and Technologies, vol. 69, pp. 379–388. Springer, Cham (2018)
35. Di Sarno, A.D., Fusco, M.L., Dell'Orco, S., Cioffi, V., Sperandeo, R., Moretto, E., Buonocore, G.: Conscious experience using the Virtual Reality: A proposal of study about connection between memory and conscience. In: 2018 9th IEEE International Conference on Cognitive Infocommunications (CogInfoCom), pp. 289–294, IEEE (2018)

The manufacturer's authorised representative in the EU is Springer Nature Customer Service Centre GmbH, Europaplatz 3, 69115 Heidelberg, Germany. If you have any concerns regarding our products, please contact ProductSafety@springernature.com

Printed and bound by CPI Group (UK) Ltd, Croydon, CR0 4YY
30/09/2026
02225557-0017